Effects of Phytochemicals on Metabolic Disorders and Human Health

Effects of Phytochemicals on Metabolic Disorders and Human Health

Guest Editors

María Angeles Martín
Sonia Ramos Rivero

Guest Editors
María Angeles Martín
Institute of Food Science,
Technology and Nutrition
(ICTAN-CSIC)
Madrid
Spain

Sonia Ramos Rivero
Institute of Food Science,
Technology and Nutrition
(ICTAN-CSIC)
Madrid
Spain

Editorial Office
MDPI AG
Grosspeteranlage 5
4052 Basel, Switzerland

This is a reprint of the Special Issue, published open access by the journal *Nutrients* (ISSN 2072-6643), freely accessible at: https://www.mdpi.com/journal/nutrients/special_issues/641KCSF5MZ.

For citation purposes, cite each article independently as indicated on the article page online and as indicated below:

Lastname, A.A.; Lastname, B.B. Article Title. *Journal Name* **Year**, *Volume Number*, Page Range.

ISBN 978-3-7258-3361-0 (Hbk)
ISBN 978-3-7258-3362-7 (PDF)
https://doi.org/10.3390/books978-3-7258-3362-7

Contents

About the Editors

María Angeles Martín

María Angeles Martín Arribas is a Tenured Scientist at the Institute of Food Science, Technology and Nutrition (ICTAN) of the Spanish National Research Council (CSIC). She obtained her B.S. degree in Biology from the University Complutense of Madrid in 1991 and her Ph.D. in Biochemistry and Molecular Biology from the same university in 1995. She started her research career in the Institute of Biochemistry (IB, CSIC-UCM), and in 2005, she moved to the ICTAN (CSIC). Since 2007, she participates as a Researcher at the Centro de Investigación Biomédicas en Red de Diabetes y Enfermedades Metabólicas Asociadas (CIBERDEM), which brings together 29 top Spanish Research Groups in the field of diabetes and associated diseases. She is currently the Head Researcher of the NUTRALMET (Nutrition and Metabolic Alterations) Group. Her research activity is focused on the study of natural bioactive compounds, mainly polyphenols, with favourable effects against diseases associated with oxidative stress and inflammation, with special attention paid to diabetes and its associated complications. Her interest is also focused on the characterization of the specific molecular mechanisms involved in the beneficial effects of these compounds, including their relationship with the gut microbiome and their potential impact on metabolomes.

Sonia Ramos Rivero

Sonia Ramos Rivero holds a degree in Pharmacy (1994) and a Ph.D. from the University Complutense of Madrid (1999). She is a postdoctoral research fellow at the Institute of Biochemistry, CSIC-UCM, Madrid (2000), and at the institute of Biochemistry I, University of Cologne (2000–2003). Following this, she has been working at ICTAN-CSIC (formerly Instituto del Frío) thanks to a "Ramón y Cajal" contract and also, since 2008, as a Tenured Scientist. She also participates as a Researcher at the Centro de Investigación Biomédicas en Red de Diabetes y Enfermedades Metabólicas Asociadas (CIBERDEM). Her research activity in the field of nutritional biochemistry is focused on the study of the molecular mechanisms of action of bioactive food components, mainly phenolic compounds and their metabolites, and their potential effect on health and on the prevention of chronic diseases related to oxidative stress (cancer, diabetes, cardiovascular disease, and obesity). Her expertise includes the development of different experimental approaches in cultured cells, experimental animals (pre-clinical studies), and, more recently, human clinical trials both in healthy and risk population groups, with special interest in diabetes and its complications.

Preface

Phytochemicals include a wide range of compounds belonging to different groups such as polyphenols, carotenoids, alkaloids, etc. They are commonly found in the diet, mainly in plant foods and derived products. Phytochemicals have been gaining attention in recent years because of their their health-promoting properties. However, it is still difficult to define their impact on health maintenance and the prevention or even treatment of chronic diseases, including metabolic disorders. Nevertheless, at present, it is generally accepted that phytochemicals exert anti-inflammatory, anti-obesity, and anti-diabetic effects, among others, and that these effects are related to their modulatory role on redox status, inflammation, and microbiota. In all of these ways, phytochemicals may be useful in the prevention and treatment of high prevalence metabolic diseases worldwide, such as obesity, cardiovascular disease, non-alcoholic fatty liver disease, and diabetes.

María Angeles Martín and Sonia Ramos Rivero

Guest Editors

nutrients

MDPI

ial

ects of Phytochemicals on Metabolic Diseases and .man Health

a Ángeles Martín [1,2,*] and Sonia Ramos [1,2,*]

1 Department of Metabolism and Nutrition, Institute of Food Science and Technology and Nutrition (ICTAN), Consejo Superior de Investigaciones Científicas (CSIC), José Antonio Novais 10, Ciudad Universitaria, 28040 Madrid, Spain
2 CIBER de Diabetes y Enfermedades Metabólicas Asociadas (CIBERDEM), Instituto de Salud Carlos III (ISCIII), 28029 Madrid, Spain
* Correspondence: amartina@ictan.csic.es (M.Á.M.); s.ramos@ictan.csic.es (S.R.)

Metabolic diseases constitute a worldwide health concern because of their increasing prevalence and associated mortality [1–4]. Actually, pathologies such as type 2 diabetes mellitus, obesity, metabolic syndrome, non-alcoholic fatty liver disease, and cardiovascular diseases are the most common chronic illnesses in almost all countries. Moreover, these diseases are responsible for causing disability in the affected population. All this results in a substantial world economic burden that could be reduced and prevented by lowering the associated risk factors. In this context, metabolic diseases can result from a combination of genetic, physiological, environmental, and behavioral factors. Consequently, a cost-effective solution is to reduce the modifiable behavioral risk factors, with physical activity and a healthy diet being very valuable tools.

Phytochemicals are commonly found in the diet and are abundant in fruits, vegetables, olive oil, legumes, cocoa, and beverages such as tea, coffee, and wine [5]. In fact, because of this wide distribution and their potential effects on health, their consumption may represent a promising protective and therapeutic approach to prevent metabolic diseases and improve health and, therefore, constitutes an important area of research. Notably, phytochemicals have received rising attention for their health-promoting properties in many chronic diseases, including metabolic disorders and their associated complications.

The pathogenesis of the metabolic diseases is very different, but dysregulation of key metabolic signaling pathways (i.e., lipid and glucose homeostasis, insulin pathway, etc.), as well as enhanced oxidative stress and inflammation and alterations in the gut microbiota, are present in all of them [6–10]. In this context, it could be highlighted that the health benefits of the phytochemicals have been related to different biological activities, such as their well-known antioxidant property, but also their anti-diabetic, anti-obesity, and anti-inflammatory actions, among others [11]. It is therefore becoming clear that phytochemicals can modulate different signaling pathways that are responsible for these final actions and, hence, their health benefits.

on: Martín, M.Á.; Ramos, S.
s of Phytochemicals on
olic Diseases and Human
h. *Nutrients* **2024**, *16*, 4323.
//doi.org/10.3390/
244323

ved: 27 November 2024
oted: 3 December 2024
shed: 14 December 2024

This Editorial refers to the Special Issue "Effects of Phytochemicals on Metabolic Diseases and Human Health". The Special Issue highlights the role of oxidative stress, inflammation, and microbiota in the pathophysiology of the different metabolic diseases and comprises thirteen new studies and two reviews. All contributions are listed below:

In this line, Godos et al. have reviewed the clinical studies carried out with flavanols, mainly from cocoa, to expose their impact on vascular health (contribution 1). This review highlights the effects of this family of flavonoids in reducing blood pressure and improving endothelial function, as well as the relevant role that their microbial metabolites play in these final effects by modulating gene expression and protein functions. Likewise, Huang and colleagues (contribution 2) review the potential effects of *Vaccinium* L. on diabetes and microvascular complications, such as nephropathy and retinopathy. In addition, the of plants on reducing blood glucose, oxidative stress, and inflammation, which make t excellent candidates for the management of diabetes and diabetic vascular complicati although further studies on the mechanisms of action and larger randomized blinded t are needed.

Similarly, related to the effects on diabetes, Gallardo-Villanueva et al. report benefits of a flavanol-rich cocoa–carob blend (CCB) in suppressing islet inflammatio preventing macrophage infiltration into the islets and overproduction of pro-inflamma cytokines, along with the inactivation of nuclear factor kappa B, which was associated a prevented beta-cell apoptosis and the loss of beta cells in Zucker Diabetic Fatty (Z rats (contribution 3). In addition, García-Díez and coworkers (contribution 4) demonst that this supplemented diet also alleviates damage to the liver and kidneys, key org for the maintenance of glucose homeostasis, in ZDF rats by ameliorating the oxida stress, downregulating apoptosis, and improving insulin signaling and glucose homeost in both tissues. Interestingly, in both contributions, additive effects were demonstra when the CCB was combined with metformin, highlighting its potential as an adju therapy to delay the progression of type 2 diabetes. In addition, Zhou and colleag (contribution 5) proved the mechanism of the hypoglycemic effect of epimedin C, most important flavonoid of the edible herb *Epimedium*, in diabetic mice. Thus, by usi proteomic approach, the authors demonstrated that epimedin C downregulated key hep proteins involved in gluconeogenesis and insulin signaling, such as phosphoenolpyru carboxykinase, and upregulated crucial proteins for lipid digestion and degradation (nar group XIIB secretory phospholipase A2-like protein, apolipoprotein B-100 [Apob], cytochrome P450 4A14), resulting in alleviation of lipotoxicity and protection agai oxidative stress in the liver of the diabetic animals.

Arterial hypertension constitutes a cause of premature morbidity and mortality. I randomized, triple-blind controlled trial, Serrano and colleagues (contribution 6) show that an optimized extract of aged black garlic with low doses of S-allyl cysteine indu an extra-significant reduction in blood pressure in a population of grade I hypertens patients receiving antihypertensive drug treatment. Importantly, hypercholesterolemia central risk factor for atherosclerosis, which is the leading cause of mortality worldwide In this regard, Langhi et al. explore the mechanism of action of a polyphenol-rich pl extract named Totum-070 in human colonic cells and in hamsters fed a high-fat d Totum-070 demonstrated a reduction in various lipid and inflammatory gene expressi and in the intestinal lipid absorption (contribution 7). In addition, this polyphenol-r extract regulated microbial alpha- and beta-diversity. Indeed, the modulation of both microbiota and gut barrier plays a relevant role in metabolic diseases [12]. In this cont Khalil et al. have demonstrated that polyphenolic leave extracts from *Thymbra spicata* Mediterranean plant, contributed to maintaining a balanced microbial environment a beneficial metabolites, and it can potentially counteract pathogenic bacterial overgrow during dysbiosis (contribution 8). Moreover, *T. spicata* extract enhanced intestinal barr resistance, which is crucial for preserving intestinal integrity and preventing conditio such as leaky gut syndrome. Similarly, Balderas and colleagues (contribution 9) report an increase in beneficial bacteria in rats fed onion- and apple-enriched diets. Furthermo a relevant effect of onion and apple intake on major types of microbial-derived molecul such as short-chain fatty acids (SCFAs) and bile acids (BAs), was demonstrated in obe Zucker rats. Thus, boosted levels of SCFAs and decreased excretion of mainly prima and secondary BAs, as well as increased amounts of taurine- and glycine-conjugat BAs compared to the obese rats, were shown. This differential regulation of bioacti compounds and their metabolites on the microbiota and SCFA and BA formation, as well their relationship with some diseases, deserves further studies. In this regard, Nederve et al. in their randomized, double-blind, placebo-controlled clinical trial explore the effe of a novel multi-ingredient supplement on weight loss and body composition in overweig and obese volunteers and demonstrate the effectiveness of this nutritional supplement reducing weight, fat mass, and markers of liver health and metabolism, such as extracellul vesicle-associated miRNA profiles (contribution 10). At the molecular level, Cierniak and colleagues (contribution 11) show the role of 1,2-dicinnamoyl-*sn*-glycero-3-phosphocholin in increasing the number of mitochondria and the expression levels of genes encoded by mtDNA to ameliorate the insulin resistance in cultured mature 3T3-L1 cells exposed to palmitic acid.

Importantly, the development and progression of liver diseases associated with inappropriate diets is growing worldwide [13], being connected to multiple metabolic diseases. In this regard, Park and coworkers (contribution 12) prove that a homoisoflavonoid isolated from *Portulaca oleracea* named (*E*)-5-hydroxy-7-methoxy-3-(2-hydroxybenzyl)-4-chromanone (HMC) attenuates the hepatic steatosis by decreasing lipid accumulation and triglyceride content in free fatty acid-treated human hepatic HepG2 cells. Mechanistically, HMC promotes fatty acid oxidation via AMPK and PPARα and inhibits key proteins involved in lipid synthesis. Similarly, Donghia et al. in their study cohort report the protective role of lycopene extracts from cooked and fresh tomatoes and conclude that the potent anti-inflammatory and antioxidant properties may serve as a protective factor against the development and progression of steatosis and liver injury (contribution 13). Aiello and coworkers, in a single-arm longitudinal interventional pilot study, also demonstrate that the administration of a dietary supplement based on a liquid residue produced during olive oil extraction has the potential to improve anthropometric, hematological, and metabolic parameters in volunteers at risk of metabolic syndrome (contribution 14).

Anxiety, depression, and cognitive dysfunction are also highly prevalent diseases in our society [14,15]. Wang et al. demonstrate that evodiamine, a major alkaloid from dry unripened fruit *Evodia fructus*, reverses anxiety- and depression-like behavior in rodents (contribution 14). Moreover, evodiamine protected against apoptosis of hippocampal neurons in the post-traumatic stress disorder by modulating the renin–angiotensin pathway.

Overall, this Special Issue provides an update on recent advances in the study of the health benefits of phytochemicals, phytochemical-rich foods, their extracts, and health-promoting benefits in the context of metabolic diseases, as well as revealing novel molecular mechanisms and testing potential novel therapies. This Special Issue highlights the role of oxidative stress, inflammation, and microbiota in the pathophysiology of the different metabolic diseases and the necessity of deepening knowledge about the probable benefits of the phytochemicals against metabolic diseases. In addition, the fact that further in vivo studies are needed before recommending phytochemicals as an alternative/adjuvant approach in the treatment of metabolic diseases is emphasized.

Conflicts of Interest: The authors declare no conflicts of interest.

List of Contributions

1. Godos, J.; Romano, G.L.; Laudani, S.; Gozzo, L.; Guerrera, I.; Dominguez Azpíroz, I.; Martínez-Diaz, R.; Quiles, J.L.; Battino, M.; Drago, F. et al. Flavan-3-ols and Vascular Health: Clinical Evidence and Mechanisms of Action. *Nutrients* **2024**, *16*, 2471. https://doi.org/2410.3390/nu16152471.
2. Huang, H.; Luo, Y.; Wang, Q.; Zhang, Y.; Li, Z.; He, R.; Chen, X.; Dong, Z. Vaccinium as Potential Therapy for Diabetes and Microvascular Complications. *Nutrients* **2023**, *15*, 2031. https://doi.org/2010.3390/nu15092031.
3. Gallardo-Villanueva, P.; Fernández-Marcelo, T.; Villamayor, L.; Valverde, A.M.; Ramos, S.; Fernández-Millán, E.; Martín, M.A. Synergistic effect of a flavonoid-rich cocoa-carob blend and metformin in preserving pancreatic beta cells in Zucker diabetic fatty rats. *Nutrients* **2024**, *16*, 273. https://doi.org/210.3390/nu16020273.
4. García-Díez, E.; López-Oliva, M.E.; Pérez-Jiménez, J.; Martín, M.Á.; Ramos, S. Cocoa and Carob Supplementation, Alone or in Combination with Metformin, Protects against Hepatorenal Injury in Zucker Diabetic Fatty Rats. *Nutrients* **2024**, *16*, 3087. https://doi.org/3010.3390/nu16183087.
5. Zhou, X.; Liu, Z.; Yang, X.; Feng, J.; Gins, M.S.; Yan, T.; Han, L.; Zhang, H. The Mechanism Underlying the Hypoglycemic Effect of Epimedin C on Mice with Type 2 Diabetes Mellitus Based on Proteomic Analysis. *Nutrients* **2024**, *16*, 25. https://doi.org/10.3390/nu16010025.

6. Serrano, J.C.E.; Castro-Boqué, E.; García-Carrasco, A.; Morán-Valero, M.I.; González-Hedst D.; Bermúdez-López, M.; Valdivielso, J.M.; Espinel, A.E.; Portero-Otín, M. Antihyperter Effects of an Optimized Aged Garlic Extract in Subjects with Grade I Hypertension and hypertensive Drug Therapy: A Randomized, Triple-Blind Controlled Trial. *Nutrients* **202** 3691. https://doi.org/3610.3390/nu1517369.
7. Langhi, C.; Vallier, M.; Otero, Y.F.; Maura, M.; Le Joubioux, F.; Groult, H.; Achour, O.; Pebr R.B.; Giera, M.; Guigas, B. et al. Totum-070, a Polyphenol-Rich Plant Extract, Prevents Hy holesterolemia in High-Fat Diet-Fed Hamsters by Inhibiting Intestinal Cholesterol Absorp *Nutrients* **2023**, *15*, 5056. https://doi.org/5010.3390/nu15245056.
8. Khalil, M.; Piccapane, F.; Vacca, M.; Celano, G.; Mahdi, L.; Perniola, V.; Apa, C.A.; Annunz A.; Iacobellis, I.; Procino, G. et al. Nutritional and Physiological Properties of Thymbra spi In Vitro Study Using Fecal Fermentation and Intestinal Integrity Models. *Nutrients* **2024**, *16*, https://doi.org/510.3390/nu16050588.
9. Balderas, C.; de Ancos, B.; Sánchez-Moreno, C. Bile Acids and Short-Chain Fatty Acids Modulated after Onion and Apple Consumption in Obese Zucker Rats. *Nutrients* **2023**, *15*, https://doi.org/3010.3390/nu15133035.
10. Nederveen, J.P.; Mastrolonardo, A.J.; Xhuti, D.; Di Carlo, A.; Manta, K.; Fuda, M.R.; Tarnopo M.A. Novel Multi-Ingredient Supplement Facilitates Weight Loss and Improves Body Com tion in Overweight and Obese Individuals: A Randomized, Double-Blind, Placebo-Contro Clinical Trial. *Nutrients* **2023**, *15*, 3693. https://doi.org/3610.3390/nu15173693.
11. Cierzniak, A.; Gliszczynska, A.; Małodobra-Mazur, M. 1,2-Dicinnamoyl-sn-glyce phosphocholine Improves Insulin Sensitivity and Upregulates mtDNA Encoded Gene Insulin-Resistant 3T3-L1 Adipocytes: A Preliminary Study. *Nutrients* **2024**, *16*, 3163. ht //doi.org/3110.3390/nu16183163.
12. Park, J.-E.; Han, J.-S. (E)-5-hydroxy-7-methoxy-3-(2-hydroxybenzyl)-4-chromanone, a M Homoisoflavonoid, Attenuates Free Fatty Acid-Induced Hepatic Steatosis by Activating AN and PPARαPathways in HepG2 Cells. *Nutrients* **2024**, *16*, 3475. https://doi.org/3410.3390/ 6203475.
13. Donghia, R.; Campanella, A.; Bonfiglio, C.; Cuccaro, F.; Tatoli, R.; Giannelli, G. Protec Role of Lycopene in Subjects with Liver Disease: NUTRIHEP Study. *Nutrients* **2024**, *16*, https://doi.org/510.3390/nu16040562.
14. Aiello, A.; Calabrone, L.; Noonan, D.M.; Corradino, P.; Nofri, S.; Cristoni, S.; Accardi, Candore, G.; Caruso, C.; Zinellu, A.; et al. Effect of a Phytochemical-Rich Olive-Deri Extract on Anthropometric, Hematological, and Metabolic Parameters. *Nutrients* **2024**, *16*, 3 https://doi.org/10.3390/nu16183068.
15. Wang, Z.; Lai, C.; Shen, B.; Li, B.; Chen, J.; Shen, X.; Huang, Z.; Yang, C.; Gao, Y. Effects of E diamine on Behavior and Hippocampal Neurons through Inhibition of Angiotensin-Convert Enzyme and Modulation of the Renin Angiotensin Pathway in a Mouse Model of Post-Traum Stress Disorder. *Nutrients* **2024**, *16*, 1957. https://doi.org/1910.3390/nu16121957.

References

1. Saklayen, M.G. The Global Epidemic of the Metabolic Syndrome. *Curr. Hypertens. Rep.* **2018**, *20*, 12. [CrossRef]
2. WHO. 2024. Available online: https://www.who.int/news-room/fact-sheets/detail/diabetes (accessed on 2 December 202
3. WHO. 2024. Available online: https://www.who.int/news-room/fact-sheets/detail/obesity-and-overweight (accessed o December 2024).
4. WHO. 2024. Available online: https://www.who.int/news-room/fact-sheets/detail/cardiovascular-diseases-(cvds) (access on 2 December 2024).
5. Guan, R.; Van Le, Q.; Yang, H.; Zhang, D.; Gu, H.; Yang, Y.; Sonne, C.; Lam, S.S.; Zhong, J.; Jianguang, Z.; et al. A review of diet phytochemicals and their relation to oxidative stress and human diseases. *Chemosphere* **2021**, *271*, 129499. [CrossRef] [PubMe
6. Ghaben, A.L.; Scherer, P.E. Adipogenesis and Metabolic Health. *Nat. Rev. Mol. Cell Biol.* **2019**, *20*, 242–258. [CrossRef] [PubMe
7. Hurrle, S.; Hsu, W.H. The Etiology of Oxidative Stress in Insulin Resistance. *Biomed. J.* **2017**, *40*, 257–262. [CrossRef] [PubMec
8. Martín, M.A.; Goya, L.; Ramos, S. Antidiabetic actions of cocoa flavanols. *Mol. Nutr. Food Res.* **2016**, *60*, 1756–1769. [CrossRef
9. Martín, M.A.; Ramos, S. Impact of Dietary Flavanols on Microbiota, Immunity and Inflammation in Metabolic Diseases. *Nutrie* **2021**, *13*, 850. [CrossRef] [PubMed]
10. Van Treuren, W.; Dodd, D.M. Microbial Contribution to the Human Metabolome: Implications for Health and Disease. *Ann. R Pathol. Mech. Dis.* **2020**, *15*, 345–369. [CrossRef] [PubMed]
11. Cote, B.; Elbarbry, F.; Bui, F.; Su, J.W.; Seo, K.; Nguyen, A.; Lee, M.; Rao, D.A. Mechanistic Basis for the Role of Phyto-chemicals Inflammation-Associated Chronic Diseases. *Molecules* **2022**, *27*, 781. [CrossRef] [PubMed]

Fan, Y.; Pedersen, O. Gut microbiota in human metabolic health and disease. *Nat. Rev. Microbiol.* **2021**, *10*, 55–71. [CrossRef] [PubMed]
Han, S.K.; Baik, S.K.; Kim, M.Y. Non-alcoholic fatty liver disease: Definition and subtypes. *Clin. Mol. Hepatol.* **2023**, *29* (Suppl. 29), S5–S16. [CrossRef]
WHO. 2024. Available online: https://www.who.int/news-room/fact-sheets/detail/mental-disorders (accessed on 2 December 2024).
Suddell, S.; Mahedy, L.; Skirrow, C.; Penton-Voak, I.S.; Munafò, M.R.; Wootton, R.E. Cognitive functioning in anxiety and depression: Results from the ALSPAC cohort. *R. Soc. Open Sci.* **2023**, *10*, 221161. [CrossRef] [PubMed]

nutrients

MD

Review

Flavan-3-ols and Vascular Health: Clinical Evidence and Mechanisms of Action

Justyna Godos [1], Giovanni Luca Romano [2], Samuele Laudani [1], Lucia Gozzo [3], Ida Guerrera [1], Irma Dominguez Azpíroz [4,5,6], Raquel Martínez Diaz [4,7,8], José L. Quiles [4,9,10], Maurizio Battino [4,11,12], Filippo Drago [1], Francesca Giampieri [4,11,*], Fabio Galvano [1] and Giuseppe Grosso [1,13]

1 Department of Biomedical and Biotechnological Sciences, University of Catania, 95123 Catania, Italy
2 Department of Medicine and Surgery, University of Enna "Kore", 94100 Enna, Italy
3 Clinical Pharmacology Unit/Regional Pharmacovigilance Centre, Azienda Ospedaliero Universitaria Policlinico "G. Rodolico-S. Marco", 95123 Catania, Italy
4 Research Group on Food, Nutritional Biochemistry and Health, Universidad Europea del Atlántico, Isabel Torres 21, 39011 Santander, Spain
5 Research Group on Food, Nutritional Biochemistry and Health, Universidade Internacional do Cuanza, Cuito EN250, Angola
6 Research Group on Food, Nutritional Biochemistry and Health, Universidad de La Romana, La Romana 22000, Dominican Republic
7 Research Group on Food, Nutritional Biochemistry and Health, Universidad Internacional Iberoamerica Campeche 24560, Mexico
8 Research Group on Food, Nutritional Biochemistry and Health, Universidad Internacional Iberoamerica Arecibo, PR 00613, USA
9 Department of Physiology, Institute of Nutrition and Food Technology "José Mataix Verdú", Biomedical Research Centre, University of Granada, 18016 Armilla, Spain
10 Research and Development Functional Food Centre (CIDAF), Health Science Technological Park, Avenida Conocimiento 37, 18016 Granada, Spain
11 Department of Clinical Sciences, Università Politecnica delle Marche, 60131 Ancona, Italy
12 International Joint Research Laboratory of Intelligent Agriculture and Agri-Products Processing, Jiangsu University, Zhenjiang 212013, China
13 Center for Human Nutrition and Mediterranean Foods (NUTREA), University of Catania, 95123 Catania,
* Correspondence: f.giampieri@staff.univpm.it

Citation: Godos, J.; Romano, G.L.; Laudani, S.; Gozzo, L.; Guerrera, I.; Dominguez Azpíroz, I.; Martínez Diaz, R.; Quiles, J.L.; Battino, M.; Drago, F.; et al. Flavan-3-ols and Vascular Health: Clinical Evidence and Mechanisms of Action. *Nutrients* **2024**, *16*, 2471. https://doi.org/10.3390/nu16152471

Academic Editors: Lindsay Brown and Motoyuki Iemitsu

Received: 1 June 2024
Revised: 22 July 2024
Accepted: 22 July 2024
Published: 30 July 2024

Abstract: Cardiovascular diseases (CVDs) are one of the main causes of mortality and morbic worldwide. A healthy diet rich in plant-derived compounds such as (poly)phenols appears to h a key role in improving cardiovascular health. Flavan-3-ols represent a subclass of (poly)pher of great interest for their possible health benefits. In this review, we summarized the results clinical studies on vascular outcomes of flavan-3-ol supplementation and we focused on the role the microbiota in CVD. Clinical trials included in this review showed that supplementation w flavan-3-ols mostly derived from cocoa products significantly reduces blood pressure and impro endothelial function. Studies on catechins from green tea demonstrated better results when involvi healthy individuals. From a mechanistic point of view, emerging evidence suggests that microb metabolites may play a role in the observed effects. Their function extends beyond the previous bel of ROS scavenging activity and encompasses a direct impact on gene expression and protein funct Although flavan-3-ols appear to have effects on cardiovascular health, further studies are needed clarify and confirm these potential benefits and the rising evidence of the potential involvement the microbiota.

Keywords: flavan-3-ols; polyphenols; inflammation; vascular; gut microbiota

1. Introduction

Cardiovascular diseases (CVDs) are among the leading causes of death globally [Different risk factors are involved in CVD development, such as age, high blood pressu smoking, dyslipidemia, and diet [2]. The latest estimates from the Global Burden of Disease study reported that dietary risks represent the leading factors in affecting CVD burden [3], with an estimated 10 million CVD deaths in 2017 due to diets low in fruits, vegetables, whole grains, nuts, and seeds, while high in sodium [4]. Mechanistically, the processes leading to vascular and endothelial function consist of the activation of different pathways involved in inflammatory responses (including the nuclear factor-kappa B (NF-κB), p38, protein kinase B (PKB/Akt), and extracellular signal-regulated kinases (ERK)1/2) which impair nitric oxide (NO) synthesis and the expression of different adhesion molecules (i.e., intercellular adhesion molecule-1 (ICAM-1) and vascular cell adhesion protein-1 (VCAM-1)) [5]. Chronic protraction of such pro-inflammatory insults, especially when promoted through a suboptimal diet, may lead to substantial impairment of vascular function, the establishment of hypertension, and ultimately an increased risk of CVD [6].

There is consistent evidence that plant-based dietary patterns are associated with a reduced risk of CVD [7,8]. While adherence to a healthy diet is ideal for reducing health risks, an increasing body of research suggests the key role played by bioactive compounds [9,10]. Among the most studied, (poly)phenols achieved great interest for their beneficial effects on cardiovascular health [11]. Flavanols (also known as flavan-3-ols) represent one of the most abundant polyphenols in the human diet, mainly found in fruits, green tea, red wine, and cocoa [12]. Chemically, flavan-3-ols represent a complex subclass of flavonoids that include a variety of monomeric, oligomeric, and polymeric compounds. The main monomeric forms are (+)-catechin, (−)-epicatechin, (+)-gallocatechin, (−)-epigallocatechin, (−)-epicatechin-3-*O*-gallate, and (−)-epigallocatechin-3-*O*-gallate. Oligomers or polymers of flavan-3-ols are known as proanthocyanidins, while polymers composed exclusively of epicatechin or catechin are named procyanidins. Previous evidence agreed that flavan-3-ols might exert beneficial effects toward the vascular system [13,14], by regulating various cellular signaling pathways [15]. Additionally, epidemiological studies showed an association between higher flavan-3-ol intake and a lower risk of CVD and mortality [16,17].

An intriguing hypothesis recently investigated over the last decade involves the role of gut microbiota in mediating, at least in part, the positive effects of dietary (poly)phenols on human health [18]. The gut microbiota consists of bacteria, viruses, fungi, and eukarya which together contribute to the healthy/unhealthy state of the host [19]. Flavan-3-ols introduced by diet mainly in the form of aglycone are not completely absorbed as such but rather metabolized by the microbiota residing in the gut [20], affecting gut microbiota population and gut homeostasis [21]. A recent study highlighted intra- and inter-source differences in flavan-3-ol bioavailability, suggesting the role of substantial individual differences when considering the absorption and bioavailability of such compounds [22]. This raised a growing interest in the interindividual diversity of microbiota composition that leads to differences in the microbial-derived metabolites with great impact on health outcomes [23]. The aim of this review is to provide an overview of the results coming from the clinical trials found in the literature on flavan-3-ols and their effects on vascular health and the potential role of the gut microbiota. Moreover, we will discuss the main molecular mechanisms of flavan-3-ols and their microbial-derived metabolites.

2. Gut Microbiota and Cardiovascular Health

The gut microbiota consists of the totality of microorganisms cohabiting the digestive tract including bacteria, fungi, archaea, viruses, and protozoa, which communicate bidirectionally with the host through different routes including the vagus nerve, the immune system, and different microbial-derived metabolites [24,25]. The composition of the microbiota can be influenced by different factors like drug administration, stress, mode of delivery, and diet [25,26]. Gut microbiota alterations, namely dysbiosis, can lead to impaired intestinal barrier integrity (leaky gut) leading to the translocation of bacteria products into the host circulation system and the onset of an inflammatory state which have been correlated with an augmented incidence of different pathologies including cardiovascular disease [27]. Many studies investigated microbiota composition and its correlation with CVD. Comparing the gut microbiota of atherosclerotic patients with a control gr an increase in *Escherichia coli*, *Klebsiella* spp., *Enterobacter aerogenes*, and *Ruminococcus gn* was observed, along with a reduction in different short-chain fatty acid producers inc ing *Roseburia intestinalis* and *Faecalibacterium prausnitzii* [28]. Similar results have b observed in works conducted by other authors in which when comparing the microbio patients with chronic heart failure and control participants, they demonstrated, even in case, a significant reduction in *Faecalibacterium prausnitzii* and an increase in *Ruminoco gnavus* [29]. In another study, the microbiota of pre-hypertensive and hypertensive pati was characterized by a reduced richness and diversity compared to the control gr These alterations were accompanied by an increase in the genera Klebsiella and *Prevo* and a reduced abundance of beneficial bacteria including Oscillibacter, *Faecalibacter* Bifidobacterium, *Roseburia*, Butyrivibrio, and *Coprococcus*. Fecal transplantation fron tients to germ-free mice led to an increase in blood pressure compared to control-recip mice [30]. The establishment of a dysbiosis state has been associated with altered duction of microbial-derived metabolites, which could exert detrimental effects on l health. Among the different metabolites produced by the gut microbiota, trimethylan N-oxide (TMAO) has been largely associated with CVD [31]. TMAO derives from liver oxidation of trimethylamine (TMA), a metabolite produced by the metabolisn foods containing L-carnitine and choline, mainly contained in animal-derived products Increased levels of TMAO were correlated with augmented cholesterol deposition ar higher risk of atherosclerosis [32]. In addition to TMAO, a longitudinal study demonstra the association of other microbial-derived metabolites, such as p-cresol, indoxyl sul and phenylacetylglutamine, with the incidence of major adverse cardiovascular events poor survival risk [33]. Diet represents an easier way to modulate microbiota composi Flavan-3-ol intake was demonstrated to possess prebiotic and antimicrobial proper resulting in an increased abundance of beneficial bacteria, such as *Lactobacillus*, and *dobacterium*, and a reduction in *Clostridium perfringens* [34]. The microbiota changes indu by flavan-3-ol intake were demonstrated to have beneficial effects on microbial-deri metabolite production, leading to a reduction in TMAO plasmatic levels [35]. Moreo flavan-3-ols promote beneficial effects on intestinal barrier function through the inhibi of inflammatory signals, reduced barrier permeability, and immune modulation [36].

3. Clinical Evidence of Flavan-3-ols on Vascular Outcomes

3.1. An Overview of Randomized Clinical Trials Evaluating the Effect of Flavan-3-ol Intake or Vascular Health

Flavan-3-ols are derivatives of flavans that include a variety of compounds commo found in several plant foods, such as cocoa products, tea, apples, pears, and berries [A summary of RCTs providing flavan-3-ol doses for the intervention on endothelial a vascular outcomes is presented in Table 1.

Author, Year, Country	Study Design	Participants (Mean Age)	Duration	Treatment	Polyphenol Constituent (Daily Intake)	Comparison	Main Findings
Murphy, 2003, Australia [38]	Double-blind, placebo-controlled	32 healthy subjects (treatment group: 40 y; placebo group: 47 y)	28 d	Tablets containing flavanols and procyanidins	234 mg flavanols and procyanidins	Placebo tablets (≤6 mg flavanols and procyanidins)	No divergences were seen between groups after treatment.
ngler, 2004, USA [39]	Double-blind, placebo-controlled	21 healthy volunteers (38 y)	2 wk	Dark chocolate bars	213 mg total procyanidins and 46 mg epicatechin	Placebo dark chocolate bars (traces of procyanidins and epicatechin)	High-flavonoid chocolate intake was associated with a significant amelioration of FMD ($p = 0.024$).
Fraga, 2005, Argentina [40]	Crossover, placebo-controlled	28 healthy participants (19 y)	2 × 14 d	Flavanol-containing milk chocolate	168 mg flavanols (39 mg epicatechin and catechin and 126 mg of procyanidins)	Cocoa butter chocolate (<5 mg/d of flavanols)	FCMC consumption was associated with a significant decrease in DBP ($p = 0.01$) and the mean BP ($p = 0.008$).
rassi, 2005, Italy [41]	Crossover, placebo-controlled	15 normotensive control (34 y) and 20 never-treated, grade I essential hypertensive individuals (43 y)	2 × 15 d (7 d washout)	Dark chocolate bars	88 mg flavanols (21.91 mg catechin, 65.97 mg epicatechin, 0.59 mg quercetin, 0.03 mg kaempferol, and 0.31 mg isorhamnetin)	Flavanol-free white chocolate bars	SBP and DBP significantly decreased after dark chocolate consumption compared to white chocolate and baseline values ($p < 0.0001$). FMD increased after dark chocolate consumption but not after white chocolate consumption ($p < 0.0001$).
Wang-Polagruto, 2006, USA [42]	Double-blind, placebo-controlled	32 postmenopausal hypercholesterolemic women (high-flavanol group: 57 y; low-flavanol group: 55 y)	6 wk	High-flavanol cocoa beverage	446 mg flavanols	Placebo cocoa beverage (43 mg flavanols)	Low-flavanol consumption led to a significant decrease in SBP and DBP ($p < 0.05$). High-flavanol treatment led to an increase in brachial artery hyperemic blood flow after ($p < 0.05$) compared to the baseline.

Table 1. *Cont.*

Author, Year, Country	Study Design	Participants (Mean Age)	Duration	Treatment	Polyphenol Constituent (Daily Intake)	Comparison	Main Findings
Baba, 2007, Japan [43]	Double-blind, placebo-controlled	25 healthy male subjects (38 y)	12 wk	Cocoa powder + sugar	98.02 mg epicatechins, 35.1 mg catechins, 41.08 procyanidin B2, and 24.9 mg procyanidin C1	Sugar	No significant results were found in SBP and DBP.
Taubert, 2007, Germany [44]	Investigator-blinded, placebo-controlled	44 participants (64 y)	18 wk	Dark chocolate	30 mg polyphenols (catechin 1.7 mg, epicatechin 5.1 mg, epicatechin-gallate 0.3 mg, procyanidin dimer 6.8 mg, procyanidin dimer-gallate 1.8 mg, procyanidin tetramer 3.7 mg, procyanidin pentamer 2.6 mg, and flavonols < 0.05 mg)	Polyphenol-free white chocolate	Dark chocolate consumption reduced SBP and DBP ($p < 0.001$).
Abdullah A, 2008, Saudi Arabia [45]	Placebo-controlled	89 healthy female (21 y)	15 d	(i) Dark chocolate	1.89 mg gallic acid, 1.08 mg ECG, 3.99 mg catechin, caffeine 16.99 mg, 4.01 mg EC, epigallocatechin 3-gallate 1.40 mg, and epicatechin 3-gallate 0.14 mg	(ii) White chocolate or no chocolate; (iii) no chocolate	Dark chocolate intake was associated with decreased SBP and DBP ($p < 0.05$).
Crews, 2008, USA [46]	Double-blind, placebo-controlled	90 healthy participants (≥60 y)	6 wk	Dark chocolate bar and artificially sweetened cocoa beverage	754.71 mg total proanthocyanidins/g	Placebo bar (0.20 mg/g proanthocyanin) and beverage (40.87 mg/g proanthocyanin)	The dark chocolate and cocoa groups displayed an increased pulse rate at the midpoint and at the end of treatment ($p = 0.007$).

Author, Year, Country	Study Design	Participants (Mean Age)	Duration	Treatment	Polyphenol Constituent (Daily Intake)	Comparison	Main Findings
Davison, 2008, Australia [47]	Double-blind, placebo-controlled	49 participants (41 y)	12 wk	High-flavanol cocoa drink	902 mg flavanols	Low-flavanol cocoa drink (36 mg/d flavanols)	High-flavanol cocoa intake increased FMD ($p < 0.01$) and reduced DBP and mean arterial BP ($p < 0.05$).
rassi, 2008, Italy [48]	Placebo-controlled, crossover	19 essential hypertension participants (44 y)	2 × 15 d (7 d washout)	Flavanol-rich dark chocolate	1008 mg total phenols (110.9 mg EC, 36.12 mg catechin, 2.5 mg quercetin, 0.03 mg kaempferol, and 0.2 mg isorhamnetin)	Flavanol-free white chocolate	Flavanol-rich chocolate reduced the SBP, DBP and ambulatory pressure which was inversely correlated with an increase in FMD with FRDC ($p < 0.0001$).
Iuniyappa, 2008, USA [49]	Double-blind, placebo-controlled, crossover	20 participants with mild-to-moderate hypertension (43 y)	2 × 2 wk (1 wk washout)	Flavanol-rich cocoa drink	902 mg cocoa polyphenols (174 mg EC, 62 mg catechin, and 676 mg procyanidins)	Flavanol-poor placebo (28 mg cocoa polyphenols: 2 mg epicatechin, 8 mg catechin, and 16 mg procyanidins)	Cocoa treatment for 2 wk was able to increase insulin-stimulated brachial arterial dilatation compared to placebo ($p = 0.028$).
Shiina, 2009, Japan [50]	Single-blind, placebo-controlled	39 healthy men (29 y)	2 wk	Flavonoid-rich dark chocolate	Catechin, epicatechin, and procyanidin (550 mg)	White chocolate	Dark chocolate consumption was associated with increased change in CFVR compared to control group ($p < 0.01$).
Ieiss, 2010, USA [51]	Double-blind, placebo-controlled, crossover	16 participants CAD (64 y)	2 × 30 d (1 wk washout)	Cocoa drink	750 mg flavanols, monomers–decamers, 130 mg monomers, 118 mg EC, 12 mg catechin, 106 mg dimers, and 516 mg trimers–decamers	Placebo cocoa drink (18 mg flavanols, monomers–decamers, 6 mg monomers, 2 mg EC, 4 mg catechin, 4 mg dimers, and 6 mg trimers–decamers)	High-flavanol intervention led to a significant increase in FMD ($p < 0.05$) and a decrease in SBP ($p = 0.013$).

Table 1. *Cont.*

Author, Year, Country	Study Design	Participants (Mean Age)	Duration	Treatment	Polyphenol Constituent (Daily Intake)	Comparison	Main Findings
Njike, 2011, USA [52]	Double-blind, placebo-controlled crossover	37 healthy participants (52 y)	2 × 6 wk (4 wk washout)	(i) Sugar-free cocoa beverage; (ii) sugar-sweetened cocoa beverage	21 mg catechin, 48 mg epicatechin, 92 mg procyanidin dimer, 98 mg procyanidin trimer, 31 mg procyanidin tetramer, 55 mg procyanidin pentamer and hexamer, and 805 mg total procyanidin	Sugar-sweetened cocoa-free beverage	Cocoa ingestion improved FMD relative to the control group ($p < 0.01$).
Almoosawi, 2012, UK [53]	Single-blind placebo-controlled, crossover	42 women (21 normal BMI, 13 overweight, and 8 obese)	2 × 4 wk (2 wk washout)	Dark chocolate	500 mg total polyphenols, 18.99 mg epicatechin and catechin	Placebo dark chocolate	Decrease in SBP ($p = 0.007$) and DBP ($p = 0.003$) after dark chocolate intake.
Desideri, 2012, Italy [54]	Double-blind	90 elderly individuals	8 wk	(i) High-flavanol cocoa drink; (ii) intermediate-flavanol cocoa drink	(i) 990 mg flavanols; (ii) 520 mg flavanols	Low flavanols drink (45 mg flavanols)	High and intermediate flavanol intake was associated with reduced BP after 8 wk of treatment ($p < 0.0001$) and compared to the low group ($p < 0.05$).
Flammer, 2012, Switzerland [55]	Double-blind, placebo-controlled	20 participants with congestive heart failure (64 y)	4 wk	Flavanol-rich chocolate	10.8 mg catechin, 36 mg epicatechin	Placebo chocolate	Flavanol-rich chocolate administration led to significant increase in FMD ($p = 0.045$). No changes were observed for BP and HR.
Mogollon, 2013, Canada [56]	Double-blind, placebo-controlled	42 healthy, pregnant women (29 y)	12 wk	High-flavanol dark chocolate	400 mg total flavanols, 64 mg total catechin and epicatechin	Low-flavanol chocolate	Results showed no differences.
Neufingerl, 2013, The Netherlands [57]	Double-blind, placebo-controlled	143 healthy participants (theobromine + cocoa group: 55 y; theobromine group: 53 y; cocoa group: 55 y; placebo group: 55 y)	4 wk	(i) Cocoa drink; (ii) theobromine drink; (iii) theobromine + cocoa drink	(i) 325 mg flavanols, 150 mg theobromine; (ii) 850 mg theobromine; (iii) 325 mg flavanols, 1000 mg theobromine	Placebo drink	No changes were reported in SBP, DBP, and HR among the groups.

Author, Year, Country	Study Design	Participants (Mean Age)	Duration	Treatment	Polyphenol Constituent (Daily Intake)	Comparison	Main Findings
ɛsser, 2014, UK [58]	Double-blind, placebo-controlled, crossover	41 overweight, middle-aged men (57 y)	4 wk (4 wk washout)	High-flavanol chocolate	1078 mg flavanols, 349 mg epicatechins	Normal flavanol chocolate	An increase in FMD by 1% and a decrease in the AIX independent of the type of chocolate consumed.
lbero-Baraibar, 2014, Spain [59]	Double-blind, placebo-controlled	47 healthy participants with BMI of 30.59 ± 2.33 kg/m^2 (57 y)	4 wk	Cocoa extract	414.26 mg total flavanols, 153.44 mg epicatechin, 14.56 mg catechin, 99.40 mg dimer B2, 13.44 mg dimer B1, and 133.53 mg oligomeric procyanidins	Control meal	Both groups showed an improvement in SBP and DBP after 4 weeks of intervention.
Nickols-ichardson, 2014, USA [60]	Placebo-controlled	51 overweight/obese postmenopausal women (35 y)	18 wk	Sugar-free natural cocoa beverage + chocolate snack	270 mg flavanols	Non-cocoa beverage and non-cocoa snack	Both groups observed a significant reduction in SBP and DBP independent of intervention.
Sarriá, 2014, Spain [61]	Placebo-controlled, crossover	24 normocholesterolemic (M: 28 y; F: 26 y) and 20 moderately hypercholesterolemic (M: 35 y; F: 25 y) volunteers	2 × 2 wk	Soluble cocoa powder + milk	416.4 mg polyphenols, 44.1 mg flavanols, 9.3 mg epicatechins, 18 mg catechins, and 16.5 mg procyanidin B2	Milk	No significant changes were observed in SBP, DBP, and HR between groups.
West, 2014, USA [62]	Placebo-controlled, crossover	30 middle-aged, overweight, and moderately obese adults (51 y)	2 × 4 wk (2 wk washout)	Dark chocolate + sugar-free cocoa beverage	814 mg total flavanols	Low-flavanol chocolate bar and cocoa-free beverage with no added sugar (3 mg/d total flavanols)	Increase in brachial artery diameter before hyperemia ($p = 0.001$) and at peak dilation following cuff release ($p = 0.0001$). Resting ($p = 0.04$) and peak ($p = 0.03$) hyperemic blood flow increased after treatment. Substantial decreases in the AIX were observed only in women ($p = 0.01$).

Table 1. *Cont.*

Author, Year, Country	Study Design	Participants (Mean Age)	Duration	Treatment	Polyphenol Constituent (Daily Intake)	Comparison	Main Findings
Heiss, 2015, Germany. [63]	Double-blind, placebo-controlled	42 healthy participants (22 young (26 y); 20 elderly (60 y)	14 d	CF drink	900 mg flavanols, 146 mg monomers, 128 mg (-)-epicatechin, 14 mg (-)-catechin, 4 mg (+)-catechin, and 754 mg dimers–decamers	CF free	Improvement in FMD ($p < 0.001$), reduced PWV and lowered total peripheral resistance, and increased arteriolar and microvascular vasodilatory capacity, and central DBP in both groups. In elderly group, CF decreased the aortic AIX and office peripheral SBP ($p < 0.05$).
Koli, 2015, Finland [64]	Crossover	22 participants with mild hypertension (45 y)	2 × 8 wk (4 wk washout)	Dark chocolate	602.7 mg flavanols	Snack restriction	No changes were observed after dark chocolate intake.
Massee, 2015, Australia [65]	Double-blind, placebo-controlled	38 healthy participants (24 y)	4 wk	Cocoa tablet	250 mg catechin	Placebo tablets (inert cellulose powder)	No differences were observed compared to baseline values.
Mastroiacovo, 2015, Italy [66]	Double-blind, placebo-controlled	90 elderly participants (69 y)	8 wk	(i) High flavanol drink; (ii) intermediate-flavanol drink	(i) 993 mg total flavanols, 185 mg epicatechin, 62 mg catechin, 182 mg dimers, 141 mg trimers, 126 mg tetramers, 297 mg pentamers–decamers; (ii) 520 mg total flavanols, 95 mg epicatechin, 35 mg catechin, 96 mg dimers, 72 mg trimers, 64 mg tetramers, and 158 mg pentamers–decamers	Low flavanol drink containing 48 mg total flavanols, 5 mg epicatechin, 8 mg catechin, 10 mg dimers, 4 mg trimers, 2 mg tetramers, and 17 mg pentamers–decamers	SBP was significantly reduced in HF and IF after treatment ($p < 0.0001$) and compared to the LF group ($p < 0.0001$).
Ottaviani, 2015, USA [67]	Double-blind, placebo-controlled	59 healthy adults (41 y)	12 wk	Cocoa extract capsules	Up to 2000 mg cocoa flavanols (248 mg flavanol monomers, 1752 mg procyanidins)	Placebo capsules, cocoa-flavanol-free	Treatment was not associated with significant changes in BP or platelet function.

Author, Year, Country	Study Design	Participants (Mean Age)	Duration	Treatment	Polyphenol Constituent (Daily Intake)	Comparison	Main Findings
Rassaf, 2016, Germany [68]	Double-blind, placebo-controlled	49 participants with ESRD (65 y)	30 days	Cocola-flavanol-rich beverage	900 mg total cocoa flavanols	Placebo beverage	A significant increase in FMD ($p < 0.001$), a reduction in DBP ($p = 0.03$), and an increase in HR ($p = 0.01$) compared to the placebo group.
ull, 2015, UK [69]	Double-blind placebo-controlled, crossover	32 healthy participants (55 years)	2 × 6 wk	High-flavanol dark chocolate	1064 mg total flavanols, 223 mg monomers, 192 mg procyanidin dimers, and 649 mg procyanidin trimers to decamers	Low-flavanol dark chocolate (88 mg total flavanols, 23 mg monomers, 24 mg procyanidin dimers, and 41 mg procyanidin trimers to decamers)	The LFDC group showed an increase in 24 h ($p = 0.008$) and daily HR ($p = 0.001$) compared to pre-study values.
Sansone, 2015, Germany [70]	Double-blind, placebo-controlled	100 healthy individuals (47 y)	1 mo	Cocoa-flavanol drink	900 mg total flavanols, 146 mg monomers, 128 mg (-)-epicatechin, 14 mg (-)-catechin, 4 mg (+)-catechin, and 754 mg dimers–decamers	Cocoa-flavanol-free drink	The CF group showed an increase in FMD and a decrease in office and central SBP and DBP. PWV and the AIX decreased in the CF group compared to the control group.
Njike, 2016, USA [71]	Double-blind, placebo-controlled	101 adults with stage 1 hypertension (140–159/90–99 mmHg) on no more than one BP medication (53 y)	2 × 8 wk (4 wk washout)	(i) 10 g cocoa powder; (ii) 5 g cocoa powder	(i) 261.8 mg, flavanols, 45.8 mg epicatechin; (ii) 130.9 mg flavonols, 22.9 mg epicatechin	Placebo chocolate	Treatment reduced BP after subgroup analysis (24 h SBP, $p = 0.038$, 24 h DBP, $p = 0.023$). It reduced BP ($p = 0.009$) and improved endothelial function ($p = 0.031$) in participants on beta blockers. Participants on diuretics daily consumption showed significant increase in 24 h ambulatory SBP ($p = 0.022$).

Table 1. *Cont.*

Author, Year, Country	Study Design	Participants (Mean Age)	Duration	Treatment	Polyphenol Constituent (Daily Intake)	Comparison	Main Findings
Garcia-Yu, 2020, Spain [72]	Controlled	137 postmenopausal women (intervention group 57 y; control group 57 y)	6 mo	Chocolate (99% cocoa)	0.58 mg protocatechuic acid, 1.76 mg procyanidin dimer (B3), 10.4 mg catechin, 14.4 mg procyanidin dimer (B2), 26.1 mg epicatechin,8.53 mg procyanidin trimer (C1), 3.54 mg procyanidin A hexoside, 0.02 mg quercetin glucoside, and 0.03 mg quercetin arabinoside	No intervention	No notable changes were observed between groups for SBP, DBP, or brachial–ankle PWV. For PP, in contrast to the increase observed in the control group, there was a decrease in the intervention group ($p = 0.048$).

Abbreviations: AIX (augmentation index); BP (blood pressure); CAD (coronary heart disease); CFVR (coronary flow velocity reserve); d (day); DBP (diastolic blood pressure); EC (epicatechin); ECG (epicatechin gallate); ESRD (end-stage renal disease); F (female); FMD (flow-mediated dilation); FRDC (flavanol-rich dark chocolate); HF (high flavanol); HR (heart rate); IF (intermediate flavanol); LF (low flavanol); LFDC (low-flavanol dark chocolate); M (male); mo (month); PWV (pulse wave velocity); SBP (systolic blood pressure); wk (week); y (year).

Only an early, double-blind, RCT examined the consequences of the ingestion of cocoa flavan-3-ol and procyanidin tablets, containing 234 mg flavan-3-ols and procyanidins, or placebo tablets on the platelet function of 32 healthy individuals. At the end of the trial, a decrease in P-selectin ($p = 0.06$) and adenosine diphosphate (ADP)-induced aggregation ($p = 0.042$) was observed between groups [38]. Further studies have been conducted using cocoa products. Among the studies conducted on healthy adults, a double-blind, RCT conducted on 21 healthy adults (21–55 y) investigated the effects of high-flavonoid (213 mg procyanidins and 46 mg of epicatechin) or low-flavonoid dark chocolate bar intake (46 g/day) showing an improvement in the flow-mediated dilation (FMD) of the brachial artery ($1.3 \pm 0.7\%$; $p = 0.024$) in high-flavonoid chocolate consumers likened to the control group after 2 weeks of treatment, while no differences were observed in blood pressure (BP) in both groups [39]. Similarly, in a crossover RCT conducted on 28 healthy male adults, aged between 18 and 20 years, the effects of flavan-3-ol-containing milk chocolate (providing 168 mg flavan-3-ols, 39 mg epicatechin, and catechin and 126 mg of procyanidins) was evaluated compared to white chocolate with a negligible amount of flavan-3-ol, which crossed over after 14 days: the intervention resulted in a reduction in systolic blood pressure (SBP), mean BP (from 89 ± 2 to 84 ± 2 mmHg, $p = 0.008$), and diastolic blood pressure (DBP) (from 72 ± 2 to 67 ± 2 mmHg, $p = 0.01$), while no differences were observed for white chocolate consumption [40]. Another single-blind, parallel RCT involving 44 participants (56–73 y) investigated the effects of dark chocolate supplementation, containing 30 mg of polyphenols, or polyphenol-free white chocolate. At the end of the treatment (18 weeks) a reduction in SBP (-2.9 ± 1.6 mmHg, $p < 0.001$) and DBP (-1.9 ± 1.0 mmHg, $p < 0.001$) from baseline in the intervention group was observed [44]. Similarly, an RCT conducted on 89 healthy females (18–25 y) administered with either (i) 100 g of dark chocolate (providing 3.99 catechin, 4.01 epicatechin, and 1.89 gallic acid) daily, (ii) 90 g/day of white chocolate (containing no polyphenols), or (iii) no chocolate, showed that only dark chocolate supplementation was associated with a lowering in SBP (from 115.9 ± 12.6 to 107 ± 8.6 mmHg; $p < 0.05$) and DBP (from 73 ± 9.9 to 67.7 ± 9.7 mmHg; $p < 0.05$) [45]. Another study including 39 healthy men (23–40 y) evaluated the effects of flavonoid-rich dark chocolate (providing 500 mg of catechin, epicatechin, and procyanidin) compared to white chocolate consumption on coronary flow velocity reserve (CFVR) measured by noninvasive transthoracic doppler echocardiography. The results demonstrated an increase in CFVR in the treated group (from 3.38 ± 0.49 to 4.28 ± 0.85, $p < 0.01$) likened to the non-flavonoid group [50]. In a crossover, double-blind RCT, the effect of daily consumption of 22 g/day sugar-free cocoa, 22 g/day sugar-sweetened cocoa (providing 21 mg catechin, 48 mg epicatechin, 92 mg procyanidin dimer, 98 mg procyanidin trimer, 31 mg procyanidin tetramer, 55 mg procyanidin pentamer, and hexamer, 805 mg total procyanidin) or placebo (no cocoa with 100 g/day of sugar) beverages for 6 weeks on 37 healthy participants was tested. Results displayed a protective effect of cocoa ingestion, independent of sugar content, on endothelial function measured as FMD (2.4%, 95% CI: 1.5%, 3.2% and 1.5%, 95% CI: 0.6%, 2.4%, respectively), which was increased compared to the placebo group ($p < 0.01$), while no differences were observed concerning BP in any groups [52]. Another crossover, single-blind RCT compared the effect of 20 g of polyphenol-rich dark chocolate (providing 18.99 mg epicatechin and catechin) on 42 women volunteers (21 with a normal BMI, 13 overweight, and 8 obese). The results showed a decreased SBP ($p = 0.007$) and DBP ($p = 0.003$) after 4 weeks of polyphenol-rich dark chocolate administration compared to pretreatment [53]. In another double-blind, parallel-arm RCT, 90 elderly individuals consumed a cocoa drink containing 990 mg (high flavan-3-ols), 520 mg (intermediate flavan-3-ols), or 45 mg (low flavan-3-ols) of flavan-3-ols once daily for 8 weeks. Results displayed a reduction in SBP (high cocoa flavanol (CF) from 141.1 to 131 mmHg, intermediate CF from 142.4 to 134.2 mmHg, $p < 0.0001$) and DBP (high CF from 84.5 to 79.7 mmHg, intermediate CF from 86.4 to 83 mmHg, $p < 0.0001$) after 8 weeks of treatment, as well as a notable decrease in SBP ($p = 0.0018$) and DBP ($p = 0.007$) comparing high and intermediate CF versus low CF [54]. In another double-masked parallel-group RCT, the effect of flavan-3-ol intake in 22 young (<35 y) and 20 eld (50–80 y) healthy males allocated to consume a cocoa-flavan-3-ol drink (providing 900 of flavan-3-ols, 128 mg (−)-epicatechin, 14 mg (−)-catechin, 4 mg (+)-catechin, and 754 dimers–decamers) or a nutrient-matched, cocoa-flavan-3-ol-free control drink was e uated. The results displayed an improvement in FMD (young: 7.8 ± 0.2% and elde 6.3 ± 0.3%; $p < 0.05$), reduced pulse wave velocity (PWV) (in young: from 6.0 ± 0. 5.6 ± 0.1 m/s; in elderly: from 9.3 ± 0.5 to 8.5 ± 0.4 m/s; $p < 0.05$) and total periph resistance, as well as increased arteriolar and microvascular vasodilatory capacity central DBP ($p < 0.05$) in both groups after 2 weeks, with a decreased aortic augmenta index (AIX) and SBP ($p < 0.05$) in elderly participants [63]. In a parallel-arm, double-b RCT conducted on 90 elderly participants (mean age 70 y) without any significant clin medical conditions, the effects of high-, intermediate-, or low-flavan-3-ol drinks (provid 993 mg total flavan-3-ols, 520 mg total flavan-3-ols, or 48 mg total flavan-3-ols, res tively) for 8 weeks were evaluated, resulting in a reduction in SBP in participants a treatment with high (from 138 ± 1.28 to 130 ± 1.21 mmHg; $p < 0.0001$) and intermed (from 137 ± 1.37 to 130.2 ± 1.28 mmHg; $p < 0.0001$) flavan-3-ol cocoa compared with low-treated group ($p < 0.05$), while DBP was reduced in all three groups without any nificant changes between them [66]. Another double-blind, parallel-group RCT, involv 100 healthy individuals aged between 35 and 60 years and divided into two group consume a cocoa-flavan-3-ol-containing drink (providing 900 mg total flavan-3-ols, 146 monomers, 128 mg (−)-epicatechin, 14 mg (−)-catechin, 4 mg (+)-catechin, and 754 dimers–decamers) or a placebo, for 1 month, showed an increase in FMD (1.2, 95% CI: 1.4%) and a decrease in SBP (−4.4, 95% CI: −7.9, −0.9 mmHg) and DBP (−3.9, 95% −6.7, −1.1 mmHg) and PWV (−0.4, 95% CI: −0.8, −0.04) in the intervention group [7

Various studies, conducted in hypertensive individuals and high-CVD-risk indivi als, resulted in promising results as well. A crossover RCT explored the impacts of d chocolate consumption in 20 never-treated, grade I essential hypertensive individu (mean age 43 y) and normotensive control individuals (mean age 34 y) receiving 100 g/ of dark chocolate (containing 88 mg of flavan-3-ols, 21.91 mg catechin, 65.97 mg epic chin, 0.59 mg quercetin, 0.03 mg kaempferol, and 0.31 mg isorhamnetin) or 80 g of wh chocolate. The treatment led to a reduction in SBP (−11.0 ± 6.3 mmHg; $p = 0.0001$) DBP (−6.2 ± 4.2 mmHg; $p = 0.0001$) in comparison to the baseline. A response of and DBP to treatment was observed also in the control group, with a significant ef of treatment ($p < 0.0001$) and time ($p < 0.0001$). Furthermore, an augmentation of FI likened to the baseline and to the white chocolate group was observed in both the hy tensive and control groups ($p < 0.0001$) [41]. In a double-blind RCT, 32 postmenopau hypercholesterolemic women were recruited to evaluate the effects of 6-week supplem tation with a high-flavan-3-ol cocoa beverage, containing 446 mg of flavan-3-ols, c low-flavan-3-ol cocoa beverage, containing 43 mg of flavan-3-ols: at the end of the stu brachial artery hyperemic blood flow increased by 76% after the high-flavan-3-ol tr ment ($p < 0.05$) compared to the baseline, but a decrease in SBP and DBP by 9.3% 6.5%, respectively ($p < 0.05$) in the low-flavan-3-ol group compared to the high-flavan- group was reported [42]. Another crossover RCT was conducted on 19 essential hy tensive participants with impaired glucose tolerance (mean age of 44 y) receiving eit 100 g/day of chocolate rich in flavan-3-ols (providing 110.9 mg epicatechin, 36.12 mg c chin, 2.5 mg quercetin, 0.03 mg kaempferol, and 0.2 mg isorhamnetin) or flavan-3-ol- white chocolate for 15 days followed by washout and crossover. The results showed t flavan-3-ol-rich chocolate was able to reduce both SBP (−3.82 ± 2.40 mmHg) and D (−3.92 ± 1.98 mmHg) and increase FMD ($p < 0.0001$) while no effects were observed for flavan-3-ol-free chocolate [48]. A double-blind, crossover RCT investigated the effects of coa drink supplementation (150 mL twice a day, 900 mg flavan-3-ol/day) on 20 participa with essential hypertension for 2 weeks of treatment (followed by 1 week of wash before crossover), resulting in an increase in insulin-stimulated changes in brachial art diameter compared to the placebo group (8.1 ± 1.2 versus 5.0 ± 1.3; $p = 0.028$), but v not able to significantly reduce BP [49]. A double-blind, parallel RCT, was conducted on 49 overweight/obese participants, randomly assigned to intake high-flavan-3-ol (902 mg flavan-3-ol) or low-flavan-3-ol (36 mg flavan-3-ols) cocoa, in combination or not with exercise for 12 weeks. Compared to the low-flavan-3-ol group, the high-flavan-3-ol group showed an increase in FMD chronically (over 12 weeks) by 1.6% ($p < 0.01$) and a reduction in DBP by 1.6 mmHg and mean arterial BP by 1.2 mmHg ($p < 0.05$), independent of exercise [47]. Another double-blind, crossover RCT, involving 16 patients with coronary artery disease (mean age 64 y) receiving dietary high-flavan-3-ol intervention (providing 750 mg flavan-3-ols) and low-flavan-3-ol intervention for 30 days showed an improvement in brachial artery FMD by 4.6% ($p < 0.05$) and a reduction in SBP (-4.2 ± 2.7 mmHg; $p = 0.013$) in the treated group compared to the control group [51]. A double-blind, RCT conducted on 20 participants with congestive heart failure evaluated the effects of 40 g of flavan-3-ol-rich chocolate (providing 10.8 mg catechin, 36 mg epicatechin) or 28.4 g of control chocolate. The treatment led to an improvement in FMD after 4 weeks of daily intake of flavan-3-ol-rich chocolate (from 5.15 ± 1.23 mm to 5.25 ± 1.22 mm; $p = 0.02$), while no differences were observed in BP and heart rate in either group [55]. Finally, a study conducted on another category of patients (hemodialyzed), including 57 participants supplemented with chocolate-flavan-3-ol rich beverages (900 mg of flavan-3-ols per day) or a placebo for 30 days, showed an increase in FMD (from $3.4 \pm 0.9\%$ to $3.9 \pm 0.8\%$ vs. placebo, from $3.5 \pm 0.7\%$ to $3.5 \pm 0.7\%$; $p < 0.001$), a reduction in DBP (from 73 ± 12 to 69 ± 11 mmHg vs. placebo, from 70 ± 11 to 73 ± 13 mmHg; $p = 0.03$) and an increase in heart rate (from 70 ± 12 to 74 ± 13 bpm vs. placebo, from 75 ± 15 to 74 ± 13 bpm; $p = 0.01$) [68].

In contrast, some studies did not detect substantial differences after supplementation. A crossover, double-blind, RCT investigated the effects of 70 g/day of flavan-3-ol-enriched chocolate consumption (providing 1078 mg flavan-3-ols, 349 mg epicatechins) or normal-flavan-3-ol chocolate on increased endothelium-dependent vasodilatation in 41 healthy overweight men (45–70 y) leading to an increase in FMD ($p = 0.01$) and the AIX ($p = 0.024$) by about 1% as well as soluble ICAM (sICAM) 1, sICAM3, and leukocyte adhesion markers ($p < 0.05$), independent of the type of chocolate consumed [58]. In another parallel-arm RCT, 60 overweight or obese premenopausal women (25–45 y) were instructed to consume dark chocolate snacks along with a once-daily sugar-free cocoa beverage (providing 270 mg of flavan-3-ols) or non-chocolate snacks plus a sugar-free non-cocoa beverage for 18 weeks: results showed that both groups experienced reductions in SBP and DBP with no significant changes between groups [60]. Among other studies with contrasting results, an RCT investigated the effects of 10 g or 5 g of cocoa consumption (providing 261.8 mg flavan-3-ols, 45.8 mg epicatechin, 130.9 mg flavonols, and 22.9 mg epicatechin, respectively) on BP and endothelial function n 101 adults with stage 1 hypertension for 8 weeks: daily cocoa intake did not ameliorate BP and endothelial function but led to BP reduction in a subgroup of participants in angiotensin-converting enzyme inhibitors (24 h SBP: -3.2 ± 9.3 vs. 3.6 ± 8.5; $p = 0.038$, 24 h DBP: -2.0 ± 5.0 vs. 2.3 ± 5.6 mm Hg; $p = 0.023$) whereas daily consumption of 5 g of cocoa-containing products reduced SBP (-4.6 ± 3.2 vs. 1.8 ± 2.8 mm Hg; $p = 0.009$) and ameliorated endothelial activity (3.1 ± 2.3 vs. $-3.4 \pm 7.4\%$; $p = 0.031$) [71]. Also, a placebo-controlled, crossover RCT was conducted in 30 middle-aged individuals who were overweight and moderately obese to evaluate the effect of 37 g/day dark chocolate and a sugar-free cocoa beverage (total flavan-3-ols = 814 mg/day) consumption against a low-flavan-3-ol chocolate bar and a beverage mix without cocoa and no added sugar (total flavan-3-ols = 3 mg/day) on arterial stiffness and endothelial function. After 4 weeks of treatment, no changes were observed in fasting BP or heart rate, but some positive acute responses after the last intervention administration were seen [i.e., augmentation in the diameter of the brachial artery before reactive hyperemia ($p = 0.001$) and at the peak dilation following cuff release ($p = 0.0001$), as well as an increase in resting ($p = 0.04$) and peak ($p = 0.03$) hyperemic blood flow, and increases in SBP ($p = 0.02$), DBP ($p = 0.005$) and heart rate ($p = 0.02$)] compared to the control treatment [62]. Moreover, some other studies showed no important changes after supplementation. In an RCT in which 25 male subj (mean age 38 y), were supplemented with 26 g/day of cocoa powder and 12 g/day of su (providing 98 mg epicatechins, 35 mg catechins, 41 mg procyanidin B2, and 24 mg cyanidin C1) or only with 12 g/day of sugar for 12 weeks, did not show significant eff on BP in both groups [43]. Similarly, a double-blind, fixed-dose, parallel-group RCT inv ing 90 healthy participants, showed that dark chocolate and cocoa (providing 754.71 total proanthocyanidins) were not associated with beneficial effects on BP. However, d chocolate and cocoa groups displayed an increased pulse rate at the midpoint and at end of treatment [46]. An RCT investigated the effects of 20 g/day of flavan-3-ol-rich ch late (providing 400 mg of flavan-3-ol) or low-flavan-3-ol chocolate (containing < 60 of flavan-3-ol) on BP and endothelial function in 42 healthy pregnant women. No ferences neither on FMD or on BP were observed [56]. In another study designed a double-blind, parallel RCT and involving 143 healthy participants, the effects of the take of a 200 mL/day cocoa drink, a 200 mL/day theobromine drink, or a 200 mL/ theobromine + cocoa drink (providing 325 mg flavan-3-ols, 150 mg theobromine, 850 theobromine, 325 mg flavan-3-ols, and 1000 mg theobromine, respectively) for 4 we were evaluated, without reporting significant differences for SBP, DBP, and HR [57]. double-blind, placebo-controlled, parallel RCT involving 47 healthy participants, sup mentation with 1.4 g of cocoa extract (providing 414.26 mg total flavan-3-ols, 153.44 epicatechin, 14.56 mg catechin, 99.40 mg dimer B2, 13.44 mg dimer B1, and 133.53 oligomeric procyanidins) or controlled meals for 4 weeks led to an improvement in BP an decrease in sICAM-1 in both groups [59]. In a crossover RCT, 24 normocholesterolemic 20 moderately hypercholesterolemic volunteers received 30 g/day of a cocoa product in polyphenols (providing 416.4 mg of total polyphenols, 44,1 mg flavan-3-ols, 9.3 mg catechins, 18 mg catechins, and 16.5 mg procyanidin B2) plus milk or a control, consist of milk. At the end of the treatment period, no significant changes were reported in BP heart rate among the groups [61]. A double-blind, parallel RCT evaluated the effects o acute and subchronic daily intake of cocoa tablets for 4 weeks, supplementing 250 mg/ of catechin in 38 young healthy participants aged between 18 and 40 years. Neithe acute nor in subchorionic administration were any changes in cardiovascular functi observed [65]. In a double-blind, crossover RCT, a total of 32 individuals (mean age 5 diagnosed with hypertension, were divided to be supplemented with high-flavan-3-ol d chocolate (1064 mg flavan-3-ols/day for 6 weeks) or low-flavan-3-ol dark chocolate (88 flavan-3-ols/day for 6 weeks), resulting in no changes on BP, although low-flavan-3-ol d chocolate showed a significant increase in HR [69]. In another double-blind, two-para arm RCT, the influence of cocoa extract capsule supplementation (providing 2000 mg co flavan-3-ols) on BP and platelet function was evaluated, but the results showed no relev changes in SBP, DBP, and platelet function after 2 weeks of intervention [67]. In a crosso RCT involving 22 adults aged 33–64 years with mild hypertension, the effect of 49 g/ dark chocolate (providing 602.7 mg flavan-3-ols) intake on cardiovascular risk factors evaluated. Results showed that after 8 weeks, no effects by chocolate intake on cardiova lar risk factors were observed [64]. Another RCT investigated the consequences of intak 10 g of cocoa-rich chocolate (providing 10.4 mg catechin, 14.4 mg procyanidin dimer (26.1 mg epicatechin, and 8.53 mg procyanidin trimer (C1)) for 6 months, showing no r vant changes comparing groups for SBP ($p = 0.391$), DBP ($p = 0.622$) or brachial–ankle P ($p = 0.263$), but a decrease in pulse pressure (PP) was observed in the intervention gr (−2.05 mm Hg; 95% CI: −4.08, −0.02; $p = 0.048$) compared to the control group [72].

It is noteworthy to mention some limitations of the trials included. First, there no consistency in terms of length, dose, type of intervention (tablets vs. food), hea status of participants, and study design; thus, a formal meta-analysis of results is h to interpret. Second, from a clinical standpoint, the formulations themselves were univocal in terms of molecules included; hence, it should be taken into account that effects solely attributed to simple flavan-3-ols may also depend on other components u

3.2. A Summary of Randomized Clinical Trials Investigating the Impact of Catechin Supplementation on Vascular Health

Another group of studies were conducted focusing their attention on green tea catechins. Studies conducted in both healthy and unhealthy individuals resulted in contrasting results presented in Table 2.

A double-blind, parallel RCT, conducted on 111 healthy adult volunteers aged between 21 and 70 years old and administered with two capsules per day of *Camellia sinensis* (containing 400 mg decaffeinated catechin green tea extract) or two placebo capsules for 3 months, resulted in a 3 mmHg reduction in SBP ($p < 0.05$) in the treated group (from 131 ± 6.3 to 128 ± 6.3 mmHg) with no effects on DPB [73]. Another three-arm parallel, double-blind RCT was conducted on 49 healthy adult men aged 18–35 years who were administered 450 mg green tea (about 240 mg catechins) or sour tea (at least 250 mg of anthocyanin) and one placebo group consuming 450 mg of maltodextrin for 6 weeks and showed a reduction in SDP (from 124.4 ± 5.5, after 4 weeks: 114.4 ± 7.4, $p = 0.004$) and DBP ($p = 0.007$) in the sour tea group when compared with pre-intervention values, whereas no significant differences were observed in the green tea and control groups [74]. However, a double-blind, parallel RCT involving 22 healthy postmenopausal women (62–73 y) who were administered one bottle of green tea beverage with 615 mg of total catechins or a placebo beverage with 92 mg of catechins for 4 weeks showed no changes compared to the other groups [75]. Another study included 51 subjects were randomly divided to receive beverages containing high catechin (400 mg/day) or low catechin (100 mg/day) for 9 weeks, followed by 2 weeks of washout, resulting in no significant difference between the high and low catechin group in any CVD risk factors [76]. Similarly, 46 overweight women aged 19–57 years old received a low-calorie diet with three capsules of green tea, three times a day (total intake of tea catechins 1125 mg/day), or three capsules of placebo (maltodextrins) total for 3 months, and showed a decrease in SBP and DBP in both the intervention and control group, with no change and no differences in heart rate [77]. Another RCT conducted on 33 healthy male volunteers who received six capsules per day containing green tea extracts (714 mg/day of green tea polyphenols) or placebo, displayed no effects in the intervention on endothelium-dependent or independent vascular function, as well as no changes in plasma concentrations of asymmetric dimethylarginine [78]. Similar results were detected in another double-blind RCT involving 50 older adults assigned to receive 350 mL of green tea containing 630.9 mg of catechins or a beverage without catechins for 14 weeks; also in this study, no significant differences in the change between the two groups were found [79]. Moreover, an RCT involving 114 healthy subjects receiving either barley extract without catechin (placebo), a "Sunrouge" extract containing 11.2 mg anthocyanin and 323.6 mg (−)-epigallo-catechin-3-gallate (EGCG), or a "Yabukita" extract containing 322.2 mg EGCG for 12 weeks showed an increase following intake of the "Sunrouge" extract ($p < 0.05$) [80].

Table 2. Main characteristics of randomized clinical trials concerning catechins supplementation and cardiovascular risk factors.

Author, Year, Country	Study Design	Participants (Mean Age)	Duration	Treatment	Polyphenol Constituent (Daily Intake)	Comparison	Main Findings
Fukino, 2005, Japan [81]	Controlled	66 patients with borderline diabetes or diabetes (53 y)	2 mo	One packet of GTE/powder	544 mg polyphenols, 456 mg catechins	No intervention	No significant changes were observed between groups.
Diepvens, 2005, The Netherlands [77]	Double-blind, placebo-controlled	46 overweight women (placebo: 41 y; green tea: 41 y)	87 d	GTE capsules with low-calorie diet	1125 mg tea catechins	Maltodextrin capsules	No changes were observed between groups.
Hill, 2007, Australia [82]	Placebo-controlled	38 overweight or obese postmenopausal women (58 y)	12 wk	EGCG capsules + walking for 45 min at 75% of age-predicted maximum heart rate	300 mg EGCG	2 lactose capsules/d + walking for 45 min at 75% of age-predicted maximum heart rate 3/wk	A significant reduction in HR was observed in the EGCG group ($p < 0.01$).
Nagao, 2007, Japan [83]	Double-blind, placebo-controlled	240 women and men with visceral fat-type obesity (41 y)	12 wk	GTE beverage high in catechins	583 mg total catechins, 42.84 mg catechin, 40.12 mg catechin gallate, 127.5 mg gallocatechin, 139.74 mg gallocatechin gallate, 32.3 mg epicatechin, 30.94 mg epicatechin gallate, 60.36 mg epigallocatechin, and 100.3 mg epigallocatechin gallate	Control beverage (96 mg catechins)	Compared with the control group, in the catechin group, a greater decrease in SBP for subjects with initial SBP values of 130 mm Hg or higher was observed ($p < 0.05$).
Fukino, 2008, Japan [84]	Crossover	60 volunteers with fasting blood glucose levels of ≥6.1 mmol/L or nonfasting blood glucose levels of ≥7.8 mmol/L (52 y)	2 × 2 mo (no washout)	GTE powder	544 mg polyphenols, 456 mg catechins	No intervention	No significant changes were observed after intervention.
Frank, 2009, UK [78]	Double-blind, placebo-controlled	33 healthy men (treatment group: 41 y; control group 40 y)	3 wk	GTE capsules	714 mg of green tea polyphenols, ~670 mg flavanols	Placebo capsules (maltodextrin and caffeine)	No changes were observed after intervention.

Author, Year, Country	Study Design	Participants (Mean Age)	Duration	Treatment	Polyphenol Constituent (Daily Intake)	Comparison	Main Findings
Isu, 2008, Taiwan [85]	Double-blind, placebo-controlled	78 obese women (green tea: 43 y; placebo: 43 y)	12 wk	Dried, powdered GTE capsules	491 mg catechins, 302 mg EGCG	Placebo capsules (cellulose)	No significant differences were observed when comparing the two groups.
own, 2009, UK [86]	Double-blind, placebo-controlled	88 overweight or obese male subjects (53 y)	8 wk	EGCG capsules	800 mg EGCG	Placebo capsule (lactose)	Compared to the placebo, EGCG treatment reduced DBP ($p = 0.014$).
lagao, 2009, Japan [87]	Double-blind, placebo-controlled	43 patients with T2DM (catechin group: 64 y; control group 62 y)	12 wk	Catechin-rich beverage	582.8 mg catechins	Control beverage with 96.3 mg catechins	No significant differences were observed between groups.
Nantz, 2009, USA [73]	Double-blind, placebo-controlled	111 healthy adult volunteers (29 y)	3 mo	Camellia sinensis compounds capsules	polyphenols > 80%, catechins > 80%, EGCG > 45%	Placebo capsules (microcrystalline cellulose)	Treatment led to a significant reduction in SBP ($p < 0.05$).
su, 2011, USA [88]	Single-blind, placebo-controlled	35 obese subjects with MetS (42 y)	8 wk	(i) Green tea; (ii) GTE	(i) 440 mg EGCG, 220 mg EGC, 180 mg ECG, and 88 mg EC; (ii) 460 mg EGCG, 240 mg EGC, and 120 mg ECG e 50 mg EC	Water	No effects were observed after treatment.
own, 2011, UK [89]	Double-blind, placebo-controlled, crossover	64 overweight and obese sedentary males (placebo: 49 y; intervention: 49 y)	2 × 6 wk (2 wk washout)	DGT capsules	EGCG 40.71%, EGC 16.27%, EC 8.74%, ECG 6.02%, GC 2.02%, GCG 1.27%, catechin 1.16%, gallic acid 0.75%, and catechin gallate 0.03% in each capsule	Placebo capsules (lactose)	No effect of treatment on any of the ambulatory BP-monitoring parameters.

Table 2. *Cont.*

Author, Year, Country	Study Design	Participants (Mean Age)	Duration	Treatment	Polyphenol Constituent (Daily Intake)	Comparison	Main Findings
Hsu, 2011, Taiwan [90]	Double-blind, placebo-controlled	68 subjects with BMI > 25 kg/m2 and T2DM for more than one year (GTE: 50 y; placebo: 52 y)	16 wk	DGT capsules	856.8 mg EGCG, 236.1 mg ECG, 115.5 mg EGC, 71.9 mg EC, 63.7 mg GCG, and <1.05 mg GC	Placebo capsules (cellulose)	No statistically significant differences were observed between groups.
Sone, 2011, Japan [76]	Placebo-controlled	51 individuals (high-catechin group: 43 y; low-catechin group: 48 y)	9 wk treatment	Catechin-enriched green tea beverage	400 mg catechins	Control beverage with 100 mg catechins	No significant differences were observed between groups.
Bogdanski, 2012, Poland [91]	Double-blind, placebo-controlled	56 obese, hypertensive subjects (GTE group: 49 y; placebo group 51 y)	3 mo	GTE capsules	208 mg EGCG	Placebo capsules (microcrystalline cellulose)	In the treated group, both SBP and DBP significantly decreased ($p < 0.01$).
Suliburska, 2012, Poland [92]	Double-blind, placebo-controlled	46 obese patients (GTE group: 48 y; placebo group: 52 y)	3 mo	GTE capsules	208 mg EGCG	Placebo capsules (microcrystalline cellulose)	No significant changes in SBP and DBP.
Miyazaki, 2013, Japan [79]	Double-blind, placebo-controlled	50 older adults participating in a pedometer-based walking program (69 y)	14 wk	Green tea catechins beverage	630.9 mg total catechin, 125.7 mg GC, 114 mg EGC, 30 mg catechin, 34.7 mg EC, 143.2 mg EGCG, 112.6 mg GCG, 45.8 mg ECG, and 24.8 mg CG	Control beverage (88.7 mg total catechin)	No significant changes were shown when comparing groups.
Liu, 2014, Taiwan [93]	Double-blind, placebo-controlled	77 subjects with type 2 diabetes and lipid abnormalities (GTE group: 55 y; cellulose group: 53 y)	16 wk	GTE capsules	856.8 mg EGCG, 236.1 mg ECG, 115.5 mg EGC, 71.9 mg EC, 63.7 mg GCG, and <1.05 mg GC	Placebo capsules (cellulose)	Treatment with GTE did not show significant changes in BP.
Takahashi, 2014, Japan [75]	Double-blind, placebo-controlled	22 healthy postmenopausal women (placebo: 66 y; green tea: 66 y)	4 wk	Green tea beverage	615 mg catechins	Placebo beverage (92 mg catechins)	No differences in BP between the groups were observed.

Author, Year, Country	Study Design	Participants (Mean Age)	Duration	Treatment	Polyphenol Constituent (Daily Intake)	Comparison	Main Findings
Dower, 2015, The Netherlands [94]	Double-blind, placebo-controlled, crossover	33 healthy (pre)hypertensive men and women (SBP between 125 and 160 mmHg) (66 y)	3 × 4 wk (2 × 4 wk washout)	(i) Epicatechin capsules; (ii) quercetin capsules	(i) 100 mg epicatechin; (ii) 160 mg quercetin-3-glucoside	Placebo capsules	Epicatechin and quercetin supplementation significantly decreased soluble endothelial selectin ($p = 0.03$ and $p = 0.03$, respectively).
hen, 2016, Taiwan [95]	Double-blind, placebo-controlled	77 women with central obesity (GTE: 44 y; placebo: 44 y)	12 wk	DGT capsules	EGCG 856.8 mg, ECG 236.1 mg, EGC 115.5 mg, EC 71.9 mg, GCG 63.7 mg, and GC < 1.05 mg	Placebo capsules (microcrystalline cellulose)	There were no significant percentage reductions in BP between groups.
utiérrez-Salmeàn, 2016, India [96]	Double-blind, placebo-controlled	30 subjects with hypertriglyceridemia (37 y)	4 wk	Epicatechin capsules	100 mg epicatechin	Placebo capsules	No significant changes in SBP and DBP were observed when comparing groups.
Lu and Hsu, 2016, Taiwan [97]	Double-blind, placebo-controlled	64 women with moderate or severe acne vulgaris (GTE 28 y, placebo 30 y)	4 wk	DGT capsules	EGCG 856.8 mg ECG 236.1 mg, EGC 115.5 mg, EC 71.9 mg GCG 63.7 mg, and GC < 1.05 mg	Placebo capsules (microcrystalline cellulose)	No changes were observed after intervention.
Kafeshani, 2017, Iran [74]	Double-blind, placebo-controlled	49 healthy adult men (green tea: 20 y; sour tea: 20 y; placebo: 21 y)	6 wk	(i) Green tea tablets; (ii) sour tea tables	(i) ~240 mg catechins; (ii) at least 250 mg anthocyanins	Placebo tablets (maltodextrin)	A reduction in DBP in the sour tea group compared to the baseline ($p = 0.007$) and a significant reduction in SBP ($p = 0.004$) compared to the placebo.

Table 2. *Cont.*

Author, Year, Country	Study Design	Participants (Mean Age)	Duration	Treatment	Polyphenol Constituent (Daily Intake)	Comparison	Main Findings
Saarenhovi, 2017, Finland [98]	Double-blind, placebo-controlled, crossover	57 otherwise healthy subjects with borderline hypertension (130–139/85–89 mmHg) or unmedicated mild hypertension (140–165/90–95 mmHg) (55 y)	2 × 4 wk (4 to 5 wk washout)	Epicatechin capsules	100 mg epicatechin	Placebo capsules (microcrystalline cellulose)	Treatment induced significant increase in FMD% at the first visit ($p = 0.04$), last visit ($p = 0.02$), and for both visits combined ($p < 0.01$) but not significantly compared to placebo. DBP at end of treatment was −3.3 mmHg ($p = 0.008$).
Kirch, 2018, Germany [99]	Double-blind, placebo-controlled, crossover	47 overweight or obese nonsmokers with clear signs of MetS (M: 36 y; F: 35 y)	2 × 2 wk (2 wk washout)	Epicatechin capsules	25 mg epicatechin	Placebo capsules (mannitol)	No differences were observed between treatments.
Maeda-Yamamoto, 2018, Japan [80]	Double-blind, placebo-controlled	114 healthy subjects, with an SBP value of ≤125 and <155 and a DBP value < 95, or a DBP of ≤75 mmHg and <95 mmHg and an SBP < 155 mmHg (40 y)	12 wk	(i) Green tea cultivar Sunrouge extract; (ii) green tea cultivar Yabukita extract	(i) 11.2 mg anthocyanin, 323.6 mg EGCG, 85.7 mg EGC, 12.3 mg EGCG3″Me, 40.4 mg flavonols; (ii) 322.2 mg EGCG, 413.1 mg EGC, and 54.8 mg flavonols	Barley extract without catechin	After "Sunrouge" extract treatment, participants showed increased blood pressure ($p < 0.05$).

Abbreviations: BP (blood pressure); d (day); DBP (diastolic blood pressure); DGT (decaffeinated green tea extract); EC (epicatechin); ECG (epicatechin gallate); EGC (epigallocatechin); EGCG (epigallocatechin gallate); EGCG3″Me (epigallocatechin-3-=-(3-O-methyl)-gallate; F (female); GC (Gallocatechin); GCG (Gallocatechin gallate); GTE (green tea extract); HR (heart rate); MetS (metabolic syndrome); M (male); mo (month); SBP (systolic blood pressure); T2DM (type 2 diabetes mellitus); wk (week); y (year).

Studies conducted on unhealthy individuals included mostly patients with cardiometabolic risk factors. Most studies provided interventions with green tea as a drink or an extract. A 12-week intervention study investigated the effect of green tea extract on 240 individuals (25–55 y) with visceral fat-type obesity who were administered either green tea (583 mg/day) or a control drink (96 mg catechins/day), and showed that individuals with higher baseline SBP had a better effect of the treatment compared to controls ($p < 0.05$) with no effects on DBP [83]. The same group performed a similar study on 43 patients with type 2 diabetes mellitus, administering for 12 weeks (plus a 4-week follow-up period) green tea containing either 582.8 mg of catechins or 96.3 mg of catechins (control group). Results displayed a marginal reduction in SBP in the treated group during the trial but no significant differences between the two groups [87]. In contrast, a more recent study conducted in 77 women with central obesity (20–60 y) randomly assigned to receive high-dose green tea (consumption of 856.8 mg EGCG/day) or a placebo for 12 weeks reported no reductions between groups and in other measurements analyzed [95].

Among studies on green tea extracts in obese individuals, most studies reported null findings. A double-blind RCT, conducted on 78 obese women, examined the consequence of green tea extract intake (about 500 mg of catechins/day) three times a day for 12 weeks compared to the placebo group. Results showed no significant difference between SBP and DBP in either group nor between them after the follow-up [85]. A similar RCT was conducted on 35 obese participants with metabolic syndrome (mean age 42.5 ± 1.7 y), who were randomly assigned to intake green tea (four cups/day), green tea extract (two capsules), or four cups water/day, and showed that the supplementation did not have any effect on metabolic syndrome features, including SBP ($p = 0.94$ for green tea, $p = 0.86$ for green tea extract vs. no treatment) and DBP ($p = 0.96$ for green tea, $p = 0.48$ for green tea extract vs. no treatment) [88]. Another double-blind, crossover RCT conducted on 64 sedentary males diagnosed as overweight and obese (40–69 y) who were allocated to receive decaffeinated green tea extract for 6 weeks and a crossover, did not lead to any effect on the ambulatory parameters [89]. Another study conducted on 46 slightly younger obese individuals (30–60 y) with similar intervention and duration did not display any significant differences comparing groups [92]. Finally, in another double-blind RCT, 56 obese and hypertensive subjects, aged 30–60 years old, were randomly assigned to ingest a daily dose of one capsule of 379 mg green tea extract containing 208 mg of EGCG or a matching placebo capsule of pure microcrystalline cellulose for 3 months. Results reported a reduction in both SBP (from 145 ± 10 to 141 ± 8 mmHg, $p = 0.004$) and DBP (from 88 ± 4 to 84 ± 3 mmHg, $p < 0.001$) in the treated group compared with the placebo group [91]. Also, studies on diabetic individuals led to mostly null results. Only the first study, conducted on 66 diabetic or prediabetic patients (aged 32–73 y) who received a packet of green tea extract containing 456 mg catechins for 2 months or no intervention, resulted in an improvement in both SBP (from 139.3 ± 15.7 to 131.6 ± 20.8 mmHg) and DBP (from 92.5 ± 11.1 to 83.3 ± 12,6), although with no significant differences with the control group [81]. Then, a crossover RCT was conducted in 60 volunteers (32–73 y) with fasting blood glucose levels of ≥6.1 mmol/L or nonfasting blood glucose levels of ≥7.8 mmol/L, who consumed a packet of green tea-extract powder containing 456 mg catechins daily for 2 months and a 2-month non-intervention period, and showed no significant changes in SBP ($p = 0.43$) whereas the reduction in DBP associated with the intervention was of borderline statistical significance ($p = 0.06$) [84]. A double-blind RCT aimed to examine the effect of a decaffeinated green tea extract providing a daily dose of 856 mg of epigallocatechin gallate or a placebo (cellulose) on 68 subjects with a BMI of >25 kg/m^2 and type 2 diabetes for 16 weeks and did not display a statistically significant difference between groups for any measured variable [90]. Similarly, a study conducted on 77 patients aged between 20 and 65 years and diagnosed with type 2 diabetes mellitus and lipid abnormalities explored the effects of 16 weeks of treatment with decaffeinated green tea extract (856.8 mg of EGCG) compared with a placebo. The results did not show any significant changes in the parameters investigated [93]. The same null results were obtained by a double-blind

RCT conducted on 64 women between 25 to 45 years old with moderate or severe a supplemented with decaffeinated green tea extract, providing a daily dose of 856 m epigallocatechin gallate upon or a placebo (cellulose) for 4 weeks [97].

Studies using purified catechin extracts resulted in more significant, yet not f consistent results. A study comparing the effect of 8 weeks of dietary supplementation v epicatechin (400 mg capsules twice daily) vs. placebo lactose on overweight or obese n subjects (40–65 y) resulted in a reduction in DBP (mean change: placebo −0.058 (SE 0 mmHg vs. EGCG −2·68 (SE 0.72) mmHg; $p = 0.014$) but no significant effects on SBP [Another double-blind RCT conducted on 33 healthy prehypertensive men and wom with a mean age of 66.4 ± 7.9 years, and receiving epicatechin (100 mg/day), querceti glucoside (160 mg/day), or placebo capsules for 4 weeks showed that either epicatec and quercetin supplementation decreased soluble endothelial selectin (−7.7 ng/mL, CI: 14.5, 0.83; $p = 0.03$ and −7.4 ng/mL, 95% CI: 14.3, 0.56; $p = 0.03$, respectively), whe all other markers of endothelial dysfunction and the derived z score for endothelial dysf tion did not change significantly [94]. Another study conducted on 38 overweight or ob postmenopausal women (45–70 y) instructed to intake two capsules containing 300 mg/ of epigallocatechin gallate or lactose (placebo) for 12 weeks resulted in a reduction in h rate in the treated group ($p < 0.01$) although no significant differences were observe arterial compliance, endothelial function, and BP in either treatment [82]. On the contrar double-blind, multicenter RCT including 30 subjects diagnosed with hypertriglyceride (18–50 y), instructed to take two capsules of 25 mg epicatechin or a placebo for 4 we twice a day (total daily dose of 100 mg) [96] and a double-blind, crossover RCT conduc on 47 overweight or obese nonsmokers with clear signs of metabolic syndrome admi tered with 25 mg epicatechin/day or a placebo in random order for 14 consecutive d (with a 2 wk washout) [99] resulted in no significant changes in blood pressure.

Among other food sources of catechins, in a placebo-controlled, repeated-dose, dou blind crossover trial, 57 volunteers with borderline hypertension (130–139/85–89 mm or unmedicated mild hypertension (140–165/90–95 mmHg) were allocated to receive ap polyphenol extract rich in epicatechin and flavan-3-ol oligomers (100 mg of epicatec or a placebo followed by crossover. The results showed that treatment led to a decre in DBP by −3.3 mmHg (95% CI: −5.6, −0.9, $p = 0.008$), and an acute improvement in mean change in maximum FMD% at the first visit of 1.16 (95% CI: 0.04, 2.28 $p = 0.04$), in last visit of 1.37 (95% CI: 0.22, 2.52, $p = 0.02$), and for both visits combined, 1.29 (95% 0.40, 2.18, $p < 0.01$), although the statistical significance was lost when compared with placebo. Moreover, the study did not identify any long-term change in either FMD% endothelium-independent, nitrate-mediated vasodilatation (NMD) [98].

4. Mechanisms of Action of Flavan-3-ols on Vascular Health

4.1. Antioxidant Activity

Cardiovascular disease is associated with impairment in endothelial functions a the assessment of a low-grade inflammatory state resulting in an increased productio cytokines [100,101]. The basic inflammatory process may start following a cellular ins due to pro-oxidant stimuli by reactive oxygen species (ROS) (abnormally generated a by-product of mitochondrial metabolism or by specific enzymes) that exceeds endo nous antioxidant capacity and leads to cell death [102]. The main beneficial effects flavan-3-ols were attributed to their ROS scavenger activity which is exerted through a direct or (2) an indirect mechanism [103] through the reduction in oxidative dam and lipid peroxidation in vascular smooth muscle cells [104]. The first one consists of capability of flavan-3-ols to donate an electron and stop ROS production [105,106]. Th antioxidant activity is also ascribed to their ability to chelate metal ions implicated in rad production [107]. Otherwise, the indirect mechanism consists of the ability to regulate expression of different enzymes involved in ROS scavenging including catalase (CA superoxide dismutase (SOD), and glutathione peroxidase (GSH) [108–110].

4.2. Anti-Inflammatory Regulation

Although the beneficial effect of (poly)phenols on the cardiovascular system have been generally attributed to their antioxidant activity [111], recent knowledge has changed this point of view [112]. Indeed, it is increasingly becoming evident that the protective function of polyphenols may occur through the activation of cellular signaling and the regulation of gene expression [112]. A bioinformatic analysis conducted on nutrigenomic data was performed to investigate the influence of flavan-3-ols in the modulation of gene expression involved in cardiometabolic health [113]. Results identified many genes modulated by flavan-3-ols, not only in endothelial cells and smooth muscle cells but also in hepatocytes, adipocytes, and immune cells that are all involved in the development of the pathology. Here, it was observed that flavan-3-ols mainly modulated different genes implicated in the inflammatory response. In particular, flavan-3-ols acted on transcriptional factors, such as endothelial transcription factor GATA-2 (GATA2), the nuclear factor NF-kappa-B p105 subunit (NF-κB1), forkhead box C1 (FOXC1), peroxisome proliferator-activated receptor-gamma (PPARG), and different miRNA such as let-7b-5p, mir-335-5p, mir-16-5por, and mir-26b-5p [113]. In support of these results, a multi-omic study demonstrated that epicatechin-derived microbial metabolites gamma-valerolactones can regulate the expression of different mRNA, miRNA, and proteins, resulting in the modulation of different cellular pathways involved in cell adhesion, signaling pathways, and interactions with the immune system [114]. In particular, miRNA targets have been found to play a key role in the modulation of different pathways such as endothelial cell functions (adherens junction, gap junction, and focal adhesion), cell signaling (phosphatidylinositol 3-kinase (PI3K)/protein kinase B (PKB/AKT), Wnt, forkhead box O (FOXO), mitogen-activated protein kinases (MAPK), and PPAR signaling), or cell metabolism [114]. Finally, cocoa flavan-3-ols metabolites are able to increase *ApoAI* expression through the involvement of Sites A and B of the *ApoAI* promoter, the major component of HDL which exerts its antiatherogenic properties through the transport of cholesterol to the liver [115].

4.3. Flavan-3-ol Microbial-Derived Metabolites and Cardiovascular Health

Flavan-3-ols and proanthocyanidins maintain relative stability during gastric transit [116]. In the small intestine, flavan-3-ol monomers are rapidly converted to generate phase II sulfated, methylated, and glucuronidated derivatives, by the action of sulfotransferases, catechol-*O*-methyltransferases, and uridine-5′-diphosphate-glucuronosyltransferases, respectively [117]. However, most of the ingested flavan-3-ols are not absorbed in the small intestine, reaching the large intestine and metabolized by the resident microbiota to produce two of their major microbial-derived metabolites, namely the hydroxyphenyl-gamma-valerolactone (PVLs) and their derived hydroxy-phenylvaleric acids (PVAs), which account for 42% of plasma metabolites following flavan-3-ol intake [117]. Other non-specific flavan-3-ol metabolites include vanillic acid, homovanillic acid, hippuric acid, or p-coumaric acid [118]. Interestingly, unlike the original monomeric molecules, colonic metabolites are more available in the human organism [119]. However, the interindividual differences in microbial composition, influenced by age, sex, and diet, correlate with changes in the ability to metabolize (poly)phenols [120,121]. Information on specific bacteria strains and the enzymes involved in flavan-3-ol metabolism to PVLs/PVAs is very limited. *Flavonifractor plautii* can convert 1-(3′,4′-dihydroxyphenyl)-3-(2″,4″,6″-trihydroxyphenyl)-propan-2-ol to 5-(3′,4′-dihydroxyphenyl)-gamma-valerolactone and 4-hydroxy-5-(dihydroxyphenyl)valeric acid [122]. However, another study investigating the interindividual difference in flavan-3-ol metabolism demonstrated that the production of dihydroxyphenyl-gamma-valerolactone conjugates was influenced by interindividual microbial composition with a positive correlation with Clostridia and Actinobacteria class, including the species *F. plautii*, Clostridium leptum, *Sporobacter termitidis*, Ruminococcus bromii, and *Eubacterium ramulus*, and the genus *Propionibacterium* [123]. Similar results have been reported in another study in which high levels of PVLs and PVAs in the plasma and a negative correlation was observed with bacteria *Anaerostipes, Turicibacter, Blautia,* a *Lachnospiraceae* in the levels of these metabolites in the plasma and urine, respectively [1

Furthermore, an in vitro study demonstrated the involvement of *Akkermansia mucinip* in the degradation of EGCG. However, it was demonstrated that *A. muciniphila* did metabolize EGCG alone but in the presence of either mucin or glucose substrate [125] a converted EGCG into gallic acid, epigallocatechin, and (−)-epicatechin [125], promoting own grown as demonstrated by many other works [126,127].

The cardioprotective properties attributed to flavan-3-ols could be associated w their microbial-derived metabolites PVLs and PVAs through their anti-inflammatory tivity, hypotensive properties, and ability to reduce the adhesion of monocyte to endo lial cells, preventing atherosclerosis development as represented in Figure 1. 5-(3 dihydroxyphenyl)-gamma-valerolactone has been tested for its anti-inflammatory pro ties, showing the capacity to inhibit the expression of iNOS [128] which increased acti has been correlated with heart failure [129]. Moreover, two microbial-derived metabol 5-(3,4,5-trihydroxyphenyl)-gamma-valerolactone and 5-(3,5-dihydroxyphenyl)-gam valerolactone showed hypotensive properties in an in vivo study conducted on rats [1 Hydroxyphenyl valeric acids, also derived by flavan-3-ols metabolism, demonstrate strong inhibitory activity of angiotensin I-converting enzymes (ACEs) [130]. The car protective effect of the microbial-derived metabolite 5-(3′,4′-dihydroxyphenyl)-gam valerolactone has been demonstrated by another in vitro study, demonstrating that the tr ment prevents monocyte adhesion to endothelial cells through downregulation of VCA in tumor necrosis factor-alpha (TNF-alpha)-treated cells [131]. This mechanism could attributed to a downregulation of phosphorylation and the activation of NF-κB [131]. Mi bial catabolism of flavan-3-ols leads also to the production of phenolic acids [132] show anti-inflammatory, antioxidant, and cardioprotective effects. Among catabolic metabol derived by flavan-3-ols, *trans*-caffeic acid and 3,4-dihydroxyphenylpropionic acid (DHPP) inhibited NO production in lipopolysaccharides (LPS)-treated RAW264.7 cells [1 Furthermore, in vitro treatment with benzoic acid-sulfate, vanillic acid-glucuronide, a protocatechuic acid-3-sulfate leads to a reduction in TNF-alpha secretion induced by I stimulation [134]. Similar results demonstrated a dose-dependent response to protoc chuic acid treatment, leading to a reduction in TNF-alpha, interleukin 6 (IL-6), IL-1 beta, a prostaglandin E2 production as well as a reduced expression of toll-like receptor 4 (TL activation of NF-κB, and MAPKs [135]. In another study, protocatechuic acid used a concentration between 5 and 25 uM showed a significant reduction in platelet aggregat which was mediated by blocking the interaction between the von Willebrand factor (v and the platelet receptor glycoprotein Ib [136]. Furthermore, it was demonstrated that pr catechuic acid administration to type 1 diabetic rats decreased plasma glycated hemoglo (HbA1c) and cardiac malondialdehyde (MDA) levels, indicating an improvement of card mitochondrial function, and increased anti-apoptotic B-cell lymphoma 2 (BCL2) exp sion [137]. Protocatechuic acid also showed hypoglycemic and insulin-sensitizing acti exerted via the stimulation of the poly (ADP-ribose) polymerase (PARP)/protein kinas (PKC)/NF-κB pathway in myocardial tissue [138]. 3-hydroxyphenylpropionic acid (3-H was also reported to have beneficial effects on the endothelium through the increas phosphorylation of eNOS and Akt, a pathway that has been greatly demonstrated to protective for endothelial function [139].

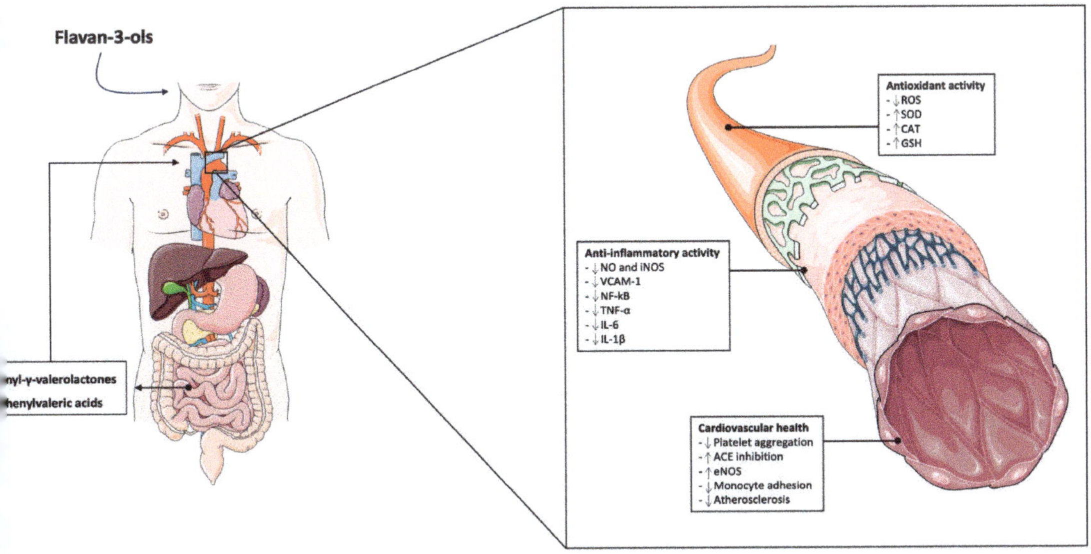

Figure 1. A schematic overview of the primary activities of microbial-derived metabolite flavan-3-ols. ACE (angiotensin-converting enzyme); CAT (catalase); eNOS (endothelial nitric oxide synthase); GSH (glutathione peroxidase); IL-1β (interleukin-1β); IL-6 (interleukin-6); iNOS (inducible nitric oxide synthase); NF-κB (nuclear factor-kappa B); NO (nitric oxide); ROS (reactive oxygen species); SOD (superoxide dismutase); TNF-α (tumor necrosis factor-alpha).

5. Future Perspectives

This review summarized the main results from clinical studies exploring the effects of flavan-3-ols on vascular outcomes. Although most trials reported significant effects on vascular health, including reductions in blood pressure and improved endothelial function, not all reported results in the literature are univocal. Thus, the evidence from reviewed studies should be considered in light of some limitations. First, the heterogeneity among the reviewed trials related to differences in the health status of the target population and characteristics of exposure, including differences in the doses, supplement composition, and trial duration, may at least partially explain the discrepancies among the explored studies. Second, the limitations comprise the potential interactions, including accumulating, synergistic, and antagonistic effects, with other supplement components as well as food matrix components. Finally, the variations in the actual exposure to flavan-3-ol metabolites related to the differences in gut microbiota composition cannot be ruled out. Therefore, future clinical studies exploring variables that may influence the intra- and inter-individual variations in the bioavailability of flavan-3-ols to better elucidate the role of flavan-3-ols and their metabolites in vascular functioning are warranted.

6. Conclusions

Clinical trials demonstrated that supplementation with flavan-3-ols, primarily derived from cocoa products, exhibits significant effects in reducing blood pressure and ameliorating endothelial function. Studies focusing on catechins have shown enhanced outcomes when the supplementation occurred through green tea drinks in healthy individuals. Moreover, quantitative analyses support the efficacy of green tea supplementation, including extracts, in reducing blood pressure, particularly in hypertensive individuals or in those at high risk of CVD. From a mechanistic point of view, emerging evidence suggests that microbial metabolites may play a role in the observed effects. Their function extends beyond the previous belief of ROS scavenging activity and encompasses a direct impact on gene the role of the microbiota and the inter-individual variations in gut microbiota composit aiming to gain a comprehensive understanding of how differences in microbiota-deri metabolites may influence cardiovascular risk.

Author Contributions: Conceptualization, J.G. and G.G.; investigation, J.G., S.L., I.G. and C writing—original draft preparation, J.G., G.L.R., S.L., L.G., I.G. and G.G.; writing—review editing, J.G., G.L.R., S.L., L.G., I.G., I.D.A., R.M.D., J.L.Q., M.B., F.D., F.G. (Francesca Giampieri), (Fabio Galvano) and G.G.; table visualization, J.G., S.L. and I.G.; figure visualization, S.L.; supervis M.B., F.D., F.G. (Fabio Galvano) and G.G.; project administration, G.G. All authors have read agreed to the published version of the manuscript.

Funding: This research received no external funding.

Acknowledgments: J.G. was supported by the co-financing of the European Union—FSE-REA EU, PON Research and Innovation 2014-2020 DM1062/2021; CUP: E65F21002560001. The ures were generated by using Servier Medical Art available at smart.servier.com (accessed 15 November 2023).

Conflicts of Interest: The authors declare no conflicts of interest.

Abbreviations

3,4-DHPP (3,4-dihydroxyphenylpropionic acid); 3-HPP (3-hydroxyphenylpropionic); A (angiotensin-converting enzyme); ADP (adenosine diphosphate); AIX (augmentation index); A (protein kinase B); (ApoAI) apolipoprotein AI; BCL2 (B-cell lymphoma 2); BMI (body mass index) (blood pressure); bpm (beats per minute); CAT (catalase); CFVR (flow velocity reserve); CI (confide interval); CVD (cardiovascular disease); DBP (diastolic blood pressure); EGCG (epigallo-catechi gallate); eNOS (endothelial nitric oxide synthase); ERK1/2 (extracellular signal-regulated kina FMD (flow-mediated dilation); FOXC1 (forkhead box C1); Foxo (forkhead box o); GSH (glutathi peroxidase); HbA1c (Glycated hemoglobin); HDL (high-density lipoprotein); HR (heart rate); ICA (intercellular adhesion molecule-1); IL (interleukin); iNOS (inducible nitric oxide synthase); (lipopolysaccharides); MAPK (mitogen-activated protein kinase); MDA (malondialdehyde); NF (nuclear factor-kappa B); NMD (nitrate-mediated vasodilatation); NO (nitric oxide); PI3K-Akt (p phatidylinositol 3-kinase-Akt); PKC (protein kinase C); PARP (poly (ADP-ribose) polymerase); PI (peroxisome proliferator-activated receptor); PPARG (peroxisome proliferator-activated recep gamma); PVAs (hydroxy-phenylvaleric acids); PVLs (hydroxy-phenyl-gamma-valerolactone); P (pulse wave velocity); RCT (randomized clinical trial); ROS (reactive oxygen species); SBP (syst blood pressure); sICAM1 (soluble intercellular cell adhesion molecule-1); sICAM3 (soluble interce lar cell adhesion molecule-3); SOD (superoxide dismutase); TLR4 (toll-like receptor 4); TNF-al (tumor necrosis factor-alpha); VCAM-1 (vascular cell adhesion protein-1); vWF (von Willebr factor); wk (week); wnt (wingless/integrated); y (year).

References

1. GBD 2019 Diseases and Injuries Collaborators. Global burden of 369 diseases and injuries in 204 countries and territo 1990–2019: A systematic analysis for the Global Burden of Disease Study 2019. *Lancet* **2020**, *396*, 1204–1222. [CrossRef] [PubM
2. Crowe-White, K.M.; Evans, L.W.; Kuhnle, G.G.C.; Milenkovic, D.; Stote, K.; Wallace, T.; Handu, D.; Senkus, K.E. Flavan-3-ols Cardiometabolic Health: First Ever Dietary Bioactive Guideline. *Adv. Nutr.* **2022**, *13*, 2070–2083. [CrossRef] [PubMed]
3. GBD 2019 Risk Factors Collaborators Global burden of 87 risk factors in 204 countries and territories, 1990–2019: A system analysis for the Global Burden of Disease Study 2019. *Lancet* **2020**, *396*, 1223–1249. [CrossRef] [PubMed]
4. GBD 2017 Diet Collaborators Health effects of dietary risks in 195 countries, 1990–2017: A systematic analysis for the Glc Burden of Disease Study 2017. *Lancet* **2019**, *393*, 1958–1972. [CrossRef] [PubMed]
5. Figueiredo, C.S.; Roseira, E.S.; Viana, T.T.; Silveira, M.A.D.; de Melo, R.M.V.; Fernandez, M.G.; Lemos, L.M.G.; Passos, L. Inflammation in Coronary Atherosclerosis: Insights into Pathogenesis and Therapeutic Potential of Anti-Inflammatory Dr *Pharmaceuticals* **2023**, *16*, 1242. [CrossRef] [PubMed]
6. Grosso, G.; Laudisio, D.; Frias-Toral, E.; Barrea, L.; Muscogiuri, G.; Savastano, S.; Colao, A. Anti-Inflammatory Nutrients Obesity-Associated Metabolic-Inflammation: State of the Art and Future Direction. *Nutrients* **2022**, *14*, 1137. [CrossRef] [PubM

Grosso, G.; Marventano, S.; Yang, J.; Micek, A.; Pajak, A.; Scalfi, L.; Galvano, F.; Kales, S.N. A comprehensive meta-analysis on evidence of Mediterranean diet and cardiovascular disease: Are individual components equal? *Crit. Rev. Food Sci. Nutr.* **2017**, *57*, 3218–3232. [CrossRef] [PubMed]

Wang, T.; Kroeger, C.M.; Cassidy, S.; Mitra, S.; Ribeiro, R.V.; Jose, S.; Masedunskas, A.; Senior, A.M.; Fontana, L. Vegetarian Dietary Patterns and Cardiometabolic Risk in People with or at High Risk of Cardiovascular Disease: A Systematic Review and Meta-analysis. *JAMA Netw. Open* **2023**, *6*, e2325658. [CrossRef] [PubMed]

Poulsen, N.B.; Lambert, M.N.T.; Jeppesen, P.B. The Effect of Plant Derived Bioactive Compounds on Inflammation: A Systematic Review and Meta-Analysis. *Mol. Nutr. Food Res.* **2020**, *64*, e2000473. [CrossRef]

Grosso, G. Effects of Polyphenol-Rich Foods on Human Health. *Nutrients* **2018**, *10*, 1089. [CrossRef]

Grosso, G.; Godos, J.; Currenti, W.; Micek, A.; Falzone, L.; Libra, M.; Giampieri, F.; Forbes-Hernández, T.Y.; Quiles, J.L.; Battino, M.; et al. The effect of dietary polyphenols on vascular health and hypertension: Current evidence and mechanisms of action. *Nutrients* **2022**, *14*, 545. [CrossRef] [PubMed]

Pinto, P.; Santos, C.N. Worldwide (poly)phenol intake: Assessment methods and identified gaps. *Eur. J. Nutr.* **2017**, *56*, 1393–1408. [CrossRef] [PubMed]

Raman, G.; Shams-White, M.; Avendano, E.E.; Chen, F.; Novotny, J.A.; Cassidy, A. Dietary intakes of flavan-3-ols and cardiovascular health: A field synopsis using evidence mapping of randomized trials and prospective cohort studies. *Syst. Rev.* **2018**, *7*, 100. [CrossRef] [PubMed]

Godos, J.; Vitale, M.; Micek, A.; Ray, S.; Martini, D.; Del Rio, D.; Riccardi, G.; Galvano, F.; Grosso, G. Dietary Polyphenol Intake, Blood Pressure, and Hypertension: A Systematic Review and Meta-Analysis of Observational Studies. *Antioxidants* **2019**, *8*, 152. [CrossRef] [PubMed]

Mozaffarian, D.; Wu, J.H.Y. Flavonoids, dairy foods, and cardiovascular and metabolic health: A review of emerging biologic pathways. *Circ. Res.* **2018**, *122*, 369–384. [CrossRef] [PubMed]

Micek, A.; Godos, J.; Del Rio, D.; Galvano, F.; Grosso, G. Dietary Flavonoids and Cardiovascular Disease: A Comprehensive Dose-Response Meta-Analysis. *Mol. Nutr. Food Res.* **2021**, *65*, e2001019. [CrossRef]

Grosso, G.; Micek, A.; Godos, J.; Pajak, A.; Sciacca, S.; Galvano, F.; Giovannucci, E.L. Dietary Flavonoid and Lignan Intake and Mortality in Prospective Cohort Studies: Systematic Review and Dose-Response Meta-Analysis. *Am. J. Epidemiol.* **2017**, *185*, 1304–1316. [CrossRef]

Mena, P.; Bresciani, L. Dietary fibre modifies gut microbiota: What's the role of (poly)phenols? *Int. J. Food Sci. Nutr.* **2020**, *71*, 783–784. [CrossRef]

Zmora, N.; Suez, J.; Elinav, E. You are what you eat: Diet, health and the gut microbiota. *Nat. Rev. Gastroenterol. Hepatol.* **2019**, *16*, 35–56. [CrossRef]

Stalmach, A.; Mullen, W.; Steiling, H.; Williamson, G.; Lean, M.E.J.; Crozier, A. Absorption, metabolism, and excretion of green tea flavan-3-ols in humans with an ileostomy. *Mol. Nutr. Food Res.* **2010**, *54*, 323–334. [CrossRef]

Baky, M.H.; Elshahed, M.; Wessjohann, L.; Farag, M.A. Interactions between dietary flavonoids and the gut microbiome: A comprehensive review. *Br. J. Nutr.* **2022**, *128*, 577–591. [CrossRef] [PubMed]

Di Pede, G.; Mena, P.; Bresciani, L.; Achour, M.; Lamuela-Raventós, R.M.; Estruch, R.; Landberg, R.; Kulling, S.E.; Wishart, D.; Rodriguez-Mateos, A.; et al. Revisiting the bioavailability of flavan-3-ols in humans: A systematic review and comprehensive data analysis. *Mol. Asp. Med.* **2023**, *89*, 101146. [CrossRef]

Anhê, F.F.; Choi, B.S.Y.; Dyck, J.R.B.; Schertzer, J.D.; Marette, A. Host-Microbe Interplay in the Cardiometabolic Benefits of Dietary Polyphenols. *Trends Endocrinol. Metab.* **2019**, *30*, 384–395. [CrossRef]

Adak, A.; Khan, M.R. An insight into gut microbiota and its functionalities. *Cell. Mol. Life Sci.* **2019**, *76*, 473–493. [CrossRef] [PubMed]

Cryan, J.F.; O'Riordan, K.J.; Cowan, C.S.M.; Sandhu, K.V.; Bastiaanssen, T.F.S.; Boehme, M.; Codagnone, M.G.; Cussotto, S.; Fulling, C.; Golubeva, A.V.; et al. The Microbiota-Gut-Brain Axis. *Physiol. Rev.* **2019**, *99*, 1877–2013. [CrossRef]

Góralczyk-Bińkowska, A.; Szmajda-Krygier, D.; Kozłowska, E. The Microbiota-Gut-Brain Axis in Psychiatric Disorders. *Int. J. Mol. Sci.* **2022**, *23*, 1245. [CrossRef] [PubMed]

Di Vincenzo, F.; Del Gaudio, A.; Petito, V.; Lopetuso, L.R.; Scaldaferri, F. Gut microbiota, intestinal permeability, and systemic inflammation: A narrative review. *Intern. Emerg. Med.* **2024**, *19*, 275–293. [CrossRef]

Jie, Z.; Xia, H.; Zhong, S.-L.; Feng, Q.; Li, S.; Liang, S.; Zhong, H.; Liu, Z.; Gao, Y.; Zhao, H.; et al. The gut microbiome in atherosclerotic cardiovascular disease. *Nat. Commun.* **2017**, *8*, 845. [CrossRef]

Cui, X.; Ye, L.; Li, J.; Jin, L.; Wang, W.; Li, S.; Bao, M.; Wu, S.; Li, L.; Geng, B.; et al. Metagenomic and metabolomic analyses unveil dysbiosis of gut microbiota in chronic heart failure patients. *Sci. Rep.* **2018**, *8*, 635. [CrossRef]

Li, J.; Zhao, F.; Wang, Y.; Chen, J.; Tao, J.; Tian, G.; Wu, S.; Liu, W.; Cui, Q.; Geng, B.; et al. Gut microbiota dysbiosis contributes to the development of hypertension. *Microbiome* **2017**, *5*, 14. [CrossRef]

Thomas, M.S.; Fernandez, M.L. Trimethylamine N-Oxide (TMAO), Diet and Cardiovascular Disease. *Curr. Atheroscler. Rep.* **2021**, *23*, 12. [CrossRef]

Rahman, M.M.; Islam, F.; -Or-Rashid, M.H.; Mamun, A.A.; Rahaman, M.S.; Islam, M.M.; Meem, A.F.K.; Sutradhar, P.R.; Mitra, S.; Mimi, A.A.; et al. The gut microbiota (microbiome) in cardiovascular disease and its therapeutic regulation. *Front. Cell. Infect. Microbiol.* **2022**, *12*, 903570. [CrossRef] [PubMed]

33. Nemet, I.; Li, X.S.; Haghikia, A.; Li, L.; Wilcox, J.; Romano, K.A.; Buffa, J.A.; Witkowski, M.; Demuth, I.; König, M.; et al. Atla gut microbe-derived products from aromatic amino acids and risk of cardiovascular morbidity and mortality. *Eur. Heart J.* **2** *44*, 3085–3096. [CrossRef] [PubMed]
34. Martín, M.Á.; Ramos, S. Impact of dietary flavanols on microbiota, immunity and inflammation in metabolic diseases. *Nutr.* **2021**, *13*, 850. [CrossRef] [PubMed]
35. García-Cordero, J.; Martinez, A.; Blanco-Valverde, C.; Pino, A.; Puertas-Martín, V.; San Román, R.; de Pascual-Teres Regular consumption of cocoa and red berries as a strategy to improve cardiovascular biomarkers via modulation of microb metabolism in healthy aging adults. *Nutrients* **2023**, *15*, 2299. [CrossRef] [PubMed]
36. Xiong, H.-H.; Lin, S.-Y.; Chen, L.-L.; Ouyang, K.-H.; Wang, W.-J. The Interaction between Flavonoids and Intestinal Microbe Review. *Foods* **2023**, *12*, 320. [CrossRef] [PubMed]
37. Aron, P.M.; Kennedy, J.A. Flavan-3-ols: Nature, occurrence and biological activity. *Mol. Nutr. Food Res.* **2008**, *52*, 79– [CrossRef] [PubMed]
38. Murphy, K.J.; Chronopoulos, A.K.; Singh, I.; Francis, M.A.; Moriarty, H.; Pike, M.J.; Turner, A.H.; Mann, N.J.; Sinclair, A.J. Die flavanols and procyanidin oligomers from cocoa (*Theobroma cacao*) inhibit platelet function. *Am. J. Clin. Nutr.* **2003**, *77*, 1466–1 [CrossRef] [PubMed]
39. Engler, M.B.; Engler, M.M.; Chen, C.Y.; Malloy, M.J.; Browne, A.; Chiu, E.Y.; Kwak, H.-K.; Milbury, P.; Paul, S.M.; Blumber et al. Flavonoid-rich dark chocolate improves endothelial function and increases plasma epicatechin concentrations in hea adults. *J. Am. Coll. Nutr.* **2004**, *23*, 197–204. [CrossRef]
40. Fraga, C.G.; Actis-Goretta, L.; Ottaviani, J.I.; Carrasquedo, F.; Lotito, S.B.; Lazarus, S.; Schmitz, H.H.; Keen, C.L. Reg consumption of a flavanol-rich chocolate can improve oxidant stress in young soccer players. *Clin. Dev. Immunol.* **2005**, *12*, 11 [CrossRef]
41. Grassi, D.; Necozione, S.; Lippi, C.; Croce, G.; Valeri, L.; Pasqualetti, P.; Desideri, G.; Blumberg, J.B.; Ferri, C. Cocoa redu blood pressure and insulin resistance and improves endothelium-dependent vasodilation in hypertensives. *Hypertension* **2005** 398–405. [CrossRef] [PubMed]
42. Wang-Polagruto, J.F.; Villablanca, A.C.; Polagruto, J.A.; Lee, L.; Holt, R.R.; Schrader, H.R.; Ensunsa, J.L.; Steinberg, F.M.; Schr H.H.; Keen, C.L. Chronic consumption of flavanol-rich cocoa improves endothelial function and decreases vascular cell adhe molecule in hypercholesterolemic postmenopausal women. *J. Cardiovasc. Pharmacol.* **2006**, *47* (Suppl. S2), S177–S186, discuss S206. [CrossRef] [PubMed]
43. Baba, S.; Osakabe, N.; Kato, Y.; Natsume, M.; Yasuda, A.; Kido, T.; Fukuda, K.; Muto, Y.; Kondo, K. Continuous intak polyphenolic compounds containing cocoa powder reduces LDL oxidative susceptibility and has beneficial effects on pla HDL-cholesterol concentrations in humans. *Am. J. Clin. Nutr.* **2007**, *85*, 709–717. [CrossRef] [PubMed]
44. Taubert, D.; Roesen, R.; Lehmann, C.; Jung, N.; Schömig, E. Effects of low habitual cocoa intake on blood pressure and bioac nitric oxide: A randomized controlled trial. *JAMA* **2007**, *298*, 49–60. [CrossRef] [PubMed]
45. Abdullah, A.N. Short-Term Consumption of a Dark Chocolate Containing Flavanols is Followed by a Significant Decreas Normotensive Population. *Pak. J. Nutr.* **2008**, *7*, 773–781. [CrossRef]
46. Crews, W.D.; Harrison, D.W.; Wright, J.W. A double-blind, placebo-controlled, randomized trial of the effects of dark choc and cocoa on variables associated with neuropsychological functioning and cardiovascular health: Clinical findings fro sample of healthy, cognitively intact older adults. *Am. J. Clin. Nutr.* **2008**, *87*, 872–880. [CrossRef] [PubMed]
47. Davison, K.; Coates, A.M.; Buckley, J.D.; Howe, P.R.C. Effect of cocoa flavanols and exercise on cardiometabolic risk factor overweight and obese subjects. *Int. J. Obes.* **2008**, *32*, 1289–1296. [CrossRef] [PubMed]
48. Grassi, D.; Desideri, G.; Necozione, S.; Lippi, C.; Casale, R.; Properzi, G.; Blumberg, J.B.; Ferri, C. Blood pressure is redu and insulin sensitivity increased in glucose-intolerant, hypertensive subjects after 15 days of consuming high-polyphenol d chocolate. *J. Nutr.* **2008**, *138*, 1671–1676. [CrossRef]
49. Muniyappa, R.; Hall, G.; Kolodziej, T.L.; Karne, R.J.; Crandon, S.K.; Quon, M.J. Cocoa consumption for 2 wk enhances insu mediated vasodilatation without improving blood pressure or insulin resistance in essential hypertension. *Am. J. Clin. Nutr.* **2** *88*, 1685–1696. [CrossRef]
50. Shiina, Y.; Funabashi, N.; Lee, K.; Murayama, T.; Nakamura, K.; Wakatsuki, Y.; Daimon, M.; Komuro, I. Acute effect of flavonoid-rich dark chocolate intake on coronary circulation, as compared with non-flavonoid white chocolate, by transthor Doppler echocardiography in healthy adults. *Int. J. Cardiol.* **2009**, *131*, 424–429. [CrossRef]
51. Heiss, C.; Jahn, S.; Taylor, M.; Real, W.M.; Angeli, F.S.; Wong, M.L.; Amabile, N.; Prasad, M.; Rassaf, T.; Ottaviani, J.I.; e Improvement of endothelial function with dietary flavanols is associated with mobilization of circulating angiogenic cell patients with coronary artery disease. *J. Am. Coll. Cardiol.* **2010**, *56*, 218–224. [CrossRef] [PubMed]
52. Njike, V.Y.; Faridi, Z.; Shuval, K.; Dutta, S.; Kay, C.D.; West, S.G.; Kris-Etherton, P.M.; Katz, D.L. Effects of sugar-sweetened sugar-free cocoa on endothelial function in overweight adults. *Int. J. Cardiol.* **2011**, *149*, 83–88. [CrossRef] [PubMed]
53. Almoosawi, S.; Tsang, C.; Ostertag, L.M.; Fyfe, L.; Al-Dujaili, E.A.S. Differential effect of polyphenol-rich dark chocolate biomarkers of glucose metabolism and cardiovascular risk factors in healthy, overweight and obese subjects: A randomi clinical trial. *Food Funct.* **2012**, *3*, 1035–1043. [CrossRef] [PubMed]

Desideri, G.; Kwik-Uribe, C.; Grassi, D.; Necozione, S.; Ghiadoni, L.; Mastroiacovo, D.; Raffaele, A.; Ferri, L.; Bocale, R.; Lechiara, M.C.; et al. Benefits in cognitive function, blood pressure, and insulin resistance through cocoa flavanol consumption in elderly subjects with mild cognitive impairment: The Cocoa, Cognition, and Aging (CoCoA) study. *Hypertension* **2012**, *60*, 794–801. [CrossRef] [PubMed]
Flammer, A.J.; Sudano, I.; Wolfrum, M.; Thomas, R.; Enseleit, F.; Périat, D.; Kaiser, P.; Hirt, A.; Hermann, M.; Serafini, M.; et al. Cardiovascular effects of flavanol-rich chocolate in patients with heart failure. *Eur. Heart J.* **2012**, *33*, 2172–2180. [CrossRef] [PubMed]
Mogollon, J.A.; Bujold, E.; Lemieux, S.; Bourdages, M.; Blanchet, C.; Bazinet, L.; Couillard, C.; Noël, M.; Dodin, S. Blood pressure and endothelial function in healthy, pregnant women after acute and daily consumption of flavanol-rich chocolate: A pilot, randomized controlled trial. *Nutr. J.* **2013**, *12*, 41. [CrossRef]
Neufingerl, N.; Zebregs, Y.E.M.P.; Schuring, E.A.H.; Trautwein, E.A. Effect of cocoa and theobromine consumption on serum HDL-cholesterol concentrations: A randomized controlled trial. *Am. J. Clin. Nutr.* **2013**, *97*, 1201–1209. [CrossRef] [PubMed]
Esser, D.; Mars, M.; Oosterink, E.; Stalmach, A.; Müller, M.; Afman, L.A. Dark chocolate consumption improves leukocyte adhesion factors and vascular function in overweight men. *FASEB J.* **2014**, *28*, 1464–1473. [CrossRef]
Ibero-Baraibar, I.; Abete, I.; Navas-Carretero, S.; Massis-Zaid, A.; Martinez, J.A.; Zulet, M.A. Oxidised LDL levels decreases after the consumption of ready-to-eat meals supplemented with cocoa extract within a hypocaloric diet. *Nutr. Metab. Cardiovasc. Dis.* **2014**, *24*, 416–422. [CrossRef]
Nickols-Richardson, S.M.; Piehowski, K.E.; Metzgar, C.J.; Miller, D.L.; Preston, A.G. Changes in body weight, blood pressure and selected metabolic biomarkers with an energy-restricted diet including twice daily sweet snacks and once daily sugar-free beverage. *Nutr. Res. Pract.* **2014**, *8*, 695–704. [CrossRef]
Sarriá, B.; Martínez-López, S.; Sierra-Cinos, J.L.; García-Diz, L.; Mateos, R.; Bravo, L. Regular consumption of a cocoa product improves the cardiometabolic profile in healthy and moderately hypercholesterolaemic adults. *Br. J. Nutr.* **2014**, *111*, 122–134. [CrossRef] [PubMed]
West, S.G.; McIntyre, M.D.; Piotrowski, M.J.; Poupin, N.; Miller, D.L.; Preston, A.G.; Wagner, P.; Groves, L.F.; Skulas-Ray, A.C. Effects of dark chocolate and cocoa consumption on endothelial function and arterial stiffness in overweight adults. *Br. J. Nutr.* **2014**, *111*, 653–661. [CrossRef]
Heiss, C.; Sansone, R.; Karimi, H.; Krabbe, M.; Schuler, D.; Rodriguez-Mateos, A.; Kraemer, T.; Cortese-Krott, M.M.; Kuhnle, G.G.C.; Spencer, J.P.E.; et al. Impact of cocoa flavanol intake on age-dependent vascular stiffness in healthy men: A randomized, controlled, double-masked trial. *Age* **2015**, *37*, 9794. [CrossRef]
Koli, R.; Köhler, K.; Tonteri, E.; Peltonen, J.; Tikkanen, H.; Fogelholm, M. Dark chocolate and reduced snack consumption in mildly hypertensive adults: An intervention study. *Nutr. J.* **2015**, *14*, 84. [CrossRef]
Massee, L.A.; Ried, K.; Pase, M.; Travica, N.; Yoganathan, J.; Scholey, A.; Macpherson, H.; Kennedy, G.; Sali, A.; Pipingas, A. The acute and sub-chronic effects of cocoa flavanols on mood, cognitive and cardiovascular health in young healthy adults: A randomized, controlled trial. *Front. Pharmacol.* **2015**, *6*, 93. [CrossRef]
Mastroiacovo, D.; Kwik-Uribe, C.; Grassi, D.; Necozione, S.; Raffaele, A.; Pistacchio, L.; Righetti, R.; Bocale, R.; Lechiara, M.C.; Marini, C.; et al. Cocoa flavanol consumption improves cognitive function, blood pressure control, and metabolic profile in elderly subjects: The Cocoa, Cognition, and Aging (CoCoA) Study--a randomized controlled trial. *Am. J. Clin. Nutr.* **2015**, *101*, 538–548. [CrossRef]
Ottaviani, J.I.; Balz, M.; Kimball, J.; Ensunsa, J.L.; Fong, R.; Momma, T.Y.; Kwik-Uribe, C.; Schroeter, H.; Keen, C.L. Safety and efficacy of cocoa flavanol intake in healthy adults: A randomized, controlled, double-masked trial. *Am. J. Clin. Nutr.* **2015**, *102*, 1425–1435. [CrossRef] [PubMed]
Rassaf, T.; Rammos, C.; Hendgen-Cotta, U.B.; Heiss, C.; Kleophas, W.; Dellanna, F.; Floege, J.; Hetzel, G.R.; Kelm, M. Vasculoprotective Effects of Dietary Cocoa Flavanols in Patients on Hemodialysis: A Double-Blind, Randomized, Placebo-Controlled Trial. *Clin. J. Am. Soc. Nephrol.* **2016**, *11*, 108–118. [CrossRef] [PubMed]
Rull, G.; Mohd-Zain, Z.N.; Shiel, J.; Lundberg, M.H.; Collier, D.J.; Johnston, A.; Warner, T.D.; Corder, R. Effects of high flavanol dark chocolate on cardiovascular function and platelet aggregation. *Vascul. Pharmacol.* **2015**, *71*, 70–78. [CrossRef]
Sansone, R.; Rodriguez-Mateos, A.; Heuel, J.; Falk, D.; Schuler, D.; Wagstaff, R.; Kuhnle, G.G.C.; Spencer, J.P.E.; Schroeter, H.; Merx, M.W.; et al. Cocoa flavanol intake improves endothelial function and Framingham Risk Score in healthy men and women: A randomised, controlled, double-masked trial: The Flaviola Health Study. *Br. J. Nutr.* **2015**, *114*, 1246–1255. [CrossRef]
Njike, V.Y.; Hamburg, N.; Kellogg, M.; Annapureddy, A.; Vita, J. Dose and response to cocoa (DARC): A randomized double-blind controlled trial. *Clin. Trials Regul. Sci. Cardiol.* **2016**, *23–24*, 9–15. [CrossRef]
Garcia-Yu, I.A.; Garcia-Ortiz, L.; Gomez-Marcos, M.A.; Rodriguez-Sanchez, E.; Agudo-Conde, C.; Gonzalez-Sanchez, J.; Maderuelo-Fernandez, J.A.; Recio-Rodriguez, J.I. Effects of Cocoa-Rich Chocolate on Blood Pressure, Cardiovascular Risk Factors, and Arterial Stiffness in Postmenopausal Women: A Randomized Clinical Trial. *Nutrients* **2020**, *12*, 1758. [CrossRef] [PubMed]
Nantz, M.P.; Rowe, C.A.; Bukowski, J.F.; Percival, S.S. Standardized capsule of *Camellia sinensis* lowers cardiovascular risk factors in a randomized, double-blind, placebo-controlled study. *Nutrition* **2009**, *25*, 147–154. [CrossRef] [PubMed]
Kafeshani, M.; Entezari, M.H.; Karimian, J.; Pourmasoumi, M.; Maracy, M.R.; Amini, M.R.; Hadi, A. A comparative study of the effect of green tea and sour tea on blood pressure and lipid profile in healthy adult men. *ARYA Atheroscler.* **2017**, *13*, 109–116.

75. Takahashi, M.; Miyashita, M.; Suzuki, K.; Bae, S.-R.; Kim, H.-K.; Wakisaka, T.; Matsui, Y.; Takeshita, M.; Yasunaga, K. A ingestion of catechin-rich green tea improves postprandial glucose status and increases serum thioredoxin concentratior postmenopausal women. *Br. J. Nutr.* **2014**, *112*, 1542–1550. [CrossRef] [PubMed]
76. Sone, T.; Kuriyama, S.; Nakaya, N.; Hozawa, A.; Shimazu, T.; Nomura, K.; Rikimaru, S.; Tsuji, I. Randomized controlled tria an effect of catechin-enriched green tea consumption on adiponectin and cardiovascular disease risk factors. *Food Nutr. Res.* 2 *55*, 8326. [CrossRef] [PubMed]
77. Diepvens, K.; Kovacs, E.M.R.; Nijs, I.M.T.; Vogels, N.; Westerterp-Plantenga, M.S. Effect of green tea on resting energy expendi and substrate oxidation during weight loss in overweight females. *Br. J. Nutr.* **2005**, *94*, 1026–1034. [CrossRef] [PubMed]
78. Frank, J.; George, T.W.; Lodge, J.K.; Rodriguez-Mateos, A.M.; Spencer, J.P.E.; Minihane, A.M.; Rimbach, G. Daily consumptio an aqueous green tea extract supplement does not impair liver function or alter cardiovascular disease risk biomarkers in hea men. *J. Nutr.* **2009**, *139*, 58–62. [CrossRef] [PubMed]
79. Miyazaki, R.; Kotani, K.; Ayabe, M.; Tsuzaki, K.; Shimada, J.; Sakane, N.; Takase, H.; Ichikawa, H.; Yonei, Y.; Ishii, K. Minor eff of green tea catechin supplementation on cardiovascular risk markers in active older people: A randomized controlled *Geriatr. Gerontol. Int.* **2013**, *13*, 622–629. [CrossRef]
80. Maeda-Yamamoto, M.; Nishimura, M.; Kitaichi, N.; Nesumi, A.; Monobe, M.; Nomura, S.; Horie, Y.; Tachibana, H.; Nishil J. A Randomized, Placebo-Controlled Study on the Safety and Efficacy of Daily Ingestion of Green Tea (*Camellia sinensis* L. "Yabukita" and "Sunrouge" on Eyestrain and Blood Pressure in Healthy Adults. *Nutrients* **2018**, *10*, 569. [CrossRef]
81. Fukino, Y.; Shimbo, M.; Aoki, N.; Okubo, T.; Iso, H. Randomized controlled trial for an effect of green tea consumption on ins resistance and inflammation markers. *J. Nutr. Sci. Vitaminol.* **2005**, *51*, 335–342. [CrossRef] [PubMed]
82. Hill, A.M.; Coates, A.M.; Buckley, J.D.; Ross, R.; Thielecke, F.; Howe, P.R.C. Can EGCG reduce abdominal fat in obese subject *Am. Coll. Nutr.* **2007**, *26*, 396S–402S. [CrossRef]
83. Nagao, T.; Hase, T.; Tokimitsu, I. A green tea extract high in catechins reduces body fat and cardiovascular risks in hum *Obesity* **2007**, *15*, 1473–1483. [CrossRef] [PubMed]
84. Fukino, Y.; Ikeda, A.; Maruyama, K.; Aoki, N.; Okubo, T.; Iso, H. Randomized controlled trial for an effect of green tea-ext powder supplementation on glucose abnormalities. *Eur. J. Clin. Nutr.* **2008**, *62*, 953–960. [CrossRef] [PubMed]
85. Hsu, C.-H.; Tsai, T.-H.; Kao, Y.-H.; Hwang, K.-C.; Tseng, T.-Y.; Chou, P. Effect of green tea extract on obese women: A randomi double-blind, placebo-controlled clinical trial. *Clin. Nutr.* **2008**, *27*, 363–370. [CrossRef] [PubMed]
86. Brown, A.L.; Lane, J.; Coverly, J.; Stocks, J.; Jackson, S.; Stephen, A.; Bluck, L.; Coward, A.; Hendrickx, H. Effects of die supplementation with the green tea polyphenol epigallocatechin-3-gallate on insulin resistance and associated metabolic factors: Randomized controlled trial. *Br. J. Nutr.* **2009**, *101*, 886–894. [CrossRef] [PubMed]
87. Nagao, T.; Meguro, S.; Hase, T.; Otsuka, K.; Komikado, M.; Tokimitsu, I.; Yamamoto, T.; Yamamoto, K. A catechin-rich beve improves obesity and blood glucose control in patients with type 2 diabetes. *Obesity* **2009**, *17*, 310–317. [CrossRef] [PubMed
88. Basu, A.; Du, M.; Sanchez, K.; Leyva, M.J.; Betts, N.M.; Blevins, S.; Wu, M.; Aston, C.E.; Lyons, T.J. Green tea minimally aff biomarkers of inflammation in obese subjects with metabolic syndrome. *Nutrition* **2011**, *27*, 206–213. [CrossRef] [PubMed]
89. Brown, A.L.; Lane, J.; Holyoak, C.; Nicol, B.; Mayes, A.E.; Dadd, T. Health effects of green tea catechins in overweight and ob men: A randomised controlled cross-over trial. *Br. J. Nutr.* **2011**, *106*, 1880–1889. [CrossRef]
90. Hsu, C.-H.; Liao, Y.-L.; Lin, S.-C.; Tsai, T.-H.; Huang, C.-J.; Chou, P. Does supplementation with green tea extract improve ins resistance in obese type 2 diabetics? A randomized, double-blind, and placebo-controlled clinical trial. *Altern. Med. Rev.* **2011** 157–163.
91. Bogdanski, P.; Suliburska, J.; Szulinska, M.; Stepien, M.; Pupek-Musialik, D.; Jablecka, A. Green tea extract reduces blood press inflammatory biomarkers, and oxidative stress and improves parameters associated with insulin resistance in obese, hyperten patients. *Nutr. Res.* **2012**, *32*, 421–427. [CrossRef]
92. Suliburska, J.; Bogdanski, P.; Szulinska, M.; Stepien, M.; Pupek-Musialik, D.; Jablecka, A. Effects of green tea supplementatior elements, total antioxidants, lipids, and glucose values in the serum of obese patients. *Biol. Trace Elem. Res.* **2012**, *149*, 315– [CrossRef] [PubMed]
93. Liu, C.-Y.; Huang, C.-J.; Huang, L.-H.; Chen, I.-J.; Chiu, J.-P.; Hsu, C.-H. Effects of green tea extract on insulin resista and glucagon-like peptide 1 in patients with type 2 diabetes and lipid abnormalities: A randomized, double-blinded, placebo-controlled trial. *PLoS ONE* **2014**, *9*, e91163. [CrossRef]
94. Dower, J.I.; Geleijnse, J.M.; Gijsbers, L.; Schalkwijk, C.; Kromhout, D.; Hollman, P.C. Supplementation of the Pure Flavonc Epicatechin and Quercetin Affects Some Biomarkers of Endothelial Dysfunction and Inflammation in (Pre)Hypertensive Ad A Randomized Double-Blind, Placebo-Controlled, Crossover Trial. *J. Nutr.* **2015**, *145*, 1459–1463. [CrossRef] [PubMed]
95. Chen, I.-J.; Liu, C.-Y.; Chiu, J.-P.; Hsu, C.-H. Therapeutic effect of high-dose green tea extract on weight reduction: A randomi double-blind, placebo-controlled clinical trial. *Clin. Nutr.* **2016**, *35*, 592–599. [CrossRef] [PubMed]
96. Gutiérrez-Salmeán, G.; Meaney, E.; Lanaspa, M.A.; Cicerchi, C.; Johnson, R.J.; Dugar, S.; Taub, P.; Ramírez-Sánchez, I.; Villar F.; Schreiner, G.; et al. A randomized, placebo-controlled, double-blind study on the effects of (-)-epicatechin on the trig eride/HDLc ratio and cardiometabolic profile of subjects with hypertriglyceridemia: Unique in vitro effects. *Int. J. Cardiol.* 2 *223*, 500–506. [CrossRef] [PubMed]
97. Lu, P.H.; Hsu, C.H. Does supplementation with green tea extract improve acne in post-adolescent women? A randomi double-blind, and placebo-controlled clinical trial. *Complement. Ther. Med.* **2016**, *25*, 159–163. [CrossRef]

Saarenhovi, M.; Salo, P.; Scheinin, M.; Lehto, J.; Lovró, Z.; Tiihonen, K.; Lehtinen, M.J.; Junnila, J.; Hasselwander, O.; Tarpila, A.; et al. The effect of an apple polyphenol extract rich in epicatechin and flavan-3-ol oligomers on brachial artery flow-mediated vasodilatory function in volunteers with elevated blood pressure. *Nutr. J.* **2017**, *16*, 73. [CrossRef] [PubMed]
Kirch, N.; Berk, L.; Liegl, Y.; Adelsbach, M.; Zimmermann, B.F.; Stehle, P.; Stoffel-Wagner, B.; Ludwig, N.; Schieber, A.; Helfrich, H.-P.; et al. A nutritive dose of pure (-)-epicatechin does not beneficially affect increased cardiometabolic risk factors in overweight-to-obese adults-a randomized, placebo-controlled, double-blind crossover study. *Am. J. Clin. Nutr.* **2018**, *107*, 948–956. [CrossRef]
Henein, M.Y.; Vancheri, S.; Longo, G.; Vancheri, F. The role of inflammation in cardiovascular disease. *Int. J. Mol. Sci.* **2022**, *23*, 2906. [CrossRef]
Violi, F.; Cammisotto, V.; Bartimoccia, S.; Pignatelli, P.; Carnevale, R.; Nocella, C. Gut-derived low-grade endotoxaemia, atherothrombosis and cardiovascular disease. *Nat. Rev. Cardiol.* **2023**, *20*, 24–37. [CrossRef] [PubMed]
Dubois-Deruy, E.; Peugnet, V.; Turkieh, A.; Pinet, F. Oxidative stress in cardiovascular diseases. *Antioxidants* **2020**, *9*, 864. [CrossRef] [PubMed]
Bernatoniene, J.; Kopustinskiene, D.M. The role of catechins in cellular responses to oxidative stress. *Molecules* **2018**, *23*, 965. [CrossRef] [PubMed]
Brånén, L.; Hovgaard, L.; Nitulescu, M.; Bengtsson, E.; Nilsson, J.; Jovinge, S. Inhibition of tumor necrosis factor-alpha reduces atherosclerosis in apolipoprotein E knockout mice. *Arterioscler. Thromb. Vasc. Biol.* **2004**, *24*, 2137–2142. [CrossRef] [PubMed]
Fan, F.-Y.; Sang, L.-X.; Jiang, M. Catechins and their therapeutic benefits to inflammatory bowel disease. *Molecules* **2017**, *22*, 484. [CrossRef] [PubMed]
Bors, W.; Heller, W.; Michel, C.; Saran, M. Flavonoids as antioxidants: Determination of radical-scavenging efficiencies. *Meth. Enzymol.* **1990**, *186*, 343–355.
Fraga, C.G.; Galleano, M.; Verstraeten, S.V.; Oteiza, P.I. Basic biochemical mechanisms behind the health benefits of polyphenols. *Mol. Asp. Med.* **2010**, *31*, 435–445. [CrossRef] [PubMed]
Rodríguez-Ramiro, I.; Martín, M.A.; Ramos, S.; Bravo, L.; Goya, L. Comparative effects of dietary flavanols on antioxidant defences and their response to oxidant-induced stress on $CaCO_2$ cells. *Eur. J. Nutr.* **2011**, *50*, 313–322. [CrossRef] [PubMed]
Meng, Q.; Velalar, C.N.; Ruan, R. Effects of epigallocatechin-3-gallate on mitochondrial integrity and antioxidative enzyme activity in the aging process of human fibroblast. *Free Radic. Biol. Med.* **2008**, *44*, 1032–1041. [CrossRef]
Khan, S.G.; Katiyar, S.K.; Agarwal, R.; Mukhtar, H. Enhancement of antioxidant and phase II enzymes by oral feeding of green tea polyphenols in drinking water to SKH-1 hairless mice: Possible role in cancer chemoprevention. *Cancer Res.* **1992**, *52*, 4050–4052.
Naderi, G.A.; Asgary, S.; Sarraf-Zadegan, N.; Shirvany, H. Anti-oxidant effect of flavonoids on the susceptibility of LDL oxidation. *Mol. Cell. Biochem.* **2003**, *246*, 193–196. [CrossRef] [PubMed]
Ruskovska, T.; Maksimova, V.; Milenkovic, D. Polyphenols in human nutrition: From the in vitro antioxidant capacity to the beneficial effects on cardiometabolic health and related inter-individual variability—An overview and perspective. *Br. J. Nutr.* **2020**, *123*, 241–254. [CrossRef]
Ruskovska, T.; Massaro, M.; Carluccio, M.A.; Arola-Arnal, A.; Muguerza, B.; Vanden Berghe, W.; Declerck, K.; Bravo, F.I.; Calabriso, N.; Combet, E.; et al. Systematic bioinformatic analysis of nutrigenomic data of flavanols in cell models of cardiometabolic disease. *Food Funct.* **2020**, *11*, 5040–5064. [CrossRef]
Corral-Jara, K.F.; Nuthikattu, S.; Rutledge, J.; Villablanca, A.; Morand, C.; Schroeter, H.; Milenkovic, D. Integrated Multi-Omic Analyses of the Genomic Modifications by Gut Microbiome-Derived Metabolites of Epicatechin, 5-(4′-Hydroxyphenyl)-γ-Valerolactone, in TNFalpha-Stimulated Primary Human Brain Microvascular Endothelial Cells. *Front. Neurosci.* **2021**, *15*, 622640. [CrossRef]
Oleaga, C.; Ciudad, C.J.; Izquierdo-Pulido, M.; Noé, V. Cocoa flavanol metabolites activate HNF-3β, Sp1, and NFY-mediated transcription of apolipoprotein AI in human cells. *Mol. Nutr. Food Res.* **2013**, *57*, 986–995. [CrossRef] [PubMed]
Rios, L.Y.; Bennett, R.N.; Lazarus, S.A.; Rémésy, C.; Scalbert, A.; Williamson, G. Cocoa procyanidins are stable during gastric transit in humans. *Am. J. Clin. Nutr.* **2002**, *76*, 1106–1110. [CrossRef]
Mena, P.; Bresciani, L.; Brindani, N.; Ludwig, I.A.; Pereira-Caro, G.; Angelino, D.; Llorach, R.; Calani, L.; Brighenti, F.; Clifford, M.N.; et al. Phenyl-γ-valerolactones and phenylvaleric acids, the main colonic metabolites of flavan-3-ols: Synthesis, analysis, bioavailability, and bioactivity. *Nat. Prod. Rep.* **2019**, *36*, 714–752. [CrossRef] [PubMed]
Monagas, M.; Urpi-Sarda, M.; Sánchez-Patán, F.; Llorach, R.; Garrido, I.; Gómez-Cordovés, C.; Andres-Lacueva, C.; Bartolomé, B. Insights into the metabolism and microbial biotransformation of dietary flavan-3-ols and the bioactivity of their metabolites. *Food Funct.* **2010**, *1*, 233–253. [CrossRef] [PubMed]
Márquez Campos, E.; Stehle, P.; Simon, M.-C. Microbial Metabolites of Flavan-3-Ols and Their Biological Activity. *Nutrients* **2019**, *11*, 2260. [CrossRef]
Manach, C.; Milenkovic, D.; Van de Wiele, T.; Rodriguez-Mateos, A.; de Roos, B.; Garcia-Conesa, M.T.; Landberg, R.; Gibney, E.R.; Heinonen, M.; Tomás-Barberán, F.; et al. Addressing the inter-individual variation in response to consumption of plant food bioactives: Towards a better understanding of their role in healthy aging and cardiometabolic risk reduction. *Mol. Nutr. Food Res.* **2017**, *61*, 1600557. [CrossRef]

121. van Duynhoven, J.; Vaughan, E.E.; Jacobs, D.M.; Kemperman, R.A.; van Velzen, E.J.J.; Gross, G.; Roger, L.C.; Possemier Smilde, A.K.; Doré, J.; et al. Metabolic fate of polyphenols in the human superorganism. *Proc. Natl. Acad. Sci. USA* **2011**, (Suppl. S1), 4531–4538. [CrossRef]
122. Kutschera, M.; Engst, W.; Blaut, M.; Braune, A. Isolation of catechin-converting human intestinal bacteria. *J. Appl. Microbiol.* **2** *111*, 165–175. [CrossRef] [PubMed]
123. van Velzen, E.J.J.; Westerhuis, J.A.; Grün, C.H.; Jacobs, D.M.; Eilers, P.H.C.; Mulder, T.P.; Foltz, M.; Garczarek, U.; Kemperr R.; Vaughan, E.E.; et al. Population-based nutrikinetic modeling of polyphenol exposure. *Metabolomics* **2014**, *10*, 1059–1 [CrossRef]
124. Trošt, K.; Ulaszewska, M.M.; Stanstrup, J.; Albanese, D.; De Filippo, C.; Tuohy, K.M.; Natella, F.; Scaccini, C.; Mattivi, F. H Microbiome co-metabolic processing of dietary polyphenols—An acute, single blinded, cross-over study with different dose apple polyphenols in healthy subjects. *Food Res. Int.* **2018**, *112*, 108–128. [CrossRef]
125. Xia, Y.; Zhang, X.; Jiang, M.; Zhang, H.; Wang, Y.; Zhang, Y.; Seviour, R.; Kong, Y. In vitro co-metabolism of epigallocatechi gallate (EGCG) by the mucin-degrading bacterium *Akkermansia muciniphila. PLoS ONE* **2021**, *16*, e0260757. [CrossRef] [PubM
126. Sheng, L.; Jena, P.K.; Liu, H.-X.; Hu, Y.; Nagar, N.; Bronner, D.N.; Settles, M.L.; Bäumler, A.J.; Wan, Y.-J.Y. Obesity treatmen epigallocatechin-3-gallate-regulated bile acid signaling and its enriched *Akkermansia muciniphila. FASEB J.* **2018**, *32*, fj2018003 [CrossRef]
127. Patial, V.; Katoch, S.; Chhimwal, J.; Dadhich, G.; Sharma, V.; Rana, A.; Joshi, R.; Padwad, Y. Catechins prevent obesity-indu kidney damage by modulating PPARγ/CD36 pathway and gut-kidney axis in rats. *Life Sci.* **2023**, *316*, 121437. [Cross] [PubMed]
128. Uhlenhut, K.; Högger, P. Facilitated cellular uptake and suppression of inducible nitric oxide synthase by a metabolite of marit pine bark extract (Pycnogenol). *Free Radic. Biol. Med.* **2012**, *53*, 305–313. [CrossRef]
129. Schiattarella, G.G.; Altamirano, F.; Tong, D.; French, K.M.; Villalobos, E.; Kim, S.Y.; Luo, X.; Jiang, N.; May, H.I.; Wang, Z.V.; e Nitrosative stress drives heart failure with preserved ejection fraction. *Nature* **2019**, *568*, 351–356. [CrossRef]
130. Takagaki, A.; Nanjo, F. Effects of Metabolites Produced from (-)-Epigallocatechin Gallate by Rat Intestinal Bacteria on Angiote I-Converting Enzyme Activity and Blood Pressure in Spontaneously Hypertensive Rats. *J. Agric. Food Chem.* **2015**, *63*, 8262–8 [CrossRef]
131. Lee, C.C.; Kim, J.H.; Kim, J.S.; Oh, Y.S.; Han, S.M.; Park, J.H.Y.; Lee, K.W.; Lee, C.Y. 5-(3′,4′-Dihydroxyphenyl-γ-valerolacton Major Microbial Metabolite of Proanthocyanidin, Attenuates THP-1 Monocyte-Endothelial Adhesion. *Int. J. Mol. Sci.* **2017** 1363. [CrossRef]
132. Urpi-Sarda, M.; Monagas, M.; Khan, N.; Llorach, R.; Lamuela-Raventós, R.M.; Jáuregui, O.; Estruch, R.; Izquierdo-Pulido, Andrés-Lacueva, C. Targeted metabolic profiling of phenolics in urine and plasma after regular consumption of cocoa by lic chromatography-tandem mass spectrometry. *J. Chromatogr. A* **2009**, *1216*, 7258–7267. [CrossRef] [PubMed]
133. Yang, X.-W.; Wang, N.; Li, W.; Xu, W.; Wu, S. Biotransformation of 4,5-O-dicaffeoylquinic acid methyl ester by human intest flora and evaluation on their inhibition of NO production and antioxidant activity of the products. *Food Chem. Toxicol.* **2013** 297–303. [CrossRef]
134. di Gesso, J.L.; Kerr, J.S.; Zhang, Q.; Raheem, S.; Yalamanchili, S.K.; O'Hagan, D.; Kay, C.D.; O'Connell, M.A. Flavonoid metabo reduce tumor necrosis factor-α secretion to a greater extent than their precursor compounds in human THP-1 monocytes. *Nutr. Food Res.* **2015**, *59*, 1143–1154. [CrossRef]
135. Wang, H.-Y.; Wang, H.; Wang, J.-H.; Wang, Q.; Ma, Q.-F.; Chen, Y.-Y. Protocatechuic Acid Inhibits Inflammatory Response LPS-Stimulated BV2 Microglia via NF-κB and MAPKs Signaling Pathways. *Neurochem. Res.* **2015**, *40*, 1655–1660. [Cross [PubMed]
136. Kim, K.; Bae, O.-N.; Lim, K.-M.; Noh, J.-Y.; Kang, S.; Chung, K.Y.; Chung, J.-H. Novel antiplatelet activity of protocatechuic a through the inhibition of high shear stress-induced platelet aggregation. *J. Pharmacol. Exp. Ther.* **2012**, *343*, 704–711. [CrossR
137. Semaming, Y.; Kumfu, S.; Pannangpetch, P.; Chattipakorn, S.C.; Chattipakorn, N. Protocatechuic acid exerts a cardioprotec effect in type 1 diabetic rats. *J. Endocrinol.* **2014**, *223*, 13–23. [CrossRef] [PubMed]
138. Bhattacharjee, N.; Dua, T.K.; Khanra, R.; Joardar, S.; Nandy, A.; Saha, A.; De Feo, V.; Dewanjee, S. Protocatechuic Acid, a Phen from Sansevieria roxburghiana Leaves, Suppresses Diabetic Cardiomyopathy via Stimulating Glucose Metabolism, Ameliora Oxidative Stress, and Inhibiting Inflammation. *Front. Pharmacol.* **2017**, *8*, 251. [CrossRef]
139. Qian, Y.; Babu, P.V.A.; Symons, J.D.; Jalili, T. Metabolites of flavonoid compounds preserve indices of endothelial cell nitric ox bioavailability under glucotoxic conditions. *Nutr. Diabetes* **2017**, *7*, e286. [CrossRef]

nutrients

MDPI

'w

ccinium as Potential Therapy for Diabetes and icrovascular Complications

Huang [1,2], Yayong Luo [1,2], Qian Wang [1], Yihan Zhang [3], Zhongxia Li [3], Ruikun He [3], Xiangmei Chen [1,2,*] Zheyi Dong [1,*]

1 National Clinical Research Center for Kidney Diseases, State Key Laboratory of Kidney Diseases, Beijing Key Laboratory of Kidney Disease Research, First Medical Center of Chinese PLA General Hospital, Nephrology Institute of the Chinese People's Liberation Army, Beijing 100853, China
2 School of Clinical Medicine, Guangdong Pharmaceutical University, Guangzhou 510006, China
3 BYHEALTH Institute of Nutrition & Health, No. 3 Kehui 3rd Street, No. 99 Kexue Avenue Central, Huangpu District, Guangzhou 510663, China
* Correspondence: xmchen301@126.com (X.C.); shengdai26@163.com (Z.D.)

Abstract: Diabetes mellitus is one of the most critical global health concerns, with a fast-growing prevalence. The incidence of diabetic vascular complications is also rapidly increasing, exacerbating the burden on individuals with diabetes and the consumption of public medical resources. Despite the overall improvements in the prevention, diagnosis, and treatment of diabetic microvascular complications in recent years, safe and effective alternative or adjunctive therapies are urgently needed. The mechanisms underlying diabetic vascular complications are complex, with hyperglycemia-induced oxidative stress and inflammation being the leading causes. Therefore, glycemic control, antioxidation, and anti-inflammation are considered the main targets for the treatment of diabetes and its vascular comorbidities. *Vaccinium* L. (Ericaceae) is a genus of plants enriched with polyphenolic compounds in their leaves and fruits. *Vaccinium* and its extracts have demonstrated good bioactivity in reducing blood glucose, oxidative stress, and inflammation, making them excellent candidates for the management of diabetes and diabetic vascular complications. Here, we review recent preclinical and clinical studies on the potential effect of *Vaccinium* on ameliorating diabetes and diabetic complications, particularly diabetic kidney disease and diabetic retinopathy.

Keywords: diabetes mellitus; diabetic kidney disease; diabetic retinopathy; Ericaceae; *Vaccinium*

›n: Huang, H.; Luo, Y.; Wang, ang, Y.; Li, Z.; He, R.; Chen, X.; Z. *Vaccinium* as Potential ›y for Diabetes and ›ascular Complications. *›nts* **2023**, *15*, 2031. https:// ›/10.3390/nu15092031

mic Editors: Francesca ›ieri, Sonia Ramos Rivero and Angeles Martín

›ed: 15 March 2023
›d: 15 April 2023
›ted: 20 April 2023
hed: 23 April 2023

1. Introduction

Due to rapid urbanization and population aging, people are changing their dietary and nutritional habits, leading to an increase in the global prevalence of diabetes mellitus (DM) [1,2]. Diabetes is one of the world's greatest public health challenges [3], particularly in developing countries [4]. Type 2 diabetes accounts for approximately 90–95% of all diabetes cases and is strongly associated with an unhealthy lifestyle, such as energy-rich diets and sedentary behaviors [5]. Patients with diabetes are prone to microangiopathic complications and have a higher risk of mortality [6]. Microvascular complications of diabetes continue to compromise the quality of life of patients with diabetes. The most representative of these are diabetic kidney disease (DKD) and diabetic retinopathy (DR). However, the inclusion of diabetic neuropathy as a microvascular complication has recently been challenged. In China, an estimated 24.3 million people with diabetes have chronic kidney disease (CKD) [7]. In the United States, the number of people with diabetes who have started treatment for end-stage renal disease (ESRD) has increased considerably from more than 40,000 in 2000 to more than 50,000 in 2014 [8]. Mortality is approximately 30 times higher in patients with diabetic nephropathy compared to in those without [9]. DR is the leading cause of low vision and blindness in patients with diabetes, with an annual incidence of up to 12.7% [10]. These microvascular lesions also increase the risk of all-cause and cardiovascular disease (CVD) mortality [11,12]. Owing to the increasing incidenc diabetes and the associated socioeconomic impact of diabetes-related complications, th is an urgent need to improve the management of patients with diabetes.

Inflammation and oxidative stress play key roles in the pathogenesis of diabetes its microvascular complications. Metabolic abnormalities in diabetes lead to the excess production of mitochondrial superoxide in the endothelium of small blood vessels. T increase in superoxide production activates five major pathways: overactivity of the osamine pathway, polyol pathway flux, activation of protein kinase C isoforms, increa formation of advanced glycation end products (AGEs), and increased expression of AGE receptor and its activating ligands [13]. This review focuses on the managemen two microvascular complications, DKD and DR. Although angiotensin-converting enzy inhibitors (ACEI)/angiotensin receptor blockers (ARB) and sodium–glucose cotranspo 2 (SGLT2) have been widely used for DKD treatment, as well as laser and injection ther for DR, safer and more acceptable adjuvant treatments are urgently needed.

Natural compounds in plants provide a safer and more effective way to treat diab and its microvascular complications than drugs [14]. The *Vaccinium* L. (Ericaceae) ge which includes a range of shrubs and dwarf shrubs, contains a wide diversity of econ ically and culturally important berry crop species [15,16]. These fruits contain vari volatile organic compounds [17]. *Vaccinium* reportedly contains phenolic compour pectin, vitamins, and sugars [18,19]. Anthocyanins (ANT) are found in the genus *Vaccin* L. (Ericaceae) [20]. ANT is metabolized in the body as follows: it is absorbed in the and then enters the liver through the portal vein, where it is metabolized, secreted, reabsorbed [21].

Vaccinium, including its fruits, flowers, and leaves, is widely used in traditic medicine, especially for the empirical treatment of diabetes, vision-related ailments, and eral cardiovascular disorders [19,22,23]. Phytoconstituents of *Vaccinium* have antioxid anti-inflammatory, antibacterial, anti-obesity, anti-cancer, anti-diabetic, eye-cardioprotec and neuroprotective activities [18,24–29]. The *Vaccinium* extract provides an avenue for development of new drugs.

In this review, we aimed to describe the current knowledge of the anti-diabetic a anti-microvascular complications (DKD and DR) properties of some *Vaccinium* spe (focusing on four popular species of *Vaccinium*, referring to the fruits, if not specifi including the mechanisms of action, experimental study results, and some positive clin evidence.

2. Materials and Methods

The Web of Science Core Collection (WoSCC), Scopus, and PubMed databases w used for literature searches. The search ended on 1 February 2023. Combinations of seve search terms such as "*Vaccinium*" "bilberry," "anthocyanins," "flavonoids," "polyp nols," "diabetes," "diabetic kidney disease," "diabetic retinopathy," "in vitro," "in viv "clinical trial," "antioxidant," "anti-inflammatory," and "immune-modulatory" were us Bibliometric and visual analyses were performed using a scholarly search on the webs

3. Research Status of *Vaccinium* in Treating Diabetes Mellitus and Its Microvascular Complications

The number of publications focusing on *Vaccinium*, diabetes, and its complicati has rapidly increased over the last 10 years (Figure 1a). High-frequency terms focused *Vaccinium*, DM, metabolic syndrome (MS) or DR, plant extracts or juice, anthocyanins, and mouse animal experiments, clinical trials, anti-diabetic, and oxidative stress (Figure In terms of regional distribution, the Americas, Asia, and Europe, represented by United States, Canada, China, Iran, Russia, and Finland, respectively, have made signific contributions to the research in this field (Figure 1(c-1,c-2)).

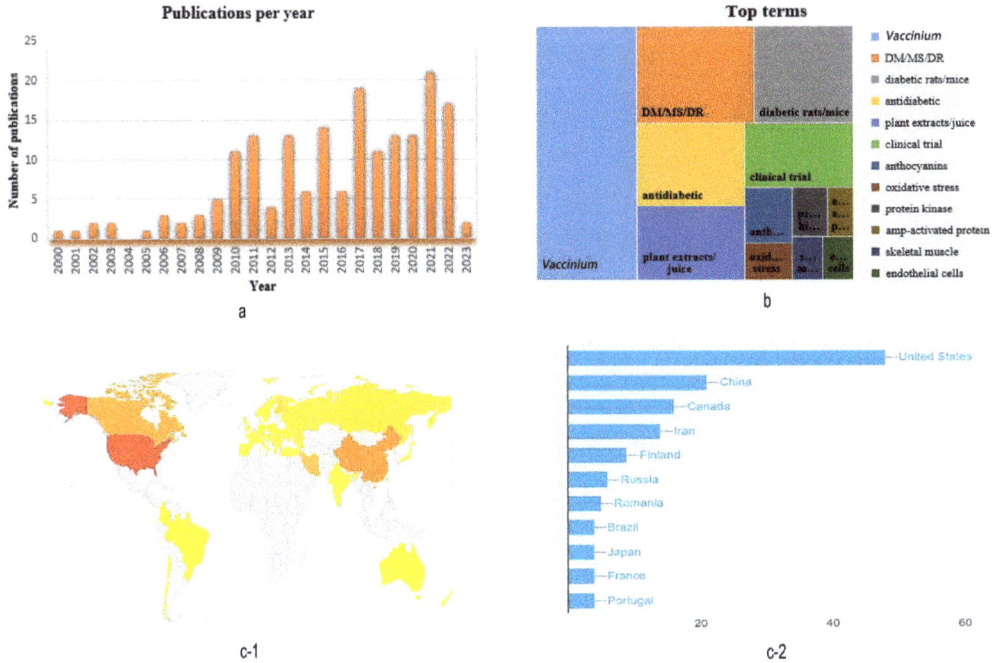

Figure 1. Research status of *Vaccinium* in treating diabetes mellitus and its microvascular complications. (**a**) Publications per year; (**b**) Top terms; (**c-1**,**c-2**) Regional distribution.

4. Vaccinium

4.1. General

The *Vaccinium* L. (Ericaceae) genus, consisting of approximately 450 species, contains a range of terrestrial or epiphytic shrubs and dwarf shrubs that mainly grow in cooler areas across Europe, Southeast and Central Africa, North and Central America, and Asia [15,30]. Most *Vaccinium* fruits are edible, and some have a long history of human consumption. *V. corymbosum* (blueberry), *V. oxycoccos* (cranberry), *V. macrocarpon* (American cranberry), *V. myrtillus* (bilberry), *V. Arctostaphylos* (bearberry), and *V. vitis idaea* (lingonberry) are the species of *Vaccinium* most investigated [17]. Arevka illustrated the differences between four common species: bilberry, blueberry, lingonberry, and cranberry [15].

Leaves and fruits have been widely used in traditional medicine for the treatment of stomatitis; diabetes; renal stones; and intestinal, liver, and urinary tract disorders, as early as the 18th century [31]. Some *Vaccinium* species were domesticated in the 20th century and are now cultured on a large scale in several regions worldwide as economic fruits.

4.2. Chemical Profile

The chemical components of *Vaccinium* have been extensively investigated in several studies. Polyphenolic components, including ANT (cyanidin, malvidin, and delphinidin), flavonoids (quercetin, isoquercetin, kaempferol, apigenin, and myricetin), phenolic acids (gallic, *p*-coumaric, cinnamic, syringic, ferulic, and caffeic acids), and ellagitannins, are considered the main bioactive compounds of *Vaccinium* [19].

ANT is the primary phytochemical characteristic of *Vaccinium*. To date, more than 35 anthocyanin glycosides have been identified in *Vaccinium*, with the principal anthocyanins varying among species (Table 1). In addition, more than 50 other flavonoids, mainly flavanols and pro-anthocyanidins, have been identified in *Vaccinium*, and their profiles vary between species [30].

Table 1. The main flavonoids in four popular *Vaccinium* species.

Species	Geographical Sources	Anthocyanins	Other Flavonoids
V. myrtillus (bilberry) [19,30]	Central and northern parts of Europe	cyanidin 3-galactoside, cyanidin 3-glucoside, cyanidin 3-arabinoside, delphinidin 3-galactoside, delphinidin 3-arabinoside, delphinidin 3-glucoside, malvidin 3-galactoside, malvidin 3-arabinoside, malvidin 3-glucoside, petunidin 3-galactoside, petunidin 3-arabinoside, petunidin 3-acetylglucoside, petunidin 3-glucoside, peonidin 3-galactoside, peonidin 3-arabinoside, cyanidin 3-xyloside, cyanidin 5-glucoside, cyanidin 3,5-diglucoside, cyanidin 3-(6″-O-2-rhamnopyranpsyl-2″-O-β-xylopranosyl-β-glucopyranoside), cyanidin 3-sambubioside, delphinidin 3-sambuobiside, peonidin-3-glycoside	myricetin 3-glucoside, myricetin 3-arabinoside, myricetin3-rhamnoside, quercetin 3-arabinoside, quercetin 3-rhamnoside, quercetin 3-galactoside, quercetin 3-glucoside, quercetin 3-rutinoside, apigenin, chrysoeriol, myricetin, myricetin-3-xyloside, quercetin 3-glucuronide, quercetin 3-xyloside, isorhamnetin 3-glucoside, isorhamnetin 3-xyloside isorhamnetin, laricitrin, syringetin, luteolin, kaempferol isorhamnetin 3-galactoside, myricetin 3-glucuronide, laricitrin 3-glucoside, syringetin 3-glucoside, kaempferol 3-glucoside, myricetin 3-galactoside,
V. corymbosum (blueberry) [19,32,33]	Parts of Asia and North America	delphinidin 3-galactoside, delphinidin 3-glucoside, cyanidin 3-galactoside, delphinidin 3-arabinoside, cyanidin 3-glucoside, petunidin 3-galactoside cyanidin 3-arabinoside, petunidin 3-glucoside, peonidin 3-galactoside, petunidin 3-arabinoside malvidin 3-galactoside, malvidin 3-glucoside peonidin 3-arabinoside, malvidin 3-arabinoside, delphinidin 3-acetylglucoside, petunidin 3-acetylglucoside, malvidin 3-acetylglucoside, petunidin 3-glucoside	myricetin 3-galactoside, myricetin 3-glucoside, myricetin 3-rhamnoside, quercetin 3-galactoside, quercetin 3-glucoside, quercetin 3-xylosylglucuronide, quercetin 3-glucosylxyloside, quercetin 3-rutinoside, quercetin 3-acetylrhamnoside, quercetin 3-xyloside
V. vitis idaea (lingonberry) [30,34]	North Eurasia and North America	cyanidin 3-glucoside, delphinidin 3-glucoside, cyanidin 3-arabinoside, peonidin 3-arabinoside, peonidin 3-glucoside, peonidin 3-galactoside, delphinidin 3-arabinoside, delphinidin 3-galactoside, petunidin 3-galactoside, petunidin 2-glucoside, malvidin 3-galactoside, malvidin 3-glucoside	kaempferol, quercetin, myricetin, rutin myricetin 3-glucoside, quercetin 3-glucoside, quercetin 3-galactoside, quercetin 3-arabinoside, quercetin 3-xyloside, kaempferol 3-rhamnoside, quercetin 3-rhamnoside, isorhamnetin 3-galactoside, isorhamnetin 3-glucoside, syringetin 3-glucoside, kaempferol 3-glucoside
V. macrocarpon (cranberry) [33,35]	Eastern US and Canada	cyanidin-3-glucoside, cyanidin-3-galactoside, cyanidin-3-arabinoside, peonidin-3-glucoside, peonidin-3-galactoside, peonidin-3-arabinoside, pelargonidin-3-galactoside, pelargonidin-3-arabinoside, malvidin-3-galactoside, malvidin-3-arabinoside delphinidin-3-arabinoside, petunidin-3-galactoside	kaempferol-3-glucoside, quercetin-3-galactoside, quercetin-3-arabinoside, quercetin-3-rhamnoside

4.3. Bioactivity

Phytochemicals from several *Vaccinium* species exhibit good activity in multiple biofunctions. The enrichment of polyphenolic compounds leads to a strong antioxidant effect, which is the most acknowledged bioactivity of these berries [20]. Similarly, high concentrations of ANT and flavonoids contribute to the anti-inflammatory effects of *Vaccinium*. As many types of tissue damage are closely associated with oxidative stress and inflammation, *Vaccinium* demonstrates therapeutic potential under multiple pathological conditions, such as diabetes and diabetic vascular damage [36,37].

Moreover, *Vaccinium* has antimicrobial, anticarcinogenic, cardiovascular protective, vision improvement, and anti-neurodegenerative effects, which have been described in detail elsewhere [23,31,38–40]. Berries have the potential to reduce metabolic and cardiovascular risk [40,41]. Similarly, the intake of blueberries has been associated with a reduced risk of cardiovascular disease, death, and type 2 diabetes (T2D), as well as improved weight maintenance and neuroprotection in some epidemiological studies [42,43]. In addition, cranberries have special effect against urinary tract inflammation, tooth decay, periodontitis, and *Helicobacter pylori* infection of the stomach [44].

5. Effects of Oxidative Stress and Inflammation on Diabetes and Its Microvascular Complications

5.1. Abnormalities of Glucose and Lipid Metabolism

Type 2 diabetes mellitus (T2DM) accounts for more than 90% of diabetes cases and is typically characterized by abnormally high blood glucose levels and insulin resistance [5]. The maintenance of hyperglycemia leads to the production of mitochondrial superoxide in endothelial cells of small vessels. As a lack of insulin stimulates malonyl coenzyme A (CoA) production, the insulin receptor (IR) increases the oxidation of free fatty acids (FFAs) in endothelial cells, leading to increased superoxide production by the mitochondrial electron transport chain [13]. Mitochondrial superoxide production activates five pathways involved in diabetic microvascular pathogenesis [13]. During the development of diabetic complications, hyperglycemia acts synergistically with other risk factors (obesity, hypertension, and dyslipidemia) to accelerate the presentation of histopathological features of diabetes [45].

In contrast, cellular oxidative stress and disequilibrium of REDOX homeostasis are common features in patients with DM [46–48]. The bioavailability of nitric oxide (NO) and uncoupling endothelial nitric oxide synthase (eNOS) are two major factors that lead to changes in vascular reactivity and the production of reactive oxygen species (ROS) and reactive nitrogen species (RNS) [49–51]. In addition, the main factor in oxidative stress is the imbalance between promoting enzymes (e.g., NADPH oxidase complex (Nox), cytochrome 450, xanthine oxidase, and myeloperoxidase) and antioxidant enzymes (e.g., superoxide dismutase, catalase, and glutathione (GSH) peroxidase) [52–54]. Kinases and transcription factors involved in many inflammatory and oxidative stress responses activate intracellular signaling pathways that lead to the production of pro-oxidative, pro-vascular and pro-inflammatory factors, such as chemokines, cytokines, pro-oxidant enzymes, extracellular matrix proteins, growth factors, and adhesion molecules [55–57].

In conclusion, oxidative stress and inflammation induced by abnormal glucose and lipid metabolism are the major pathogenic factors associated with diabetic complications.

5.2. Diabetic Kidney Disease

DKD is characterized by thickening of the glomerular basement membrane (GBM), dilation of the mesangial matrix, the formation of characteristic Kimmelstiel–Wilson nodules, and a progressive decline in albuminuria and glomerular filtration rate (GFR) [58]. It is characterized by metabolic disturbances and hemodynamic abnormalities caused by hyperglycemia. Some of these pathways involve the formation of AGEs, renin-angiotensin-aldosterone system (RAAS), aldol reductase activation, polyol pathway activation, protein kinase C (PKC), ROS, an increase in some cytokines, connective tissue growth factor (CTGF), and the activation of transforming growth factor beta 1 (TGF-β1) [59–61]. Elevated levels due to hyperglycemia are central to the pathogenesis of DKD. In diabetes, the n sources of ROS are NOX, AGE, and polyol chains [62]. Oxidative stress can directly dam the podocytes, mesangial and endothelial cells, leading to albuminuria and tubuloi stitial fibrosis [63,64]. Innate immunity is involved in the occurrence and developmer DKD. Mechanistically, TLR4 is overexpressed in DKD and is negatively correlated renal function and positively correlated with HbA1c levels [65,66]. In addition, TLR2 [6 and NLRP3 inflammasome activation of interleukin-1β (IL-1β) [65,68,69] play a major in metabolic stress in DKD. The pathogenesis of diabetic nephropathy is influenced combination of multiple factors, and there is a large amount of overlap between pathw and intermediaries. For example, oxidative stress can indirectly activate other pathw to cause damage, while other pathogenic pathways can cause damage through oxida stress. Therefore, the exact pathogenic and molecular mechanisms of DKD remain unc

At present, strict control of blood pressure and blood glucose and inhibition of RAAS by ACEI or ARB are the main methods for the treatment of DKD. The introduc of new glucose-lowering agents, finerenone, and sodium-dependent glucose transpo 2 (SGLT-2) has dramatically changed the treatment landscape of T2D [70,71]. The clir advantages associated with the use of SGLT2 inhibitors include antifibrotic effects to the correction of oxidative stress and inflammation, autophagy, and modulatio mitochondrial function [72]. The EMPA-REG outcome study (Empagliflozin Cardiovasc Outcome Event Trial in Type 2 Diabetes Mellitus Patients) showed that empaglifl reduces cardiovascular death or worsens kidney disease by 39% [73]. The CANVAS st (Canagliflozin Cardiovascular Assessment Study) has also shown that canagliflozin redu cardiovascular and renal outcomes [74].

However, these methods only delay the progression of DKD and do not preven reverse its progression to ESRD [61]. Therefore, new drugs targeting the patholog mechanisms of DKD, such as oxidative stress and inflammation, have become the m focus of new therapies to treat DKD [75].

5.3. *Diabetic Retinopathy*

DR is classically divided into two major categories: non-proliferative (NPDR) proliferative (PDR). NPDR is characterized by defects in the retinal vasculature, incluc hemorrhage, such as punctate bleeding and spot bleeding [76–78], hard exudation [77 flocculent spot [76,77,79,80], microaneurysms [76–79], and vascular leakage [79,80]. main feature of PDR is pathological retinal angiogenesis [81], which includes the gro of new abnormal blood vessels from the existing vascular network and the generatio retinal neovascularization [77,82,83]. The progression of NPDR to PDR is characteri by increased expression of ischemia, hypoxia, and pro-angiogenesis. Angiogenic gro factors include vascular endothelial growth factor (VEGF), fibroblast growth factor-2 (F 2), platelet-derived growth factor (PDGF), and angiopoietin-2 (Ang-2), which can activ abnormal retinal vessels, causing them to protrude into the preretinal space [81–85]. Infl mation has been described as a mechanism of DR [86,87]. The key cytokines in this pro are IL-1β, IL-1-dependent IL-6, IL-8, and TNF-α, as well as monocyte adhesion to endothelial wall and then chemotaxis to the subendothelial space [65,86,88]. In end-st diabetic retinopathy, severe hypoxia leads to neovascularization, vitreous hemorrhage, retinal detachment [89].

Although the mechanism behind this has not been fully elucidated, oxidative st has been shown to be a key factor in this process [90]. In the ischemic state, oxidative st GSH, lipid peroxide, malondialdehyde and superoxide dismutase levels increase, w antioxidant levels decrease, thereby inducing oxidative damage to the retina [91]. Accord to in vitro experiments, elevated superoxide levels were observed under hyperglyce conditions and increased hydrogen peroxide content was observed in retinal cells [92 Oxidative stress can damage cell membranes and induce apoptosis, microvascular dam and barrier damage, ultimately leading to the development of DR.

Intraocular treatments for DR include laser photocoagulation, intravitreal injections of anti-vascular endothelial growth factor (VEGF) agents and steroids, and vitreoretinal surgery [77,94]. Although these treatments may slow the progression of DR blindness, they are not effective in treating the disease and have considerable side effects [95,96]. Therefore, there is an urgent need to identify alternative or adjuvant treatments to prevent DR and slow its progression.

6. Experimental Study on Diabetes Mellitus and Diabetic Microvascular Complications Treatment with *Vaccinium* Extract

6.1. Regulation of Glucose and Lipid Metabolism Disorders

Phenolics target the key pathways involved in carbohydrate metabolism and hepatic glucose homeostasis, including glycogenesis, glycolysis, and gluconeogenesis [30]. The mechanism of the hypoglycemic action of *Vaccinium* may be mediated in part by interference with enzyme action, and polyphenols in *Vaccinium*, such as flavonoids and tannins, can inhibit α-amylase and α-glucosidase [97,98]. Intestinal α-glucosidase breaks down oligosaccharides and disaccharides into monosaccharides suitable for absorption, and *Vaccinium* slows the release of glucose into the bloodstream [99]. A study in prediabetic and diabetic mice showed that bilberry extracts inhibited the activities of α-glucosidase and α-amylase and prevented postprandial hyperglycemia by slowing the rate of carbohydrate digestion [100]. Several polyphenols, such as quercetin, resveratrol, and epigallocatechin-3-gallate, are transported to the plasma membrane mainly through the activation of the protein kinase (AMPK) pathway, thereby enhancing glucose uptake in muscle and adipocytes [98].

There is scientific evidence that the intake of polyphenol-rich fruits, which improve diet-induced insulin resistance, is beneficial for the health of obese animals [101]. Mice treated with bilberry extract (BBE) show a considerable reduction in blood glucose levels [102]. One study showed that rats treated with bilberry extract (nonacylated anthocyanins extract from bilberries: NAAB) for 8 weeks had decreased fasting plasma glucose levels [103]. Male mice treated with lingonberry for 8 weeks showed reduced fasting and postprandial hyperinsulinism, improved insulin sensitivity, and enhanced hepatic insulin clearance. In a diet-induced obesity (DIO) mouse model, bilberry treatment at 125, 250, and 500 mg/kg/day significantly reduced blood glucose levels by 28%, 25%, and 17%, respectively. In this model, lingonberry considerably reduced blood glucose and insulin levels [104]. In another study, mice fed lingonberry showed improvements in blood sugar and liver function, along with a reduction in inflammation [105].

Vaccinium is beneficial in reducing adipose tissue inflammation in models of metabolic syndrome [106–109]. In addition, dried bilberry (*Vaccinium myrtillus* L.) slowly reduces serum cholesterol and delays the adverse consequences of high-fat diet-induced lipid and glucose metabolism caused by high-fat diet [110]. Lingonberry (*Vaccinium vitis-idaea* L.) treatment of hypertrophic adipocytes resulted in reduced lipid accumulation and triglyceride (TG) content, and downregulated expression of lipogenic genes for fatty acid and TC synthesis, such as fatty acid synthase (FAS), adipocyte protein 2 (aP2), and diacylglycerol acyltransferase-1 (DGAT1) [111]. Bilberries, lingonberries, cranberries, and blueberries can regulate glucose and lipid metabolism [112–121]. Experimental studies on diabetes mellitus treatment with *Vaccinium* extracts are listed in Table 2.

Table 2. Experimental studies on the treatment of diabetes mellitus with *Vaccinium* extracts.

	Animal	Intervention	Duration	Results
Takács et al. 2020 [100]	Normal (control), obese, prediabetic, and streptozotocin-induced diabetic mice	Wild strawberry (Fragaria vesca), blackberry (Rubus fruticosus), and European blueberry (*Vaccinium* myrtillus) leaf extracts	/	Inhibit α-glucosidase and α-amylase enzyme activity in vitro, attenuated the starch-induced rise of blood glucose levels.
Mykkänen et al. 2014 [122]	C57BL mice fed with a high-fat diet (HFD)	The effects of 5% or 10% (*w*/*w*) of whole bilberries (BB)	3 months	Prevented ameliorated type 1 pro-inflammatory responsiveness, blood pressure
Feshani et al. 2011 [123]	Alloxan-diabetic male Wistar rats	Fruit of *Vaccinium* arctostaphylos L. (Ericaceae)	3 weeks	Decreased postprandial blood glucose and TG, increased erythrocyte superoxide dismutase, GPX, and catalase.
Prior et al. 2010 [112]	Male C57BL/6J mice (25 days of age) fed with HFD or LFD (low-fat diet)	Blueberry juice or purified blueberry ANT (0.2 or 1.0 mg/mL) in the drinking water	72 days	Blueberry juice was not as effective as the low dose of ANT in the drinking water in preventing obesity.
Zhong et al. 2020 [113]	HFD-fed mice	BBJ, and FBJ with homemade probiotic starter or CFBJ	17 weeks	All decreased fat accumulation and LDL-C levels. BBJ and FBJ treatments regulated the liver mRNA and protein expression levels involved in lipid and glucose metabolism.
Anhê et al. 2015 [118]	C57BL/6J mice were fed either a chow or a high fat/high sucrose diet	Daily either with vehicle (water) or cranberry extract (200 mg/kg)	8 weeks	Cranberry extract reduced weight gain and visceral obesity, improve insulin sensitivity, lowered intestinal TG content, increased the proportion of the mucin-degrading bacterium Akkermansia.
Brader et al. 2013 [124]	Male Zucker Diabetic Fatty (ZDF) rats	A control, bilberry-enriched, blackcurrant-enriched, or fiber-enriched diet	8 weeks	Bilberry enrichment ameliorated total and LDL but not HDL
Eid et al. 2014 [104]	C57BL/6 mice fed a HFD	Lingonberry extract to HFD at three different concentrations (125, 250, and 500 mg/Kg)	8 weeks	Decreased glycemia and strongly tended to decrease insulin levels, improved hepatic steatosis by decreasing hepatic TG levels and significantly activated liver AMPK and Akt pathways.
Al Hamimi et al. 2017 [105].	C57BL/6J fed a HFD	Control, two of which containing lingonberries (L1D and L2D) from different sources,	/	Glycemia was improved only in mice fed L1D, both L1D and L2D liver function was improved, and inflammation reduced. Increased phosphatidylcholines and lysophosphatidylcholines, decreased serine and sphingomyelins

	Animal	Intervention	Duration	Results
Madduma Hewage et al. 2020 [125]	Mice (C57BL/6J) fed a HFD	Dietary supplementation of lingonberry	12 weeks	Decreased BUN, KIM-1, NGAL, NF-κB, MCP-1, TNF-α, IL-6.
Vendrame et al. 2015 [114]	The obese Zucker rat (OZR)	Fed an 8% enriched wild blueberry diet or a control (C) diet	8 weeks	Decreased plasma concentrations of glucose, insulin, glycated hemoglobin GHbA1c, resistin, and retinol-binding protein 4 (RBP4), compared to control diet.
Ryyti et al. 2021 [126]	C57BL/6N male mice	Fed with either a high-fat (HF) or low-fat (LF) diet or HF diet supplemented with air-dried lingonberry powder (HF + LGB).	6 weeks	Lingonberry supplementation prevented the effect of HF diet on an array of genes, such a Mogat1, Plin4, Igfbp2, Lcn2, Saa1, Saa2, Cxcl14, Gcp1, S100a10, Cdkn1a, Tubb2a, and Tubb6.
Khanal et al. 2010 [115]	Rats fed a high fructose diet	Dietary treatments were control (starch based), high fructose (HF), and HF containing either 3.3, 6.6, or 33 g cranberry powder/kg diet.	/	Fed with cranberry powder decreased plasma glucose and triglycerides, lower fasting plasma insulin.
Chen et al. 2020 [103]	ZDF rats	Fed with the nonacylated anthocyanin extract from bilberries (NAAB) or the acylated anthocyanin extract from purple potatoes (AAPP)	Daily doses of 25 and 50 mg/kg body weight for 8 weeks	NAAB reduced fasting plasma glucose level, the levels of branched-chain amino acids and improved lipid profiles.
Pemmari et al. 2022 [110]	HFD-induced mouse model of obesity	Air-dried bilberry powder	/	The bilberry supplementation was unable to modify the weight gain, prevented the increase in the hepatic injury marker ALT and many inflammatory factors like SAA, MCP1, and CXCL14, prevented the increase in serum cholesterol, glucose, and insulin levels.
Petersen et al. 2022 [116]	Seven-week-old diabetic db/db mice	Standard diet (db/db) or a diet supplemented with 3.8% freeze-dried blueberry (db/db + BB)	10 weeks	Blueberry supplementation reduces NOX4 and IκKβ, increases commensal microbes.
Takikawa et al. 2010 [127]	T2D mice	The effect of dietary bilberry extract (BBE)	/	Ameliorates hyperglycemia and insulin sensitivity via activation of AMP-activated protein kinase (AMPK).
Medina-Larqué et al. 2022 [119]	C57BL6 male mice fed an obesogenic high-fat and high-sucrose (HFHS) diet	Cranberry polyphenols (CP), agavins (AG), CP + AG	9 weeks	Both CP and AG can shape gut microbiota composition and regulate key mucosal markers involved in the repair of epithelial barrier integrity.

Table 2. *Cont.*

	Animal	Intervention	Duration	Results
Zhou et al. 2020 [120]	Male C57BL/6J mice fed with normal diet or HFD	Cranberry polyphenolic extract	16 weeks	CPE reduced but did not normalize HFD-induced body weight gain.
Singh et al. 2018 [121]	Male Swiss albino mice were fed normal chow or HFD	Administered either cranberry extract (CRX) (200 mg/kg) alone or in combination with isomalto-oligosaccharides (IMOs) (1 g/kg)	/	Combination of CRX and IMOs prevented systemic and tissue inflammation, glucose intolerance, and systemic obesity-associated metabolic changes in adipose tissue and liver.
Nair et al. 2014 [117]	Five-week-old lean and obese Zucker rats (LZR and OZR)	Fed a blueberry-enriched diet or an isocaloric control diet	15 weeks	Blueberry (BB) protects by inhibiting TLR4.
Lee et al. 2018 [128]	Twenty-four male Wistar rats	Fed low-fat (LF; 10% fat), HF or HF with 10% by weight blueberry powder diets	8 weeks	Increase Gammaproteobacteria abundance. Ileal villus height, tumor necrosis factor α (Tnfa) and interleukin 1β (Il1b) gene expression normalized by blueberry supplementation. improved markers of insulin sensitivity.
Morissette et al. 2020 [129]	Sixty-eight C57BL/6 male mice	Balanced diet (Chow); high-fat, high-sucrose diet (HFHS); or HFHS supplemented with whole blueberry powder (BB), anthocyanidin (ANT)-rich extract, or proanthocyanidin (PAC)-rich extract	12 weeks	PAC-treated mice were leaner, improved insulin responses during OGTT.
Seymour et al. 2011 [130]	Zucker Fatty and Zucker Lean rats	Fed a HFD or LFD containing 2% (*wt*/*wt*) freeze-dried whole highbush blueberry powder	/	The addition of blueberry reduced triglycerides, fasting insulin, homeostasis model index of insulin resistance, and glucose area under the curve, reduced abdominal fat mass.
Nunes et al. 2021 [131]	Hypercaloric diet-induced prediabetic rat model	Blueberry juice (BJ)	14 weeks	Counteracted diet-evoked metabolic deregulation, improving glucose tolerance, insulin sensitivity, and hypertriglyceridemia, along with systemic and hepatic antioxidant properties, alleviated hepatic steatosis and mitochondrial dysfunction.
Elks et al. 2015 [132]	Four-week-old female C57BL/6J mice after induction of menopause	Fed a high-fat diet or the same diet supplemented with 4% blueberries (BB) powder	12 weeks	BB supplementation prevents the glucose intolerance and hepatic steatosis, and these effects are independent of body weight.

	Animal	Intervention	Duration	Results
Wu et al. 2018 [133]	HFD fed C57BL/6 mice	LFD, HFD, or HFD plus orlistat, and blackberry anthocyanins (BLA) or blueberry anthocyanins (BBA) in their daily food	12 weeks	Reduced serum and hepatic lipid levels and increased hepatic superoxide dismutase and GPx, activities, attenuated expression of tumor necrosis factor TNF-α, interleukin-6, and nuclear factor-kappaB genes.
Ryyti et al. 2020 [134]	Thirty male C57BL/6N mice	Receive LF, HF and lingonberry-supplemented high-fat (HF + LGB) diet	6 weeks	Lingonberry supplementation prevented adverse changes blood cholesterol and glucose levels, restrained proinflammatory adipocytokine leptin, increase serum amyloid A.
Vuong et al. 2009 [135]	KKA(y) mice	Biotransformed blueberry juice (BJ)	/	BJ decreases hyperglycemia, in part by reversing adiponectin levels, protects young pre-diabetic mice from developing obesity and diabetes.
Wu et al. 2016 [136]	HFD induced obese male C57BL/6 mice	Blueberry anthocyanin (BA) at doses of 50, 100, and 200 mg/kg	8 weeks	BA at high doses reduced body weight, low and middle doses did not affect. A high dose could effectively decrease serum glucose, attenuate epididymal adipocytes, improve lipid profiles, and significantly down-regulate expression levels of TNFα, IL-6 PPARγ, and FAS genes.

In summary, evidence from preclinical studies suggests that the *Vaccinium* extr is beneficial for controlling blood glucose, lipids, insulin resistance, oxidative stress, a inflammation.

6.2. Treating DR and DKD

The results of experimental studies on the treatment of DR and DKD with *Vaccini* extracts are listed in Table 3. An increasing number of consumers and scientists are realiz the visual benefits of ANT-rich Vaccinium, and ANT is currently used in ophthalmology prevent diabetic retinopathy and improve vision [137]. Blueberries contain abundant A which is beneficial for eye health. Blueberry extracts from northeast China, in which C glu is the most abundant, ameliorated oxidative stress-induced blood retinal barrier (B damage in the retina [138]. The total anthocyanin content at the optimal dose was estima to be 36 mg/kg [138]. In another study, blueberry anthocyanin extract (BAE) prevented progression of DR via molecular regulation of ROS/endoplasmic reticulum stress (ERS) a the miR-182/8-oxoguanine-DNA glycosylase (OGG1) axis [139]. One study showed t blueberry ANT can protect retinal cells from diabetes-induced inflammation and oxidat stress through the regulation of Nrf2/HO-1 signaling [140]. ANT in blueberries prote human retinal capillary endothelial cells through anti-inflammatory and anti-oxidat mechanisms because malvidin-3-glucoside can reduce angiogenesis in a DR-induced model by inhibiting the Akt pathway and reducing VEGF levels, inhibits the protein kin B pathway and decreases the level of VEGF to reduce angiogenesis in the DR-induced mode [141]. Blueberries and bilberries can be used to develop nutritional supplements the prevention of diabetic retinopathy. Bilberry anthocyanosides promote the synthesis a regeneration of rhodopsin, increase the sensitivity of the retina to changes in light inten improve the blood supply to the retina, visual acuity and dark adaptation [142]. Bilbe extract treatment also decreased the expression of DR markers, such as degradation zonula occludens-1, occludin, claudin-5, and retinal VEGF, and prevented or delayed onset of early diabetic retinopathy in diabetic rats [143].

There are few experimental studies of Vaccinium in the treatment of DKD. DIAVI natural sea buckthorn and Vaccinium myrtillus extract, manipulates gene splicing and pression to treat type II mouse model of diabetic nephropathy in mice. DIAVIT, particula delphinidin, changes vascular endothelial growth factor A (VEGF-A) splicing and resc the diabetic nephropathy (DN) phenotype [144]. One study showed that key indicator renal failure, such as urine color, turbidity, and total protein, were considerably reduce cats with chronic kidney disease receiving a nutritious diet containing 0.0371% cran ries [145]. The chemical components of Vaccinium include ANT, flavonoids, ellagitann and phenolic acids. ANT are polyphenolic compounds present in various foods and play important role in treating DKD. A study showed that prevention of the progression of D by ANT could be related to the regulation of amino acid metabolism. After treatment w ANT, fasting blood glucose levels, glomerular fibrosis scores, glomerular lesion perimet and kidney function (urine creatinine and Cystatin C) were considerably alleviated in D mice [146]. Another study found that body weight, systolic blood pressure, C-pepti serum insulin, glycosylated hemoglobin A1c, and elevated fasting blood glucose lev in diabetic mice were remarkably reduced by ANT [147]. One of the main mechanis by which ANT plays a protective role in DN is by inhibiting the inflammatory respo induced by the LXRα pathway and blocking cholesterol deposition [148]. Flavonoids c stitute a major class of polyphenolic compounds with diverse pharmacological activi Flavonoids also have antifibrotic and antiapoptotic properties and play an important rol renoprotective effects in CKD by interfering with TGF-β1/Smad signaling and inhibit the epithelial-to-mesenchymal transition [149]. Tannins, polyphenolic compounds fr bilberries, play an important role in controlling the progression of diabetic microvasc complications. This will help researchers find ways to develop new cost-effective thera for managing the complications of diabetes [14].

	Animal/Cell	Intervention	Duration	Results
Wang et al. 2015 [137]	Visible light-induced damage in human retinal pigment epithelial (RPE) cells	Four ACNs, pelargonidin-3-glucoside (Pg-3-glu), cyanidin-3-glucoside (Cy-3-glu), delphinidin-3-glucoside, and malvidin-3-glucoside (Mv-3-glu) from blueberry, blackberry and strawberry	/	Cy-3-glu exhibited the highest reactive oxygen species inhibitory capacity, Cy-3-glu and Mv-3-glu decrease VEGF, Cy-3-glu and Pg-3-glu inhibited the increase in β-galactosidase.
Li et al. 2022 [138]	HG-treated ARPE-19 cells	10 μM Cy-3-glu (blueberry extracts from northeast China)	/	Cy-3-glu ameliorating oxidative stress-induced BRB damage via the Nrf2 pathways.
Wang et al. 2022 [139].	Human retinal pigment epithelium cell line ARPE-19 cells were exposed to high concentration glucose (H-Glu) with 25 mM for 24 h	Blueberry anthocyanin extract (BAE)		The increase of apoptosis, ROS level and ERS in ARPE-19 cells induced by H-Glu was notably restored by BAE.
Song et al. 2016 [140]	Intraperitoneal injection of streptozotocin (STZ, 60 mg/kg) was used to induce a rat diabetes model.	Blueberry anthocyanins at 20, 40, and 80 mg/kg were given orally	12 weeks	Blueberry anthocyanins prevent diabetes-induced weight loss and increased blood glucose, increased GSH and GPx, decreased MDA, ROS, VEGF and IL-1β, increased the mRNA levels of Nrf2 and HO-1.
Huang et al. 2018 [141]	high glucose- (HG-) induced injury in human retinal capillary endothelial cells (HRCECs)	Blueberry anthocyanin extract and its predominant constituents, malvidin (Mv), malvidin-3-glucoside (Mv-3-glc), and malvidin-3-galactoside (Mv-3-gal),	24 h	All increased cell viability, SOD he enzyme activity of catalase; decreased ROS.
Kim et al. 2015 [143]	streptozotocin-induced diabetic rats	Bilberries extract (100 mg/kg)	6 weeks	Bilberries extract did not affect the blood glucose levels and body weight; reduced the fluorescein leakage; decreased markers of diabetic retinopathy, such as retinal VEGF expression and degradation of zonula occludens-1, occludin, and claudin-5.
Stevens et al. 2019 [144]	a model of type II DN. Diabetic db/db mice	Administered DIAVIT in their drinking water	14 weeks	DIAVIT prevented albuminuria and glomerular water permeability; alters VEGF-A splicing in type II DN, rescuing the DN phenotype.

Table 3. *Cont.*

	Animal/Cell	Intervention	Duration	Results
Di Cerbo et al. 2018 [145]	34 client-owned, neutered cats with II-III CKD	Control diet (n = 17) or a nutraceutical diet (n = 17, contain 0.0371% cranberry)	90 days	creatinine, blood urea nitrogen, total proteins, aspartate aminotransferase, urine turbidity score, color score, and total proteins decreased in cats that received the ND.
Li et al. 2022 [146]	6-week-old male C57BLKS/J-Lepr^{db}/Lepr^{db} mice	10 mg/kg Cyanidin-3-*O*-glucoside per day by oral gavage	12 weeks	The fasting blood glucose level, perimeter of glomerular lesions, perimeter of glomerular lesions and kidney function (Cystatin C, urine creatinine) alleviated after ANT treatment compared to untreated; upregulated taurine, hypotaurine metabolism pathway tryptophan metabolism and tyrosine metabolism.
Qin et al. 2018 [147]	DN in db/db mice	Cyanidin 3-glucoside	/	Cyanidin 3-glucoside reduced body weight, the levels of blood urea nitrogen (BUN), serum creatinine, urinary albumin content and albumin/creatinine ratio (ACR);reduced the surface area of Bowman's capsule, glomerular tuft, Bowman's space, and decreased renal expression of collagen IV, fibronectin, transforming growth factor β 1 (TGFβ1), matrix metalloprotein 9 (MMP9) and α-smooth muscle actin (α-SMA), the Lee's index, perirenal white adipose tissue weight, and high levels of blood and renal triglyceride and cholesterol, reduced systemic levels and renal expression of TNFα, IL-1α, and monocyte chemotactic protein-1 (MCP-1); increased GSH; decreased GSSG level.
Du et al. 2015 [148]	High-glucose (HG)-stimulated HK-2 cells	ANTs:cyanidin-3-O-β-glucoside chloride [C3G] or cyanidin chloride [Cy]		Enhanced cholesterol efflux and ABCA1 expression; increased peroxisome proliferator-activated receptor alpha (PPARα) and liver X receptor alpha (LXRα) expression and decreased the HG-induced expression of the proinflammatory cytokines intercellular adhesion molecule-1 (ICAM1), monocyte chemoattractant protein-1 (MCP1), and transforming growth factor-β1 (TGFβ1), as well as NFκB activation, blocking cholesterol deposition and inhibiting the LXRα pathway-induced inflammatory response.

7. Clinical Evidence for the Effect of *Vaccinium* on Diabetes and Diabetic Microvascular Complications

Owing to the powerful antioxidant effects of *Vaccinium*, the therapeutic potential of these fruits and their extracts has been evaluated for several chronic diseases, including diabetes mellitus, cancer, and neurodegenerative and cardiovascular diseases. This review focuses on *Vaccinium* extracts for the treatment of diabetes and diabetic microvascular complications (DR and DKD).

7.1. Effect of Vaccinium on Type 2 Diabetes Mellitus Treatment

There are many clinical studies on the treatment of diabetes with *Vaccinium*, which can lower blood glucose levels. Clinical evidence for the anti-diabetic effects of *Vaccinium.* is listed in Table 4. Whole blueberry and soluble fiber supplementation prevents gestational weight gain, improves inflammation, and controls blood glucose levels in obese women [150]. In addition, in adults, pancreatic polypeptide (PP) concentrations were remarkably higher when 140 g of whole blueberries were administered [151]. The consumption of 22 g of freeze-dried blueberries for 8 weeks was beneficial to the hearts of men with T2D [152]. In addition to blueberries, bilberries, cranberries, and whortleberries have a similar effect on blood sugar control, and some studies have recommended the use of bilberries to regulate blood glucose levels in patients with T2D or metabolic syndrome [153–155]. In addition, one study showed that bilberries lower postprandial blood glucose and insulin levels [155]. One study showed that cranberries could improve postprandial glucose management [156]. In addition, dried cranberries [27] and cranberry juice [157–159] have similar effects and that whortleberry extract considerably decreases HbA1c, fasting glucose, and 2 h postprandial glucose levels [160]. ANT are chemicals found in *Vaccinium* species. Purified ANT favorably affects glycemic control and the lipid profile [161,162].

A recent meta-analysis showed that consumption of blueberries and cranberries remarkably reduced fasting blood glucose and glycated hemoglobin levels in patients with diabetes is highly credible. In individuals with diabetes, the consumption of cranberries or blueberries considerably reduced fasting blood glucose [MD: −17.72 mg/dL; 95% CI: −29.62, −5.82; $p = 0.03$; I2 = 57%] and glycated hemoglobin [MD: −0.32; 95% CI: −0.57, −0.07; $p = 0.15$; I2 = 39%]; however, there was no effect on insulin resistance [37]. Similarly, another meta-analysis, including seven randomized controlled trials, involved 270 adult patients with T2D, who consumed cranberry juice (240 mL) daily for 12 weeks and were supplemented with powder or blueberry extract (9.1–9.8 mg of ANT) for 8 to 12 weeks to control blood glucose in patients with T2D, despite the heterogeneity in the form of dose, administration (natural, extract, dried, preparation-juice), duration of intervention, and type of population studied involving these two berries [36]. Grohmann et al. [39] showed that interventions with lingonberry and blackcurrant extracts resulted in a mean reduction in HbA1c and fasting glucose levels of 4.7% and 3%, respectively, and that lingonberry and blackcurrant extracts were beneficial for glucose metabolism, although the current evidence is supported by only a few studies in Chinese subjects with T2DM.

In clinical trials using *Vaccinium* specifically, oral administration of the fruit and its extracts has shown mixed results. Owing to the high amount of sugar present in *Vaccinium*, extracts without sugar tend to show better anti-diabetic effects than the whole fruit or juice because of the higher content of bioactive substances [23]. A study in which patients consumed 400 g of fresh bilberries for eight weeks showed a negative correlation between the dietary intake of lingonberries and fasting plasma glucose levels; however, insulin sensitivity remained unchanged [163].

Table 4. Clinical evidence on anti-diabetic effects of *Vaccinium*.

	Sample	Age (y)	Intervention	Duration	Results
Basu et al. 2021 [150]	34 women at high risk of developing GDM	27 ± 5	280 g whole blueberries and 12 g soluble fiber daily or standard prenatal care	18 weeks	Lower maternal weight gain, C-reactive protein, and blood glucose based on GCT in intervention group, compared to the control group.
Mirfeizi et al. 2016 [153]	105 T2DM patients	30–65	Bilberry supplements 1 g or placebo daily	90 days	Reduced FBG, 2 h blood postprandial glucose and homeostasis model assessment of insulin resistance (HOMA-IR) scores in the bilberry group, compared with placebo group.
De Mello et al. 2017 [154]	47 individuals with metabolic syndrome	25–60	200 g of bilberry purée and 40 g of dried bilberries (altogether eq. 400 g of fresh bilberries) or control	8 weeks	Significant increase in fasting serum hippuric acid in intervention group, compared to the control group.
Wilson et al. 2010 [27]	13 noninsulin-dependent diabetics	61.6 ± 2.3	Sweetened dried cranberries (40 g, 113 cal, 1.8 g fiber, 10 g polydextrose)	/	Favorable glycemic and insulinemic response in intervention group.
Chan. et al. 2021 [28]	20 T2DM patients	55.8 ± 9.5	Bilberry supplementation (1.4 g of extract) or placebo daily	4 weeks	Tendency of improved glycemic control in intervention group, compared to the placebo group.
Stote et al. 2019 [151]	17 healthy adults	22–65	140 g of whole blueberries or placebo daily	/	Significant increase in pancreatic polypeptide (PP) concentrations in intervention group, compared to the placebo group.
Schell et al. 2017 [156]	25 T2DM patients	56 ± 6	Fast-food style high-fat breakfast (70 g fat, 974 kcal) with or without cranberries (40 g).	/	Lower postprandial increases of glucose at 2 and 4 h in the cranberry group, compared to control group.
Kianbakht. et al. 2013 [160]	37 T2DM patients	40–60	1050 mg of Caucasian whortleberry fruit hydroalcoholic extract or placebo daily, in combination with anti-hyperglycemic drugs	2 months	Lower blood levels of fasting glucose, 2 h postprandial glucose, and HbA1c in intervention group, compared to the placebo group.
Novotny et al. 2015 [157]	56 individuals	25–65	480 mL of low-calorie cranberry juice or placebo daily	8 weeks	Reduced circulating TGs, CRP, and glucose, insulin resistance, and diastolic BP in intervention group, compared to the placebo group.

	Sample	Age (y)	Intervention	Duration	Results
Hoggard et al. 2013 [155]	8 male volunteers with T2D	62 ± 5	A single capsule of 0.47 g standardized bilberry extract (36% (*w*/*w*) anthocyanins) (eq. 50 g of fresh bilberries) or placebo followed by a polysaccharide drink (eq. 75 g glucose)	/	Reduced postprandial glycaemia and insulin levels in intervention group, compared to the placebo group.
Wilson et al. 2008 [158]	12 T2DM patients	65.3 ± 2.3	Unsweetened low-calorie cranberry juice (LCCBJ; 19 Cal/240 mL) and control	/	Favorable metabolic response in intervention group, compared to the control group.
Shidfar et al. 2012 [159]	58 male volunteers with T2D	54.8 ± 9.1	240 mL of cranberry juice or placebo daily	12 weeks	Significant decrease in serum glucose and apo B; and significant increase in serum apoA-1 and PON-1 activity in intervention group, compared to the placebo group.
Yang et al. 2017 [161]	160 participants with prediabetes or early untreated diabetes	40–75	Purified anthocyanins (320 mg/day) or placebo	12 weeks	Reduced HbA1c, low-density lipoprotein-c, apolipoprotein A-1, apolipoprotein B in intervention group, compared to the placebo group.
Li et al. 2015 [162]	58 diabetic patients	56–67	160 mg of anthocyanins twice daily or placebo	24 weeks	Decreased serum LDL cholesterol, triglycerides, apolipoprotein B, and apo C-III; increased HDL cholesterol; higher total radical-trapping antioxidant parameter and ferric ion reducing antioxidant power values; lower fasting plasma glucose, homeostasis model assessment for insulin resistance index, and elevated serum adiponectin and b-hydroxybutyrate in intervention group, compared to the placebo group.
Stote et al. 2020 [152]	58 male volunteers with T2D	51–75	22 g freeze-dried blueberries or placebo daily	8 weeks	Lower hemoglobin A1c, fructosamine, triglycerides, aspartate transaminase, and alanine transaminase levels in intervention group, compared to the placebo group.

However, other clinical studies have shown no significant differences in fasting g cose levels between treatment and control groups after 12 [164] or 24 weeks [165] of diet anthocyanin supplementation, or 2 months of daily intake of 400 g of fresh lingonb ries [166]. Even in the latest clinical study in Chinese patients with T2DM, using 1. of bilberry extract daily for 6 weeks, HbA1c decreased by 0.31 ± 0.58% while taking supplement; however, this change was not considerably different compared to place and there was also no considerable difference between lingonberry extract and placebo antioxidant status, oxidative stress and inflammatory status treatment [28].

7.2. Research for the Treatment of DR and DKD

There are few clinical studies on *Vaccinium* and its extracts in the treatment of DR the first open-label placebo-controlled study of bilberry extract in DR, a combination 200 mg bilberry extract and 10 mg carotene administered thrice a day reduced vascu permeability and improved retinal vascularity [167]. In another study, in patients w diabetic and hypertensive retinopathy, 160 mg of bilberry extract containing 25% A taken twice daily showed a 77–90% improvement in fundus examination and fluorosco angiographic abnormalities compared with placebo [168]. One study tested the effec bilberry fruit extract on patients with diabetic retinopathy at a dose of 510 mg/day for year, with gradual improvement in contrast sensitivity, but other measured paramet (corrected visual acuity, microaneurysms, hard exudates, and leaking points) remair unchanged for the entire duration of the study [169]. In a randomized, double-bli monocentric, prospective study, supplementation with Macuprev (containing bilberries 3 and anthocyanosides 90 mg) increased the function of macular preganglional compone which helped to decrease inflammation in DR lesions [170]. Bilberries are also used treat diabetes and microvascular complications [171]. However, high HDL levels are a associated with diabetic retinopathy [172]. Therefore, more basic experiments are needec understand the mechanisms by which HDL affects DR. Further clinical trials are requi

Although there are few clinical studies on *Vaccinium* in the treatment of DKD, *V cinium* and its active components have shown promising results in the clinical intervent of CKD. *Vaccinium* is an important component of local diets in many countries. It is popu because of its pleasant taste and is often processed into alcoholic beverages, preservati jams, pies, and juices. Plant-based diets may help manage and prevent some of the syr toms and metabolic complications of CKD [173]. In a meta-analysis of cohort studies CKD, seven studies including 15,285 participants showed that a plant-based diet redu the risk of CKD [174]. There is growing evidence that an entire plant-based diet may sl the progression of CKD, reduce the incidence of cardiovascular disease, and lower rates of obesity and diabetes, which, in turn, may delay the onset of kidney failure a dialysis [175–177].

In addition, some clinical studies have demonstrated the vascular protective effe of *Vaccinium* (Table 5), indicating its potential application in the prevention of diab microvascular complications. One study showed the first sustained improvements lipid status, vascular function, and underlying NO bioactivity following consumpt of one cup of blueberries per day [178]. These findings suggest that blueberries ex immunomodulatory effects and reduce oxidative stress and inflammation in patients w metabolic syndrome [179]. Among the *Vaccinium* species, blueberry, cranberry, and bilbe have this vascular protective function. Cranberries decrease atherosclerotic choleste profiles, including total and LDL-C levels, and the total-to-HDL cholesterol ratio [1 These findings suggest that daily consumption of cranberry beverages for 8 weeks n help reduce lipid status and alter certain biomarkers of oxidative stress in individuals w obesity and a pro-inflammatory state [181].

	Sample	Age (y)	Intervention	Duration	Results
Curtis et al. 2019 [178]	115 adults with MetS	63 ± 7	Blueberries (75 g or 150 g) or placebo daily	6 months	Improvements in vascular function, lipid status, and underlying NO bioactivity in intervention group, compared to the placebo group.
Nair et al. 2017 [179]	27 adults with metabolic syndrome	/	Blueberries (45 g freeze-dried) or placebo daily	6 weeks	Decreased superoxide and total ROS in whole blood and monocytes; increased myeloid DC; decreased monocyte gene expression of TNFα, IL-6, TLR4 and reduced serum GMCSF in intervention group, compared to the placebo group.
Hsia et al. 2020 [181]	35 individuals with obesity and with elevated fasting glucose or impaired glucose tolerance	/	450 mL of low-energy cranberry beverage or placebo daily	8 weeks	Levels of 8-isoprostane (biomarker of lipid peroxidation) decreased in the cranberry group but increased in the placebo group.
Lehtonen et al. 2011 [182]	110 female overweight and obese women	44.2 ± 6.2	Bilberry diets (equivalent to an average daily dose of 100 g fresh bilberries)	33–35 days	Decrease in waist circumference, weight, and Vascular cell adhesion molecule (VCAM).
Lee et al. 2008 [180]	30 T2D subjects taking oral glucose-lowering drugs	65 ± 1	Three capsules (500 mg/capsule) of cranberry extracts or placebo daily	12 weeks	Reduced atherosclerotic cholesterol profiles, including LDL-C, total cholesterol, and total: HDL cholesterol ratio in intervention group, compared to placebo group.

Despite progress in studies on the hypolipidemic and hypoglycemic effects of *V cinium* and improvement in DR, further studies with larger cohorts, longer follow- periods, and more reliable endpoints (for example, proteinuria, glomerular filtration r and disease progression) are required to evaluate the use of lingonberry extract as add-on therapy for T2D, diabetic retinal disease, and glycogenic kidneys.

8. Conclusions and Future Perspectives

Diabetes mellitus and its microvascular complications require effective dietary sup ment adjuvant therapy. Because *Vaccinium* fruit contains many antioxidant compoun clinical application of *Vaccinium* extract as nutritional health products may be beneficial diabetes-related microvascular complications, especially DKD and DR. Compared w common drugs, the *Vaccinium* extract is currently safe and mostly has no side effects. *V cinium* extracts offer a means to discover and develop new drugs; however, they have so drawbacks. *Vaccinium* does not deliver the intense and potent therapeutic effects of dr such as SGLT2. *Vaccinium* is generally studied as a whole, the exact components that af the disease are unknown, and the understanding of the underlying mechanism is still in infancy. The limitation of clinical usefulness is poor bioavailability. In clinical trials, ju is available in 240 mL [159,160] or 480 mL [157] daily, bilberry supplements are availa in doses of 1 g [153] or 1.4 g [28] daily, and the dosage of the whole fruit of blueberrie 140 to 300 g [150–152,178] daily. The best dose of *Vaccinium* is unknown; it is usually fou in juice, fruits, extract, and other forms, and more experiments are needed to determ the dosage. Although ANT are rapidly absorbed, they are also rapidly metabolized a excreted from the body [183,184].

Current technological developments in the pharmaceutical industry are driving development of *Vaccinium* extracts for T2D and its associated microvascular complicatio allowing for increased purity percentages and optimized formulations to obtain grea in vivo stability and target tissue bioavailability, thereby prolonging their therape effects.

This makes *Vaccinium* a promising treatment for diabetes and diabetic microvascu complications. However, further studies on the mechanisms involved, as well as lar randomized blinded trials, are urgently required.

Author Contributions: Conceptualization, Z.D. and X.C.; methodology, Z.D.; writing—original d preparation, H.H. and Y.L.; writing—review and editing, H.H., Q.W., Z.L., Y.Z. and R.H. All auth have read and agreed to the published version of the manuscript.

Funding: This research was funded by the Science and Technology Project of Beijing (Z2211000074221 Beijing Natural Science Foundation (L222133), and Natural Science Foundation of China (62250001

Institutional Review Board Statement: Not applicable.

Informed Consent Statement: Not applicable.

Data Availability Statement: Not applicable.

Conflicts of Interest: The authors declare no conflict of interest.

Abbreviations

ACEI, Angiotensin-converting enzyme inhibitors; AGEs, advanced glycation end produ AMPK, adenosine monophosphate-activated protein kinase; Ang-2, Angiopoietin-2; ANT, An cyanins; aP2, adipocyte protein 2; ARB, Angiotensin Receptor Blocker; BA, blueberry anthocya BAE, blueberry anthocyanin extract; BBE, bilberry extract; BBJ, blueberry juice; BRB, blood ret barrier; CANVAS, Canagliflozin Cardiovascular Assessment Study; CFBJ, commercial fermen blueberry juice; CKD, chronic kidney disease; CoA, coenzyme A; CTGF, connective tissue gro factor; CVD, cardiovascular disease; DGAT1, diacylglycerol acyltransferase-1; DIO, diet-indu obesity; DKD, diabetic kidney disease; DM, diabetes mellitus; DN, diabetic nephropathy; DR, diab retinopathy; EMPA-REG Outcomes, Empagliflozin, Cardiovascular Outcomes, and Mortality

Type 2 Diabetes; eNOS, endothelial nitric oxide synthase; ERS, endoplasmic reticulum stress; ESRD, end-stage renal disease; FAS, fatty acid synthase; FBJ, fermented blueberry juice; FFA, free fatty acid; FGF-2, fibroblast growth factor-2; GBM, glomerular basement membrane; GFR, glomerular filtration rate; GPx, glutathione peroxidase; GSH, glutathione; HbA1, Hemoglobin alpha 1; HDL, High density liptein; HFD, high-fat diet; HO-1, heme oxygenase 1; LCCBJ, low-calorie cranberry juice; LDL, low density lipoprotein; LDL-C, low density lipoprotein cholesterol; LF, low-fat diet; IL-1, interleukin-1; IL-1β, interleukin-1β; IL-6, interleukin-6; IL-8, interleukin-8; LXRα, Liver X Receptor Alpha; IR, Insulin receptor; IRS, Insulin receptor substrate; MDA, malondialdehyde; MS, metabolic syndrome; NAAB, Nonacylated anthocyanins extract from bilberries; NLRP3, NOD-like receptor thermal protein domain associated protein 3; NO, nitric oxide; Nox, NADPH oxidase complex; NOX4, NADPH oxidase 4; NPDR, non-proliferative diabetic retinopathy; Nrf2, nuclear factor erythroid 2-related factor 2; OGG1, 8-oxoguanine-DNA glycosylase; OZR, obese Zucker rat; PDGF, Platelet-derived growth factor; PDR, proliferative diabetic retinopathy; PKC, protein kinase C; PP, pancreatic polypeptide; RAAS, Renin-angiotensin-aldosterone system; RNS, reactive nitrogen species; ROS, reactive oxygen species; SGLT2, sodium–glucose cotransporter 2; T2D, type 2 diabetes; T1D, type 1 diabetes; TG, triglycerides; TGF-β1, transforming growth factor beta 1; TLR2, Toll-like receptor 2; TLR4, Toll-like receptor 4; TNFα, tumor necrosis factor alpha; VEGF, vascular endothelial growth factor; VEGF-A, Vascular endothelial growth factor A; WoSCC, Web of Science Core Collection; ZDF, Zucker Diabetic Fatty.

erences

Misra, A.; Gopalan, H.; Jayawardena, R.; Hills, A.P.; Soares, M.; Reza-Albarrán, A.A.; Ramaiya, K.L. Diabetes in developing countries. *J. Diabetes* **2019**, *11*, 522–539. [CrossRef] [PubMed]

Ali, M.K.; Pearson-Stuttard, J.; Selvin, E.; Gregg, E.W. Interpreting global trends in type 2 diabetes complications and mortality. *Diabetologia* **2022**, *65*, 3–13. [CrossRef] [PubMed]

Sun, H.; Saeedi, P.; Karuranga, S.; Pinkepank, M.; Ogurtsova, K.; Duncan, B.B.; Stein, C.; Basit, A.; Chan, J.C.N.; Mbanya, J.C.; et al. IDF Diabetes Atlas: Global, regional and country-level diabetes prevalence estimates for 2021 and projections for 2045. *Diabetes Res. Clin. Pract.* **2022**, *183*, 109119. [CrossRef] [PubMed]

Xue, R.; Gui, D.; Zheng, L.; Zhai, R.; Wang, F.; Wang, N. Mechanistic Insight and Management of Diabetic Nephropathy: Recent Progress and Future Perspective. *J. Diabetes Res.* **2017**, *2017*, 1839809. [CrossRef]

Chatterjee, S.; Khunti, K.; Davies, M.J. Type 2 diabetes. *Lancet* **2017**, *389*, 2239–2251. [CrossRef]

Barrett, E.J.; Liu, Z.; Khamaisi, M.; King, G.L.; Klein, R.; Klein, B.E.K.; Hughes, T.M.; Craft, S.; Freedman, B.I.; Bowden, D.W.; et al. Diabetic Microvascular Disease: An Endocrine Society Scientific Statement. *J. Clin. Endocrinol. Metab.* **2017**, *102*, 4343–4410. [CrossRef]

Zhang, L.; Long, J.; Jiang, W.; Shi, Y.; He, X.; Zhou, Z.; Li, Y.; Yeung, R.O.; Wang, J.; Matsushita, K.; et al. Trends in Chronic Kidney Disease in China. *N. Engl. J. Med.* **2016**, *375*, 905–906. [CrossRef]

Burrows, N.R.; Hora, I.; Geiss, L.S.; Gregg, E.W.; Albright, A. Incidence of End-Stage Renal Disease Attributed to Diabetes Among Persons with Diagnosed Diabetes—United States and Puerto Rico, 2000–2014. *MMWR Morb. Mortal. Wkly. Rep.* **2017**, *66*, 1165–1170. [CrossRef]

Sagoo, M.K.; Gnudi, L. Diabetic Nephropathy: An Overview. *Methods Mol. Biol.* **2020**, *2067*, 3–7. [CrossRef]

Sabanayagam, C.; Banu, R.; Chee, M.L.; Lee, R.; Wang, Y.X.; Tan, G.; Jonas, J.B.; Lamoureux, E.L.; Cheng, C.Y.; Klein, B.E.K.; et al. Incidence and progression of diabetic retinopathy: A systematic review. *Lancet Diabetes Endocrinol.* **2019**, *7*, 140–149. [CrossRef]

Sabanayagam, C.; Chee, M.L.; Banu, R.; Cheng, C.Y.; Lim, S.C.; Tai, E.S.; Coffman, T.; Wong, T.Y. Association of Diabetic Retinopathy and Diabetic Kidney Disease With All-Cause and Cardiovascular Mortality in a Multiethnic Asian Population. *JAMA Netw. Open.* **2019**, 2, e191540. [CrossRef] [PubMed]

Cao, K.; Wang, B.; Friedman, D.S.; Hao, J.; Zhang, Y.; Hu, A.; Wang, N. Diabetic Retinopathy, Visual Impairment, and the Risk of Six-Year Death: A Cohort Study of a Rural Population in China. *Ophthalmic Res.* **2021**, *64*, 983–990. [CrossRef]

Giacco, F.; Brownlee, M. Oxidative stress and diabetic complications. *Circ. Res.* **2010**, *107*, 1058–1070. [CrossRef]

Laddha, A.P.; Kulkarni, Y.A. Tannins and vascular complications of Diabetes: An update. *Phytomedicine* **2019**, *56*, 229–245. [CrossRef]

Edger, P.P.; Iorizzo, M.; Bassil, N.V.; Benevenuto, J.; Ferrão, L.F.V.; Giongo, L.; Hummer, K.; Lawas, L.M.F.; Leisner, C.P.; Li, C.; et al. There and back again; historical perspective and future directions for Vaccinium breeding and research studies. *Hortic. Res.* **2022**, *9*, uhac083. [CrossRef] [PubMed]

Bläsing, D. A Review of Vaccinium Research and the Vaccinium Industry of the Federal Republic of Germany. *Acta Hortic.* **1989**, *241*, 101–109. [CrossRef]

17. Sater, H.M.; Bizzio, L.N.; Tieman, D.M.; Muñoz, P.D. A Review of the Fruit Volatiles Found in Blueberry and Other Vaccini Species. *J. Agric. Food Chem.* **2020**, *68*, 5777–5786. [CrossRef]
18. Sharma, A.; Lee, H.J. Anti-Inflammatory Activity of Bilberry (*Vaccinium myrtillus* L.). *Curr. Issues Mol. Biol.* **2022**, *44*, 4570–4. [CrossRef] [PubMed]
19. Martău, G.A.; Bernadette-Emőke, T.; Odocheanu, R.; Soporan, D.A.; Bochiș, M.; Simon, E.; Vodnar, D.C. Vaccinium Spe (*Ericaceae*): Phytochemistry and Biological Properties of Medicinal Plants. *Molecules* **2023**, *28*, 1533. [CrossRef]
20. Hasanloo, T.; Jafarkhani Kermani, M.; Dalvand, Y.A.; Rezazadeh, S. A Complete Review on the Genus Vaccinium and Iran Ghareghat. *J. Med. Plants* **2019**, *18*, 46–65. [CrossRef]
21. Sapian, S.; Taib, I.S.; Katas, H.; Latip, J.; Zainalabidin, S.; Hamid, Z.A.; Anuar, N.N.M.; Budin, S.B. The Role of Anthocyani Modulating Diabetic Cardiovascular Disease and Its Potential to Be Developed as a Nutraceutical. *Pharmaceuticals* **2022**, *15*, 1 [CrossRef] [PubMed]
22. Ştefănescu Braic, R.; Vari, C.; Imre, S.; Huţanu, A.; Fogarasi, E.; Todea, T.; Groşan, A.; Eşianu, S.; Laczkó-Zöld, E.; Dogaru Vaccinium Extracts as Modulators in Experimental Type 1 Diabetes. *J. Med. Food* **2018**, *21*, 1106–1112. [CrossRef] [PubMed]
23. Vaneková, Z.; Rollinger, J.M. Bilberries: Curative and Miraculous—A Review on Bioactive Constituents and Clinical Resea *Front. Pharmacol.* **2022**, *13*, 909914. [CrossRef]
24. Thibado, S.P.; Thornthwaite, J.T.; Ballard, T.K.; Goodman, B.T. Anticancer effects of Bilberry anthocyanins compared w NutraNanoSphere encapsulated Bilberry anthocyanins. *Mol. Clin. Oncol.* **2018**, *8*, 330–335. [CrossRef] [PubMed]
25. Satoh, Y.; Ishihara, K. Investigation of the antimicrobial activity of Bilberry (*Vaccinium myrtillus* L.) extract against periodontopa bacteria. *J. Oral. Biosci.* **2020**, *62*, 169–174. [CrossRef]
26. Osada, H.; Okamoto, T.; Kawashima, H.; Toda, E.; Miyake, S.; Nagai, N.; Kobayashi, S.; Tsubota, K.; Ozawa, Y. Neuroprotec effect of bilberry extract in a murine model of photo-stressed retina. *PLoS ONE* **2017**, *12*, e0178627. [CrossRef]
27. Wilson, T.; Luebke, J.L.; Morcomb, E.F.; Carrell, E.J.; Leveranz, M.C.; Kobs, L.; Schmidt, T.P.; Limburg, P.J.; Vorsa, N.; Singh, Glycemic responses to sweetened dried and raw cranberries in humans with type 2 diabetes. *J. Food Sci.* **2010**, *75*, H218–H [CrossRef]
28. Chan, S.W.; Chu, T.T.W.; Choi, S.W.; Benzie, I.F.F.; Tomlinson, B. Impact of short-term bilberry supplementation on glyce control, cardiovascular disease risk factors, and antioxidant status in Chinese patients with type 2 diabetes. *Phytother. Res.* **2** *35*, 3236–3245. [CrossRef]
29. Ulbricht, C.; Basch, E.; Basch, S.; Bent, S.; Boon, H.; Burke, D.; Costa, D.; Falkson, C.; Giese, N.; Goble, M.; et al. An evidence-ba systematic review of bilberry (*Vaccinium myrtillus*) by the Natural Standard Research Collaboration. *J. Diet. Suppl.* **2009**, *6*, 162– [CrossRef] [PubMed]
30. Tundis, R.; Tenuta, M.C.; Loizzo, M.R.; Bonesi, M.; Finetti, F.; Trabalzini, L.; Deguin, B. Vaccinium Species (*Ericaceae*): F Chemical Composition to Bio-Functional Activities. *Appl. Sci.* **2021**, *11*, 5655. [CrossRef]
31. Ștefănescu, R.; Laczkó-Zöld, E.; Ősz, B.E.; Vari, C.E. An Updated Systematic Review of *Vaccinium myrtillus* Leaves: Phytochemi and Pharmacology. *Pharmaceutics* **2022**, *15*, 16. [CrossRef]
32. Cho, M.J.; Howard, L.R.; Prior, R.L.; Clark, J.R. Flavonoid glycosides and antioxidant capacity of various blackberry, bluebe and red grape genotypes determined by high-performance liquid chromatography/mass spectrometry. *J. Sci. Food Agric.* **2** *84*, 1771–1782. [CrossRef]
33. Skrovankova, S.; Sumczynski, D.; Mlcek, J.; Jurikova, T.; Sochor, J. Bioactive Compounds and Antioxidant Activity in Differ Types of Berries. *Int. J. Mol. Sci.* **2015**, *16*, 24673–24706. [CrossRef]
34. Ștefănescu, B.E.; Szabo, K.; Mocan, A.; Crişan, G. Phenolic Compounds from Five *Ericaceae* Species Leaves and Their Rela Bioavailability and Health Benefits. *Molecules* **2019**, *24*, 2046. [CrossRef]
35. Rodriguez-Bonilla, L.; Williams, K.A.; Rodríguez Bonilla, F.; Matusinec, D.; Maule, A.; Coe, K.; Wiesman, E.; Diaz-Garcia Zalapa, J. The Genetic Diversity of Cranberry Crop Wild Relatives, Vaccinium macrocarpon Aiton and V. oxycoccos L., in the with Special Emphasis on National Forests. *Plants* **2020**, *9*, 1446. [CrossRef] [PubMed]
36. Rocha, D.; Caldas, A.P.S.; da Silva, B.P.; Hermsdorff, H.H.M.; Alfenas, R.C.G. Effects of blueberry and cranberry consumptio type 2 diabetes glycemic control: A systematic review. *Crit. Rev. Food Sci. Nutr.* **2019**, *59*, 1816–1828. [CrossRef]
37. Delpino, F.M.; Figueiredo, L.M.; Gonçalves da Silva, T.; Flores, T.R. Effects of blueberry and cranberry on type 2 diab parameters in individuals with or without diabetes: A systematic review and meta-analysis of randomized clinical trials. N *Metab. Cardiovasc. Dis.* **2022**, *32*, 1093–1109. [CrossRef]
38. Bujor, O.C.; Tanase, C.; Popa, M.E. Phenolic Antioxidants in Aerial Parts of Wild Vaccinium Species: Towards Pharmaceutical Biological Properties. *Antioxidants* **2019**, *8*, 649. [CrossRef] [PubMed]
39. Grohmann, T.; Litts, C.; Horgan, G.; Zhang, X.; Hoggard, N.; Russell, W.; de Roos, B. Efficacy of Bilberry and Grape Seed Ext Supplement Interventions to Improve Glucose and Cholesterol Metabolism and Blood Pressure in Different Population Systematic Review of the Literature. *Nutrients* **2021**, *13*, 1692. [CrossRef] [PubMed]
40. Basu, A. Role of Berry Bioactive Compounds on Lipids and Lipoproteins in Diabetes and Metabolic Syndrome. *Nutrients* **2019** 1983. [CrossRef]
41. Chan, S.W.; Tomlinson, B. Effects of Bilberry Supplementation on Metabolic and Cardiovascular Disease Risk. *Molecules* **2020** 1653. [CrossRef] [PubMed]

Kalt, W.; Cassidy, A.; Howard, L.R.; Krikorian, R.; Stull, A.J.; Tremblay, F.; Zamora-Ros, R. Recent Research on the Health Benefits of Blueberries and Their Anthocyanins. *Adv. Nutr.* **2020**, *11*, 224–236. [CrossRef] [PubMed]
de Oliveira, M.S.; Pellenz, F.M.; de Souza, B.M.; Crispim, D. Blueberry Consumption and Changes in Obesity and Diabetes Mellitus Outcomes: A Systematic Review. *Metabolites* **2022**, *13*, 19. [CrossRef]
Nemzer, B.V.; Al-Taher, F.; Yashin, A.; Revelsky, I.; Yashin, Y. Cranberry: Chemical Composition, Antioxidant Activity and Impact on Human Health: Overview. *Molecules* **2022**, *27*, 1503. [CrossRef]
Fowler, M.J. Microvascular and Macrovascular Complications of Diabetes. *Clin. Diabetes* **2011**, *29*, 116–122. [CrossRef]
Zoja, C.; Benigni, A.; Remuzzi, G. The Nrf2 pathway in the progression of renal disease. *Nephrol. Dial. Transplant.* **2014**, *29* (Suppl. S1), i19–i24. [CrossRef]
Tan, S.M.; de Haan, J.B. Combating oxidative stress in diabetic complications with Nrf2 activators: How much is too much? *Redox Rep.* **2014**, *19*, 107–117. [CrossRef] [PubMed]
Uruno, A.; Yagishita, Y.; Yamamoto, M. The Keap1-Nrf2 system and diabetes mellitus. *Arch. Biochem. Biophys.* **2015**, *566*, 76–84. [CrossRef]
Cheng, H.; Harris, R.C. Renal endothelial dysfunction in diabetic nephropathy. *Cardiovasc. Hematol. Disord. Drug. Targets* **2014**, *14*, 22–33. [CrossRef]
Capellini, V.K.; Celotto, A.C.; Baldo, C.F.; Olivon, V.C.; Viaro, F.; Rodrigues, A.J.; Evora, P.R. Diabetes and vascular disease: Basic concepts of nitric oxide physiology, endothelial dysfunction, oxidative stress and therapeutic possibilities. *Curr. Vasc. Pharmacol.* **2010**, *8*, 526–544. [CrossRef]
Santiago, A.R.; Boia, R.; Aires, I.D.; Ambrósio, A.F.; Fernandes, R. Sweet Stress: Coping With Vascular Dysfunction in Diabetic Retinopathy. *Front. Physiol.* **2018**, *9*, 820. [CrossRef]
Lindblom, R.; Higgins, G.; Coughlan, M.; de Haan, J.B. Targeting Mitochondria and Reactive Oxygen Species-Driven Pathogenesis in Diabetic Nephropathy. *Rev. Diabet. Stud.* **2015**, *12*, 134–156. [CrossRef]
Miranda-Díaz, A.G.; Pazarín-Villaseñor, L.; Yanowsky-Escatell, F.G.; Andrade-Sierra, J. Oxidative Stress in Diabetic Nephropathy with Early Chronic Kidney Disease. *J. Diabetes Res.* **2016**, *2016*, 7047238. [CrossRef] [PubMed]
Santilli, F.; Cipollone, F.; Mezzetti, A.; Chiarelli, F. The role of nitric oxide in the development of diabetic angiopathy. *Horm. Metab. Res.* **2004**, *36*, 319–335. [CrossRef] [PubMed]
Shikata, K.; Makino, H. Microinflammation in the pathogenesis of diabetic nephropathy. *J. Diabetes Investig.* **2013**, *4*, 142–149. [CrossRef] [PubMed]
Moreno, J.A.; Gomez-Guerrero, C.; Mas, S.; Sanz, A.B.; Lorenzo, O.; Ruiz-Ortega, M.; Opazo, L.; Mezzano, S.; Egido, J. Targeting inflammation in diabetic nephropathy: A tale of hope. *Expert. Opin. Investig. Drugs* **2018**, *27*, 917–930. [CrossRef] [PubMed]
Garibotto, G.; Carta, A.; Picciotto, D.; Viazzi, F.; Verzola, D. Toll-like receptor-4 signaling mediates inflammation and tissue injury in diabetic nephropathy. *J. Nephrol.* **2017**, *30*, 719–727. [CrossRef]
Akpoveso, O.P.; Ubah, E.E.; Obasanmi, G. Antioxidant Phytochemicals as Potential Therapy for Diabetic Complications. *Antioxidants* **2023**, *12*, 123. [CrossRef]
Pérez-Morales, R.E.; Del Pino, M.D.; Valdivielso, J.M.; Ortiz, A.; Mora-Fernández, C.; Navarro-González, J.F. Inflammation in Diabetic Kidney Disease. *Nephron* **2019**, *143*, 12–16. [CrossRef]
Tavafi, M. Diabetic nephropathy and antioxidants. *J. Nephropathol.* **2013**, *2*, 20–27. [CrossRef]
Samsu, N. Diabetic Nephropathy: Challenges in Pathogenesis, Diagnosis, and Treatment. *Biomed. Res. Int.* **2021**, *2021*, 1497449. [CrossRef] [PubMed]
Krishan, P.; Chakkarwar, V.A. Diabetic nephropathy: Aggressive involvement of oxidative stress. *J. Pharm. Educ. Res.* **2011**, *2*, 35.
Khazim, K.; Gorin, Y.; Cavaglieri, R.C.; Abboud, H.E.; Fanti, P. The antioxidant silybin prevents high glucose-induced oxidative stress and podocyte injury in vitro and in vivo. *Am. J. Physiol. Renal Physiol.* **2013**, *305*, F691–F700. [CrossRef] [PubMed]
Duni, A.; Liakopoulos, V.; Roumeliotis, S.; Peschos, D.; Dounousi, E. Oxidative Stress in the Pathogenesis and Evolution of Chronic Kidney Disease: Untangling Ariadne's Thread. *Int. J. Mol. Sci.* **2019**, *20*, 3711. [CrossRef] [PubMed]
Rohm, T.V.; Meier, D.T.; Olefsky, J.M.; Donath, M.Y. Inflammation in obesity, diabetes, and related disorders. *Immunity* **2022**, *55*, 31–55. [CrossRef]
Lin, M.; Yiu, W.H.; Wu, H.J.; Chan, L.Y.; Leung, J.C.; Au, W.S.; Chan, K.W.; Lai, K.N.; Tang, S.C. Toll-like receptor 4 promotes tubular inflammation in diabetic nephropathy. *J. Am. Soc. Nephrol.* **2012**, *23*, 86–102. [CrossRef]
Ma, J.; Wu, H.; Zhao, C.Y.; Panchapakesan, U.; Pollock, C.; Chadban, S.J. Requirement for TLR2 in the development of albuminuria, inflammation and fibrosis in experimental diabetic nephropathy. *Int. J. Clin. Exp. Pathol.* **2014**, *7*, 481–495.
Tang, S.C.W.; Yiu, W.H. Innate immunity in diabetic kidney disease. *Nat. Rev. Nephrol.* **2020**, *16*, 206–222. [CrossRef] [PubMed]
Shahzad, K.; Bock, F.; Al-Dabet, M.M.; Gadi, I.; Kohli, S.; Nazir, S.; Ghosh, S.; Ranjan, S.; Wang, H.; Madhusudhan, T.; et al. Caspase-1, but Not Caspase-3, Promotes Diabetic Nephropathy. *J. Am. Soc. Nephrol.* **2016**, *27*, 2270–2275. [CrossRef]
Alicic, R.Z.; Rooney, M.T.; Tuttle, K.R. Diabetic Kidney Disease: Challenges, Progress, and Possibilities. *Clin. J. Am. Soc. Nephrol.* **2017**, *12*, 2032–2045. [CrossRef]
Alicic, R.Z.; Cox, E.J.; Neumiller, J.J.; Tuttle, K.R. Incretin drugs in diabetic kidney disease: Biological mechanisms and clinical evidence. *Nat. Rev. Nephrol.* **2021**, *17*, 227–244. [CrossRef] [PubMed]
Salvatore, T.; Galiero, R.; Caturano, A.; Rinaldi, L.; Di Martino, A.; Albanese, G.; Di Salvo, J.; Epifani, R.; Marfella, R.; Docimo, G.; et al. An Overview of the Cardiorenal Protective Mechanisms of SGLT2 Inhibitors. *Int. J. Mol. Sci.* **2022**, *23*, 3651. [CrossRef]

73. Wanner, C.; Inzucchi, S.E.; Lachin, J.M.; Fitchett, D.; von Eynatten, M.; Mattheus, M.; Johansen, O.E.; Woerle, H.J.; Broedl, U Zinman, B. Empagliflozin and Progression of Kidney Disease in Type 2 Diabetes. *N. Engl. J. Med.* **2016**, *375*, 323–334. [CrossI [PubMed]
74. Mahaffey, K.W.; Neal, B.; Perkovic, V.; de Zeeuw, D.; Fulcher, G.; Erondu, N.; Shaw, W.; Fabbrini, E.; Sun, T.; Li, Q.; e Canagliflozin for Primary and Secondary Prevention of Cardiovascular Events: Results From the CANVAS Program (Canaglifle Cardiovascular Assessment Study). *Circulation* **2018**, *137*, 323–334. [CrossRef]
75. Zac-Varghese, S.; Winocour, P. Managing diabetic kidney disease. *Br. Med. Bull.* **2018**, *125*, 55–66. [CrossRef] [PubMed]
76. Stitt, A.W.; Lois, N.; Medina, R.J.; Adamson, P.; Curtis, T.M. Advances in our understanding of diabetic retinopathy. *Clin.* **2013**, *125*, 1–17. [CrossRef]
77. Duh, E.J.; Sun, J.K.; Stitt, A.W. Diabetic retinopathy: Current understanding, mechanisms, and treatment strategies. *JCI Ins* **2017**, *2*, e93751. [CrossRef]
78. Lin, K.Y.; Hsih, W.H.; Lin, Y.B.; Wen, C.Y.; Chang, T.J. Update in the epidemiology, risk factors, screening, and treatmen diabetic retinopathy. *J. Diabetes Investig.* **2021**, *12*, 1322–1325. [CrossRef]
79. Stitt, A.W.; Curtis, T.M.; Chen, M.; Medina, R.J.; McKay, G.J.; Jenkins, A.; Gardiner, T.A.; Lyons, T.J.; Hammes, H.P.; Simó, R.; e The progress in understanding and treatment of diabetic retinopathy. *Prog. Retin. Eye Res.* **2016**, *51*, 156–186. [CrossRef]
80. Kollias, A.N.; Ulbig, M.W. Diabetic retinopathy: Early diagnosis and effective treatment. *Dtsch. Arztebl. Int.* **2010**, *107*, 75- quiz 84. [CrossRef]
81. Biswas, S.; Feng, B.; Chen, S.; Liu, J.; Aref-Eshghi, E.; Gonder, J.; Ngo, V.; Sadikovic, B.; Chakrabarti, S. The Long Non-Cod RNA HOTAIR Is a Critical Epigenetic Mediator of Angiogenesis in Diabetic Retinopathy. *Invest. Ophthalmol. Vis. Sci.* **2021**, *62* [CrossRef]
82. Capitão, M.; Soares, R. Angiogenesis and Inflammation Crosstalk in Diabetic Retinopathy. *J. Cell. Biochem.* **2016**, *117*, 2443–2 [CrossRef]
83. Praidou, A.; Androudi, S.; Brazitikos, P.; Karakiulakis, G.; Papakonstantinou, E.; Dimitrakos, S. Angiogenic growth factors their inhibitors in diabetic retinopathy. *Curr. Diabetes Rev.* **2010**, *6*, 304–312. [CrossRef]
84. Yang, Y.; Liu, Y.; Li, Y.; Chen, Z.; Xiong, Y.; Zhou, T.; Tao, W.; Xu, F.; Yang, H.; Ylä-Herttuala, S.; et al. MicroRNA-15b Targ VEGF and Inhibits Angiogenesis in Proliferative Diabetic Retinopathy. *J. Clin. Endocrinol. Metab.* **2020**, *105*, 3404–3415. [Cross [PubMed]
85. Tang, L.; Xu, G.T.; Zhang, J.F. Inflammation in diabetic retinopathy: Possible roles in pathogenesis and potential implications therapy. *Neural Regen. Res.* **2023**, *18*, 976–982. [CrossRef] [PubMed]
86. Joussen, A.M.; Poulaki, V.; Le, M.L.; Koizumi, K.; Esser, C.; Janicki, H.; Schraermeyer, U.; Kociok, N.; Fauser, S.; Kirchhof, B.; e A central role for inflammation in the pathogenesis of diabetic retinopathy. *Faseb J.* **2004**, *18*, 1450–1452. [CrossRef] [PubMed
87. Mesquida, M.; Drawnel, F.; Fauser, S. The role of inflammation in diabetic eye disease. *Semin. Immunopathol.* **2019**, *41*, 427– [CrossRef]
88. Dong, N.; Xu, B.; Wang, B.; Chu, L. Study of 27 aqueous humor cytokines in patients with type 2 diabetes with or with retinopathy. *Mol. Vis.* **2013**, *19*, 1734–1746.
89. Wong, T.Y.; Cheung, C.M.; Larsen, M.; Sharma, S.; Simó, R. Diabetic retinopathy. *Nat. Rev. Dis. Primers* **2016**, *2*, 16012. [Cross
90. Madsen-Bouterse, S.A.; Zhong, Q.; Mohammad, G.; Ho, Y.S.; Kowluru, R.A. Oxidative damage of mitochondrial DNA in diab and its protection by manganese superoxide dismutase. *Free. Radic. Res.* **2010**, *44*, 313–321. [CrossRef]
91. Cade, W.T. Diabetes-related microvascular and macrovascular diseases in the physical therapy setting. *Phys. Ther.* **2008**, 1322–1335. [CrossRef] [PubMed]
92. Cui, Y.; Xu, X.; Bi, H.; Zhu, Q.; Wu, J.; Xia, X.; Qiushi, R.; Ho, P.C. Expression modification of uncoupling proteins and MnSOI retinal endothelial cells and pericytes induced by high glucose: The role of reactive oxygen species in diabetic retinopathy. *Eye Res.* **2006**, *83*, 807–816. [CrossRef] [PubMed]
93. Du, Y.; Miller, C.M.; Kern, T.S. Hyperglycemia increases mitochondrial superoxide in retina and retinal cells. *Free. Radic. Med.* **2003**, *35*, 1491–1499. [CrossRef]
94. Tan, T.E.; Wong, T.Y. Diabetic retinopathy: Looking forward to 2030. *Front. Endocrinol.* **2022**, *13*, 1077669. [CrossRef]
95. Wubben, T.J.; Johnson, M.W. Anti-Vascular Endothelial Growth Factor Therapy for Diabetic Retinopathy: Consequence Inadvertent Treatment Interruptions. *Am. J. Ophthalmol.* **2019**, *204*, 13–18. [CrossRef] [PubMed]
96. Zehden, J.A.; Mortensen, X.M.; Reddy, A.; Zhang, A.Y. Systemic and Ocular Adverse Events with Intravitreal Anti-VEGF Ther Used in the Treatment of Diabetic Retinopathy: A Review. *Curr. Diab Rep.* **2022**, *22*, 525–536. [CrossRef]
97. Podsędek, A.; Majewska, I.; Redzynia, M.; Sosnowska, D.; Koziołkiewicz, M. In vitro inhibitory effect on digestive enzymes antioxidant potential of commonly consumed fruits. *J. Agric. Food Chem.* **2014**, *62*, 4610–4617. [CrossRef]
98. Shahwan, M.; Alhumaydhi, F.; Ashraf, G.M.; Hasan, P.M.Z.; Shamsi, A. Role of polyphenols in combating Type 2 Diabetes insulin resistance. *Int. J. Biol. Macromol.* **2022**, *206*, 567–579. [CrossRef]
99. Sales, P.M.; Souza, P.M.; Simeoni, L.A.; Silveira, D. α-Amylase inhibitors: A review of raw material and isolated compounds fr plant source. *J. Pharm. Pharm. Sci.* **2012**, *15*, 141–183. [CrossRef]
100. Takács, I.; Szekeres, A.; Takács, Á.; Rakk, D.; Mézes, M.; Polyák, Á.; Lakatos, L.; Gyémánt, G.; Csupor, D.; Kovács, K.J.; et al. V Strawberry, Blackberry, and Blueberry Leaf Extracts Alleviate Starch-Induced Hyperglycemia in Prediabetic and Diabetic M *Planta Med.* **2020**, *86*, 790–799. [CrossRef]

Anhê, F.F.; Pilon, G.; Roy, D.; Desjardins, Y.; Levy, E.; Marette, A. Triggering Akkermansia with dietary polyphenols: A new weapon to combat the metabolic syndrome? *Gut Microbes* **2016**, *7*, 146–153. [CrossRef]
Grace, M.H.; Ribnicky, D.M.; Kuhn, P.; Poulev, A.; Logendra, S.; Yousef, G.G.; Raskin, I.; Lila, M.A. Hypoglycemic activity of a novel anthocyanin-rich formulation from lowbush blueberry, Vaccinium angustifolium Aiton. *Phytomedicine* **2009**, *16*, 406–415. [CrossRef]
Chen, K.; Wei, X.; Zhang, J.; Pariyani, R.; Jokioja, J.; Kortesniemi, M.; Linderborg, K.M.; Heinonen, J.; Sainio, T.; Zhang, Y.; et al. Effects of Anthocyanin Extracts from Bilberry (*Vaccinium myrtillus* L.) and Purple Potato (Solanum tuberosum L. var. 'Synkeä Sakari') on the Plasma Metabolomic Profile of Zucker Diabetic Fatty Rats. *J. Agric. Food Chem.* **2020**, *68*, 9436–9450. [CrossRef] [PubMed]
Eid, H.M.; Ouchfoun, M.; Brault, A.; Vallerand, D.; Musallam, L.; Arnason, J.T.; Haddad, P.S. Lingonberry (*Vaccinium vitis-idaea* L.) Exhibits Antidiabetic Activities in a Mouse Model of Diet-Induced Obesity. *Evid. Based Complement. Alternat. Med.* **2014**, *2014*, 645812. [CrossRef] [PubMed]
Al Hamimi, S.; Heyman-Lindén, L.; Plaza, M.; Turner, C.; Berger, K.; Spégel, P. Alterations in the plasma metabolite profile associated with improved hepatic function and glycemia in mice fed lingonberry supplemented high-fat diets. *Mol. Nutr. Food Res.* **2017**, *61*, 1600442. [CrossRef] [PubMed]
Kowalska, K.; Olejnik, A.; Zielińska-Wasielica, J.; Olkowicz, M. Inhibitory effects of lingonberry (*Vaccinium vitis-idaea* L.) fruit extract on obesity-induced inflammation in 3T3-L1 adipocytes and RAW 264.7 macrophages. *J. Funct. Foods* **2019**, *54*, 371–380. [CrossRef]
Suzuki, R.; Tanaka, M.; Takanashi, M.; Hussain, A.; Yuan, B.; Toyoda, H.; Kuroda, M. Anthocyanidins-enriched bilberry extracts inhibit 3T3-L1 adipocyte differentiation via the insulin pathway. *Nutr. Metab.* **2011**, *8*, 14. [CrossRef]
Kowalska, K.; Olejnik, A.; Szwajgier, D.; Olkowicz, M. Inhibitory activity of chokeberry, bilberry, raspberry and cranberry polyphenol-rich extract towards adipogenesis and oxidative stress in differentiated 3T3-L1 adipose cells. *PLoS ONE* **2017**, *12*, e0188583. [CrossRef]
Niesen, S.; Göttel, C.; Becker, H.; Bakuradze, T.; Winterhalter, P.; Richling, E. Fractionation of Extracts from Black Chokeberry, Cranberry, and Pomegranate to Identify Compounds That Influence Lipid Metabolism. *Foods* **2022**, *11*, 570. [CrossRef]
Pemmari, T.; Hämäläinen, M.; Ryyti, R.; Peltola, R.; Moilanen, E. Dried Bilberry (*Vaccinium myrtillus* L.) Alleviates the Inflammation and Adverse Metabolic Effects Caused by a High-Fat Diet in a Mouse Model of Obesity. *Int. J. Mol. Sci.* **2022**, *23*, 1021. [CrossRef]
Kowalska, K.; Dembczyński, R.; Gołąbek, A.; Olkowicz, M.; Olejnik, A. ROS Modulating Effects of Lingonberry (*Vaccinium vitis-idaea* L.) Polyphenols on Obese Adipocyte Hypertrophy and Vascular Endothelial Dysfunction. *Nutrients* **2021**, *13*, 885. [CrossRef]
Prior, R.L.; Wilkes, S.E.; Rogers, R.T.; Khanal, R.C.; Wu, X.; Howard, L.R. Purified blueberry anthocyanins and blueberry juice alter development of obesity in mice fed an obesogenic high-fat diet. *J. Agric. Food Chem.* **2010**, *58*, 3970–3976. [CrossRef] [PubMed]
Zhong, H.; Abdullah; Deng, L.; Zhao, M.; Tang, J.; Liu, T.; Zhang, H.; Feng, F. Probiotic-fermented blueberry juice prevents obesity and hyperglycemia in high fat diet-fed mice in association with modulating the gut microbiota. *Food Funct.* **2020**, *11*, 9192–9207. [CrossRef]
Vendrame, S.; Zhao, A.; Merrow, T.; Klimis-Zacas, D. The effects of wild blueberry consumption on plasma markers and gene expression related to glucose metabolism in the obese Zucker rat. *J. Med. Food* **2015**, *18*, 619–624. [CrossRef] [PubMed]
Khanal, R.C.; Rogers, T.J.; Wilkes, S.E.; Howard, L.R.; Prior, R.L. Effects of dietary consumption of cranberry powder on metabolic parameters in growing rats fed high fructose diets. *Food Funct.* **2010**, *1*, 116–123. [CrossRef] [PubMed]
Petersen, C.; Bharat, D.; Wankhade, U.D.; Kim, J.S.; Cutler, B.R.; Denetso, C.; Gholami, S.; Nelson, S.; Bigley, J.; Johnson, A.; et al. Dietary Blueberry Ameliorates Vascular Complications in Diabetic Mice Possibly through NOX4 and Modulates Composition and Functional Diversity of Gut Microbes. *Mol. Nutr. Food Res.* **2022**, *66*, e2100784. [CrossRef]
Nair, A.R.; Elks, C.M.; Vila, J.; Del Piero, F.; Paulsen, D.B.; Francis, J. A blueberry-enriched diet improves renal function and reduces oxidative stress in metabolic syndrome animals: Potential mechanism of TLR4-MAPK signaling pathway. *PLoS ONE* **2014**, *9*, e111976. [CrossRef]
Anhê, F.F.; Roy, D.; Pilon, G.; Dudonné, S.; Matamoros, S.; Varin, T.V.; Garofalo, C.; Moine, Q.; Desjardins, Y.; Levy, E.; et al. A polyphenol-rich cranberry extract protects from diet-induced obesity, insulin resistance and intestinal inflammation in association with increased Akkermansia spp. population in the gut microbiota of mice. *Gut* **2015**, *64*, 872–883. [CrossRef]
Medina-Larqué, A.S.; Rodríguez-Daza, M.C.; Roquim, M.; Dudonné, S.; Pilon, G.; Levy, É.; Marette, A.; Roy, D.; Jacques, H.; Desjardins, Y. Cranberry polyphenols and agave agavins impact gut immune response and microbiota composition while improving gut barrier function, inflammation, and glucose metabolism in mice fed an obesogenic diet. *Front. Immunol.* **2022**, *13*, 871080. [CrossRef]
Zhou, F.; Guo, J.; Han, X.; Gao, Y.; Chen, Q.; Huang, W.; Zhan, J.; Huang, D.; You, Y. Cranberry Polyphenolic Extract Exhibits an Antiobesity Effect on High-Fat Diet-Fed Mice through Increased Thermogenesis. *J. Nutr.* **2020**, *150*, 2131–2138. [CrossRef]
Singh, D.P.; Singh, S.; Bijalwan, V.; Kumar, V.; Khare, P.; Baboota, R.K.; Singh, P.; Boparai, R.K.; Singh, J.; Kondepudi, K.K.; et al. Co-supplementation of isomalto-oligosaccharides potentiates metabolic health benefits of polyphenol-rich cranberry extract in high fat diet-fed mice via enhanced gut butyrate production. *Eur. J. Nutr.* **2018**, *57*, 2897–2911. [CrossRef]

122. Mykkänen, O.T.; Huotari, A.; Herzig, K.H.; Dunlop, T.W.; Mykkänen, H.; Kirjavainen, P.V. Wild blueberries (*Vaccinium myrtil* alleviate inflammation and hypertension associated with developing obesity in mice fed with a high-fat diet. *PLoS ONE* **201** e114790. [CrossRef]
123. Feshani, A.M.; Kouhsari, S.M.; Mohammadi, S. Vaccinium arctostaphylos, a common herbal medicine in Iran: Molecular a biochemical study of its antidiabetic effects on alloxan-diabetic Wistar rats. *J. Ethnopharmacol.* **2011**, *133*, 67–74. [CrossRef]
124. Brader, L.; Overgaard, A.; Christensen, L.P.; Jeppesen, P.B.; Hermansen, K. Polyphenol-rich bilberry ameliorates total choleste and LDL-cholesterol when implemented in the diet of Zucker diabetic fatty rats. *Rev. Diabet. Stud.* **2013**, *10*, 270–282. [CrossI [PubMed]
125. Madduma Hewage, S.; Prashar, S.; Debnath, S.C.; Karmin, O.; Siow, Y.L. Inhibition of Inflammatory Cytokine Expression Preve High-Fat Diet-Induced Kidney Injury: Role of Lingonberry Supplementation. *Front. Med.* **2020**, *7*, 80. [CrossRef]
126. Ryyti, R.; Pemmari, A.; Peltola, R.; Hämäläinen, M.; Moilanen, E. Effects of Lingonberry (*Vaccinium vitis-idaea* L.) Supplementa on Hepatic Gene Expression in High-Fat Diet Fed Mice. *Nutrients* **2021**, *13*, 3693. [CrossRef]
127. Takikawa, M.; Inoue, S.; Horio, F.; Tsuda, T. Dietary anthocyanin-rich bilberry extract ameliorates hyperglycemia and insu sensitivity via activation of AMP-activated protein kinase in diabetic mice. *J. Nutr.* **2010**, *140*, 527–533. [CrossRef] [PubMed]
128. Lee, S.; Keirsey, K.I.; Kirkland, R.; Grunewald, Z.I.; Fischer, J.G.; de La Serre, C.B. Blueberry Supplementation Influences the (Microbiota, Inflammation, and Insulin Resistance in High-Fat-Diet-Fed Rats. *J. Nutr.* **2018**, *148*, 209–219. [CrossRef] [PubMe
129. Morissette, A.; Kropp, C.; Songpadith, J.P.; Junges Moreira, R.; Costa, J.; Mariné-Casadó, R.; Pilon, G.; Varin, T.V.; Dudor S.; Boutekrabt, L.; et al. Blueberry proanthocyanidins and anthocyanins improve metabolic health through a gut microbi dependent mechanism in diet-induced obese mice. *Am. J. Physiol. Endocrinol. Metab.* **2020**, *318*, E965–E980. [CrossR [PubMed]
130. Seymour, E.M.; Tanone, I.I.; Urcuyo-Llanes, D.E.; Lewis, S.K.; Kirakosyan, A.; Kondoleon, M.G.; Kaufman, P.B.; Bolling, Blueberry intake alters skeletal muscle and adipose tissue peroxisome proliferator-activated receptor activity and reduces ins resistance in obese rats. *J. Med. Food* **2011**, *14*, 1511–1518. [CrossRef]
131. Nunes, S.; Viana, S.D.; Preguiça, I.; Alves, A.; Fernandes, R.; Teodoro, J.S.; Matos, P.; Figueirinha, A.; Salgueiro, L.; André, A.; e Blueberry Counteracts Prediabetes in a Hypercaloric Diet-Induced Rat Model and Rescues Hepatic Mitochondrial Bioenerge *Nutrients* **2021**, *13*, 4192. [CrossRef]
132. Elks, C.M.; Terrebonne, J.D.; Ingram, D.K.; Stephens, J.M. Blueberries improve glucose tolerance without altering body comp tion in obese postmenopausal mice. *Obesity* **2015**, *23*, 573–580. [CrossRef] [PubMed]
133. Wu, T.; Gao, Y.; Guo, X.; Zhang, M.; Gong, L. Blackberry and Blueberry Anthocyanin Supplementation Counteract High- Diet-Induced Obesity by Alleviating Oxidative Stress and Inflammation and Accelerating Energy Expenditure. *Oxid. Med. C Longev.* **2018**, *2018*, 4051232. [CrossRef]
134. Ryyti, R.; Hämäläinen, M.; Peltola, R.; Moilanen, E. Beneficial effects of lingonberry (*Vaccinium vitis-idaea* L.) supplementation metabolic and inflammatory adverse effects induced by high-fat diet in a mouse model of obesity. *PLoS ONE* **2020**, *15*, e0232 [CrossRef]
135. Vuong, T.; Benhaddou-Andaloussi, A.; Brault, A.; Harbilas, D.; Martineau, L.C.; Vallerand, D.; Ramassamy, C.; Matar, C.; Hadc P.S. Antiobesity and antidiabetic effects of biotransformed blueberry juice in KKA(y) mice. *Int. J. Obes.* **2009**, *33*, 1166–1 [CrossRef] [PubMed]
136. Wu, T.; Jiang, Z.; Yin, J.; Long, H.; Zheng, X. Anti-obesity effects of artificial planting blueberry (*Vaccinium ashei*) anthocyani high-fat diet-treated mice. *Int. J. Food Sci. Nutr.* **2016**, *67*, 257–264. [CrossRef] [PubMed]
137. Wang, Y.; Zhang, D.; Liu, Y.; Wang, D.; Liu, J.; Ji, B. The protective effects of berry-derived anthocyanins against vis light-induced damage in human retinal pigment epithelial cells. *J. Sci. Food Agric.* **2015**, *95*, 936–944. [CrossRef]
138. Li, R.; Ye, Z.; Yang, W.; Xu, Y.J.; Tan, C.P.; Liu, Y. Blueberry Anthocyanins from Commercial Products: Structure Identification Potential for Diabetic Retinopathy Amelioration. *Molecules* **2022**, *27*, 7475. [CrossRef]
139. Wang, C.; Wang, K.; Li, P. Blueberry anthocyanins extract attenuated diabetic retinopathy by inhibiting endoplasmic reticul stress via the miR-182/OGG1 axis. *J. Pharmacol. Sci.* **2022**, *150*, 31–40. [CrossRef]
140. Song, Y.; Huang, L.; Yu, J. Effects of blueberry anthocyanins on retinal oxidative stress and inflammation in diabetes throu Nrf2/HO-1 signaling. *J. Neuroimmunol.* **2016**, *301*, 1–6. [CrossRef]
141. Huang, W.; Yan, Z.; Li, D.; Ma, Y.; Zhou, J.; Sui, Z. Antioxidant and Anti-Inflammatory Effects of Blueberry Anthocyanins High Glucose-Induced Human Retinal Capillary Endothelial Cells. *Oxid. Med. Cell. Longev.* **2018**, *2018*, 1862462. [CrossRef]
142. Vorob'eva, I.V.; Vorob'eva, I.V. Current data on the role of anthocyanosides and flavonoids in the treatment of eye diseases. *Ve Oftalmol.* **2015**, *131*, 104–110. [CrossRef]
143. Kim, J.; Kim, C.S.; Lee, Y.M.; Sohn, E.; Jo, K.; Kim, J.S. *Vaccinium myrtillus* extract prevents or delays the onset of diabetes—Indu blood-retinal barrier breakdown. *Int. J. Food Sci. Nutr.* **2015**, *66*, 236–242. [CrossRef]
144. Stevens, M.; Neal, C.R.; Craciun, E.C.; Dronca, M.; Harper, S.J.; Oltean, S. The natural drug DIAVIT is protective in a typ mouse model of diabetic nephropathy. *PLoS ONE* **2019**, *14*, e0212910. [CrossRef] [PubMed]
145. Di Cerbo, A.; Iannitti, T.; Guidetti, G.; Centenaro, S.; Canello, S.; Cocco, R. A nutraceutical diet based on Lespedeza s Vaccinium macrocarpon and Taraxacum officinale improves spontaneous feline chronic kidney disease. *Physiol. Rep.* **201** e13737. [CrossRef] [PubMed]

Li, Y.X.; Lu, Y.P.; Tang, D.; Hu, B.; Zhang, Z.Y.; Wu, H.W.; Fan, L.J.; Cai, K.W.; Tang, C.; Zhang, Y.Q.; et al. Anthocyanin improves kidney function in diabetic kidney disease by regulating amino acid metabolism. *J. Transl. Med.* **2022**, *20*, 510. [CrossRef] [PubMed]
Qin, Y.; Zhai, Q.; Li, Y.; Cao, M.; Xu, Y.; Zhao, K.; Wang, T. Cyanidin-3-O-glucoside ameliorates diabetic nephropathy through regulation of glutathione pool. *Biomed. Pharmacother.* **2018**, *103*, 1223–1230. [CrossRef]
Du, C.; Shi, Y.; Ren, Y.; Wu, H.; Yao, F.; Wei, J.; Wu, M.; Hou, Y.; Duan, H. Anthocyanins inhibit high-glucose-induced cholesterol accumulation and inflammation by activating LXRα pathway in HK-2 cells. *Drug. Des. Devel Ther.* **2015**, *9*, 5099–5113. [CrossRef] [PubMed]
Vargas, F.; Romecín, P.; García-Guillén, A.I.; Wangesteen, R.; Vargas-Tendero, P.; Paredes, M.D.; Atucha, N.M.; García-Estañ, J. Flavonoids in Kidney Health and Disease. *Front. Physiol.* **2018**, *9*, 394. [CrossRef]
Basu, A.; Feng, D.; Planinic, P.; Ebersole, J.L.; Lyons, T.J.; Alexander, J.M. Dietary Blueberry and Soluble Fiber Supplementation Reduces Risk of Gestational Diabetes in Women with Obesity in a Randomized Controlled Trial. *J. Nutr.* **2021**, *151*, 1128–1138. [CrossRef]
Stote, K.; Corkum, A.; Sweeney, M.; Shakerley, N.; Kean, T.; Gottschall-Pass, K. Postprandial Effects of Blueberry (Vaccinium angustifolium) Consumption on Glucose Metabolism, Gastrointestinal Hormone Response, and Perceived Appetite in Healthy Adults: A Randomized, Placebo-Controlled Crossover Trial. *Nutrients* **2019**, *11*, 202. [CrossRef] [PubMed]
Stote, K.S.; Wilson, M.M.; Hallenbeck, D.; Thomas, K.; Rourke, J.M.; Sweeney, M.I.; Gottschall-Pass, K.T.; Gosmanov, A.R. Effect of Blueberry Consumption on Cardiometabolic Health Parameters in Men with Type 2 Diabetes: An 8-Week, Double-Blind, Randomized, Placebo-Controlled Trial. *Curr. Dev. Nutr.* **2020**, *4*, nzaa030. [CrossRef] [PubMed]
Mirfeizi, M.; Mehdizadeh Tourzani, Z.; Mirfeizi, S.Z.; Asghari Jafarabadi, M.; Rezvani, H.R.; Afzali, M. Controlling type 2 diabetes mellitus with herbal medicines: A triple-blind randomized clinical trial of efficacy and safety. *J. Diabetes* **2016**, *8*, 647–656. [CrossRef] [PubMed]
de Mello, V.D.; Lankinen, M.A.; Lindström, J.; Puupponen-Pimiä, R.; Laaksonen, D.E.; Pihlajamäki, J.; Lehtonen, M.; Uusitupa, M.; Tuomilehto, J.; Kolehmainen, M.; et al. Fasting serum hippuric acid is elevated after bilberry (*Vaccinium myrtillus*) consumption and associates with improvement of fasting glucose levels and insulin secretion in persons at high risk of developing type 2 diabetes. *Mol. Nutr. Food Res.* **2017**, *61*, 1700019. [CrossRef]
Hoggard, N.; Cruickshank, M.; Moar, K.M.; Bestwick, C.; Holst, J.J.; Russell, W.; Horgan, G. A single supplement of a standardised bilberry (*Vaccinium myrtillus* L.) extract (36% wet weight anthocyanins) modifies glycaemic response in individuals with type 2 diabetes controlled by diet and lifestyle. *J. Nutr. Sci.* **2013**, *2*, e22. [CrossRef] [PubMed]
Schell, J.; Betts, N.M.; Foster, M.; Scofield, R.H.; Basu, A. Cranberries improve postprandial glucose excursions in type 2 diabetes. *Food Funct.* **2017**, *8*, 3083–3090. [CrossRef]
Novotny, J.A.; Baer, D.J.; Khoo, C.; Gebauer, S.K.; Charron, C.S. Cranberry juice consumption lowers markers of cardiometabolic risk, including blood pressure and circulating C-reactive protein, triglyceride, and glucose concentrations in adults. *J. Nutr.* **2015**, *145*, 1185–1193. [CrossRef]
Wilson, T.; Meyers, S.L.; Singh, A.P.; Limburg, P.J.; Vorsa, N. Favorable glycemic response of type 2 diabetics to low-calorie cranberry juice. *J. Food Sci.* **2008**, *73*, H241–H245. [CrossRef]
Shidfar, F.; Heydari, I.; Hajimiresmaiel, S.J.; Hosseini, S.; Shidfar, S.; Amiri, F. The effects of cranberry juice on serum glucose, apoB, apoA-I, Lp(a), and Paraoxonase-1 activity in type 2 diabetic male patients. *J. Res. Med. Sci.* **2012**, *17*, 355–360.
Kianbakht, S.; Abasi, B.; Dabaghian, F.H. Anti-hyperglycemic effect of Vaccinium arctostaphylos in type 2 diabetic patients: A randomized controlled trial. *Forsch. Komplementmed* **2013**, *20*, 17–22. [CrossRef]
Yang, L.; Ling, W.; Yang, Y.; Chen, Y.; Tian, Z.; Du, Z.; Chen, J.; Xie, Y.; Liu, Z.; Yang, L. Role of Purified Anthocyanins in Improving Cardiometabolic Risk Factors in Chinese Men and Women with Prediabetes or Early Untreated Diabetes-A Randomized Controlled Trial. *Nutrients* **2017**, *9*, 1104. [CrossRef]
Li, D.; Zhang, Y.; Liu, Y.; Sun, R.; Xia, M. Purified anthocyanin supplementation reduces dyslipidemia, enhances antioxidant capacity, and prevents insulin resistance in diabetic patients. *J. Nutr.* **2015**, *145*, 742–748. [CrossRef] [PubMed]
de Mello, V.D.; Schwab, U.; Kolehmainen, M.; Koenig, W.; Siloaho, M.; Poutanen, K.; Mykkänen, H.; Uusitupa, M. A diet high in fatty fish, bilberries and wholegrain products improves markers of endothelial function and inflammation in individuals with impaired glucose metabolism in a randomised controlled trial: The Sysdimet study. *Diabetologia* **2011**, *54*, 2755–2767. [CrossRef]
Qin, Y.; Xia, M.; Ma, J.; Hao, Y.; Liu, J.; Mou, H.; Cao, L.; Ling, W. Anthocyanin supplementation improves serum LDL- and HDL-cholesterol concentrations associated with the inhibition of cholesteryl ester transfer protein in dyslipidemic subjects. *Am. J. Clin. Nutr.* **2009**, *90*, 485–492. [CrossRef] [PubMed]
Zhu, Y.; Ling, W.; Guo, H.; Song, F.; Ye, Q.; Zou, T.; Li, D.; Zhang, Y.; Li, G.; Xiao, Y.; et al. Anti-inflammatory effect of purified dietary anthocyanin in adults with hypercholesterolemia: A randomized controlled trial. *Nutr. Metab. Cardiovasc. Dis.* **2013**, *23*, 843–849. [CrossRef]
Kolehmainen, M.; Mykkänen, O.; Kirjavainen, P.V.; Leppänen, T.; Moilanen, E.; Adriaens, M.; Laaksonen, D.E.; Hallikainen, M.; Puupponen-Pimiä, R.; Pulkkinen, L.; et al. Bilberries reduce low-grade inflammation in individuals with features of metabolic syndrome. *Mol. Nutr. Food Res.* **2012**, *56*, 1501–1510. [CrossRef]
Scharrer, A.; Ober, M. Anthocyanosides in the treatment of retinopathies (author's transl). *Klin. Monbl Augenheilkd.* **1981**, *178*, 386–389. [CrossRef] [PubMed]

168. Perossini, M.; Guidi, G.; Chiellini, S.; Siravo, D. Diabetic and hypertensive retinopathy therapy with *Vaccinium myrti* anthocyanosides (Tegens). A double blind placebo-controlled clinical trial. *Ann. Ottalmol. Clin. Ocul.* **1987**, *113*, 1173.
169. Kim, E.S.; Yu, S.-Y.; Kwon, S.J.; Kwon, O.W.; Kim, S.-Y.; Kim, T.W.; Ahn, J.K.; Oum, B.S.; Lew, Y.J.; Lee, J.E.; et al. Clin Evaluation of Patients with Nonproliferative Diabetic Retinopathy Following Medication of Anthocyanoside: Multicenter St *J. Korean Ophthalmol. Soc.* **2008**, *49*, 1629–1633. [CrossRef]
170. Parravano, M.; Tedeschi, M.; Manca, D.; Costanzo, E.; Di Renzo, A.; Giorno, P.; Barbano, L.; Ziccardi, L.; Varano, M.; Paris Effects of Macuprev(®) Supplementation in Age-Related Macular Degeneration: A Double-Blind Randomized Morpho-Functio Study Along 6 Months of Follow-Up. *Adv. Ther.* **2019**, *36*, 2493–2505. [CrossRef]
171. Savickiene, N.; Dagilyte, A.; Lukosius, A.; Zitkevicius, V. Importance of biologically active components and plants in prevention of complications of diabetes mellitus. *Medicina* **2002**, *38*, 970–975. [PubMed]
172. Sasso, F.C.; Pafundi, P.C.; Gelso, A.; Bono, V.; Costagliola, C.; Marfella, R.; Sardu, C.; Rinaldi, L.; Galiero, R.; Acierno, C.; e High HDL cholesterol: A risk factor for diabetic retinopathy? Findings from NO BLIND study. *Diabetes Res. Clin. Pract.* **2019**, 236–244. [CrossRef] [PubMed]
173. Carrero, J.J.; González-Ortiz, A.; Avesani, C.M.; Bakker, S.J.L.; Bellizzi, V.; Chauveau, P.; Clase, C.M.; Cupisti, A.; Espinosa-Cue A.; Molina, P.; et al. Plant-based diets to manage the risks and complications of chronic kidney disease. *Nat. Rev. Nephrol.* **2** *16*, 525–542. [CrossRef]
174. Kelly, J.T.; Palmer, S.C.; Wai, S.N.; Ruospo, M.; Carrero, J.J.; Campbell, K.L.; Strippoli, G.F. Healthy Dietary Patterns and R of Mortality and ESRD in CKD: A Meta-Analysis of Cohort Studies. *Clin. J. Am. Soc. Nephrol.* **2017**, *12*, 272–279. [CrossR [PubMed]
175. Adair, K.E.; Bowden, R.G. Ameliorating Chronic Kidney Disease Using a Whole Food Plant-Based Diet. *Nutrients* **2020**, *12*, 1 [CrossRef] [PubMed]
176. Joshi, S.; McMacken, M.; Kalantar-Zadeh, K. Plant-Based Diets for Kidney Disease: A Guide for Clinicians. *Am. J. Kidney* **2021**, *77*, 287–296. [CrossRef]
177. Kim, H.; Caulfield, L.E.; Garcia-Larsen, V.; Steffen, L.M.; Grams, M.E.; Coresh, J.; Rebholz, C.M. Plant-Based Diets and Incid CKD and Kidney Function. *Clin. J. Am. Soc. Nephrol.* **2019**, *14*, 682–691. [CrossRef]
178. Curtis, P.J.; van der Velpen, V.; Berends, L.; Jennings, A.; Feelisch, M.; Umpleby, A.M.; Evans, M.; Fernandez, B.O.; Meiss, M Minnion, M.; et al. Blueberries improve biomarkers of cardiometabolic function in participants with metabolic syndrome-res from a 6-month, double-blind, randomized controlled trial. *Am. J. Clin. Nutr.* **2019**, *109*, 1535–1545. [CrossRef]
179. Nair, A.R.; Mariappan, N.; Stull, A.J.; Francis, J. Blueberry supplementation attenuates oxidative stress within monocytes a modulates immune cell levels in adults with metabolic syndrome: A randomized, double-blind, placebo-controlled trial. *F Funct.* **2017**, *8*, 4118–4128. [CrossRef]
180. Lee, I.T.; Chan, Y.C.; Lin, C.W.; Lee, W.J.; Sheu, W.H. Effect of cranberry extracts on lipid profiles in subjects with Type 2 diabe *Diabet. Med.* **2008**, *25*, 1473–1477. [CrossRef]
181. Hsia, D.S.; Zhang, D.J.; Beyl, R.S.; Greenway, F.L.; Khoo, C. Effect of daily consumption of cranberry beverage on ins sensitivity and modification of cardiovascular risk factors in adults with obesity: A pilot, randomised, placebo-controlled st *Br. J. Nutr.* **2020**, *124*, 577–585. [CrossRef] [PubMed]
182. Lehtonen, H.M.; Suomela, J.P.; Tahvonen, R.; Yang, B.; Venojärvi, M.; Viikari, J.; Kallio, H. Different berries and berry fracti have various but slightly positive effects on the associated variables of metabolic diseases on overweight and obese women. *J. Clin. Nutr.* **2011**, *65*, 394–401. [CrossRef] [PubMed]
183. Rosales, T.K.O.; Hassimotto, N.M.A.; Lajolo, F.M.; Fabi, J.P. Nanotechnology as a Tool to Mitigate the Effects of Intesti Microbiota on Metabolization of Anthocyanins. *Antioxidants* **2022**, *11*, 506. [CrossRef] [PubMed]
184. Tena, N.; Martín, J.; Asuero, A.G. State of the Art of Anthocyanins: Antioxidant Activity, Sources, Bioavailability, and Therape Effect in Human Health. *Antioxidants* **2020**, *9*, 451. [CrossRef]

nutrients MDPI

cle

·nergistic Effect of a Flavonoid-Rich Cocoa–Carob Blend and etformin in Preserving Pancreatic Beta Cells in Zucker iabetic Fatty Rats

la Gallardo-Villanueva [1,2], Tamara Fernández-Marcelo [1,2], Laura Villamayor [2,3], Angela M. Valverde [2,3], ia Ramos [2,4], Elisa Fernández-Millán [1,2,*] and María Angeles Martín [2,4,*]

1 Departamento de Bioquímica y Biología Molecular, Facultad de Farmacia, Universidad Complutense de Madrid, 28040 Madrid, Spain; paulaisg@ucm.es (P.G.-V.); tafernan@ucm.es (T.F.-M.)
2 CIBER de Diabetes y Enfermedades Metabólicas Asociadas (CIBERDEM), Instituto de Salud Carlos III (ISCIII), 28029 Madrid, Spain; lvillamayor@iib.uam.es (L.V.); avalverde@iib.uam.es (A.M.V.); s.ramos@ictan.csic.es (S.R.)
3 Instituto de Investigaciones Biomedicas Sols-Morreale (IIB-CSIC), 28029 Madrid, Spain
4 Instituto de Ciencia y Tecnología de Alimentos y Nutrición (ICTAN-CSIC), 28040 Madrid, Spain
* Correspondence: elfernan@ucm.es (E.F.-M.); amartina@ictan.csic.es (M.A.M.)

Abstract: The loss of functional beta-cell mass in diabetes is directly linked to the development of diabetic complications. Although dietary flavonoids have demonstrated antidiabetic properties, their potential effects on pancreatic beta-cell preservation and their synergistic benefits with antidiabetic drugs remain underexplored. We have developed a potential functional food enriched in flavonoids by combining cocoa powder and carob flour (CCB), which has shown antidiabetic effects. Here, we investigated the ability of the CCB, alone or in combination with metformin, to preserve pancreatic beta cells in an established diabetic context and their potential synergistic effect. Zucker diabetic fatty rats (ZDF) were fed a CCB-rich diet or a control diet, with or without metformin, for 12 weeks. Markers of pancreatic oxidative stress and inflammation, as well as relative beta-cell mass and beta-cell apoptosis, were analyzed. Results demonstrated that CCB feeding counteracted pancreatic oxidative stress by enhancing the antioxidant defense and reducing reactive oxygen species. Moreover, the CCB suppressed islet inflammation by preventing macrophage infiltration into islets and overproduction of pro-inflammatory cytokines, along with the inactivation of nuclear factor kappa B (NFκB). As a result, the CCB supplementation prevented beta-cell apoptosis and the loss of beta cells in ZDF diabetic animals. The observed additive effect when combining the CCB with metformin underscores its potential as an adjuvant therapy to delay the progression of type 2 diabetes.

Keywords: type 2 diabetes; pancreatic beta cells; macrophages; oxidative stress; inflammation; flavonoids

on: Gallardo-Villanueva, P.; ndez-Marcelo, T.; Villamayor, lverde, A.M.; Ramos, S.; ndez-Millán, E.; Martín, M.A. gistic Effect of a Flavonoid-Rich –Carob Blend and Metformin serving Pancreatic Beta Cells in r Diabetic Fatty Rats. *Nutrients* *16*, 273. https://doi.org/ 0/nu16020273

emic Editors: Maria Luz ndez and Andrea Fabbri

ved: 24 November 2023
ed: 13 January 2024
ted: 15 January 2024
hed: 17 January 2024

1. Introduction

Type 2 diabetes (T2D) is a chronic metabolic disorder characterized by dysregulation of glucose homeostasis leading to hyperglycemia. This condition is associated with a gradual decline in the ability to produce and release insulin, frequently on the background of insulin resistance [1]. In this context, pancreatic beta cells, which are responsible for insulin secretion, play a central role in maintaining glucose homeostasis [2] but their progressive deterioration and loss, exacerbated by prolonged hyperglycemia, contribute significantly to the development of diabetes and its associated complications [3]. Therefore, the preservation of beta-cell function and mass is fundamental for controlling T2D, restoring proper glucose control, and improving metabolic regulation.

Oxidative stress induced by prolonged hyperglycemia is one of the factors contributing to the destruction of beta cells in diabetes [4]. Moreover, the intrinsically low expression of antioxidant enzymes in pancreatic islets exacerbates the impact of oxidative stress, resulting in consequences for beta cells [5,6]. Similarly, pancreatic inflammation, frequently observ in individuals with obesity and/or T2D [7,8], also contributes to the progressive loss of be cell function. The increase in macrophage accumulation in the islets during T2D provi signals that primarily drive beta-cell hyperplasia, but also increase the production inflammatory cytokines, further accelerating beta-cell damage [9]. However, there is limi evidence supporting the beneficial effects of existing antidiabetic therapies in augmenti intracellular antioxidant defenses or reducing inflammation to ameliorate beta-cell dama in T2D. In the context of T2D therapeutics, metformin, the primary glucose-lowering ag has been suggested to be effective in maintaining beta-cell function. Its potential effic seems to be more pronounced when used alongside other pharmacological therapies [10, Interestingly, an increasing body of literature suggests that combining metformin w natural bioactive compounds may provide enhanced protection against diabetes-rela complications [12,13]. Nevertheless, the specific impact of metformin in combinati with dietary natural bioactive molecules on the preservation of beta-cell mass rema poorly studied.

Polyphenols are food-derived compounds that have attracted significant attention recent years due to their demonstrated antioxidant, anti-inflammatory, and antidiabe properties [14,15]. Cocoa is recognized as one of the most plentiful sources of diet polyphenols, mainly flavonoids, known for their beneficial effects against T2D and related complications [16]. In particular, cocoa flavonoids have been demonstrated protect the integrity of pancreatic beta cells against oxidative damage both in vitro [17] a in vivo in young prediabetic rats [18]. Likewise, carob, a Mediterranean legume wit high flavonoid content, has also demonstrated positive effects in the context of T2D [Accordingly, we have developed a potential functional food, rich in high-molecular-wei polyphenols, combining cocoa powder and carob flour [20]. Our recent findings ha demonstrated the effectiveness of this cocoa–carob blend (CCB) in combating T2D a shown its cardioprotective effects [13]. These results were obtained using Zucker diabe fatty (ZDF) rats, a well-recognized preclinical model of T2D. Within the frame of t previous study, herein, we have investigated whether the antidiabetic properties of CCB extend its beneficial impact on pancreatic beta cells in an established diabetic mil Furthermore, we explored the potential additive effect of this dietary supplement wh administered in combination with metformin.

2. Materials and Methods

2.1. Cocoa–Carob Blend Diet

The cocoa–carob blend (CCB) consisted of a mixture of cocoa powder generou provided by Idilia (Idilia S.L., Barcelona, Spain) and carob flour (sourced from Casa R Granel Selecto S.L., Madrid, Spain) in a 60:40 ratio. This product is notable for its hi content of flavonoid-type polyphenols (16.7 g/100 g). More detailed information regard the composition of the CCB is provided elsewhere [20]. The CCB-enriched diet (10%) w prepared by incorporating 100 g/kg of the CCB into the standard diet (AIN-93G diet), a both the control and CCB diets were adjusted to be isoenergetic.

2.2. Animals and Experimental Design

This study is an extension of a prior investigation aimed at exploring the cardiop tective effects of CCB supplementation in Zucker diabetic fatty (ZDF) rats, a preclini model of T2D. The metabolic characteristics and the pathophysiology of the ZDF rats very similar to those of humans with T2D, which makes this model highly suitable for evaluation of the consequences of a CCB-rich diet on pancreatic beta cells in a diab context. We have chosen to use this animal model because it is not possible to carry out s experiments in humans with T2D. Detailed descriptions of the animals and experimen designs have been previously documented [13]. Briefly, male Zucker diabetic fatty (Z rats and their corresponding lean Zucker (ZL) controls were used. Animals were purcha from Charles River Laboratories (L'arbresle, France) at 11 weeks of age and were acclima under standard controlled conditions for a week. The ZDF rats were randomly assorted in the following groups of rats: ZDF Group: Zucker diabetic fatty rats that received the standard diet; ZDF (M) Group: diabetic rats that received the standard diet and metformin (300 mg/kg/day); ZDF (CB) Group: diabetic rats that received a 10% CCB supplementation in their diet; ZDF (CB + M) Group: diabetic rats that received both a 10% CCB supplementation in their diet and metformin (300 mg/kg/day). The ZL non-diabetic group received the standard diet (ZL). Animals had ad libitum access to water and food throughout the 12-week study period. At 24 weeks of age, the animals were fasted overnight and then euthanized. Pancreases were removed and stored under different conditions depending on the analyses to be performed. A portion of the pancreas was fixed overnight in 4% p-formaldehyde (PFA) in 0.1 M phosphate buffer (pH 7.4) and embedded in paraffin for histological and immunohistochemical analyses. Another part of the pancreas was frozen at −80 °C for subsequent analyses.

All experiments were conducted in accordance with European standards and community legislation [21] and Spanish regulations [22] and with the approval of the Committee on Animal Care and Use for Experimental Purposes of the Community of Madrid (PROEX 079/19).

2.3. Biochemical Determinations

Blood glucose levels were measured using an Accounted Glucose Analyzer (LifeScan España, Madrid, Spain). To analyze serum insulin and HbA1c levels, commercial kits were employed (Rat Insulin, Mercodia, Uppsala, Sweden; HbA1c Kit Spinreact, BioAnalitica, Madrid, Spain). Fasting glycemia and insulinemia were employed to calculate the homeostatic model assessment indices for insulin resistance (HOMA-IR) and secretion (HOMA-B).

2.4. Pancreas Homogenates

Frozen pancreas samples were homogenized at a 1:10 (*w*/*v*) ratio in extraction buffer [18]. Subsequently, they were centrifuged at 14,000× *g* for 60 min, and the supernatants were collected. The protein content in the homogenates was determined using the Bio-Rad protein assay (Bio-Rad, Madrid, Spain), following the manufacturer's instructions.

2.5. Determination of Pancreatic Insulin Content

The total insulin content of the pancreas was determined as previously described [18]. Glands were minced and disrupted ultrasonically in acid ethanol (1.5 mL of 12 M HCl/100 mL ethanol) in a 10 mL/g pancreas ratio, extracted overnight at 4 °C and centrifuged. After that, insulin was determined in the supernatant using an ELISA kit (Rat Insulin, Mercodia, Uppsala, Sweden).

2.6. Reactive Oxygen Species (ROS) Determination

ROS were quantified using a fluorometric assay based on the oxidation of reduced dichlorofluorescein (DCFH) to oxidized dichlorofluorescein (DCF) (Sigma-Aldrich) by cellular oxidants, resulting in fluorescence emission [23]. Pancreatic homogenates (20 µg of protein) were diluted with Locke's buffer and incubated with 5 mM DCFH for 30 min in the dark. Fluorescence was measured at an excitation wavelength of 485 nm and an emission wavelength of 530 nm.

2.7. Determination of Carbonyl Group Content

Protein oxidation in pancreatic homogenates was measured as the content of carbonyl groups present in the samples [23]. Each pancreas sample (500 µg of protein) was derivatized with 0.2% 2,4-dinitrophenylhydrazine (DNPH) in 2 M HCl, with a parallel blank treated only with 2 M HCl. Proteins were precipitated using 20% trichloroacetic acid and centrifuged for 5 min at 10,500 rpm. The precipitates were washed twice with ethyl acetate:ethanol (1:1, *v*/*v*) and resuspended in 6 M guanidine. Absorbance was measured at a wavelength of 360 nm, with each sample corrected using its respective blank. Results expressed as nmol/mg of protein, using a coefficient of extinction of 22,000 nmol L^{-1} cm

2.8. Determination of Glutathione (GSH) Levels

The concentration of GSH was assessed using a fluorometric assay based on reaction of GSH with o-phthalaldehyde (Sigma-Aldrich) at pH 8.0 [23]. Pancreas samp (300 μg of protein) were mixed in 50 mM phosphate buffer at pH 7.0 and precipitated w trichloroacetic acid. They were then centrifuged at 1400 rpm for 30 min. Fluoresce was measured at 460 nm emission and 340 nm excitation in a microplate reader (Bio-T Winooski, VT, USA). The results were interpolated on a GSH standard curve (5–1000 n

2.9. Determination of Glutathione Peroxidase (GPx) and Glutathione Reductase (GR) Activity

To determine the activity of the antioxidant enzymes GPx and GR, pancreas samp (400 μg of protein) were mixed in Tris buffer 0.25 M, sucrose 0.2 M, and DTT buf 5 mM at pH 7.4, and then centrifuged at 3000× g for 15 min. GPx activity is based the oxidation of GSH by GPx (Sigma-Aldrich) using tert-butylhydroperoxide (t-BOC Sigma-Aldrich, Madrid, Spain) as a substrate, coupled to the disappearance of nicot adenine dinucleotide phosphate reduced salt (NADPH, Sigma Aldrich, Madrid, Spain) GR. GR activity was analyzed by monitoring the oxidation of NADPH, which is used the reduction of GSSG [23].

2.10. Western Blot Analysis

Pancreas samples were lysed at 4 °C in a lysis buffer. Afterward, supernatants w collected, and protein concentration was determined. These protein samples were th aliquoted and stored at −80 °C until they were utilized for Western blot analyses. S polyacrylamide gel electrophoresis was employed to separate equal amounts of prote which were subsequently transferred onto polyvinylidene difluoride filters (Bio-Ra Next, the membranes were probed with anti-TNF-α (sc-52746, Santa Cruz Biotechn ogy), anti-IL-6 (sc-57315, Santa Cruz Biotechnology, Quimigen, Madrid, Spain), anti- nuclear factor kappa B (NFκB) (8242S, Cell Signaling, Madrid, Spain) and anti-phosp (Ser 536)-p65 NFκB (3033, Cell Signaling, Madrid, Spain), followed by incubation w peroxide-conjugated anti-rabbit (A6154, Sigma-Aldrich, Madrid, Spain) or anti-mo (A4416, Sigma-Aldrich, Madrid, Spain) immunoglobulin. Protein bands were visualized ing the SuperSignal™ West Pico PLUS Chemiluminescent Substrate (34580, ThermoFis Scientific, Madrid, Spain) in an Imager2Imager CHEMI Premium (VWR). Band densito try was quantified using ImageJ Software (v1.52a, National Institute of Health, Bethes MD, USA). To ensure the Western blot's normalization, β-actin was employed.

2.11. Activity of Caspase-3

The activation of caspase-3 was assessed following a previously established pro col [18]. In brief, pancreatic tissues were lysed and 50 μg of protein for each condition w mixed with 20 mM HEPES (pH 7), 10% glycerol, 2 mM DTT, and 20 μM Ac-DEV-DAMC the substrate. Enzymatic activity was determined by measuring fluorescence (excitat wavelength 380 nm and emission wavelength 440 nm).

2.12. Histological and Immunohistochemical Analysis

Pancreas samples embedded in paraffin (Panreac, Madrid, Spain) were cut into se sections with a thickness of 5 μm using a microtome (Leica RM2125RT, Leica Biosyst Madrid, Spain) and mounted on glass slides. For immunohistochemical analysis, se sections of the samples were blocked with goat serum (S-1000, Vector Laboratories, Mad Spain) and incubated with a primary mouse antibody against insulin (I2018, Sigma-Aldri overnight at 4 °C. Subsequently, a secondary goat anti-mouse antibody conjugated w peroxidase (A4416, Sigma-Aldrich, Madrid, Spain) was added, and the sections w finally developed with DAB substrate (SK-4100, Palex Medical, Madrid, Spain) and co terstained with Harris hematoxylin (HHS32-1L). For fibrosis quantification, serial sections from samples of each condition were subjected to Masson's trichrome staining (HT15-1KT, Sigma-Aldrich, Madrid, Spain), which stains fibrotic areas in blue. Images of the sections were acquired using a digital camera connected to a microscope (NIKON Eclipse 80i, Izasa, Barcelona, Spain). The percentage of beta-cell fractional area (brown) was analyzed using ImageJ v1.8 software and is expressed relative to the total pancreatic area measured from the sections of each condition. At least 30 sections per condition were analyzed. The percentage of fibrotic area was analyzed using the same software and is expressed as the percentage of collagen area (blue) relative to the total pancreatic area measured from the sections of each condition. At least 4 sections per animal and per group were analyzed.

Beta-cell apoptosis was estimated using the TUNEL method (ApopTag Peroxidase In Situ Apoptosis Detection Kit, Millipore, Madrid, Spain) coupled to insulin immunostaining, which was developed after incubation with the alkaline phosphatase secondary antibody with an alkaline substrate kit (SK-5100, Palex Medical, Madrid, Spain). The tissue was then counterstained with Harris hematoxylin. The beta-cell apoptosis rate is expressed as the percentage of apoptotic beta cells. At least 1500 beta cells were counted per pancreas. For cluster of differentiation 68-positive (CD68+) cell staining, rabbit anti-rat CD68 (ab125212, Abcam, Madrid, Spain) antibody was used coupled to insulin immunostaining as described for the TUNEL assay. The number of CD68+ cells was counted and is expressed per mm^2 of total pancreatic surface area or as the number of CD68+ cells per islet distinguishing between islet peripheric localization, defined as CD68+ cells around the periphery of the islet, and intra-islet distribution, when CD68+ cells were located within the islet among the endocrine cells. Clusters of 6 or more beta cells were considered islets. A minimum of 60 islets were analyzed per animal.

2.13. Statistical Analysis

Data underwent statistical analysis using GraphPad Prism v8.2.1 software (GraphPad Software, Boston, MA, USA). The normality of the distribution was assessed through the Shapiro–Wilk test, and data were analyzed using one-way analysis of variance (ANOVA). Significant differences among the means of each condition were identified using the Tukey post-hoc test, with a significance level of 95% ($p < 0.05$). Results are expressed as mean ± standard error of the mean (SEM).

3. Results

3.1. Biochemical Characteristics of Diabetic Animals

At the study's outset, ZDF animals exhibited a significantly higher body weight compared to ZL animals (340.3 ± 12.1 vs. 268.4 ± 9.8 g, respectively; $p < 0.05$), confirming their obese state. Additionally, at this initial time point, ZDF animals were diabetic, as evidenced by significantly elevated fasting glycemia when compared to the ZL group (182.2 ± 12.1 vs. 94.3 ± 8.1 mg/dL, respectively; $p < 0.05$).

Administration of the CCB, metformin, or both together over 12 weeks significantly reduced glucose, insulin, and glycosylated hemoglobin (HbA1c) levels, as well as decreased insulin resistance (HOMA-IR) and increased pancreatic function (HOMA-B) in ZDF animals (Supplementary Table S1). Notably, the combination of the CCB with metformin achieved values for both fasting glucose and HbA1c similar to those observed in the non-diabetic ZL group [13]. Altogether, these results indicated that the CCB diet effectively improved glucose homeostasis in ZDF rats, with the combined treatment of the CCB and metformin exhibiting even more robust protective effects than the other treatments.

3.2. Effect of the CCB and Metformin on Oxidative Stress Markers in the Pancreas of Diabetic Rats

In the context of T2D, oxidative stress induced by hyperglycemia plays a pivotal role in beta-cell dysfunction and apoptosis. Therefore, our initial focus was on investigating the oxidative state of the pancreas in diabetic animals. As illustrated in Figure 1a, a significant increase in ROS generation was observed in the pancreas of untreated diabetic rats (ZDF).

However, treatment with metformin and the CCB diet led to a substantial reduction ROS levels, effectively restoring them to levels comparable to those in non-diabetic ZL r Furthermore, we observed evidence of oxidative damage, as indicated by elevated prot carbonyl levels, in the pancreas of untreated diabetic rats (ZDF) but not in rats treated w metformin, the CCB, or the combination of both (Figure 1b). Likewise, the levels of antioxidant molecule GSH were significantly reduced in ZDF rats compared to non-diab ZL rats, but treatment with metformin and the CCB significantly increased GSH lev with a further improvement seen with the combination of both (Figure 1c). Regardi antioxidant enzymes, a notable increase in the activity of GPx was observed in all trea diabetic groups, with the most significant effect in ZDF rats fed the CCB-enriched d (Figure 1d). The activity of the GR enzyme was significantly lower in ZDF rats, but not other treated conditions, with similar levels to those found in the non-diabetic ZL gro (Figure 1e). Collectively, these findings provide strong evidence that the CCB-enriched d has the potential to enhance the activity of antioxidant defenses in the pancreas, particula GPx, and reduce the oxidative stress induced by the diabetic condition.

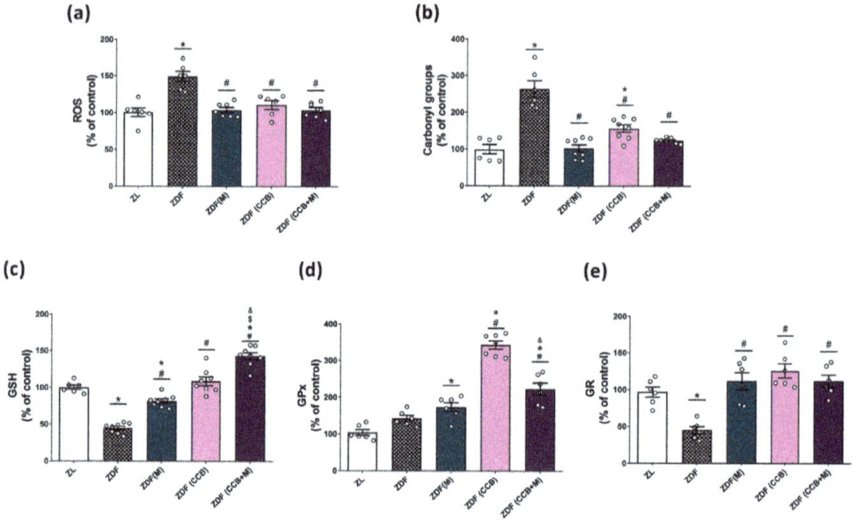

Figure 1. Effect of the CCB, metformin, and their combination on pancreatic oxidative stress. centage levels of ROS (**a**), carbonyl groups (**b**), GSH (**c**), GPx activity (**d**), and GR activity (**e**) rela to the control condition. Values are expressed as mean ± SEM of $n = 6$–8 animals. * $p < 0.05$ vs. # $p < 0.05$ vs. ZDF; $ $p < 0.05$ ZDF (CCB + M) vs. ZDF (M); & $p < 0.05$ ZDF (CCB + M) vs. ZDF (C ZL: Zucker lean; ZDF: Zucker diabetic rats; ZDF (M): Zucker diabetic rats treated with metforn ZDF (CCB): Zucker diabetic rats fed with a CCB-rich diet; ZDF (CCB + M): Zucker diabetic treated with metformin and fed with a CCB-rich diet.

3.3. *Effect of the CCB and Metformin on Inflammatory and Apoptotic Markers in the Pancreas Diabetic Rats*

The pro-oxidant environment in the diabetic pancreas can also instigate inflammat responses by activating the nuclear factor kappa B (NF-κB) signaling pathway, resulting the upregulation of specific pro-inflammatory cytokines and pancreatic apoptosis. Cor quently, our next step was to examine the impact of metformin and the CCB on mark of pancreatic inflammation (IL-6, TNF-α, and p-p65) and apoptosis (caspase-3 activi Figure 2a,b, show a marked increase in the levels of IL-6 and TNF-α in the pancreas non-treated ZDF rats. Both metformin and the CCB-rich diet were able to significar decrease the levels of pro-inflammatory cytokines with a further significant reduction their combination. Accordingly, treatment of ZDF rats with metformin and the CCB d was also able to significantly diminish the activation of the NF-κB cascade by decreasing the levels of the phosphorylated p65 subunit (Figure 2c). Similarly, the activity of caspase-3 (Figure 2d), which is involved in the final steps of cell apoptosis, was significantly elevated in the pancreas of ZDF rats. However, CCB supplementation and metformin treatment partially prevented this increase. More importantly, the combination of the CCB with metformin totally avoided the increase in caspase-3 activity in the pancreas of diabetic animals. Taken together, these results provide strong evidence for the ability of the CCB to attenuate inflammation and apoptosis in the pancreas in a diabetic environment.

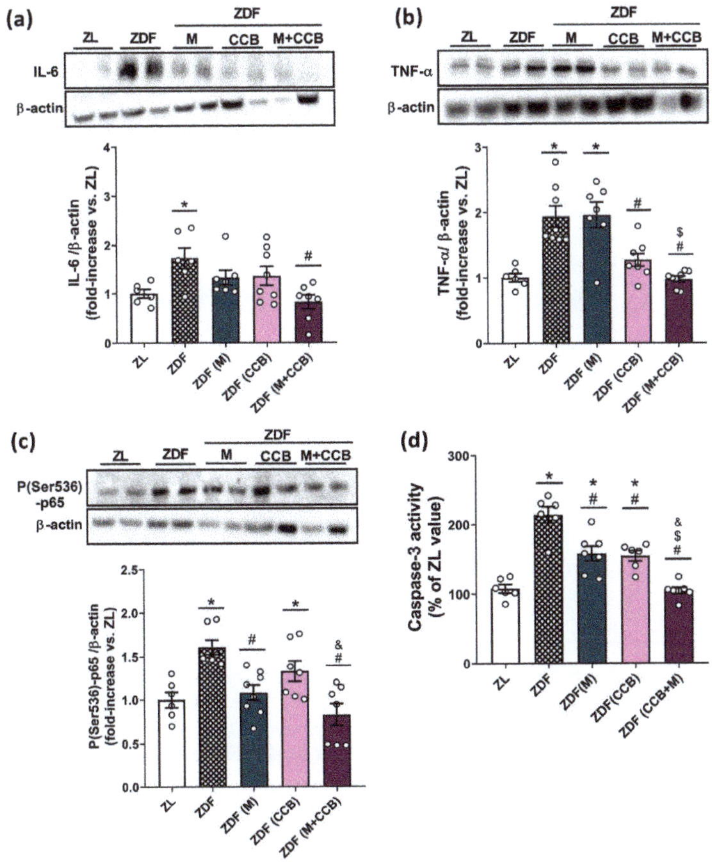

Figure 2. Effect of the CCB, metformin, and their combination on pancreatic inflammation and apoptosis. Representative Western blot analyses and levels of (**a**) IL-6, (**b**) TNF-α, and (**c**) P(Ser536)-p65 NF-κB in pancreatic tissues. Three independent experiments with two samples in each Western blot analysis were performed. (**d**) Fluorescent measurement of caspase-3 activity in pancreatic tissues. Values are expressed as mean ± SEM of n = 6–8 animals. * $p < 0.05$ vs. ZL; # $p < 0.05$ vs. ZDF; $ $p < 0.05$ ZDF (CCB + M) vs. ZDF (M); & $p < 0.05$ ZDF (CCB + M) vs. ZDF (CCB). ZL: Zucker lean; ZDF: Zucker diabetic rats; ZDF (M): Zucker diabetic rats treated with metformin; ZDF (CCB): Zucker diabetic rats fed with a CCB-rich diet; ZDF (CCB + M): Zucker diabetic rats treated with metformin and fed with a CCB-rich diet.

3.4. *Effect of the CCB and Metformin on Islet-Infiltrating Macrophages in the Pancreas of Diabetic Rats*

Macrophages are recognized as the primary immune cells involved in islet inflammation in both rodents and individuals with T2D. In fact, the infiltration of macrophages into islets during diabetes appears to mediate beta-cell destruction by secreting p inflammatory cytokines in close proximity to islet cells. For this reason, the prese of macrophages in the pancreas was investigated using CD68 staining (Figure 3a–c). In lean condition, islet macrophages were mainly localized in the periphery, surrounding islets. In contrast, the increased number of macrophages observed in non-treated ZDF p creases (Figure 3b) was linked to a shift in the distribution of islet-associated macropha with higher localization of CD68+ cells into the islets (Figure 3c,d). Metformin treatmen diabetic rats could not reduce the total number of pancreatic macrophages but substanti decreased the number of islets infiltrated with CD68+ cells. However, administration the CCB-enriched diet, alone or in combination with metformin, normalized the num and distribution of macrophages. Collectively, these findings point out the capacity of CCB to reduce islet macrophage infiltration and attenuate pancreatic inflammation i diabetic milieu.

3.5. *Effect of the CCB and Metformin on Beta-Cell Damage*

To further confirm that the antioxidant and anti-inflammatory properties of the C contribute to protect pancreatic beta cells in diabetic rats, we conducted morphometri analyses of islets in both lean and diabetic Zucker rats. As depicted in Figure 4a, is in ZDF rats exhibited a loss of structural integrity compared to those in the control rats. They showed increased extent and irregularity in their borders, with numer extensions into the exocrine tissue and a heterogeneous staining pattern characteri by intense or weak insulin immunostaining of individual beta cells. Additionally, sections analyzed in this condition revealed important peripancreatic fat accumulation a infiltrations of adipose cells into the pancreatic tissue, which could contribute to a lo lipotoxicity. Moreover, the pancreas of ZDF rats exhibited a significantly higher percent of fibrotic tissue compared to the healthy control group (Figure 4b). Treatment of diab rats with metformin alone failed to reverse the development of fibrosis. However, administration of the CCB to ZDF rats for 12 weeks resulted in a significant reduct in the area of pancreatic fibrosis compared to untreated diabetic rats. Surprisingly, t positive effect on overall fibrosis was not found when combining the dietary treatm with metformin.

Consistent with these diabetes-induced structural changes, ZDF rats had significar reduced beta-cell fractional area compared to non-diabetic ZL rats (Figure 4c). Neverthel all treatments tested resulted in a significant restoration of fractional beta cells compa to untreated diabetic rats (ZDF), reaching levels similar to those of the control group (Z Significantly, the CCB-enriched diet exerted the most pronounced effect. It is notewor that in the ZDF (CCB) and ZDF (CCB + M) groups, numerous clusters of two to six beta c distributed throughout the pancreatic tissue were observed, indicating an active proces regeneration. This is consistent with the significantly reduced total insulin content foun the pancreas of ZDF rats that was completely avoided in all treatments (Figure 4d). Th results highlight that CCB supplementation prevents beta-cell loss in ZDF rats.

To provide additional confirmation regarding the involvement of apoptosis in the of beta cells in Zucker diabetic animals, we specifically measured the apoptotic nucle beta cells in samples from the different groups by combining TUNEL staining and insu detection. As illustrated in Figure 5a,b, positive TUNEL staining signals in the pancre islets of ZDF rats were prominently present, while these signals were scarcely detectabl the pancreatic islets of control ZL rats, pointing to a substantial increase in apoptotic b cells in the pancreatic islets of ZDF diabetic animals. However, treatment with metforr or the CCB diet reduced the apoptotic effect, and the combined administration of b completely prevented it (Figure 5a,b). These results highlight the effectiveness of the C enriched diet, alone or in combination with metformin, in preventing beta-cell apopt and thereby reducing the loss of beta cells in ZDF rats.

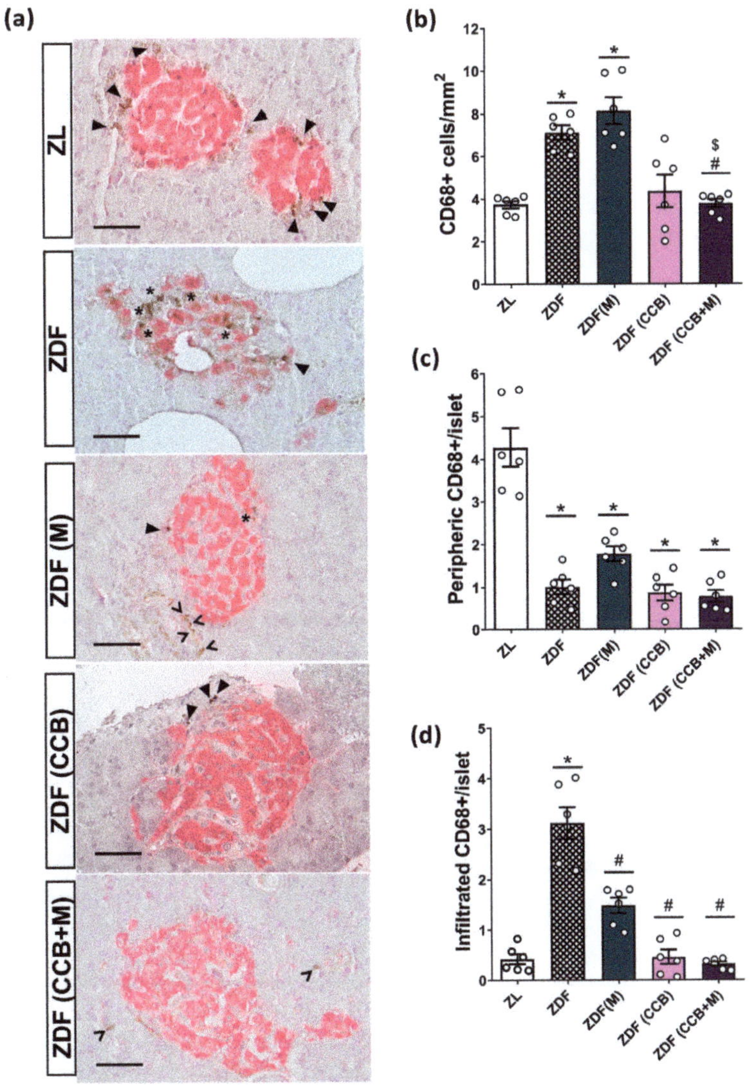

Figure 3. Effect of the CCB, metformin, and their combination on pancreatic macrophage infiltration. (**a**) Representative images of CD68+ immunohistochemical staining of islets (brown-stained). Scale bar = 50 µm. (**b**) Number of CD68+ cells expressed per mm^2 of total pancreatic surface area. (**c**) Number of peripheric CD68+ cells per islet. (**d**) Number of infiltrated CD68+ cells per islet. In the images, intra-islet macrophages are marked with stars, peripheral macrophages are marked with a closed arrowhead, and exocrine macrophages are marked with an open arrowhead. In the graphs, values are expressed as mean ± SEM of $n = 6$–8 animals. * $p < 0.05$ vs. ZL; # $p < 0.05$ vs. ZDF; $ $p < 0.05$ ZDF (CCB + M) vs. ZDF (M). ZL: Zucker lean; ZDF: Zucker diabetic rats; ZDF (M): Zucker diabetic rats treated with metformin; ZDF (CCB): Zucker diabetic rats fed with a CCB-rich diet; ZDF (CCB + M): Zucker diabetic rats treated with metformin and fed with a CCB-rich diet.

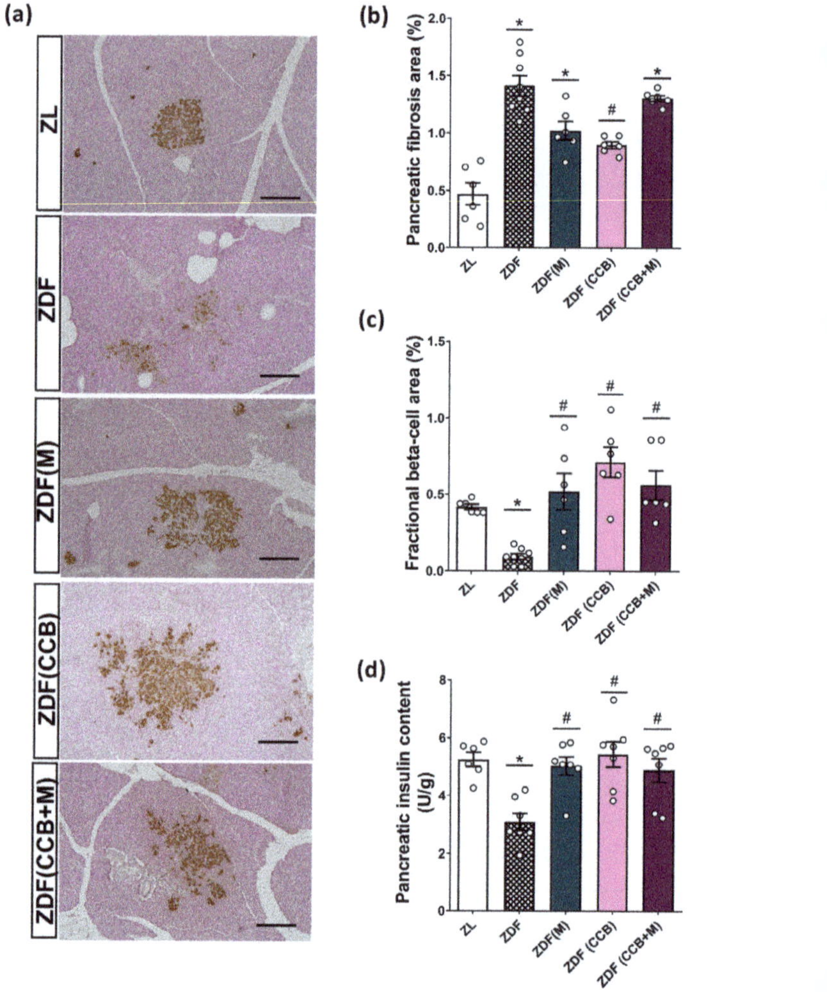

Figure 4. Effect of the CCB, metformin, and their combination on fractional beta cells. (**a**) Represe tive pancreatic tissue sections stained with insulin (brown) and hematoxylin (scale bars: 200 μ (**b**) Quantification of fibrotic area expressed as the percentage of total pancreatic tissue. (**c**) Fractic beta-cell area (stained brown) expressed as relative to the total pancreatic area (20× magnificatic (**d**) Pancreatic insulin content. Values are expressed as mean ± SEM of n = 6–8 animals. * $p < 0.05$ ZL; # $p < 0.05$ vs. ZDF. ZL: Zucker lean; ZDF: Zucker diabetic rats; ZDF (M): Zucker diabetic treated with metformin; ZDF (CCB): Zucker diabetic rats fed with a CCB-rich diet; ZDF (CCB + Zucker diabetic rats treated with metformin and fed with a CCB-rich diet.

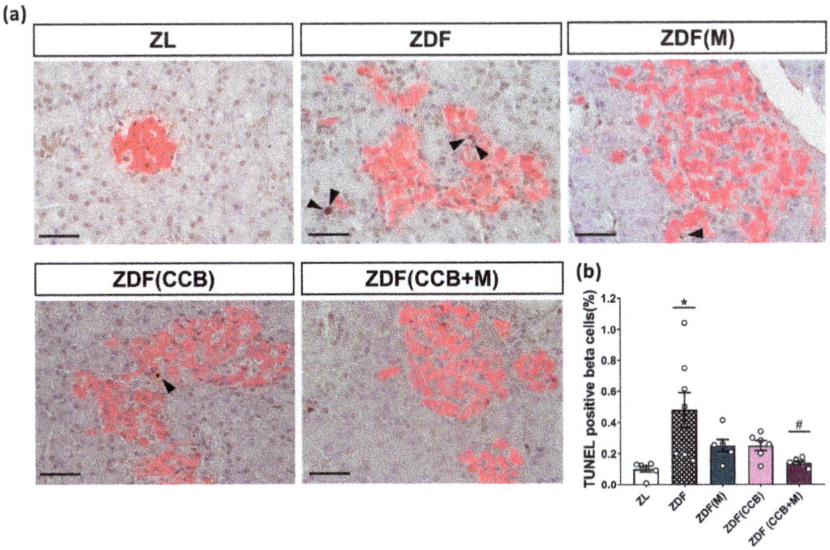

Figure 5. Effect of the CCB, metformin, and their combination on beta-cell apoptosis. (**a**) Representative immunohistochemical staining of terminal deoxynucleotidyl transferase dUTP nick end labeling-positive beta cells. Scale bar: 50 µm. Arrows indicate apoptotic nuclei (in brown) of pancreatic beta cells (insulin-positive, pink-stained cytoplasm). (**b**) Beta-cell apoptotic index expressed as the percentage of the total beta cells. Values are expressed as mean ± SEM of n = 6–8 animals. * $p < 0.05$ vs. ZL; # $p < 0.05$ vs. ZL: Zucker lean; ZDF: Zucker diabetic rats; ZDF (M): Zucker diabetic rats treated with metformin; ZDF (CCB): Zucker diabetic rats fed with a CCB-rich diet; ZDF (CCB + M): Zucker diabetic rats treated with metformin and fed with a CCB-rich diet.

4. Discussion

Numerous studies emphasize the protective role of residual beta-cell function against the progression of diabetic complications, highlighting the significant value of treatments capable of sustaining beta-cell mass and function over time [24]. In the present study, we demonstrated that supplementation of diabetic animals with a promising flavonoid-rich functional food (CCB) significantly reduced pancreatic oxidative stress and inflammation, effectively preventing beta-cell loss and supporting better glycemic control. Importantly, its combination with metformin produced a superior effect, underscoring the potential advantages of combining natural bioactive compounds with antidiabetic drugs in diabetes treatment [25].

Markers of oxidative stress, such as increased ROS generation and diminished intracellular antioxidants, are consistent with pancreatic beta-cell damage in preclinical models and individuals with T2D [26–28]. Furthermore, the naturally low expression of antioxidant enzymes in pancreatic islets renders beta cells more susceptible to oxidative stress [29]. In line with this, our findings demonstrate that the CCB can mitigate ROS generation and oxidative damage in the pancreas of diabetic animals. Moreover, CCB supplementation enhances pancreatic antioxidant capacity in diabetic animals by boosting both enzymatic and non-enzymatic antioxidant defenses. Specifically, GSH levels exhibited a significant increase, with an additional benefit observed in combination with metformin. Similarly, CCB supplementation improved the activity of antioxidant enzymes, with a substantial increase in GPx activity compared to other treatments. Previous studies have demonstrated that polyphenols increase the expression and activity of these enzymes in various tissues [13,23,30,31] and in pancreatic beta cells [32,33]. Importantly, it has been established that overexpression of GPx-1 in beta cells of diabetic mice (db/db) preserved their mass and reversed the development of hyperglycemia [34]. In the same way, early treatment diabetic ZDF rats with ebselen, a GPx mimetic agent, prevented beta-cell deterioration [. Taken together, these findings strongly indicate that a CCB-rich diet augments the activ of the antioxidant defenses in the pancreas, thereby reducing oxidative stress induced the diabetic condition.

Additionally, ZDF diabetic animals showed pancreatic inflammation evidenced increased levels of the inflammatory markers TNF-α, IL-6, and p-p65. Notably, C supplementation significantly attenuated the expression of all these pro-inflammatory diators. Macrophages are the primary contributors to immune cell-mediated inflammati in islets [9]. Several studies have demonstrated an increased macrophage infiltration T2D islets, often correlating with beta-cell dysfunction [36–38]. In our study, ZDF mals showed an increased number of macrophages primarily localized within the isl Macrophages infiltrating diabetic islets express an M1-like pro-inflammatory phenoty contributing to local overproduction of inflammatory cytokines and activation of the NF- signaling pathway, ultimately leading to beta-cell apoptosis and dysfunction [39]. though metformin treatment did not reduce macrophage numbers, it did decrease th infiltration, thereby mitigating islet inflammation. This aligns with recent findings in cating metformin's ability to inhibit islet inflammatory response and islet cell apoptosis HFD/STZ-induced diabetic mice [40]. Importantly, both the number of macrophages a their infiltration into islets significantly decreased in the pancreas of CCB supplemen animals compared to untreated ZDF rats. In line with these data, other naturally deriv compounds have been reported to attenuate inflammation in various tissues by reduc M1 polarization of macrophages in obesity and diabetes [41–46]. Notably, resveratrol, of the most studied polyphenols, attenuated macrophage infiltration in pancreatic islets T1D mice, thereby reducing beta-cell destruction [44]. More recently, it has been shown t the carotenoid lycopene may ameliorate hyperglycemia and dyslipidemia and attenu beta-cell apoptosis by regulating the toll-like receptor 4 (TLR4)/NF-κB signaling pathw both in islets of diabetic mice and in Min6 beta cells stimulated with the conditio medium collected from RAW264.7 cells treated with glucose/palmitate [40]. Collectiv these findings suggest that CCB treatment could prevent beta-cell loss in ZDF diabetic by inhibiting islet macrophage infiltration and intra-islet inflammation.

According to these results, the histological examination of the pancreas of ZDF anim revealed a reduction in the beta-cell fractional area compared to non-diabetic ZL cont This decrease was accompanied by a lower cellular insulin content and structural chan in their islets, in alignment with observations from prior studies [45,46]. Notably, the C enriched diet significantly prevented beta-cell loss, surpassing the efficacy of metforr treatment. Moreover, we observed that both the CCB and metformin treatments ha partial preventive effect on beta-cell apoptosis. These outcomes agree with prior stud demonstrating the anti-apoptotic properties of polyphenols as a significant mechanism preventing beta-cell loss [5,32]. Unexpectedly, the beneficial effect of CCB supplementat on pancreatic fibrosis was not observed when the dietary treatment was combined w metformin; the mechanistic explanation for this requires further research. Altogeth the present results highlight the effectiveness of the CCB in preventing beta-cell loss diabetic ZDF rats. Importantly, the combination of the CCB-rich diet with metformin proven to be more effective in reducing beta-cell apoptosis, suggesting that their combi use may be therapeutically beneficial in reducing beta-cell loss during the progress of diabetes. However, in this study, we focused on the effect on this treatment on whole pancreas, which is a limitation because it does not allow us to directly link treatment to the functional changes in islets. Further studies with isolated islets co provide complementary information to elucidate the molecular mechanism involved in beneficial effect of the CCB and metformin on beta cells.

It is noteworthy to mention that ZDF animals also exhibited increased insulin sistance, a condition that was significantly reduced by both CCB supplementation a metformin treatment (as indicated by HOMA-IR values). Although we did not specific evaluate the effect of CCBs on liver function in ZDF rats, the existing literature suggests that cocoa flavanols may exert an insulin-like effect on human hepatic HepG2 cells under conditions of insulin resistance. This effect involves attenuation of the blockade of the insulin signaling cascade and modulation of glucose uptake and production [47]. Additionally, in pre-diabetic rats, a cocoa-enriched diet demonstrated the ability to alleviate hepatic insulin resistance by regulating key proteins of the insulin pathway and glucose metabolism in the liver [48]. Therefore, it cannot be excluded that the observed antidiabetic properties of CCB supplementation are partly due to the effects of flavanols on liver function and insulin resistance, hallmarks of T2D development and progression.

5. Conclusions

The present study demonstrates that chronic supplementation with a potential functional food rich in flavonoids (CCB) significantly prevents beta-cell apoptosis and the loss of functional beta cells in the islets of ZDF rats, ultimately improving glucose homeostasis in diabetic animals. The protective effect of the CCB appears to be partly mediated by its ability to enhance pancreatic antioxidant defenses, thereby neutralizing oxidative stress. Additionally, the action of the CCB in preventing macrophage infiltration into islets and the subsequent local production of pro-inflammatory factors significantly contributes to its beneficial effects on beta-cell apoptosis and dysfunction. It is interesting to remark on the observed additive effect when combining CCB treatment with metformin, which highlights its potential as an adjuvant therapy to delay the progression of diabetes.

Supplementary Materials: The following supporting information can be downloaded at: https://www.mdpi.com/article/10.3390/nu16020273/s1. Supplementary Table S1. Biochemical parameters of Zucker lean rats (ZL); Zucker diabetic rats (ZDF); Zucker diabetic rats treated with metformin (ZDF (M)); Zucker diabetic rats fed with a CCB-rich diet (ZDF (CCB)), and Zucker diabetic rats treated with metformin and fed with a CCB-rich diet (ZDF (CCB + M)). Data represent the means ± SD of 6–8 animals. Means in a row without a common letter differ; $p < 0.05$.

Author Contributions: Conceptualization, T.F.-M., L.V., M.A.M. and E.F.-M.; Data curation, P.G.-V., M.A.M., E.F.-M., T.F.-M., L.V., A.M.V. and S.R.; Formal Analysis, P.G.-V., M.A.M., E.F.-M., T.F.-M., L.V., A.M.V. and S.R.; Funding Acquisition, M.A.M., E.F.-M., T.F.-M., L.V., A.M.V. and S.R.; Methodology, P.G.-V., T.F.-M., L.V. and E.F.-M.; Project administration, M.A.M., T.F.-M., L.V. and S.R.: Writing—Original Draft P.G.-V., M.A.M. and E.F.-M.; Writing—Review and Editing P.G.-V., M.A.M., E.F.-M., T.F.-M., L.V., A.M.V. and S.R.; All authors have read and agreed to the published version of the manuscript.

Funding: This work was supported by the grants RTI2018-095059-B-I00 and PID2020-116134RB-I00 (MCIN/AEI/10.13039/501100011033/ and by "ERDF A way of making Europe"), grant P2022/BMD-7227 (Comunidad de Madrid, Spain) and grant DEM2022/PIM01 by CIBERDEM (CB07/08/0013 and CB07/08/0033), Instituto de Salud Carlos III, MCIN.

Institutional Review Board Statement: All the experiments were conducted according to European Union (2010/63/EU) and Spanish (RD 53/2013) legislation and with the approval (16 May 2019) of the Animal Care and Use Committee of the Community of Madrid (PROEX 079/19).

Informed Consent Statement: Not applicable.

Data Availability Statement: Data are available upon request to the authors. The data are not publicly available due to principle of confidentiality.

Conflicts of Interest: The authors declare no conflicts of interest.

erences

American Diabetes Association. Classification and Diagnosis of Diabetes: Standards of Medical Care in Diabetes—2020. *Diabetes Care* **2020**, *43*, S14–S31. [CrossRef] [PubMed]
DeFronzo, R.A.; Ferrannini, E.; Groop, L.; Henry, R.R.; Herman, W.H.; Holst, J.J.; Hu, F.B.; Kahn, C.R.; Raz, I.; Shulman, G.I.; et al. Type 2 diabetes mellitus. *Nat. Rev. Dis. Primers* **2015**, *1*, 15019. [CrossRef] [PubMed]

3. Campbell, J.E.; Newgard, C.B. Mechanisms controlling pancreatic islet cell function in insulin secretion. *Nat. Rev. Mol. Cell B* **2021**, *22*, 142–158. [CrossRef] [PubMed]
4. Mukai, E.; Fujimoto, S.; Inagaki, N. Role of Reactive Oxygen Species in Glucose Metabolism Disorder in Diabetic Pancrea β-Cells. *Biomolecules* **2022**, *12*, 1228. [CrossRef] [PubMed]
5. García-Aguilar, A.; Guillen, C. Targeting pancreatic beta cell death in type 2 diabetes by polyphenols. *Front. Endocrinol.* **2022**, 1052317. [CrossRef] [PubMed]
6. Prasad, M.K.; Mohandas, S.; Ramkumar, K.M. Dysfunctions, molecular mechanisms, and therapeutic strategies of pancrea β-cells in diabetes. *Apoptosis* **2023**, *28*, 958–976. [CrossRef]
7. Khatua, B.; El-Kurdi, B.; Singh, V.P. Obesity and pancreatitis. *Curr. Opin. Gastroenterol.* **2017**, *33*, 374–382. [CrossRef]
8. Wang, Z.; Ni, X.; Zhang, L.; Sun, L.; Zhu, X.; Zhou, Q.; Yang, Z.; Yuan, H. Toll-Like Receptor 4 and Inflammatory Mi Environment of Pancreatic Islets in Type-2 Diabetes Mellitus: A Therapeutic Perspective. *Diabetes Metab. Syndr. Obes.* **2020**, 4261–4272. [CrossRef]
9. Ying, W.; Fu, W.; Lee, Y.S. The role of macrophages in obesity-associated islet inflammation and β-cell abnormalities. *Nat. Endocrinol.* **2020**, *16*, 81–90. [CrossRef]
10. Dludla, P.V.; Mabhida, S.E.; Ziqubu, K.; Nkambule, B.B.; Mazibuko-Mbeje, S.E.; Hanser, S.; Basson, A.K.; Pheiffer, C.; Kengne, Pancreatic β-cell dysfunction in type 2 diabetes: Implications of inflammation and oxidative stress. *World J. Diabetes* **2023**, 130–146. [CrossRef]
11. Rachdaoui, N. Early Combination Therapies for the Preservation of Pancreatic β Cells in Type 2 Diabetes Mellitus. *Endocrinol* **2023**, *164*, bqad134. [CrossRef]
12. Dludla, P.V.; Silvestri, S.; Orlando, P.; Gabuza, K.B.; Mazibuko-Mbeje, S.E.; Nyambuya, T.M.; Mxinwa, V.; Mokgalaboni, Johnson, R.; Muller, C.J.F.; et al. Exploring the Comparative Efficacy of Metformin and Resveratrol in the Managemen Diabetes-associated Complications: A Systematic Review of Preclinical Studies. *Nutrients* **2020**, *12*, 739. [CrossRef] [PubMed]
13. García-Díez, E.; López-Oliva, M.E.; Caro-Vadillo, A.; Pérez-Vizcaíno, F.; Pérez-Jiménez, J.; Ramos, S.; Martín, M.Á. Supplem tation with a Cocoa–Carob Blend, Alone or in Combination with Metformin, Attenuates Diabetic Cardiomyopathy, Card Oxidative Stress and Inflammation in Zucker Diabetic Rats. *Antioxidants* **2022**, *11*, 432. [CrossRef] [PubMed]
14. Rana, A.; Samtiya, M.; Dhewa, T.; Mishra, V.; Aluko, R.E. Health benefits of polyphenols: A concise review. *J. Food Biochem.* **2** *46*, e14264. [CrossRef]
15. Cao, H.; Ou, J.; Chen, L.; Zhang, Y.; Szkudelski, T.; Delmas, D.; Daglia, M.; Xiao, J. Dietary polyphenols and type 2 diabe Human Study and Clinical Trial. *Crit. Rev. Food Sci. Nutr.* **2019**, *59*, 3371–3379. [CrossRef]
16. Ramos, S.; Martín, M.A.; Goya, L. Effects of Cocoa Antioxidants in Type 2 Diabetes Mellitus. *Antioxidants* **2017**, *31*, 84. [Cross
17. Fernández-Millán, E.; Ramos, S.; Alvarez, C.; Bravo, L.; Goya, L.; Martín, M.A. Microbial phenolic metabolites improve gluc stimulated insulin secretion and protect pancreatic beta cells against tert-butyl hydroperoxide-induced toxicity via ERKs a PKC pathways. *Food Chem. Toxicol.* **2014**, *66*, 245–253. [CrossRef] [PubMed]
18. Fernández-Millán, E.; Cordero-Herrera, I.; Ramos, S.; Escrivá, F.; Alvarez, C.; Goya, L.; Martín, M.A. Cocoa-rich diet attenua beta cell mass loss and function in young Zucker diabetic fatty rats by preventing oxidative stress and beta cell apoptosis. *Nutr. Food Res.* **2015**, *59*, 820–824. [CrossRef]
19. Macho-González, A.; López-Oliva, M.E.; Merino, J.J.; García-Fernández, R.A.; Garcimartín, A.; Redondo-Castillejo, R.; Bast S.; Sánchez-Muniz, F.J.; Benedí, J. Carob fruit extract-enriched meat improves pancreatic beta-cell dysfunction, hepatic insu signaling and lipogenesis in late-stage type 2 diabetes mellitus model. *J. Nutr. Biochem.* **2020**, *84*, 108461. [CrossRef]
20. García-Díez, E.; Sánchez-Ayora, H.; Blanch, M.; Ramos, S.; Martín, M.A.; Pérez-Jiménez, J. Exploring a cocoa–carob blend a functional food with decreased bitterness: Characterization and sensory analysis. *LWT* **2022**, *165*, 113708. [CrossRef]
21. Directive 2010/63/EU of the European Parliament and of the Council of 22 September 2010 on the Protection of Animals U for Scientific Purposes. (OJ L 276 20.10.2010, p. 33), European Union. 2010. Available online: http://data.europa.eu/eli/dir/2 /63/2019-06-26 (accessed on 23 November 2023).
22. Real Decreto 53/2013, de 1 de febrero, por el que se establecen las normas básicas aplicables para la protección de los anima utilizados en experimentación y otros fines científicos, incluyendo la docencia. BOE-A-2013-1337, Spain. 2013. Available on https://www.boe.es/eli/es/rd/2013/02/01/53/con (accessed on 23 November 2023).
23. García-Díez, E.D.; López-Oliva, M.E.; Pérez-Vizcaíno, F.; Perez-Jimenez, J.; Ramos, S.; Martín, M.A. Dietary Supplementation w a Cocoa-Carob Blend Modulates Gut Microbiota and Prevents Intestinal Oxidative Stress and Barrier Dysfunction in Zuc Diabetic Rats. *Antioxidants* **2023**, *12*, 1519. [CrossRef]
24. Fuhri Snethlage, C.M.; McDonald, T.J.; Oram, R.D.; de Groen, P.; Rampanelli, E.; Schimmel, A.W.M.; Holleman, F.; Siegelaar Hoekstra, J.; Brouwer, C.B.; et al. Residual β-Cell Function Is Associated with Longer Time in Range in Individuals with Ty Diabetes. *Diabetes Care* **2023**, *3*, dc230776. [CrossRef] [PubMed]
25. Blahova, J.; Martiniakova, M.; Babikova, M.; Kovacova, V.; Mondockova, V.; Omelka, R. Pharmaceutical Drugs and Nat Therapeutic Products for the Treatment of Type 2 Diabetes Mellitus. *Pharmaceuticals* **2021**, *14*, 806. [CrossRef] [PubMed]
26. Schuurman, M.; Wallace, M.; Sahi, G.; Barillaro, M.; Zhang, S.; Rahman, M.; Sawyez, C.; Borradaile, N.; Wang, R. N-acety cysteine treatment reduces beta-cell oxidative stress and pancreatic stellate cell activity in a high fat diet-induced diabetic mo model. *Front. Endocrinol.* **2022**, *13*, 938680. [CrossRef] [PubMed]

Liu, J.; Zhang, Y.; Shi, D.; He, C.; Xia, G. Vitamin D Alleviates Type 2 Diabetes Mellitus by Mitigating Oxidative Stress-Induced Pancreatic β-Cell Impairment. *Exp. Clin. Endocrinol. Diabetes* **2023**, *7*, 656–666. [CrossRef] [PubMed]
Sireesh, D.; Dhamodharan, U.; Ezhilarasi, K.; Vijay, V.; Ramkumar, K.M. Association of NF-E2 Related Factor 2 (Nrf2) and inflammatory cytokines in recent onset Type 2 Diabetes Mellitus. *Sci. Rep.* **2018**, *8*, 5126. [CrossRef] [PubMed]
Eguchi, N.; Vaziri, N.D.; Dafoe, D.C.; Ichii, H. The Role of Oxidative Stress in Pancreatic β Cell Dysfunction in Diabetes. *Int. J. Mol. Sci.* **2021**, *22*, 1509. [CrossRef]
Álvarez-Cilleros, D.; López-Oliva, M.E.; Morales-Cano, D.; Barreira, B.; Pérez-Vizcaíno, F.; Goya, L.; Ramos, S.; Martín, M.A. Dietary Cocoa Prevents Aortic Remodeling and Vascular Oxidative Stress in Diabetic Rats. *Mol. Nutr. Food Res.* **2019**, *63*, e1900044. [CrossRef]
Zhang, Y.; Tan, X.; Cao, Y.; An, X.; Chen, J.; Yang, L. Punicalagin Protects against Diabetic Liver Injury by Upregulating Mitophagy and Antioxidant Enzyme Activities. *Nutrients* **2022**, *14*, 2782. [CrossRef]
Ghorbani, A.; Rashidi, R.; Shafiee-Nick, R. Flavonoids for preserving pancreatic beta cell survival and function: A mechanistic review. *Biomed. Pharmacother.* **2019**, *111*, 947–957. [CrossRef]
Bashir, N.; Manoharan, V.; Miltonprabu, S. Grape seed proanthocyanidins protects against cadmium induced oxidative pancreatitis in rats by attenuating oxidative stress, inflammation and apoptosis via Nrf-2/HO-1 signaling. *J. Nutr. Biochem.* **2016**, *32*, 128–141. [CrossRef] [PubMed]
Harmon, J.S.; Bogdani, M.; Parazzoli, S.D.; Mak, S.S.M.; Oseid, E.A.; Berghmans, M.; LeBoeuf, R.C.; Robertson, R.P. β-cell-specific overexpression of glutathione peroxidase preserves intranuclear MafA and reverses diabetes in db/db Mice. *Endocrinology* **2009**, *150*, 4855–4862. [CrossRef]
Mahadevan, J.; Parazzoli, S.; Oseid, E.; Hertzel, A.V.; Bernlohr, D.A.; Vallerie, S.N.; Liu, C.Q.; Lopez, M.; Harmon, J.S.; Robertson, R.P. Ebselen treatment prevents islet apoptosis, maintains intranuclear Pdx-1 and MafA levels, and preserves β-cell mass and function in ZDF rats. *Diabetes* **2013**, *62*, 3582–3588. [CrossRef] [PubMed]
Ying, W.; Lee, Y.S.; Dong, Y.; Seidman, J.S.; Yang, M.; Isaac, R.; Seo, J.B.; Yang, B.H.; Wollam, J.; Riopel, M.; et al. Expansion of Islet-Resident Macrophages Leads to Inflammation Affecting beta Cell Proliferation and Function in Obesity. *Cell Metab.* **2019**, *29*, 457–474. [CrossRef] [PubMed]
Marselli, L.; Bugliani, M.; Suleiman, M.; Olimpico, F.; Masini, M.; Petrini, M.; Boggi, U.; Filipponi, F.; Syed, F.; Marchetti, P. Beta-Cell inflammation in human type 2 diabetes and the role of autophagy. *Diabetes Obes. Metab.* **2013**, *3*, 130–136. [CrossRef]
Butcher, M.J.; Hallinger, D.; Garcia, E.; Machida, Y.; Chakrabarti, S.; Nadler, J.; Galkina, E.V.; Imai, Y. Association of proinflammatory cytokines and islet resident leucocytes with islet dysfunction in type 2 diabetes. *Diabetologia* **2014**, *57*, 491–501. [CrossRef]
Banu, S.; Sur, D. Role of Macrophage in Type 2 Diabetes Mellitus: Macrophage Polarization a New Paradigm for Treatment of Type 2 Diabetes Mellitus. *Endocr. Metab. Immune Disord. Drug Targets* **2022**, *23*, 2–11.
Liu, Y.; Tian, Y.; Dai, X.; Liu, T.; Zhang, Y.; Wang, S.; Shi, H.; Yin, J.; Xu, T.; Zhu, R.; et al. Lycopene ameliorates islet function and down-regulates the TLR4/MyD88/NF-κB pathway in diabetic mice and Min6 cells. *Food Funct.* **2023**, *14*, 5090. [CrossRef]
Song, Z.; Revelo, X.; Shao, W.; Tian, L.; Zeng, K.; Lei, H.; Sun, H.S.; Woo, M.; Winer, D.; Jin, T. Dietary Curcumin Intervention Targets Mouse White Adipose Tissue Inflammation and Brown Adipose Tissue UCP1 Expression. *Obesity* **2018**, *26*, 547–558. [CrossRef]
Accipe, L.; Abadie, A.; Neviere, R.; Bercion, S. Antioxidant Activities of Natural Compounds from Caribbean Plants to Enhance Diabetic Wound Healing. *Antioxidants* **2023**, *12*, 1079. [CrossRef]
Saqib, U.; Sarkar, S.; Suk, K.; Mohammad, O.; Baig, M.S.; Savai, R. Phytochemicals as modulators of M1-M2 macrophages in inflammation. *Oncotarget* **2018**, *9*, 17937–17950. [CrossRef] [PubMed]
Darwish, M.A.; Abo-Youssef, A.M.; Messiha, B.A.S.; Abo-Saif, A.A.; Abdel-Bakky, M.S. Resveratrol inhibits macrophage infiltration of pancreatic islets in streptozotocin-induced type 1 diabetic mice via attenuation of the CXCL16/NF-κB p65 signaling pathway. *Life Sci.* **2021**, *272*, 119250. [CrossRef] [PubMed]
Hallakou-Bozec, S.; Kergoat, M.; Moller, D.E.; Bolze, S. Imeglimin preserves islet β-cell mass in Type 2 diabetic ZDF rats. Endocrinol. *Diabetes Metab.* **2020**, *4*, e00193.
Lundh, M.; Galbo, T.; Poulsen, S.S.; Mandrup-Poulsen, T. Histone deacetylase 3 inhibition improves glycaemia and insulin secretion in obese diabetic rats. *Diabetes Obes. Metab.* **2015**, *17*, 703–707. [CrossRef]
Cordero-Herrera, I.; Martin, M.A.; Goya, L.; Ramos, S. Cocoa flavonoids attenuate high glucose-induced insulin signalling blockade and modulate glucose uptake and production in human HepG2 cells. *Food Chem. Toxicol.* **2014**, *64*, 10–19. [CrossRef]
Cordero-Herrera, I.; Martín, M.A.; Escrivá, F.; Álvarez, C.; Goya, L.; Ramos, S. Cocoa-rich diet ameliorates hepatic insulin resistance by modulating insulin signaling and glucose homeostasis in Zucker Diabetic Fatty rats. *J. Nutr. Biochem.* **2015**, *26*, 704–712. [CrossRef]

nutrients

MDPI

Article

Cocoa and Carob Supplementation, Alone or in Combination with Metformin, Protects against Hepatorenal Injury in Zucke Diabetic Fatty Rats

Esther García-Díez [1,†], María Elvira López-Oliva [2,3,†], Jara Pérez-Jiménez [1,4], María Ángeles Martín [1,4] and Sonia Ramos [1,4,*]

1 Instituto de Ciencia y Tecnología de Alimentos y Nutrición (ICTAN-CSIC), 28040 Madrid, Spain; esther.garciad@ictan.csic.es (E.G.-D.); jara.perez@ictan.csic.es (J.P.-J.); amartina@ictan.csic.es (M.Á.M.)
2 Sección Departamental de Fisiología, Facultad de Farmacia, Universidad Complutense de Madrid (UCM 28040 Madrid, Spain; elopez@farm.ucm.es
3 AFUSAN Group, Sanitary Research Institute of the San Carlos Clinical Hospital (IdISSC), 28040 Madrid, S
4 CIBER de Diabetes y Enfermedades Metabólicas Asociadas (CIBERDEM), Instituto de Salud Carlos III (ISCIII), 28029 Madrid, Spain
* Correspondence: s.ramos@ictan.csic.es
† The first two authors equally contributed to the article, and both should be considered as first authors.

Abstract: The liver and kidneys are crucial for glucose homeostasis and are seriously dama in diabetes. Cocoa and carob possess antidiabetic activity, but their hepatorenal protective effe especially when combined with antidiabetic drugs, are unknown. The aim of this study is investigate the effects of a cocoa–carob-supplemented diet (CC), either alone or in combination w metformin, on liver and kidney damage in Zucker diabetic fatty (ZDF) rats, a type 2 diabetes mo Male ZDF animals received a control or CC-supplemented diet, with or without metformin, Zucker lean rats were fed the control diet. The CC-supplemented diet improved glucose tolera and insulin resistance and alleviated functional and structural alterations in the diabetic liver renal cortex. The CC-supplemented diet also ameliorated oxidative stress, downregulated apopt and improved insulin signalling and glucose homeostasis. The combination of CC and metfor boosted several benefits as certain parameters related to morphological and structural alterati apoptosis, oxidative stress, glucose homeostasis, and insulin resistance, were improved in compari to animals receiving the CC-supplemented diet or metformin alone; these include the follow apoptotic index, Bax, hepatic insulin receptor or glutathione content, among others. These res demonstrate that the CC-supplemented diet alleviates the hepatorenal damage in type 2 diab ZDF rats, highlighting its potential alone or as an adjuvant therapy.

Keywords: cocoa–carob blend; glucose homeostasis; insulin resistance; oxidative stress; apopt liver and renal cortex; type 2 diabetic ZDF rats

Citation: García-Díez, E.; López-Oliva, M.E.; Pérez-Jiménez, J.; Martín, M.Á.; Ramos, S. Cocoa and Carob Supplementation, Alone or in Combination with Metformin, Protects against Hepatorenal Injury in Zucker Diabetic Fatty Rats. *Nutrients* **2024**, *16*, 3087. https://doi.org/10.3390/nu16183087

Academic Editor: Maria Teresa Guagnano

Received: 9 August 2024
Revised: 31 August 2024
Accepted: 10 September 2024
Published: 13 September 2024

1. Introduction

Type 2 diabetes is a worldwide, highly prevalent disease caused by a metabolic di der and characterised by sustained hyperglycaemia that leads to severe complications The liver and kidney are crucial organs that play an important role in maintaining syste glucose homeostasis in the body. In diabetes, the liver and kidneys are severely dama due to exacerbated oxidative stress, insulin resistance, and abnormal glucose homeosta which also contribute to maintaining hyperglycaemia in this disease [2–4]. Accordingly altered response of antioxidant defences, including glutathione (GSH) and the enzymes tathione reductase (GR), glutathione peroxidase (GPx), and superoxide dismutase (SC has been reported in patients with diabetes [5,6]. Importantly, the transcription nuc factor erythroid 2-related factor 2 (Nrf2), a key protein in the regulation of the antioxid response, and several signalling pathways involved in the cellular response to the insult, such as sirtuin-1 (SIRT-1) and apoptosis-related proteins (caspases, Bcl-2 family proteins) are also dysregulated in diabetes [3,5]. In addition, insulin resistance leads to a situation in which the insulin receptor (IR) is less responsive to insulin, leading to the blockage of this pathway and altered glycogen synthesis in the diabetic liver and kidney [2,4,7]. Similarly, glucose uptake and gluconeogenesis in the liver and renal cortex are altered due to the dysregulation of key proteins involved in these processes, e.g., glucose transporter 2 (GLUT-2) and phosphoenolpyruvate carboxykinase (PEPCK) [2,4,7]. Consequently, therapies aimed at preventing oxidative stress, abnormal glucose homeostasis, insulin resistance, and the dysregulation of closely related processes, such as apoptosis, may represent an interesting approach against diabetes.

Cocoa is a rich source of phenolic compounds, mainly flavanols, and has demonstrated beneficial effects against diabetes and its associated complications [8,9]. Cocoa possesses antioxidant and antidiabetic properties that may contribute to improving glucose homeostasis, redox status, and insulin sensitivity in this disease [8,9]. Similarly, carob, a Mediterranean legume (*Ceratonia siliqua* L.), is rich in polymeric flavanols and other polyphenols, and has attracted attention for its potential health benefits, such as the maintenance of glucose homeostasis [10]. In addition, either the combination of antidiabetic drugs with phenolic compounds or the use of mixtures of extracts enriched with these natural compounds have recently been suggested as effective therapeutic interventions that could provide an efficient treatment [11,12]. In this context, metformin (M) is the first-line antihyperglycaemic drug used in patients with type 2 diabetes; however, the evaluation of the effect of dietary flavonoids or flavonoid-rich diets in diabetes alone or in combination with antidiabetic drugs, including metformin, has been scarcely studied. Moreover, the combination of drugs with these natural compounds may bring synergistic effects and more effective therapies against diabetes [13].

Taking all this into account, a polyphenol-supplemented food was prepared by combining cocoa powder and carob flour (60:40, CC). In our recent studies, this potential functional food has shown protective effects on pancreatic beta cells and the heart in diabetes [14,15]. In this context, it was investigated whether the potential antidiabetic preventive properties of this mixture extend to the associated liver and kidney damage in the animal model of type 2 diabetes. Thus, the aim of this study was to investigate the effects of a CC-supplemented diet, alone or in combination with metformin, on liver and kidney morphology, functionality, redox status, and the molecular mechanisms connected to the modulation of apoptosis and glucose homeostasis, namely insulin signalling (IR and glycogen content), gluconeogenesis (PEPCK), and glucose transporter levels (GLUT-2) in both tissues of ZDF rats. In addition, the potential enhanced benefits derived from the co-administration of the CC-supplemented diet and metformin were studied.

2. Materials and Methods

2.1. Chemicals and Materials

3,3′-diaminobenzidine (DAB), dichlorofluorescein (DCFH), dinitrophenylhidazine (DNPH), glucose assay kit, glutathione reductase (GR), reduced and oxidised glutathione (GSH and GSSG, respectively), hydrogen peroxide, nicotine adenine dinucleotide phosphate reduced salt (NADPH), o-phthaldehyde (OPT), proteinase K, *tert*-butylhydroperoxide (t-BOOH), and SOD kit were purchased from Sigma Chemicals (Madrid, Spain). Anti-IRβ and anti-phospho-IRβ (targeting phosphorylated Tyr1150/1151 of IR), anti-Bcl-x, anti-Bax, anti-SIRT-1, and anti-PEPCK antibodies were obtained from Santa Cruz (sc-711, sc-81500, sc-634, sc-526, sc-74465, and sc-32879, respectively; Qimigen, Madrid, Spain). Antiphospho-Nrf2 antibody (recognizing phosphorylated Ser40 of Nrf2) was purchased from Signalway antibody (Qimigen, Madrid, Spain). Additionally, α-tubulin and anti-β-actin were obtained from Abcam (Madrid, Spain) and Cell Signalling Technology (Werfen, Barcelona, Spain), respectively. The anti-GLUT-2 antibody was procured from Merck (Madrid, Spain). Kits for albumin and glycated haemoglobin (HbA1c) were acquired from Spinreact (BioAnalitica,

Madrid, Spain) and the rat insulin ELISA kit was from Mercodia (AD Bioinstrumer Barcelona, Spain). The creatinine kit was purchased from Linear Chemicals S.L. (Barcelo Spain), and alanine aminotransferase (ALT) and aspartate aminotransferase (AST) kits w obtained from BioSystems (Barcelona, Spain). Chemicals and materials for electrophore were acquired from BioRad Laboratories S.A. (Madrid, Spain).

2.2. *Cocoa–Carob Diet*

The cocoa–carob diet (CC) was produced by blending pure cocoa powder (generou provided by Idilia S.L., Barcelona, Spain) with carob flour (purchased from Casa R Granel Selecto S.L., Madrid, Spain) in a 60:40 ratio. This proportion was selected af performing a preliminary sensory analysis, which showed that 40% of carob was prefer by 60% of the participants [16]. A detailed description of this cocoa–carob product is giv elsewhere [16].

Diets are based on the AIN-93G formulation, which supplies all the necessary nutrie for adult rats. The 10% cocoa–carob (CC) diet was prepared by incorporating 100 g/kg the cocoa–carob blend in the AIN-93G formulation. Briefly, AIN-93G formulation with cellulose, starch, and sucrose was prepared by including these ingredients to adjust carbohydrate and dietary fibre content based on the amount of the cocoa–carob ble added. The composition of these diets is presented in Table S1.

2.3. *Animals and Diets*

This study is an extension of a previous investigation in which animal descriptio and experimental designs were described in detail [14]. Briefly, male Zucker lean (animals and Zucker diabetic fatty [ZDF; ZDF/*crl-lepr* (fa/fa)] rats were obtained at 11 we of age from Charles River Laboratories (L' Arbresle, France). Animals were housed groups under controlled conditions, maintaining a temperature of 19–23 °C, humidity 50–60%, and a 12 h light–dark cycle. Following a one-week acclimatization period, ZDF r (n = 32) were randomly assigned to four groups: one group was fed a standard diet (ZD another group was administrated with metformin in the drinking water (300 mg/kg/d and standard diet (ZDF-M), and the remaining two groups were fed the cocoa–car supplemented diet and received metformin (ZDF-M + CC) or not (ZDF-CC). Thus, present study is focused on the effects of the combination of cocoa and carob, as well as potential boosted effect of the CC supplementation in combination with metformin a does not allow to independently assess the responses derived from cocoa or carob, wh have already demonstrated to be helpful against diabetes [8–10]. ZL rats were kept a single group and received the standard diet. This healthy ZL group was included as internal control of the diabetic situation in ZDF rats. Both diets were isocaloric, and animals had *ad libitum* access to food and water. Daily food intake was recorded, and bo weight was monitored weekly throughout the study.

At 24 weeks of age, animals were fasted overnight, and blood samples were tak for biochemical analysis. The serum was prepared by centrifugation of blood at 1000 (4 °C, 10 min) and used for further biochemical determinations. Livers and kidneys w harvested, weighed, and divided into two portions: one was fixed in paraformaldehy (PFA) for histological analysis, the other was frozen in liquid nitrogen and kept at −80 All animal procedures were approved by the Regional Committee on animal care and for experimental purposes (Comunidad de Madrid, Ref. number PROEX 079/19), and animals were treated according to the European [17] and Spanish [18] legislation on care and use of experimental animals.

2.4. *Biochemical Analysis*

Glycaemia was assessed weekly with an Accounted Glucose Analyser (LifeScan paña, Madrid, Spain), and blood HbA1c was determined through latex turbidime according to the manufacturer's instructions. Serum insulin, ALT, AST, and creatinine w assayed using commercial kits according to the manufacturer's instructions. Insulin se tivity was estimated by the homeostasis model assessment of insulin resistance (HOMA-IR) and the homeostasis model assessment of insulin sensitivity (HOMA-IS).

Urine samples were collected over a 24 h period by placing the animals in metabolic cages. Urinary glucose levels were determined using the glucose oxidase method, while albumin levels were assessed using a latex turbidimetry kit. Estimated glomerular filtration rate (eGFR, mL/min) was also calculated.

To measure hepatic and renal glycogen content, samples were boiled in 30% KOH, precipitated with 95% ethanol, and centrifuged at 840× *g* for 25 min. The glycogen was then resuspended, treated with H_2SO_4 and 5% phenol, and incubated for 30 min. The released glucose was measured at 530 nm.

2.5. Glucose Tolerance Test (GTT)

Overnight-fasted rats received an intraperitoneal injection of a 35% glucose solution (2 g/kg body weight), and blood samples were taken from the tail vein at baseline (t = 0) and at 15, 30, 60, 90, and 120 min following glucose administration. Blood glucose levels were measured using a glucometer (LifeScan España, Madrid, Spain). Overall changes in glucose levels during the GTT were assessed by calculating the area under the curve (AUC) above the baseline values.

2.6. Histological and Immunohistochemical Analysis

The liver and kidney were excised, fixed overnight in 4% PFA (dissolved in 0.1 M phosphate buffer pH 7.4), and then embedded in paraffin. Serial sections (4 μm) were mounted on glass slides, hydrated, and stained with Haematoxylin and Eosin (H&E), periodic acid–Schiff (PAS), or Masson's trichrome according to the manufacturer's instructions. To assess histological changes, 10 images of stained sections were acquired under a light microscope using a Leica DM LB2 microscope and a Leica DFC 320 digital camera (Leica, L´Hospitalet del Llobregat, Spain). Morphometric analysis was carried out with ImageJ 1.5.4 software (U. S. National Institutes of Health, Bethesda, MD, USA) and the colour deconvolution plugin. All slides were independently examined by two different researchers under blinded conditions. In the liver, the area occupied by lipid vacuoles in the H&E-stained sections from each rat was assessed across 10 fields and expressed as a percentage of the total surface area of these fields. Hepatic steatosis was graded according to the percentage of hepatocytes containing fat in accordance with the guidelines described by Kleiner et al. [19]. In the kidney, the glomerular and Bowman's space were evaluated. Measurements were taken from the average of 10 cortical fields, or 10 glomeruli cut at the vascular pole for each kidney

Masson's positive staining for fibrosis showing a blue colour to collagen and PAS-positive staining for glycogen or mucopolysaccharide deposits showing magenta-coloured granules in the cytoplasm of hepatocytes and renal tubules were determined using ImageJ 1.5.4 software (U.S. National Institutes of Health, Bethesda, MD, USA), and the colour deconvolution plugin was used to measure the stained area. The total area was determined and expressed as a percentage for statistical analysis.

Immunohistochemical analysis was performed on deparaffinised and rehydrated liver and kidney sections. After retrieving citrate antigen and quenching endogenous peroxidase, sections were incubated overnight at 4 °C with the primary antibodies to detect phospho-Ser (40)-Nrf2 and SIRT1.

The colour reaction was developed using a polymerised horseradish peroxidase-conjugated secondary antibody and counterstained with haematoxylin. Immunostaining intensity for each antibody was quantified using ImageJ 1.5.4 software (U.S. National Institutes of Health, Bethesda, MD, USA). Ten fields per section per rat were selected for analysis. Positive staining intensity was calculated as the percentage ratio of the stained area to the total field examined.

The colour reaction was developed with a polymerised horseradish peroxidase-conjugated secondary antibody and counterstained with haematoxylin. The immunostaining intensity for each antibody was measured using ImageJ 1.5.4 software (U.S. Natio Institutes of Health, Bethesda, MD, USA). For analysis, 10 fields per section per rat w selected and evaluated. The positive staining intensity was calculated as a percentage the stained area relative to the total field analysed.

2.7. Terminal Deoxynucleotidyl Transferase dUTP Nick End Labelling (TUNEL) Assay

In the liver and kidney sections, apoptotic cells were labelled *in situ* by detecting DI fragmentation on paraffin-embedded sections using the TUNEL assay. After permeabil tion and rehydration, sections were incubated with proteinase K (20 µg/mL) for 15 n at 37 °C, followed by treatment with 3% hydrogen peroxide (5 min) to quench endo nous peroxidase activity. Then, slices were incubated with equilibration buffer for 10 n before immediately applying TdT-enzyme for 1 h at 37 °C. Sections were treated w peroxidase-conjugated streptavidin, stained with DAB, and counterstained with met green. The apoptotic index, representing the percentage of cells undergoing apopto was calculated as the ratio of TUNEL-positive cells to the total number of cells coun across 10 consecutive fields using ImageJ 1.5.4 software (U.S. National Institutes of Hea Bethesda, MD, USA).

2.8. Fluorometric Analysis of Caspase-3 Activity

Activity of caspase-3 was analysed as described [20]. In brief, tissues were lysed a caspase-3 activity was assayed in a reaction mixture containing Ac-DEVD-AMC (N-ace Asp-Glu-Val-Asp-7-amino-4-methylcoumarin) as substrate. To determine the enzyma activity, fluorescence was measured at an excitation wavelength of 380 nm and an emiss wavelength of 440 nm.

2.9. Preparation of Hepatic and Renal Lysates and Western Blot Analysis

Frozen liver and kidney samples were homogenised 1:5 (*w*:*V*) in extraction buffer [21, Homogenates were then centrifuged at 14,000× *g* for 60 min. Supernatants were harves protein concentrations were measured using Bradford reagent, and samples were store −80 °C until use.

Equal amounts of protein were separated by SDS-PAGE. Then, membranes w incubated with the appropriate primary antibody, followed by treatment with peroxi conjugated secondary antibodies: anti-rabbit (GE Healthcare, Madrid, Spain) or anti-mo (Sigma, Madrid, Spain). Blots were developed using the ECL system (GE Healthc Madrid, Spain). Western blot normalisation was performed using β-actin or α-tubulin, a band quantification was carried out using a scanner and associated software.

2.10. Determination of ROS, Protein Carbonyl Groups, and GSH Content

ROS were quantified using an assay that detects the oxidation of DCFH to the fluo cent compound dichlorofluorescein (DCF) [21,22]. Briefly, homogenates of liver and re cortex tissues were diluted 1:20 (*V*/*V*) with ice-cold Locke's buffer and then incubated w 5 µM DCFH (30 min, 37 °C). Fluorescence was measured at an excitation wavelengt 485 nm and an emission wavelength of 530 nm.

Protein oxidation in liver and kidney cortex homogenates was quantified by analys carbonyl group content, as previously described [21,22]. Absorbance was measure 360 nm, and the carbonyl content was calculated as nmol/mg protein (extinction coeffic of 22,000 nmol/L/cm). Protein concentrations were analysed with the Bradford reage

GSH levels were measured using the Hissin and Hilf fluorimetric assay, as pr ously described [21,22]. Liver and renal cortex tissues were homogenised (1:20, *w*/*V* 50 mM phosphate buffer pH 7.0. Proteins were precipitated and subsequently centrifu (10,000× *g*, 30 min). The assay was based on the reaction of GSH with OPT at pH 8.0. F rescence was measured at an excitation wavelength of 340 nm and an emission wavele of 460 nm.

2.11. Analysis of GPx, GR, and SOD Activities

For analysis of antioxidant enzyme activity (GPx, GR, and SOD), frozen samples of liver and renal cortex were homogenised (1:5, w/V) in 0.25 M Tris, 0.2 M sucrose, and 5 mM DTT buffer pH 7.4 and centrifuged at $3000\times g$ for 15 min. Determination of GPx activity relies on the oxidation of GSH by GPx, using t-BOOH as a substrate, which is coupled with the depletion of NADPH by GR [21,22]. GR activity was measured by monitoring the decrease in absorbance resulting from the oxidation of NADPH during the reduction of oxidised glutathione [21,22]. SOD activity was analysed using a commercial kit following the manufacturer's instructions [21,22]. The assay relies on SOD's ability to reduce superoxide anions, which is linked to the disappearance of Dojindo's highly water-soluble tetrazolium salt (WST-1) to produce a dye. SOD activity was quantified by measuring the decrease in absorbance at 440 nm. Protein concentration was analysed using the Bradford reagent.

2.12. Statistics

Before conducting statistical analysis, data were assessed for homogeneity of variances using Levene's test. For multiple comparisons, a one-way ANOVA was performed, followed by the Bonferroni test if variances were homogeneous or the Tamhane test if variances were not homogeneous. Statistical significance was determined at $p < 0.05$. Analyses were carried out using SPSS (IBM, Madrid, Spain) version 28.0.

3. Results

3.1. CC-Supplemented Diet Alone or in Combination with Metformin Improves Body Weight, Glucose Tolerance, and Insulin Sensitivity in ZDF Rats

The initial body weight was higher in all diabetic animals compared to the ZL rats, whereas the final body weight gain increased more in the ZL and ZDF-M rats than in all other diabetic animals, which showed similar values to those of the ZDF group (Table S2). The total food intake was lower in ZL rats than in all other ZDF groups; therefore, food efficiency was reduced in all diabetic animals (Table S2). All ZDF animals displayed greater liver and kidney weights than the ZL group, which exhibited the characteristic hepatomegaly and nephromegaly of diabetes (Table 1). This increase was also maintained for the liver-to-body weight ratio in all diabetic animals, whereas for the kidney/body weight ratio, only differences between the ZL and ZDF rats were observed (Table 1).

Table 1. Liver and kidney weights of rats fed standard (ZL and ZDF) and cocoa–carob-rich (ZDF-CC) diets, and metformin-treated rats fed standard (ZDF-M) and cocoa–carob-rich diets (ZDF-CC + M) †.

	ZL	ZDF	ZDF-M	ZDF-CC	ZDF-CC + M
er weight (g)	10.15 ± 1.14 *	18.71 ± 1.49 *	19.68 ± 2.04 *	18.12 ± 1.93 *	14.89 ± 1.25 *
er-to-body weight ratio 00)	2.69 ± 0.25 *	4.77 ± 0.42 *	4.50 ± 0.67 *	4.55 ± 0.60 *	3.63 ± 0.27 *#&$
lney weight (g)	1.36 ± 0.14 *	1.78 ± 0.28 *	1.60 ± 0.22 *	1.70 ± 0.23 *	1.47 ± 0.23 ns
lney-to-body weight ratio (×100)	0.36 ± 0.03 *	0.47 ± 0.09 *	0.38 ± 0.07 ns	0.44 ± 0.05 ns	0.38 ± 0.03 ns

† Data represent the means ± SD. * $p < 0.05$ vs. ZL; # $p < 0.05$ vs. ZDF; & $p < 0.05$ ZDF-M vs. ZDF-M + CC; $ $p < 0.05$ ZDF-CC vs. ZDF-M + CC; ns, not significant.

Glucose and HbA1c levels were reduced by the administration of CC, metformin, or both agents compared to the ZDF animals (Table 2). Interestingly, the combination of CC and metformin achieved levels comparable to the ZL rats for both parameters. Consistently, insulin resistance (HOMA-IR and HOMA-IS indices) increased in the ZDF rats vs. ZL, and all other experimental groups showed intermediate values for both indices compared to their lean and diabetic littermates (Table 2). Overall, these results suggest that the CC-supplemented diet alone or in combination with metformin alleviates glucose intolerance and insulin resistance in ZDF rats, with the combined treatment having a more remarkable effect on glucose control than either CC or the drug alone.

Table 2. Biochemical parameters connected with the metabolism and liver and kidney functior rats fed the standard diet (ZL and ZDF) and treated with metformin (ZDF-M) and fed with cocoa–carob-rich (ZDF-CC) diet and treated with metformin (ZDF-CC + M) †.

	ZL	ZDF	ZDF-M	ZDF-CC	ZDF-CC + M
Serum					
Glucose (mmol/L)	5.34 ± 0.63 *	16.22 ± 1.19 *	8.58 ± 0.76 *#	9.65 ± 1.28 *#	6.42 ± 0.60 #&$
Insulin (µU/mL)	11.26 ± 0.73 *	10.56 ± 0.51 ns	12.68 ± 0.63 ns	14.53 ± 1.81 ns	13.81 ± 0.47 ns
HbA1c (%)	4.32 ± 0.21 *	20.28 ± 1.71 *	6.24 ± 0.23 #	15.75 ± 0.93 *#+	3.58 ± 0.23 #$
AUC glucose (mM/min)	1010 ± 118 *	3025 ± 227 *	1804 ± 419 *#	1758 ± 557 *#	1181 ± 472 *#
HOMA-IR	1.98 ± 0.17 *	8.98 ± 0.82 *	4.06 ± 0.36 *#	5.35 ± 0.57 *#	4.45 ± 0.40 *#
HOMA-IS	365.50 ± 31.83 *	81.65 ± 7.80 *	157.55± 8.42 *#	138.17 ± 13.55 *#	160.24 ± 15.16 *#
ALT (U/L)	71.00 ± 5.81 *	257.38 ± 19.56 *	77.66 ± 8.84 #	213.45 ± 10.83 *#+	71.50 ± 8.97 #$
AST (U/L)	240.00 ± 10.58 *	695.51 ± 38.27 *	228.57 ± 22.86 #	607.35 ± 28.36 *#+	241.50 ± 12.84 #$
Creatinine (mg/dL)	0.20 ± 0.01 *	0.33 ± 0.01 *	0.28 ± 0.02 *#	0.33 ± 0.03 *+	0.30 ± 0.02 *
Urine					
Glucose (mmol/24 h)	1.78 ± 0.20 *	147.42 ± 9.26 *	85.94 ± 16.30 *#	136.37 ± 11.51 *+	1.87 ± 0.32 #&$
Albumin (mg/24 h)	4.13 ± 0.25 *	72.34 ± 7.13 *	42.60 ± 4.69 *#	38.64 ± 6.69 *#	17.76 ± 3.67 *#&$
eGFR (mL/min)	0.48 ± 0.08 *	1.25 ± 0.11 *	0.66 ± 0.10 #	0.99 ± 0.10 *#+	0.45 ± 0.04 #$

† Data represent the means ± SD. * $p < 0.05$ vs. ZL; # $p < 0.05$ vs. ZDF; + $p < 0.05$ ZDF-M vs. ZDF-CC; & $p <$ ZDF-M vs. ZDF-M + CC; $ $p < 0.05$ ZDF-CC vs. ZDF-M + CC; ns, not significant.

3.2. CC-Supplemented Diet, Metformin, and Their Combination Ameliorate Hepatic and Rena Dysfunction in ZDF Rats

To further explore the effects of the CC-supplemented diet, alone or in combinati with metformin, on the liver and kidney of diabetic rats, hepatic ALT and AST, as well serum and urinary parameters associated with the renal function, were assayed.

Hepatic ALT and AST levels were elevated in the ZDF rats compared to the ZL a diabetic animals receiving metformin alone or in combination with the CC-supplemen diet, whilst those fed with CC alone showed values intermediate to those of the ZL a ZDF rats (Table 2). Regarding renal function, serum creatinine increased in the Z animals compared to the ZL rats, whereas it decreased in the diabetic groups treated w metformin alone or in combination with the CC-supplemented diet (Table 2). Urine gluc decreased in metformin-treated animals compared to all diabetic groups and reach levels comparable to those of the ZL group when it was administrated in combinat with the CC-supplemented diet (Table 2). Similarly, protein excretion decreased in diabetic animals treated with CC and/or metformin compared to the ZDF rats, and group receiving both agents showed the closest values to the lean littermates (Table 2). addition, the increased eGFR in the ZDF rats was improved in the ZDF-CC group, a displayed values lower than those of the ZDF animals but higher than those of the ZL metformin-treated diabetic rats (ZDF-M and ZDF-M + CC rats) (Table 2). All these chan induced by CC supplementation alone or in combination with metformin suggest that CC-supplemented diet and the antidiabetic drug might contribute to alleviate hepatic a renal dysfunction in the ZDF rats.

3.3. CC-Supplemented Diet Alone or in Combination with Metformin Alleviates Hepatic and Re Cortex Morphological Alterations in ZDF Rats

Diabetes is accompanied by clinical and morphological alterations in the liver a kidney, and to gain further insight into the potential protective effects of CC and, metformin on structural changes, histological analyses were performed using H&E, P and Masson's trichrome staining in the liver and renal cortex sections.

The ZL rats displayed a normal hepatic architecture, characterised by typical hep lobules with a thin-walled central vein and hepatic cords extending toward the peri ery, alternating with hepatic sinusoids (Figure 1A). The ZDF rats showed a signet-ri hepatocyte morphology due to extensive fat accumulation occupying between 34–66 % the liver parenchyma (steatosis score = 2.7), which appeared as focal pericentral ma and micro-vesicular fatty degeneration with the vacuolization of liver cells. In addition, the mild periportal interstitial mononuclear inflammatory infiltrates and enlargement of sinusoids affecting the centrilobular zone with a progressive loss of general organ structure were observed (Figure 1A). However, all other animal groups showed a notable restoration of normal hepatic architecture, with a decrease in the accumulation of fat droplets within the hepatocyte cytoplasm with diffuse macro-vesicular fat droplets (steatosis scores: ZDF-M = 0.9, ZDF-CC= 1.4, ZDF-CC + M= 0.6) and the regeneration of the hepatic parenchyma (Figure 1A).

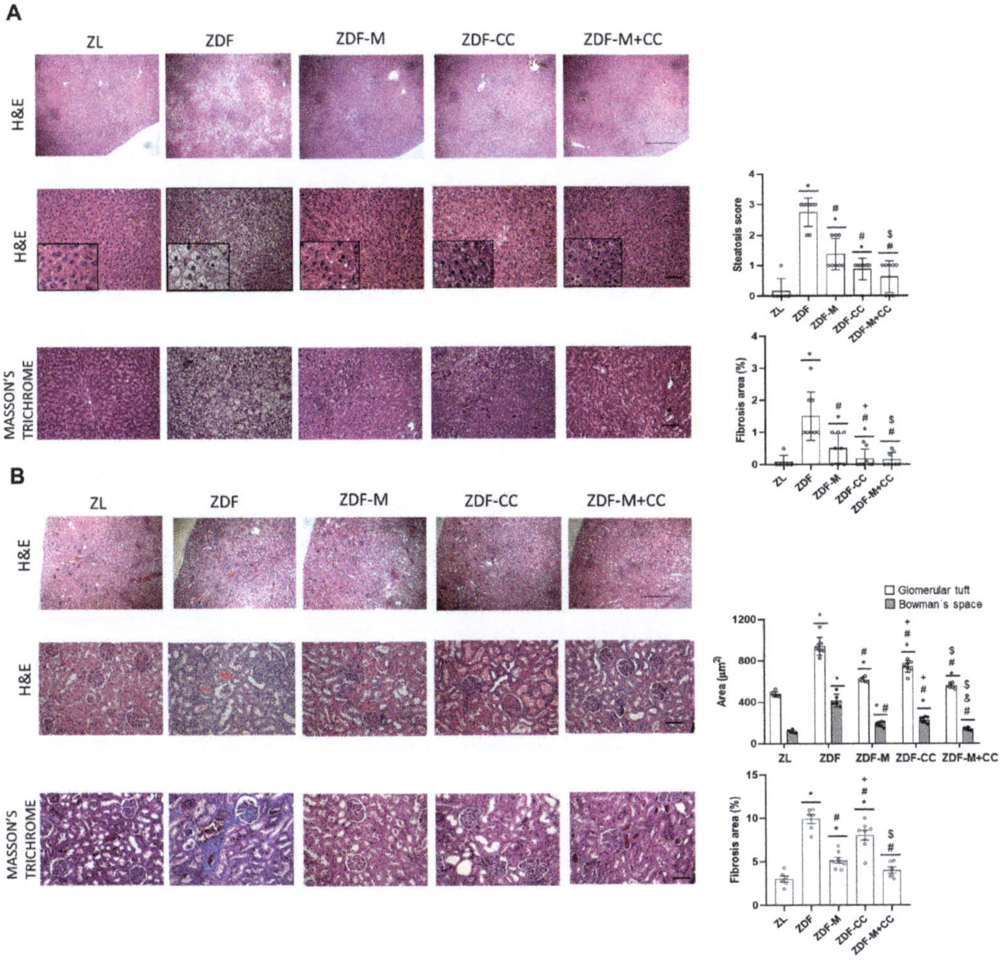

Figure 1. Effect of the cocoa–carob diet (CC), metformin (M), and their combination on the structure of the liver and renal cortex. (**A**) Representative hepatic tissue sections stained with H&E and Masson's positive staining, and representations of the steatosis grade and percentage of fibrosis area. (Scale bar: 250 μm, magnification 4×; scale bar: 50 μm, magnification 20×; inserts 20 μm, magnification 40×.) (**B**) Representative renal cortical sections stained with H&E and Masson's positive staining, and quantification of the glomerular tuft, Bowman's space, and percentage of fibrosis areas. (Scale bar: 250 μm, magnification 4×; scale bar: 50 μm, magnification 20×.) Values are depicted as mean ± SD (n = 6–8). * $p < 0.05$ vs. ZL; # $p < 0.05$ vs. ZDF; + $p < 0.05$ ZDF-M vs. ZDF-CC; & $p < 0.05$ ZDF-M vs. ZDF-M + CC; $ $p < 0.05$ ZDF-CC vs. ZDF-M + CC.

As in the liver, the ZL rats showed a normal architecture of both renal corpuscles a tubules, whilst the ZDF animals exhibited an extensive pattern of renal disease characteris by glomerular, tubular, and renal interstitial injury. As illustrated in Figure 1B, the re cortex of the ZDF rats showed glomerular hypertrophy and focal glomerulosclero characterised by the regional adhesion of the enlarged glomerular tuft to Bowman's caps associated with the expansion of the mesangial matrix and capillary filling and widening Bowman's space. Increased deposits of mucopolysaccharide and hyaline substances w also found in the cortex proximal tubules. In the ZDF rats, the renal medulla exhibi tubulointerstitial tissue with distended tubules covered by atrophic epithelial cells fil with protein casts and accompanied by increased inflammatory infiltrate around the tubu The intrarenal arterial vessel showed a mild thickening of the walls and lipid dropl inside the capillaries and collagen deposition with hyalinisation. These alterations w largely prevented when the ZDF animals were supplemented with the CC diet and/ administered with metformin (Figure 1B).

In addition, although fibrosis was not very prominent in any animal, it was enhan in the liver and kidney of the ZDF rats vs. ZL (Figure 1A,B). However, compared to the Z animals, rats receiving the CC diet and/or metformin showed ameliorated fibrotic dam (Figure 1A,B). Taken together, this may indicate that CC intake could contribute to minim the adverse morphological hepatic and renal lesions in ZDF animals. Interestingly, combination of diet and metformin treatment had a greater effect in preventing structu changes compared to the diet or drug alone.

3.4. CC-Supplemented Diet, Metformin, and Their Combination Prevent Apoptosis in the Live and Renal Cortex of ZDF Rats

Diabetes could lead to apoptosis in the liver and kidney. Therefore, to gain insight i the molecular mechanisms of the protective effect of CC and/or metformin administrat in the liver and renal cortex of the ZDF rats, apoptosis was analysed by TUNEL as Also, caspase-3 activity was measured, and key anti and pro-apoptotic Bcl-2 proteins w evaluated by Western blot.

Histological sections showed a pro-apoptotic effect in the hepatocytes and renal corti tubular cells of the ZDF animals, as assayed by TUNEL, whilst CC and/or metforn consumption remarkably prevented these changes (Figure 2A). TUNEL-positive cells w detected in the hepatocytes, and were scarcely observed in the glomeruli, with the majo of apoptotic cells localised in the renal cortical tubules (Figure 2A). In addition, the li and renal cortex of the ZDF rats showed increased values of Bax with a concomit decrease in Bcl-x_L levels (Figure 2B–E). However, in both tissues, supplementation w the CC diet and/or metformin administration reduced Bax values and raised Bcl-x_L lev (Figure 2B–E). Interestingly, Bcl-x_L levels were restored to control values (ZL animals the kidney of all treated ZDF rats and in the liver of the ZDF-CC + M group (Figure 2B-

Similarly, caspase-3 activity was increased in the liver and renal cortex of the Z group; in both tissues, these values were comparable to those of the ZL rats when anim received metformin or the CC-supplemented diet or their combination, except for liver in the ZDF-CC group, which showed levels intermediate to those of their lean a diabetic littermates (Figure 2F,G). These results suggest that the CC-supplemented d and metformin may also contribute to hepatic and renal protective effects by prevent increased apoptosis in both tissues in diabetic animals.

3.5. CC-Supplemented Diet Alone or in Combination with Metformin Ameliorate the Hepatic Renal Cortex Redox Imbalance in ZDF Rats

Oxidative stress is a major contributor to the induction of apoptosis in the liver a kidney in diabetes. To continue studying the protective effects of CC supplementat and metformin (alone or combined) in the liver and kidney of the ZDF rats, marker oxidative stress (ROS and carbonyl groups), key antioxidant defences (GSH, GPx, and SOD), and phosphorylated levels of Nrf2 and SIRT-1, which contribute to counte oxidative stress and regulate the antioxidant response, were evaluated.

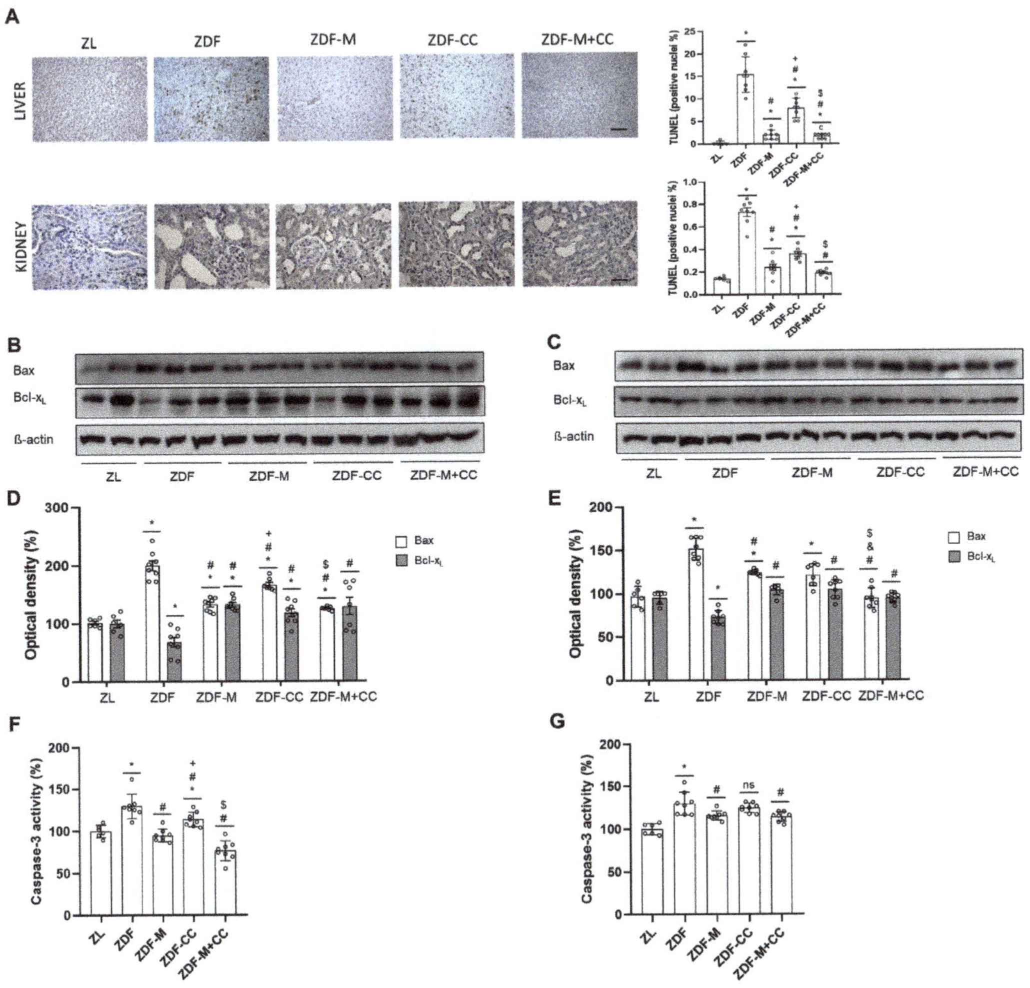

Figure 2. Effect of the cocoa–carob diet (CC), metformin (M), and their combination on apoptosis in the liver and renal cortex. (**A**) Representative TUNEL-positive hepatic and renal cells (scale bar: 20 μm, magnification 40×) and apoptotic index expressed as the percentage of positive nuclei relative to the total nuclei in the liver and renal cortex. Representative bands of Western blot analyses (**B**) in the liver and (**C**) in the renal cortex. Percentage levels of (**D**) hepatic and (**E**) renal cortex Bax and Bcl-x_L relative to the ZL animals (means ± SD, n = 6–8). Caspase-3 activity (**F**) in the liver and (**G**) renal cortex expressed as the percent of controls (ZL values). Values are shown as means ± SD of n = 6–8 animals. * $p < 0.05$ vs. ZL; # $p < 0.05$ vs. ZDF; + $p < 0.05$ ZDF-M vs. ZDF-CC; & $p < 0.05$ ZDF-M vs. ZDF-M + CC; $ $p < 0.05$ ZDF-CC vs. ZDF-M + CC; ns, not significant.

In the liver and renal cortex, ROS and protein carbonyl values were augmented in the ZDF rats when compared to all other experimental groups, whereas both parameters showed intermediate levels to those of the ZL and ZDF animals in the liver of all other diabetic rats (Figure 3A) and were similar to those of the ZL rats in the kidneys of both CC-fed groups (Figure 3B). In concert, GSH content was reduced in the ZDF animals compared to the ZL group in both tissues, but this effect was partially or completely suppressed in the liver and kidney of animals receiving the CC-supplemented diet (Figure 3A,B).

Furthermore, in the liver, all enzymatic activities were decreased in the ZDF rats compar to the lean group, whereas GPx and SOD activities were restored to the ZL values in other groups of animals, whereas GR activity was only partially recovered. Similarly, in renal cortex, SOD activity was reduced in the ZDF vs. ZL rats, but its activity was partia improved with all the treatments, although both the CC-fed animal groups showed higl values than those receiving metformin alone (Figure 2D). Cortical renal GR activity w not modified in any group of rats, while GPx activity was enhanced in the ZDF anim compared to the ZL rats, being its activity partially or totally restored to control values (animals) in ZDF-CC and ZDF-CC + M, respectively (Figure 3D).

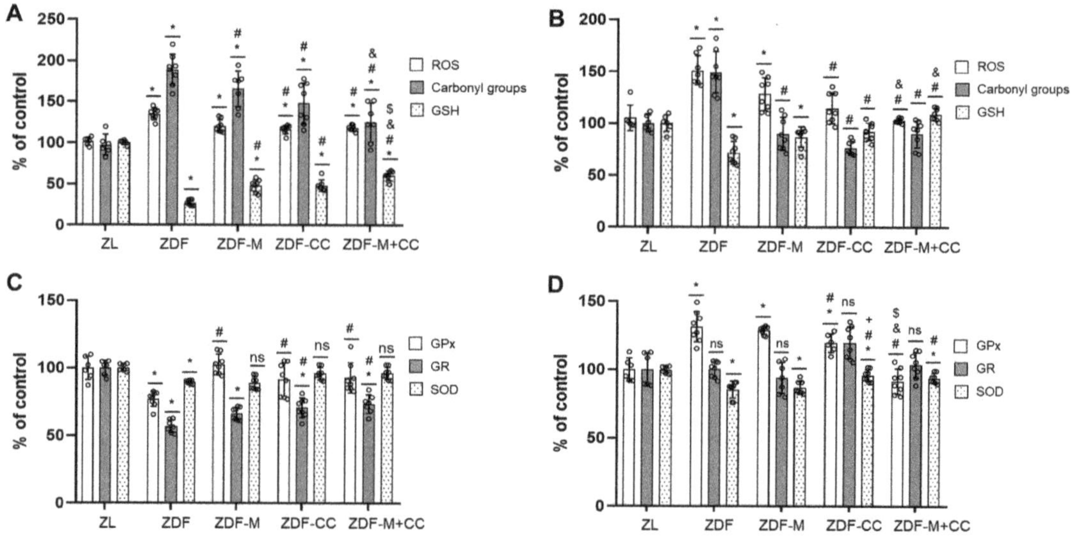

Figure 3. Effect of the cocoa–carob diet (CC), metformin (M), and their combination on the m antioxidant defences in the liver and renal cortex. Percentage of ROS levels, carbonyl groups, a GSH content (**A**) in the liver and (**B**) in the renal cortex relative to the ZL rats. Percentage of activity of antioxidant enzymes GPx, GR, and SOD (**C**) in the liver and (**D**) in the renal cortex rela to the ZL animals. Data are shown as means ± SD of n = 6–8 animals. * $p < 0.05$ vs. ZL; # p < (vs. ZDF; + $p < 0.05$ ZDF-M vs. ZDF-CC; & $p < 0.05$ ZDF-M vs. ZDF-M + CC; $ $p < 0.05$ ZDF-CC ZDF-M + CC; ns, not significant.

As shown in Figure 4, the liver and renal cortex of the ZDF rats showed the low levels of p-Nrf2 and SIRT-1 of all the animal groups, whereas rats fed with the CC diet al exhibited intermediate levels of p-Nrf2 and SIRT-1 compared to their lean and diab littermates. Interestingly, as with most of the other parameters related to redox status both tissues, the animals with values closer to or equal to those of the ZL group were receiving the CC-supplemented diet and metformin (Figures 3 and 4).

Altogether, it is suggested that the CC-supplemented diet and metformin could c tribute to restore the redox balance of the liver and renal cortex of the ZDF rats by modu ing relevant antioxidant defences and closely related proteins, with the combined treatm being more effective than either the diet or drug alone.

3.6. CC-Supplemented Diet, Metformin, and Their Combination Improve Insulin Signalling i ZDF Rat Liver and Renal Cortex

The regulation of the insulin signalling pathway in the liver and kidney contribute the maintenance of glucose homeostasis. In view of the results obtained, it was investiga whether supplementation with CC alone or in combination with metformin modulates insulin route in the liver and renal cortex; thus, the phosphorylated and total levels of the first limiting step of this pathway, the levels of IR, and glycogen content were assayed by Western blot.

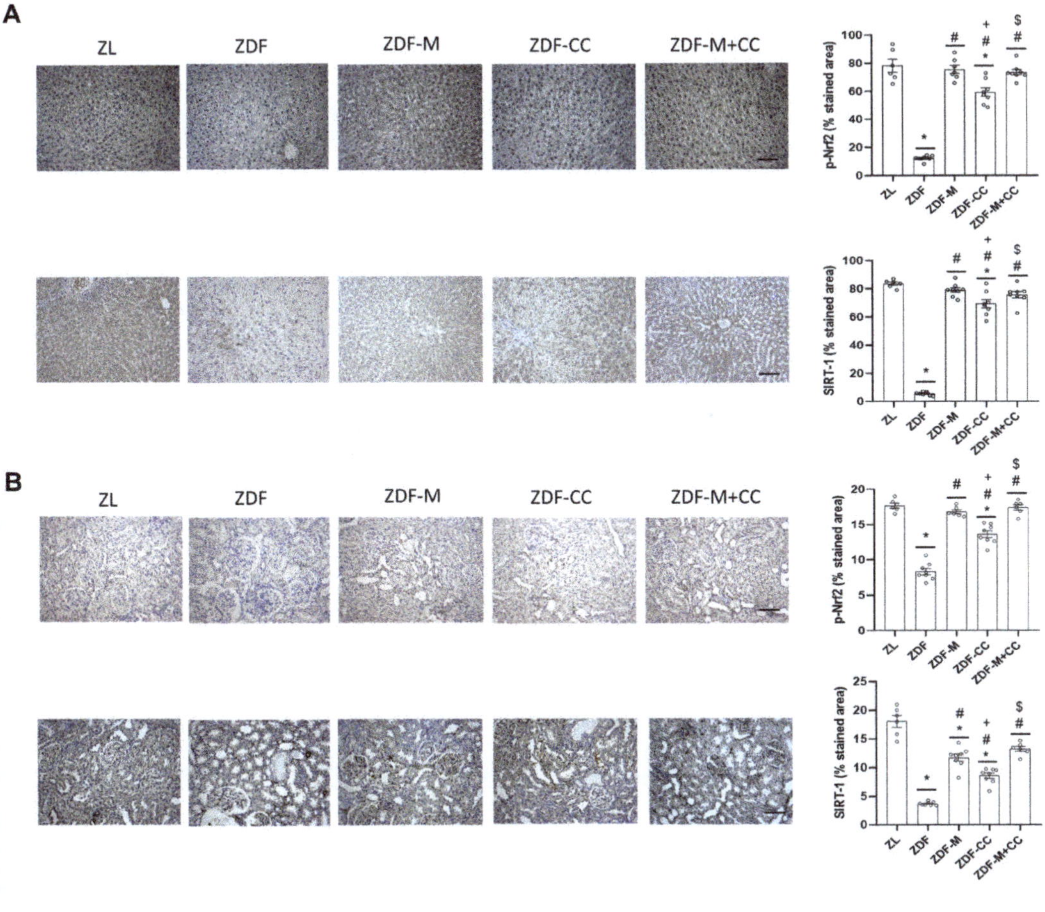

Figure 4. Effect of the cocoa–carob diet (CC), metformin (M), and their combination on levels of p-Nrf2 and SIRT-1 in the liver and renal cortex. Representative images of p-Nrf2 and SIRT-1 (scale bar: 50 µm, magnification 20×) and percentage of the stained area (**A**) in the liver and (**B**) in the renal cortex. Data are depicted as mean ± SD (n = 6–8). * $p < 0.05$ vs. ZL; # $p < 0.05$ vs. ZDF; + $p < 0.05$ ZDF-M vs. ZDF-CC; $ $p < 0.05$ ZDF-CC vs. ZDF-M + CC.

Phosphorylated-(Tyr) IR levels are associated with an early response to insulin stimulation and were decreased in the ZDF rats compared to all other experimental groups in the liver and renal cortex (Figure 5A–D). In contrast, CC and/or metformin administration contribute to restore p-IR values in both tissues, with the ZDF-CC + M animals showing the highest levels or even similar levels to those of the ZL group in the renal cortex and liver, respectively (Figure 5A–D). Total insulin receptor (IR) levels were similar in all animal groups in both tissues (Figure 5A–D).

IR activation may contribute to the modulation of glycogen content. As shown in Figure 5E,F, hepatic glycogen levels decreased in the ZDF rats compared to all other animal groups, and its content was completely restored to ZL values in the ZDF-CC + M rats, in agreement with the increased p-(Tyr) IR levels. However, glycogen content was only partially restored in the livers of animals receiving metformin or CC alone (Figure 5E On the contrary, an increase in the renal glycogen content was observed in the ZDF r whereas the ZDF-CC + M rats had values similar to those of the ZL group, and the ZI M and ZDF-CC animals showed values intermediate to those of their lean and diabe littermates, as in the liver (Figure 5E,G).

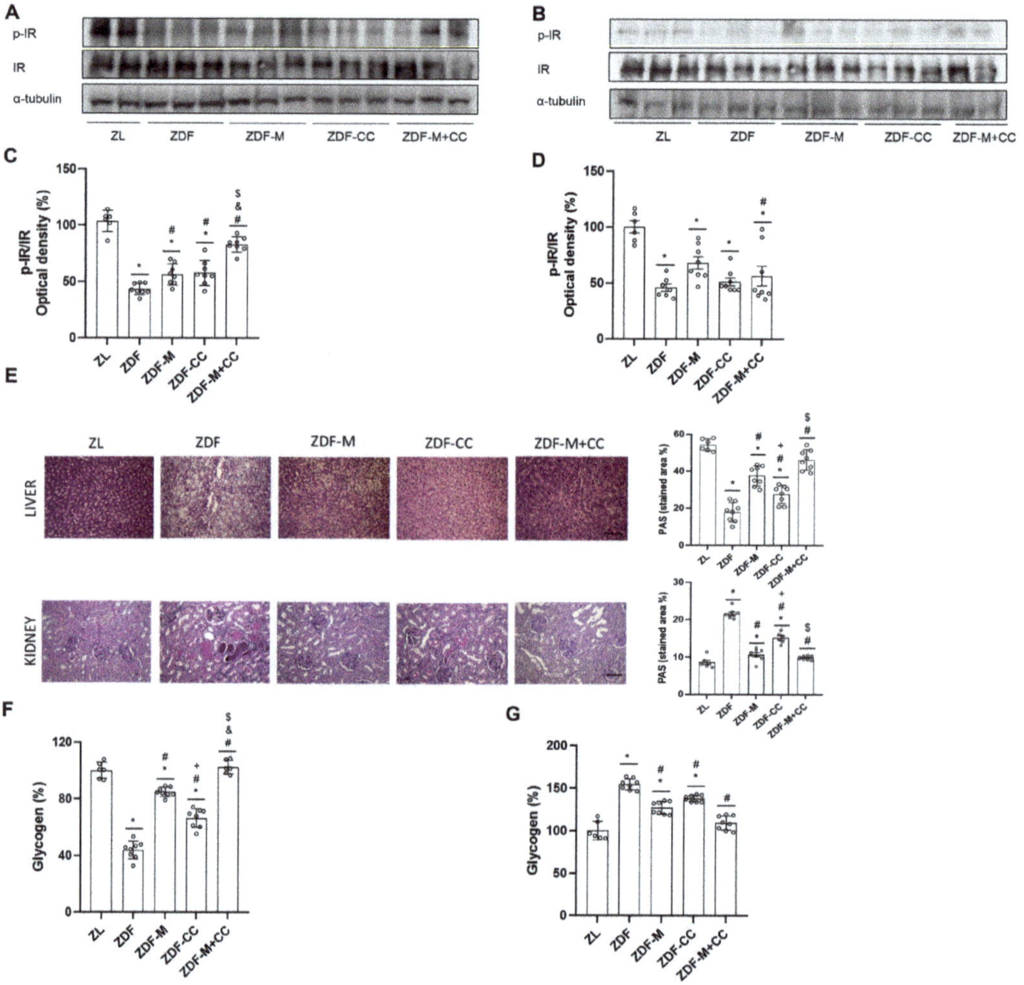

Figure 5. Effect of the cocoa–carob diet (CC), metformin (M), and their combination on ins signalling in the liver and renal cortex. Bands of representative experiments for the phosphoryla and total levels of IR (**A**) in the liver and (**B**) in the renal cortex. Densitometric quantificatio p-(Tyr)-IR and IR (**C**) in the liver and (**D**) in the renal cortex. Values (means ± SD) are expressed percentage relative to the ZL rats (n = 6–8). (**E**) Representative hepatic and renal cortex tissue secti stained with PAS (scale bar: 50 μm, magnification 20×) and the percentage of stained areas. Data shown as mean ± SD (n = 6–8). Percentage of glycogen content in the (**F**) liver and (**G**) renal co Values are expressed as a percentage relative to the ZL rats (means ± SD, n = 6–8). * $p < 0.05$ vs. # $p < 0.05$ vs. ZDF; + $p < 0.05$ ZDF-M vs. ZDF-CC; & $p < 0.05$ ZDF-M vs. ZDF-M + CC; $ $p <$ ZDF-CC vs. ZDF-M + CC.

Taken together, this suggests that in the liver and renal cortex, the CC-supplemented diet, and metformin prevent the blockade of the insulin route observed in the ZDF rats by regulating this pathway at the earliest step and contributing to the regulation of the glycogen content.

3.7. CC-Supplemented Diet Alone or in Combination with Metformin Regulates the Rate-Limiting Proteins of Gluconeogenesis and Transport of Glucose in the Liver and Renal Cortex of ZDF Rats

Increased gluconeogenesis and altered glucose uptake and reabsorption have been reported in the diabetic liver and renal cortex. To further investigate glucose homeostasis, key proteins involved in glucose production (PEPCK) and transport (GLUT-2) were analysed by Western blots in both tissues.

PEPCK levels augmented in the liver and renal cortex of the ZDF animals were compared to the lean group, and this effect was partially reversed in animals receiving the CC-supplemented diet and/or metformin (Figure 6). In addition, hepatic GLUT-2 values decreased in the ZDF animals compared to the ZL littermates, while rats fed the CC-supplemented diet and/or treated with metformin showed levels comparable to those of the ZL rats (Figure 6A,C). On the other hand, in the renal cortex, GLUT-2 levels were similar in the ZL and diabetic animals receiving the CC-supplemented diet and/or metformin, but the levels of this transporter were increased in the ZDF group (Figure 5A–C). All this suggests that the CC-supplemented diet alone or in combination with metformin may also help to preserve liver and kidney function and modulate glucose homeostasis in ZDF rats.

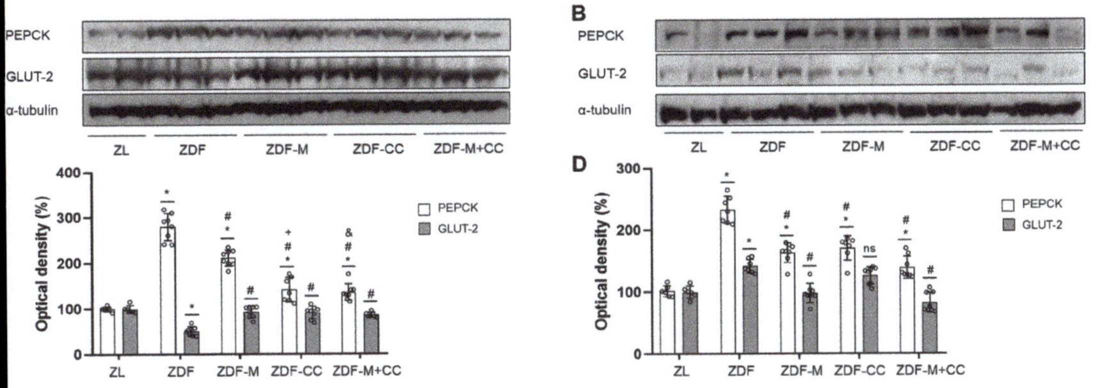

Figure 6. Effect of the cocoa–carob diet (CC), metformin (M), and their combination on levels of PEPCK and GLUT-2 in the liver and renal cortex. Bands of representative experiments (**A**) in the liver and (**B**) in the renal cortex. Densitometric quantification of PEPCK and GLUT-2 levels (**C**) in the liver and (**D**) in the renal cortex. Values (means ± SD) are expressed as a percentage relative to the ZL animal group (n = 6–8). * $p < 0.05$ vs. ZL; # $p < 0.05$ vs. ZDF; + $p < 0.05$ ZDF-M vs. ZDF-CC; & $p < 0.05$ ZDF-M vs. ZDF-M + CC; ns, not significant.

4. Discussion

Policies to prevent and treat diabetes and its associated complications are receiving increasing attention as morbidity and mortality rates continue to rise [1]. Different antidiabetic drugs are available, but their side effects and the occurrence of drug resistance are a concern. Flavonoids have been shown to have protective effects against diabetes due to their antioxidant and antidiabetic properties, among other beneficial activities [8,9]. In addition, a growing body of literature points out the synergistic effects of combined flavonoid administration [23,24]. However, there is a lack of evidence on the benefits of combining these natural compounds, especially with antidiabetic drugs, in the diabetic milieu in key organs such as the liver and kidneys. The present study demonstrates, for the first time … protective effect of a mixture of cocoa and carob (CC) and a more promising effect in combination w metformin, the first-line drug used in diabetes.

In diabetes, suppressed body weight gain has been related to an increased ener loss [25], as shown in this study. However, the absolute liver and kidney weights, and the ratio of these organ weights to the body weight increased in the diabetic animals (hepa and nephromegaly), whereas these parameters decreased when the animals receiv CC and/or metformin (Table 1). These results agree with previous reports showing improvement in the diabetic condition by reducing weight loss and decreasing hepatic a renal weights in animals receiving polyphenol-rich extracts [24,25].

Diabetes also causes changes in the structure and function of the liver and k neys [12,26,27] and importantly, at present, it is widely accepted that the alterations in c of these organs impair the function of the other [2]. In this regard, reduced renal functi has been associated with the progression of liver fibrosis in diabetic patients [28]. Th in the present study, all the recoveries in histological changes could be related to the i provement in hyperglycaemia and could be associated with the attenuation of weight lo reduced hepato- and nephromegaly, as well as with the alleviation of renal hyperfiltrati which in turn is connected with reduced levels of serum creatinine and urinary glucose a albumin [26,27]. In fact, improved glucose homeostasis delayed the progression of diabe and its complications, including nephropathy [2,29], which is in line with studies suppo ing that various hepatic and renal alterations can be reversed at a certain point by go glycaemic control [27,30,31]. Overall, the results obtained here are consistent with previo studies showing that extracts rich in flavonoids or pure phenolic compounds, as well the drug metformin alone, have protective effects against diabetes-induced functional a structural alterations in the liver and kidneys of diabetic rodents [25,26,32,33]. Howeve is worth noting that none of these studies evaluated the effects of combining a mixture polyphenol-rich foods with any antidiabetic drug, including metformin, on the liver a kidneys in type 2 diabetes, and most of them did not study both organs.

Chronic hyperglycaemia-mediated apoptosis in the liver and kidney contributes the dysfunction of both organs in diabetes [29,34]. However, various pure polyphen and flavonoid-rich extracts, as well as metformin, have been shown to attenuate apopto cell death in the liver and kidney, thus contributing to the alleviation of the diabe milieu [21,26,27], in agreement with the present results.

Oxidative stress induced by chronic hyperglycaemia is closely related to apoptosis a is an important pathophysiological mechanism underlying the structural and functio changes in the liver and kidney in diabetes [5,29,34]. In fact, hyperglycaemia increa glucose oxidation and the release of large amounts of ROS, leading to cellular oxidat damage that can anticipate to liver and kidney injury [26]. In this study, the administrat of CC and/or M to the diabetic animal groups enhanced several markers of oxidat stress, i.e., ROS generation, the oxidation of proteins, and altered antioxidant defen In addition, the levels of redox-sensitive proteins Nrf2 and SIRT-1 were enhanced by supplementation and/or M administration in comparison to the ZDF rats. Consisten several studies have demonstrated the ability of flavonoid-rich extracts and diets, as w as pure phenolic compounds, alone or combined with each other or with metformin prevent hepatic and renal oxidative stress by activating antioxidant defences via Nrf2 SIRT-1, raising the antioxidant capacity against oxidative insult and contributing to maintenance of redox balance [21–23,35], as shown here. Moreover, it is worth not that the reduction in hyperglycaemia mediated-oxidative stress could also contribute the amelioration of fibrosis and apoptosis in both organs [12,25], which is in line with present study. Interestingly, the exploration of oxidative stress-related molecular pathwa such as endoplasmic reticulum stress, whichplays a major role on main altered biologi processes in diabetes, is challenging and deserves further investigation.

Alterations in insulin signalling, gluconeogenesis, and glucose uptake in key gans critically involved in glucose homeostasis, such as the liver and kidneys, occur diabetes [4]. Thus, chronic hyperglycaemia impairs hepatic and cortical renal insu signalling at the early steps, i.e., IR recruitment and activation are inhibited [4,29], as shown in the current study. However, in agreement with the present results, it has been reported that the administration of a cocoa-rich diet, metformin, and flavonoid-rich extracts from mulberry leaves or basil ameliorated insulin resistance by increasing the total or tyrosine-phosphorylated levels of IR in the liver or renal cortex of diabetic rodents [26,30,33]. Importantly, insulin resistance has been associated with liver fibrosis and albuminuria in diabetic patients [36,37]. Similarly, the administration of CC and/or metformin improved insulin sensitivity and alleviated albuminuria and fibrosis in the liver and kidneys of the ZDF rats.

Activation of insulin signalling can lead to increased glycogen synthesis for energy storage and the suppression of glucose output by inhibiting PEPCK [2,4,29]. Thus, restoration of hepatic glycogen content has previously been shown in the liver of diabetic animals receiving metformin, flavonoid-rich foods, or mixtures of polyphenols [23,31,38], as presented here. Interestingly, the administration of flavanol-rich foods prevented glycogen accumulation in the renal tubules of diabetic rats, which is in agreement with the current study [30,32]. This differential regulation of the glycogen content in the liver and renal cortex of diabetic animals has been associated with the progression of injury induced by the disease [39]. In addition, it should be highlighted that the glycogen content in the renal cortex, as assessed here, has been scarcely studied, and that increased levels of hepatic glycogen have been reported in rats in the pre-diabetic period [31], which decrease in the later stages of the disease [38]. Taken together, CC supplementation alone or with metformin could improve both hepatic and renal insulin resistance and contribute to glucose homeostasis by modulating the glycogen content of both tissues in ZDF rats.

Another mechanism to regulate glucose homeostasis is the modulation of glucose production. Only the liver and renal cortex are capable of gluconeogenesis, and PEPCK plays a crucial role in this process as a rate-limiting enzyme [2,3]. In this line, in diabetes, low levels of this key enzyme in the liver and renal cortex could contribute to the inhibition of gluconeogenesis in both tissues, thus reducing hyperglycaemia in diabetic rats, which is in line with the present results [24,30,31]. Furthermore, it has been suggested that reduced levels of IR in hepatocytes and renal proximal tubules contribute to the maintenance of high glucose levels through enhanced gluconeogenesis [4,40]. Therefore, the current results suggest that CC and/or metformin may also contribute to a reduction in hyperglycaemia by regulating gluconeogenic PEPCK enzyme levels in the ZDF rats, with insulin signalling playing a role in this process.

Glucose uptake also contributes to the maintenance of glucose homeostasis [2,4]. Indeed, impaired glucose uptake in the liver and peripheral tissues leads to hyperglycaemia in diabetic patients [41]. Thus, reduced hepatic levels of GLUT-2 have been demonstrated in ZDF rats [42], as shown here, but the transporter levels were restored to the control values (ZL group) in ZDF animals receiving CC and/or metformin. Consistently, pure phenolic compounds, meat enriched with carob fruit extract, and antidiabetic drugs, including metformin, increased hepatic GLUT-2 levels [11,43], suggesting the normalisation of post-receptor insulin signalling and the restoration of hepatic insulin sensitivity. Similarly, the kidney also plays a relevant role in glucose homeostasis [29], although glucose transporter modulation in this organ has been disregarded. In the current study, upregulated levels of GLUT-2 were detected in the renal cortex of the ZDF animals, as previously shown [42]. This alteration contributes to the aggravation of hyperglycaemia and insulin resistance and has been associated with the accumulation of glycogen in the renal tubules [32], which is in agreement with the present results. Interestingly, a cocoa-rich diet and phlorizin supported the restoration of renal GLUT2 levels in diabetic rodents, demonstrating a renoprotective effect [30,44], as shown here. Therefore, the improved glucose homeostasis and insulin sensitivity in diabetic animals could also be related to the modulatory effect of CC and/or metformin on the hepatic and renal GLUT-2 content.

5. Conclusions

This study demonstrates that a cocoa–carob-rich diet alleviates hepatorenal dysfu tion in ZDF diabetic rats by preventing structural alterations, including fibrosis, ultimat improving glucose homeostasis. These benefits were related to the alleviation of insu resistance and to the modulation of the glucose transporter GLUT-2 and PEPCK glu neogenic enzyme levels. In addition, the CC-rich diet promoted redox balance through restoration of antioxidant defences, thereby mitigating oxidative stress and apoptosis. portantly, several of these benefits were further improved by the combination of a CC-r diet and metformin compared to either agent alone, demonstrating the efficacy of comb ing natural bioactive compounds with antidiabetic drugs in the treatment of diabetes, a also highlighting the potential of the CC-rich diet against hepatorenal complications alc or as an adjuvant therapy in the diabetic milieu.

Supplementary Materials: The following supporting information can be downloaded at: ht //www.mdpi.com/article/10.3390/nu16183087/s1. Supplementary Table S1: composition of experimental control (C) and cocoa–carob-rich (CC) diets (g/kg dry weight). Supplementary Table body weight data, food intake, and food efficiency of rats fed with standard (ZL and ZDF) a cocoa–carob-rich (ZDF-CC) diets, and treated with metformin and fed with standard (ZDF-M) a cocoa–carob-rich diets (ZDF-CC + M).

Author Contributions: Conceptualization, S.R.; data curation, E.G.-D., M.Á.M. and S.R.; for analysis, E.G.-D., M.E.L.-O., M.Á.M. and S.R.; funding acquisition, J.P.-J., M.Á.M. and S.R.; meth ology, M.E.L.-O., J.P.-J., M.Á.M. and S.R.; investigation, M.E.L.-O., J.P.-J., M.Á.M. and S.R.; pro administration, M.Á.M. and S.R.; resources, M.E.L.-O., J.P.-J., M.Á.M. and S.R.; supervision, S writing—original draft, S.R.; writing—review and editing, E.G.-D., M.E.L.-O., J.P.-J., M.Á.M. and All authors have read and agreed to the published version of the manuscript.

Funding: This research was funded by the grant RTI2018-095059-B-I00, funded by MCIN/AEI/10.130 501100011033/ and by "ERDF A way of making Europe". E.G.-D. was the recipient of a contract fr Comunidad de Madrid (PEJ-2020-AI/BIO-18529).

Institutional Review Board Statement: All the experiments were conducted according to the E pean Union (2010/63/EU) and Spanish (RD 53/2013) legislation, and with the approval (16 N 2019) of the Animal Care and Use Committee of the Community of Madrid (PROEX 079/19).

Informed Consent Statement: Not applicable.

Data Availability Statement: Data are available upon request to the authors. The data are publicly available due to principle of confidentiality.

Conflicts of Interest: The authors declare no conflicts of interest.

References

1. International Diabetes Federation. *IDF Diabetes Atlas*, 10th ed.; International Diabetes Federation: Brussels, Belgium, 2 Available online: https://www.diabetesatlas.org (accessed on 9 August 2024).
2. Artunc, F.; Schleicher, E.; Weigert, C.; Fritsche, A.; Stefan, N.; Häring, H.U. The impact of insulin resistance on the kidney vasculature. *Nat. Rev. Nephrol.* **2016**, *12*, 721–737. [CrossRef]
3. DeFronzo, R.A.; Ferrannini, E.; Groop, L.; Henry, R.R.; Herman, W.H.; Holst, J.J.; Hu, F.B.; Kahn, C.R.; Raz, I.; Shulman, G.I.; e Type 2 diabetes mellitus. *Nat. Rev. Dis. Primers* **2015**, *1*, 15019. [CrossRef]
4. Petersen, M.C.; Shulman, G.I. Mechanisms of insulin action and insulin resistance. *Physiol. Rev.* **2018**, *98*, 2133–2223. [CrossR
5. Rochette, L.; Zeller, M.; Cottin, Y.; Vergely, C. Diabetes, oxidative stress and therapeutic strategies. *Biochim. Biophys. Acta* **2** *1840*, 2709–2729. [CrossRef]
6. Arabshomali, A.; Bazzazzadehgan, S.; Mahdi, F.; Shariat-Madar, Z. Potential benefits of antioxidant phytochemicals in t 2 diabetes. *Molecules* **2023**, *28*, 7209. [CrossRef]
7. Sharma, R.; Tiwari, S. Renal gluconeogenesis in insulin resistance: A culprit for hyperglycemia in diabetes. *World J. Diabetes* **2** *12*, 556–568. [CrossRef]
8. Chen, X.; Guan, X.; Tang, Y.; Deng, J.; Zhang, X. Effects of cocoa products intake on cardiometabolic biomarkers of type 2 diab patients: A systematic review and meta-analysis based on both long-term and short-term randomised controlled trials. *Int. J. F Sci. Nutr.* **2022**, *73*, 571–587. [CrossRef]
9. Ramos, S.; Martín, M.A.; Goya, L. Effects of cocoa antioxidants in type 2 diabetes mellitus. *Antioxidants* **2017**, *31*, 84. [CrossR

Brassesco, M.L.; Brandão, T.R.S.; Silva, C.L.M.; Pintado, M. Carob bean (*Ceratonia siliqua* L.): A new perspective for functional food. *Trends Food Sci. Technol.* **2021**, *114*, 310–322. [CrossRef]

Sekar, V.; Mani, S.; Malarvizhi, R.; Nithya, P.; Vasanthi, H.R. Antidiabetic effect of mangiferin in combination with oral hypoglycemic agents metformin and gliclazide. *Phytomedicine* **2019**, *59*, 152901. [CrossRef]

Chtourou, Y.; Morjen, M.; Ammar, R.; Mhiri, R.; Jemaà, M.; ELBini-Dhouib, I.; Fetoui, H.; Srairi-Abid, N.; Marrakchi, N.; Jebali, J. Investigation of the renal protective effect of combined dietary polyphenols in streptozotocin-induced diabetic aged rats. *Nutrients* **2022**, *14*, 2867. [CrossRef]

Blahova, J.; Martiniakova, M.; Babikova, M.; Kovacova, V.; Mondockova, V.; Omelka, R. Pharmaceutical drugs and natural therapeutic products for the treatment of type 2 diabetes mellitus. *Pharmaceuticals.* **2021**, *14*, 806. [CrossRef]

García-Díez, E.; López-Oliva, M.E.; Caro-Vadillo, A.; Pérez-Vizcaíno, F.; Pérez-Jiménez, J.; Ramos, S.; Martín, M.Á. Supplementation with a cocoa-carob blend, alone or in combination with metformin, attenuates diabetic cardiomyopathy, cardiac oxidative stress and inflammation in Zucker diabetic rats. *Antioxidants* **2022**, *11*, 432. [CrossRef]

Gallardo-Villanueva, P.; Fernández-Marcelo, T.; Villamayor, L.; Valverde, A.M.; Ramos, S.; Fernández-Millán, E.; Martín, M.A. Synergistic effect of a flavonoid-rich cocoa-carob blend and metformin in preserving pancreatic beta cells in Zucker diabetic fatty rats. *Nutrients* **2024**, *16*, 273. [CrossRef]

García-Díez, E.; Sánchez-Ayora, H.; Blanch, M.; Ramos, S.; Martín, M.A.; Pérez-Jiménez, J. Exploring a cocoa-carob blend as a functional food with decreased bitterness: Characterization and sensory analysis. *LWT-Food Sci. Technol.* **2022**, *165*, 113708. [CrossRef]

Directive 2010/63/EU of the European Parliament and of the Council of 22 September 2010 on the Protection of Animals Used for Scientific Purposes. (OJ L 276 20.10.2010, p. 33), European Union. 2010. Available online: http://data.europa.eu/eli/dir/2010/63/2019-06-26 (accessed on 9 August 2024).

Real Decreto 53/2013, de 1 de Febrero, por el Que Se Establecen Las Normas Básicas Aplicables Para la Protección de Los Animales Utilizados en Experimentación y Otros Fines Científicos, Incluyendo la Docencia. BOE-A-2013-1337, Spain. 2013. Available online: https://www.boe.es/eli/es/rd/2013/02/01/53/con (accessed on 9 August 2024).

Kleiner, D.E.; Brunt, E.M.; Van Natta, M.; Behling, C.; Contos, M.J.; Cummings, O.W.; Ferrell, L.D.; Liu, Y.-C.; Torbenson, M.S.; Unalp-Arida, A.; et al. Design and validation of a histological scoring system for nonalcoholic fatty liver disease. *Hepatology* **2005**, *41*, 1313–1321. [CrossRef]

Granado-Serrano, A.; Martín, M.; Bravo, L.; Goya, L.; Ramos, S. Quercetin induces apoptosis via caspase activation, regulation of Bcl-2, and inhibition of PI-3-Kinase/Akt and ERK pathways in a human hepatoma cell line (HepG2). *J. Nutr.* **2006**, *136*, 2715–2721. [CrossRef]

Álvarez-Cilleros, D.; López-Oliva, M.E.; Martín, M.A.; Ramos, S. Cocoa ameliorates renal injury in Zucker diabetic fatty rats by preventing oxidative stress, apoptosis and inactivation of autophagy. *Food Funct.* **2019**, *10*, 7926–7939. [CrossRef]

Cordero-Herrera, I.; Martín, M.A.; Goya, L.; Ramos, S. Cocoa intake ameliorates hepatic oxidative stress in young Zucker diabetic fatty rats. *Food Res. Int.* **2015**, *69*, 194–201. [CrossRef]

Wang, Y.; Yuan, H.; Wang, Y. Treatment of diabetes nephropathy in mice by germinating seeds of *Euryale ferox* through improving oxidative stress. *Foods* **2023**, *12*, 767. [CrossRef]

Xin, C.; Zhao, M.; Wang, J.; Wang, Z. Hawthorn polyphenols, D-chiro-inositol, and epigallocatechin gallate exert a synergistic hypoglycemic effect. *J. Food Biochem.* **2021**, *45*, e13771. [CrossRef]

Salim, N.S.; Abdel-Alim, M.; Said, H.E.M.; Foda, M.F. Phenolic profiles, antihyperglycemic, anti-Diabetic, and antioxidant properties of Egyptian *Sonchus oleraceus* leaves extract: An in vivo study. *Molecules* **2023**, *28*, 6389. [CrossRef]

Othman, M.S.; Khaled, A.M.; Al-Bagawi, A.H.; Fareid, M.A.; Ghany, R.A.; Habotta, O.A.; Moneim, A.E.A. Hepatorenal protective efficacy of flavonoids from *Ocimum basilicum* extract in diabetic albino rats: A focus on hypoglycemic, antioxidant, anti-inflammatory and anti-apoptotic activities. *Biomed. Pharmacother.* **2021**, *144*, 112287. [CrossRef]

Othman, M.S.; Khaled, A.M.; Al-Bagawi, A.H.; Fareid, M.A.; Hameed, R.A.; Zahra, F.A.A.; Moneim, A.E.A. *Echinops spinosus* effect against diabetes and its hepatorenal complications: Total extract and flavonoids fraction. *Environ. Sci. Pollut. Res.* **2022**, *29*, 38606–38617. [CrossRef]

Wang, Y.; Yu, Y.; Wan, H.; Chen, Y.; Xia, F.; Zhang, W.; Zhang, K.; Gu, X.; Zhang, Y.; Lin, Z.; et al. Lower eGFR is associated with increased probability of liver fibrosis in Chinese diabetic patients. *Diabetes Metab. Res. Rev.* **2020**, *36*, e3294. [CrossRef]

Thomas, M.C.; Brownlee, M.; Susztak, K.; Sharma, K.; Jandeleit-Dahm, K.A.M.; Zoungas, S.; Rossing, P.; Groop, P.-H.; Cooper, M.E. Diabetic kidney disease. *Nat. Rev. Dis. Primer* **2015**, *1*, 15018. [CrossRef]

Álvarez-Cilleros, D.; López-Oliva, E.; Goya, L.; Martín, M.Á.; Ramos, S. Cocoa intake attenuates renal injury in Zucker Diabetic fatty rats by improving glucose homeostasis. *Food Chem. Toxicol.* **2019**, *127*, 101–109. [CrossRef]

Cordero-Herrera, I.; Martín, M.A.; Escrivá, F.; Álvarez, C.; Goya, L.; Ramos, S. Cocoa-rich diet ameliorates hepatic insulin resistance by modulating insulin signaling and glucose homeostasis in Zucker diabetic fatty rats. *J. Nutr. Biochem.* **2015**, *26*, 704–712. [CrossRef]

Renno, W.M.; Abdeen, S.; Alkhalaf, M.; Asfar, S. Effect of green tea on kidney tubules of diabetic rats. *Br. J. Nutr.* **2008**, *100*, 652–659. [CrossRef]

33. Zhang, L.; Su, S.; Zhu, Y.; Guo, J.; Guo, S.; Qian, D.; Ouyang, Z.; Duan, J.-A. Mulberry leaf active components alleviate ty 2 diabetes and its liver and kidney injury in db/db mice through insulin receptor and TGF-β/Smads signaling pathway. *Bion Pharmacother.* **2019**, *112*, 108675. [CrossRef]
34. Schattenberg, J.M.; Schuchmann, M. Diabetes and apoptosis: Liver. *Apoptosis* **2009**, *14*, 1459–1471. [CrossRef]
35. Wu, P.; Yan, Y.; Ma, L.-L.; Hou, B.-Y.; He, Y.-Y.; Zhang, L.; Niu, Z.-R.; Song, J.-K.; Pang, X.-C.; Yang, X.-Y.; et al. Effects of the N protein modulator salvianolic acid A alone or combined with metformin on diabetes-associated macrovascular and renal inj *J. Biol. Chem.* **2016**, *291*, 22288–22301. [CrossRef]
36. Sesti, G.; Fiorentino, T.V.; Arturi, F.; Perticone, M.; Sciacqua, A.; Perticone, F. Association between noninvasive fibrosis mark and chronic kidney disease among adults with non-alcoholic fatty liver disease. *PLoS ONE* **2014**, *9*, e88569. [CrossRef]
37. Manco, M.; Ciampalini, P.; DeVito, R.; Vania, A.; Cappa, M.; Nobili, V. Albuminuria and insulin resistance in children with bio proven non-alcoholic fatty liver disease. *Pediatr. Nephrol.* **2009**, *24*, 1211–1217. [CrossRef]
38. Zhou, Q.; Song, N.; Wang, S.-Q.; Wang, Y.; Zhao, Y.-K. Effect of gegen qinlian decoction on hepatic gluconeogenesis in ZDF R with type 2 diabetes mellitus based on the farnesol X receptor/ceramide signaling pathway regulating mitochondrial metaboli and endoplasmic reticulum stress. *Evid. Based Complement. Altern. Med.* **2021**, *2021*, 9922292. [CrossRef]
39. Pang, Y.-L.; Hu, J.-W.; Liu, G.-L.; Lu, S.-Y. Comparative medical characteristics of ZDF-T2DM rats during the course of devel ment to late stage disease. *Anim. Model Exp. Med.* **2018**, *1*, 203–211. [CrossRef]
40. Tiwari, S.; Singh, R.S.; Li, L.; Tsukerman, S.; Godbole, M.; Pandey, G.; Ecelbarger, C.M. Deletion of the insulin receptor in proximal tubule promotes hyperglycemia. *J. Am. Soc. Nephrol.* **2013**, *24*, 1209–1214. [CrossRef]
41. Basu, A.; Basu, R.; Shah, P.; Vella, A.; Johnson, C.M.; Nair, K.S.; Jensen, M.D.; Schwenk, W.F.; Rizz, R.A. Effects of type 2 diabe on the ability of insulin and glucose to regulate splanchnic and muscle glucose metabolism: Evidence for a defect in hepa glucokinase activity. *Diabetes* **2000**, *49*, 272–283. [CrossRef]
42. Dion, F.; Dumayne, C.; Henley, N.; Beauchemin, S.; Arias, E.B.; Leblond, F.A.; Lesage, S.; Lefrançois, S.; Cartee, G.D.; Pichette Mechanism of insulin resistance in a rat model of kidney disease and the risk of developing type 2 diabetes. *PLoS ONE* **2** *12*, e0176650. [CrossRef]
43. Macho-González, A.; Garcimartín, A.; López-Oliva, M.E.; Celada, P.; Bastida, S.; Benedí, J.; Sánchez-Muniz, F.J. Carob-fr extract-enriched meat modulates lipoprotein metabolism and insulin signaling in diabetic rats induced by high-saturated-fat *J. Funct. Food* **2020**, *64*, 103600. [CrossRef]
44. Brouwers, B.; Pruniau, V.P.; Cauwelier, E.J.; Schuit, F.; Lerut, E.; Ectors, N.; Declercq, J.; Creemers, J.W. Phlorizin pretreatm reduces acute renal toxicity in a mouse model for diabetic nephropathy. *J. Biol. Chem.* **2013**, *288*, 27200–27207. [Cross [PubMed]

nutrients

cle

ıe Mechanism Underlying the Hypoglycemic Effect Epimedin C on Mice with Type 2 Diabetes Mellitus Based on oteomic Analysis

xue Zhou [1,†], Ziqi Liu [1,†], Xiaohua Yang [2], Jing Feng [1,3], Murat Sabirovich Gins [3], Tingyu Yan [1], Lei Han [1] Huafeng Zhang [1,*]

[1] National Engineering Laboratory for Resource Development of Endangered Crude Drugs in Northwest China, Academician and Expert Workstations in Puer City of Yunnan Province, College of Food Engineering and Nutritional Science, Provincial Research Station of Se-Enriched Foods in Hanyin County of Shaanxi Province, International Joint Research Center of Shaanxi Province for Food and Health Sciences, Shaanxi Normal University, Xi'an 710119, China; guohaiwen@snnu.edu.cn (Z.L.); tyyan@snnu.edu.cn (T.Y.); lhan@snnu.edu.cn (L.H.)

[2] Research Station of Selenium-Enriched Tea of Shaanxi Province, Health Science Center, Xi'an Jiaotong University, Xi'an 710061, China

[3] Agrarian and Technological Institute, Peoples' Friendship University of Russia, Moscow 119991, Russia; anirr@bk.ru

[*] Correspondence: ijrc@snnu.edu.cn; Tel.: +86-29-85310005

[†] These authors contributed equally to this work.

Abstract: Type 2 diabetes mellitus (T2DM) has become a worldwide public health problem. Epimedin C is considered one of the most important flavonoids in *Epimedium*, a famous edible herb in China and Southeast Asia that is traditionally used in herbal medicine to treat diabetes. In the present study, the therapeutic potential of epimedin C against T2DM was ascertained using a mouse model, and the mechanism underlying the hypoglycemic activity of epimedin C was explored using a label-free proteomic technique for the first time. Levels of fasting blood glucose (FBG), homeostasis model assessment of insulin resistance (HOMA-IR), and oral glucose tolerance, as well as contents of malondialdehyde (MDA) and low-density lipoprotein cholesterol (LDL-C) in the 30 mg·kg^{-1} epimedin C group (EC30 group), were significantly lower than those in the model control group (MC group) ($p < 0.05$), while the contents of hepatic glycogen, insulin, and high-density lipoprotein cholesterol (HDL-C), as well as activities of superoxide dismutase (SOD) and glutathione peroxidase (GSH-Px) in the EC30 group were notably higher than those in the MC group ($p < 0.05$). The structures of liver cells and tissues were greatly destroyed in the MC group, whereas the structures of cells and tissues were basically complete in the EC30 group, which were similar to those in the normal control group (NC group). A total of 92 differentially expressed proteins (DEPs) were enriched in the gene ontology (GO) and Kyoto Encyclopedia of Genes and Genomes (KEGG) pathways. In the EC30 vs. MC groups, the expression level of cytosolic phosphoenolpyruvate carboxykinase (Pck1) was down-regulated, while the expression levels of group XIIB secretory phospholipase A2-like protein (Pla2g12b), apolipoprotein B-100 (Apob), and cytochrome P450 4A14 (Cyp4a14) were up-regulated. According to the KEGG pathway assay, Pck1 participated in the gluconeogenesis and insulin signaling pathways, and Pla2g12b, Apob, and Cyp4a14 were the key proteins in the fat digestion and fatty acid degradation pathways. Pck1, Pla2g12b, Apob, and Cyp4a14 seemed to play important roles in the prevention and treatment of T2DM. In summary, epimedin C inhibited Pck1 expression to maintain FBG at a relatively stable level, promoted Pla2g12b, Apob, and Cyp4a14 expressions to alleviate liver lipotoxicity, and protected liver tissues and cells from oxidant stress possibly by its phenolic hydroxyl groups.

Keywords: epimedin C; type 2 diabetes mellitus; mice; mechanism of action; label-free proteomic technique

on: Zhou, X.; Liu, Z.; Yang, X.; J.; Gins, M.S.; Yan, T.; Han, L.; ;, H. The Mechanism rlying the Hypoglycemic Effect medin C on Mice with Type 2 tes Mellitus Based on Proteomic sis. *Nutrients* **2024**, *16*, 25. //doi.org/10.3390/nu16010025

mic Editors: Sonia Ramos and María Angeles Martín

ved: 19 November 2023
d: 12 December 2023
ted: 13 December 2023
hed: 21 December 2023

1. Introduction

Diabetes mellitus (DM) is a chronic disease caused by relative or absolute decrea in insulin secretion, resulting in metabolic disorders of carbohydrates, lipids, and p teins, especially an increase in blood glucose. DM is one of the worldwide public hea problems. There were 537 million people who suffered from DM in 2021, and the num of patients was predicted to rise to 783 million by 2045 according to the Internatio Diabetes Federation [1]. Traditionally, DM was divided into four types: type 1 diabe mellitus, type 2 diabetes mellitus (T2DM), gestational diabetes mellitus, and other diabe mellitus. Among them, T2DM persons accounted for 90~95% of the total patients [2 The symptoms of T2DM are mainly hyperglycemia and lipid deposition in liver cells d to the increment in fatty acid synthesis and the reduction in fatty acid degradation Additionally, some complications such as non-alcoholic fatty liver disease (NAFLD), neu pathic diabetic retinopathy, hypertension, and cardiac disease frequently occur in T2D patients [5,6]. In recent years, the incidence of T2DM continually increased in the world p tially due to changes in dietary habits [7]. Some medicines (e.g., metformin and glimepiri are extensively used to reduce blood glucose. However, they may cause several side fects such as gastrointestinal irritation and insulin resistance, and even exacerbate dam to patients' pancreatic islet cells [8]. Thus, it is necessary to seek out safe and effici chemopreventive or anti-diabetic agents.

Recently, edible medicinal plants have received widespread attention because of l side effects and drug resistance [9,10]. *Epimedium* L. (named Yin Yang Huo in Chine a popular edible herb in China and Southeast Asia, was listed in the Raw Materials Health Foods by the Chinese National Health Commission (the former Ministry of Hea of China) in 2002. *Epimedium* was first recorded in *Shennong's Classic of Materia Med* (c.a. 184–266 AD), showing efficacy in strengthening *yang*, nourishing *qi*, and improvi *zhi* [11,12]. Traditionally, some Chinese herbalists have used *Epimedium* leaves to treat betes [13]. However, the material basis and mechanism of action remain obscure. Epime C, one of the important flavonoids widely distributed in various species of *Epimedi* (Figure S1) [11,12], has been demonstrated to possess pharmacological activities agai glucocorticoid osteoporosis and cardiovascular diseases [14], some of which were link to DM to some extent [15]. So, it is hypothesized that epimedin C possesses therapeu potential against T2DM.

In order to ascertain the hypoglycemic activity of epimedin C, the effects of epime C on T2DM mice were systematically investigated. The influence of epimedin C on bohydrate and lipid metabolisms, antioxidant systems, and structures of liver cells a tissues was discussed. Furthermore, the mechanism underlying the hypoglycemic effec epimedin C on T2DM mice was explored using a label-free proteomic technique for first time.

2. Materials and Methods

2.1. Reagents and Instruments

The main reagents included acetonitrile and formic acid (high-performance liq chromatography (LC)/mass spectrometry (MS) grade) (Thermo Fisher Scientific Co., L Waltham, MA, USA), epimedin C ($C_{39}H_{50}O_{19}$, purity $\geq$ 98%) (Shanghai YY Co., L Shanghai, China), glimepiride (Chongqing Conquer Pharmaceutical Co., Ltd., Chongq China), metformin (Bristol-Myers Squibb Co., Ltd., New York, NY, USA), streptozoto (STZ) (Sigma-Aldrich Co., Ltd., Saint Louis, MO, USA), paraformaldehyde (Beyoti Biotech Inc., Shanghai, China), a control diet (60.0% carbohydrate, 33.0% protein, 3. lipid, 3.6% amino acids, 0.3% minerals and 0.1% vitamins), a high-fat diet (66.5% con diet with 20.0% sugar, 10.0% lard, 2.5% cholesterol and 1.0% bile salt) (Beijing Bo'ai P Co., Ltd., Beijing, China), and biochemical and physiological kits (mouse insulin enzy linked immunosorbent assay (ELISA) kit, superoxide dismutase (SOD) kit, glutathi peroxidase (GSH-Px) kit, malondialdehyde (MDA) kit, high-density lipoprotein cholest

(HDL-C) kit, low-density lipoprotein cholesterol (LDL-C) kit, and hepatic glycogen assay kit) (Nanjing JC Bioengineering Institute, Nanjing, China).

The main instruments included a NanoAcuity ultra-performance liquid chromatography system (Waters Technologies Co., Ltd., Milford, DE, USA) equipped with an Acclaim PepMap C_{18} peptide analysis column (Thermo Fisher Scientific Co., Ltd., Waltham, MA, USA), an Eclipse E100 light microscope equipped with a DS-U3 imaging system (Nikon Vision Co., Ltd., Tokyo, Japan), a Q Exactive Focus Hybrid Quadrupole (Q Exactive HFX) orbitrap mass spectrometer (Thermo Fisher Scientific Co., Ltd., Waltham, MA, USA), a 5804R high-speed centrifuge (Eppendorf Co., Ltd., Hamburg, Germany), and an Onetouch UltraVue glucose meter (Johnson & Johnson Co., Ltd., New Brunswick, NJ, USA).

2.2. *Animal Experiments*

2.2.1. Experimental Design and Drug Administration

All animal experiments were approved by the Advisory Ethics Committee of the International Joint Research Center of Shaanxi Province for Food and Health Sciences (IJRC-A2020-001, approve date: 1 March 2020). Animal experiments were performed as previously described by Liu et al. [16] with some modifications. Four-week-old male specific pathogen-free (SPF) Kunming (KM) mice with an average weight of 18.0 ± 2.0 g were purchased from the Experimental Animal Center of Xi'an Jiaotong University (Certificate No.: SCXK (Shaan) 2012-003, Xi'an, China). The mice were raised at a temperature of 22 ± 2 °C and relative humidity of $60 \pm 5\%$ with a 12 h light/dark cycle. After adaptive feeding for seven days, all the mice were randomly divided into seven groups ($n = 10$ in each group): the NC (normal control) group, the MC (model control) group, the PC1 (the first positive control, 200 $mg \cdot kg^{-1}$ metformin) group, the PC2 (the second positive control, 5 $mg \cdot kg^{-1}$ glimepiride) group, the EC30 (30 $mg \cdot kg^{-1}$ epimedin C) group, the EC10 (10 $mg \cdot kg^{-1}$ epimedin C) group, and the EC5 (5 $mg \cdot kg^{-1}$ epimedin C) group (Table S1) [14,17]. Prior to oral administration (gavage) of different drugs (metformin, glimepiride, and epimedin C) or distilled water, the mice in the PC1, PC2, EC30, EC10, EC5, and MC groups were subjected to intraperitoneal injection of STZ with citrate buffer (pH 4.5). The mice in the NC group were injected with saline instead of STZ. The mice in the PC1, PC2, EC30, EC10, EC5, and MC groups were fed a high-fat diet, while the mice in the NC group were fed the control diet. Mice with the typical symptoms of diabetes (polydipsia and polyphagia) as well as blood glucose concentrations higher than 11.1 $mmol \cdot L^{-1}$ for 3 consecutive days were considered DM models [18]. In the EC30, EC10, EC5, PC1, and PC2 groups, diabetic mice were supplied with different drugs, whereas mice in the NC and MC groups were provided with sterile distilled water instead of drugs for 28 days. Water and food intake, body weight, and fasting blood glucose (FBG) levels were measured at regular intervals.

2.2.2. Biochemical and Physiological Analyses

Oral Glucose Tolerance Test

An oral glucose tolerance test was conducted according to Geidenstam et al. [19] with slight modifications. All mice were fasted overnight and supplied with glucose (2.0 $g \cdot kg^{-1}$ body weight) via oral administration. FBG levels were determined at 0 h (before glucose administration), 0.5 h, and 2.0 h (after glucose administration). The area under the curve (AUC) was calculated using the following formula:

$$AUC = [(c_1 + c_2) \times 0.5]/2 + [(c_3 + c_2) \times 1.5]/2$$

where c_1, c_2, and c_3 represent the FBG level ($mmol \cdot L^{-1}$) at 0, 0.5, and 2.0 h, respectively.

Assays of Insulin, Hepatic Glycogen, SOD, GSH-Px, MDA, HDL-C, and LDL-C and the Homeostasis Model Assessment of Insulin Resistance (HOMA-IR)

Samples were collected according to a previously reported protocol [16]. Briefly, the eyeballs of mice were treated to obtain blood samples, which were then centrifugated to

−80 °C until needed. The levels of insulin, hepatic glycogen, MDA, HDL-C, and LDL-C well as activities of SOD and GSH-Px, were analyzed with commercial kits according to manufacturer's instructions. According to Liu et al. [18], HOMA-IR was calculated usi the following formula:

$$\text{HOMA} - \text{IR} = (\text{Sg} \times \text{Si})/22.5$$

where Sg and Si represent the levels of blood glucose (mmol·L^{-1}) and serum insu (pmol·L^{-1}), respectively.

2.2.3. Histological Assessment

A histological assessment was performed using the method reported by Jack et al. [In brief, fresh liver tissues were fixed with a 4% paraformaldehyde solution and th embedded in paraffin. Tissue sections (4 μm) were cut and stained with hematoxylin a eosin (H & E) stain. The structures of liver tissues and cells were observed using a li microscope (400-fold) and were photographed with an imaging system.

2.3. Proteomic and Bioinformatic Analyses

Proteomic analysis was carried out as previously described by Yang et al. a Liu et al. [21,22]. Briefly, proteins were extracted, quantified, alkylated, digested, a desalted, and then the peptides were obtained. The resulting peptides were ionized w a voltage of 2.0 kV and were detected with ultra-performance liquid chromatograpl tandem mass spectrometry. For ultra-performance liquid chromatography, the mob phases were composed of 0.1% aqueous formic acid (phase A) and 0.1% formic acid acetonitrile (phase B). The elution program was as follows: 5~30% B for 110 min, 30~80% for 5 min, 80~5% B for 0.1 min, and 5% B for 4.9 min. For mass spectrometry, an orbit analyzer was set within a range of 350~1600 *m*/*z* for the first scanning. Ten ions with gr abundance screened with the first scanning were fragmented using higher-energy C-t dissociation (HCD). In the second scanning, the orbitrap analyzer was set at 17,500 m The raw files of mass spectrometry were treated with MaxQuant software (version 1.5. and the UniProt-Swissprot database. The expression levels of proteins were assessed w intensity-based absolute quantification (iBAQ) and normalized using the median norm ization method [23]. Data on differentially expressed proteins (DEPs) were filtered by f change (>1.20 or <0.83) and significance ($p < 0.05$), which were visualized as volcano pl (Figure S2). A gene ontology (GO) assay was applied to functional enrichment with ClusterProfiler database (version 4.0), and DEPs were classified into biological proces (BP), cellular components (CC), and molecular functions (MF). The Kyoto Encyclope of Genes and Genomes (KEGG) database (https://www.kegg.jp/) was used to clarify key pathways. Protein–protein interaction (PPI) networks were obtained using Cytosca software (version 3.9.0) and STRING database (https://cn.string-db.org/) [16].

2.4. Statistical Analysis

All data were expressed as means ± standard deviations. Figures showing the resu of the animal experiments were plotted using Origin software (version 2018), and figu showing the results of the proteomic and bioinformatic assays were plotted using above-mentioned approaches in the section "Proteomic and bioinformatic analyses". T statistical significance was assessed using SPSS software (version 20).

3. Results

3.1. Physiological and Biochemical Analysis

3.1.1. Effects of Epimedin C on the FBG, Hepatic Glycogen, Insulin, HOMA-IR, and O Glucose Tolerance of T2DM Mice

As shown in Figure 1A, the FBG levels of all mice in the T2DM groups (the M PC1, PC2, EC30, EC10, and EC5 groups) at 0 d were significantly higher than those the NC group ($p < 0.05$), and the FBG levels at 0 d in all T2DM groups were higher th 11.1 mmol·L^{-1}, indicating that theT2DM models were successfully established bef administration. From 0 to 28 d, the FBG levels of all T2DM mice were higher than those in the NC group (Figure 1B). After 14 days of treatment, the FBG levels in the MC group increased continuously, while the FBG levels of T2DM mice in the drug intervention groups (the PC1, PC2, EC30, EC10, and EC5 groups) decreased to different extents, all of which were higher than 11.1 mmol·L^{-1} (Figure 1B). At 28 d, the FBG levels of T2DM mice treated with epimedin C (the EC30, EC10, and EC5 groups) were notably lower than those in the MC group and higher than those in the NC group ($p < 0.05$). Furthermore, FBG levels exhibited a dose-dependent reduction with the increase in epimedin C concentration (Figure 1C). Compared with the FBG levels at 0 d, they decreased respectively by 27.81% in the EC30 group ($p < 0.05$), 17.14% in the EC10 group ($p < 0.05$), and 6.86% in the EC5 group ($p > 0.05$) at 28 d (Table S2). According to Figure 1D, the hepatic glycogen content in the MC group was significantly lower than that in the NC group and the drug intervention groups ($p < 0.05$). The contents of hepatic glycogen in the drug intervention groups were significantly lower than those in the NC group ($p < 0.05$). In addition, the hepatic glycogen content in the EC30 group was 1.38 and 1.52 times higher than those in the EC10 and EC5 groups, respectively ($p < 0.05$). That is to say, the contents of hepatic glycogen increased in a dose-dependent manner with the increase in epimedin C concentration.

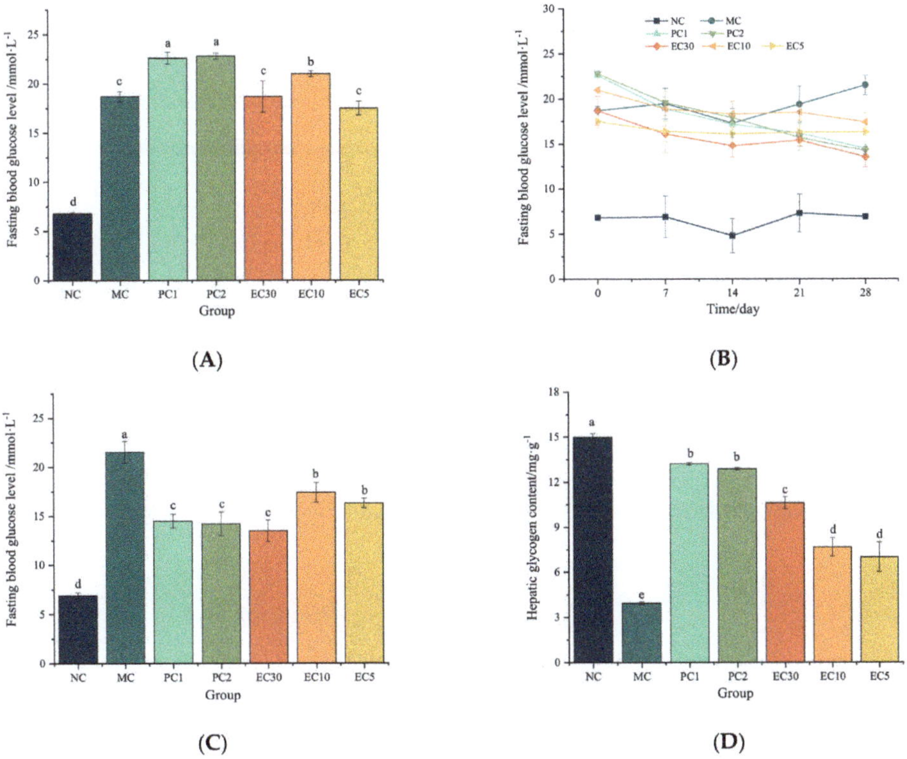

Figure 1. Effects of epimedin C on the fasting blood glucose (FBG) level and the hepatic glycogen content. (**A**) FBG levels at 0 d (before administration); (**B**) changes in FBG levels from 0 to 28 d; (**C**) FBG levels at 28 d; (**D**) hepatic glycogen content at 28 d. Different letters on the top of columns indicate significant differences between groups ($p < 0.05$).

As seen in Figure 2A, the insulin contents in the NC group were significantly higher than those in the MC group ($p < 0.05$), while there was no significant difference in the insulin contents between the NC, PC1, PC2, and EC30 groups ($p > 0.05$) (Table S3). Moreover, the

HOMA-IR levels in the MC group were noticeably higher than those in the NC and groups ($p < 0.05$) (Figure 2B and Table S3).

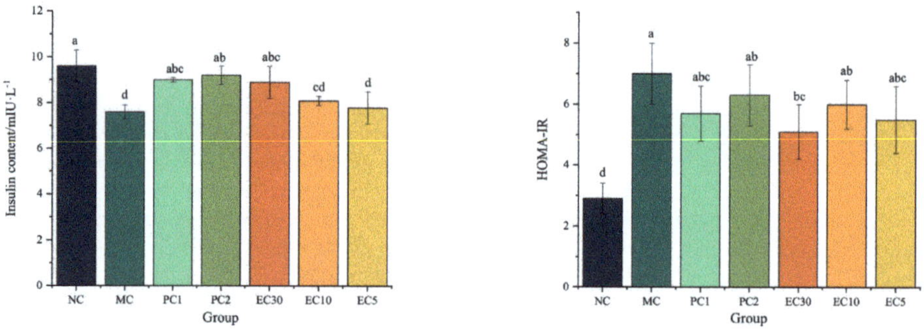

(A) **(B)**

Figure 2. Effects of epimedin C on insulin content in serum (**A**) and homeostasis model assessm of insulin resistance (HOMA-IR) (**B**). Different letters on the top of columns indicate signific differences between groups ($p < 0.05$).

Figure 3A illustrates the effects of epimedin C on the oral glucose tolerance of T2D mice. At 0 h (before glucose gavage), the FBG levels in the MC group were significan higher than those in the drug intervention groups and the NC group ($p < 0.05$), and the F levels of T2DM mice were remarkably higher than that in the NC group ($p < 0.05$). Howe there were no significant differences in the FBG levels between the PC1, PC2, and EC groups ($p > 0.05$). The FBG levels in mice in all groups rose at 0.5 h (after glucose gava whereas the FBG levels in the NC and EC30 groups decreased at 2.0 h (after glucose gava According to Figure 3B, the AUC in the MC group was significantly higher than that in NC group and the drug intervention groups ($p < 0.05$). In particular, the AUC decreased a dose-dependent manner with the increase in epimedin C concentration.

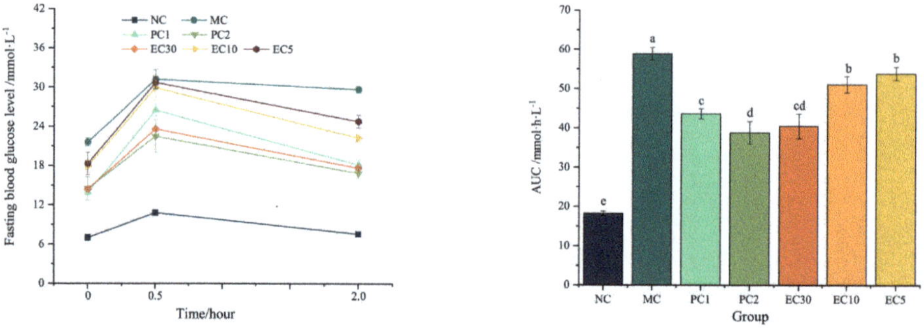

(A) **(B)**

Figure 3. Effects of epimedin C on oral glucose tolerance. (**A**) FBG levels from 0 to 2 h and (**B**) a under the curve (AUC). Different letters on the top of columns indicate significant differences betw groups ($p < 0.05$).

3.1.2. Effects of Epimedin C on Oxidative Stress

The MDA contents in the MC group were significantly higher than those in the group ($p < 0.05$) in the serums and livers, whereas the activities of SOD and GSH-Px w markedly lower than those in the NC group ($p < 0.05$) (Figure 4). The activities of S and GSH-Px in the PC1, PC2, and EC30 groups were significantly higher than those the MC group ($p < 0.05$) (Figure 4A,C,D,F), while the MDA contents in the PC1, PC2, and EC30 groups were notably lower than those in the MC group ($p < 0.05$) (Figure 4B,E). In particular, the GSH-Px activity in livers and serums rose in a dose-dependent manner with the improvement in epimedin C concentration.

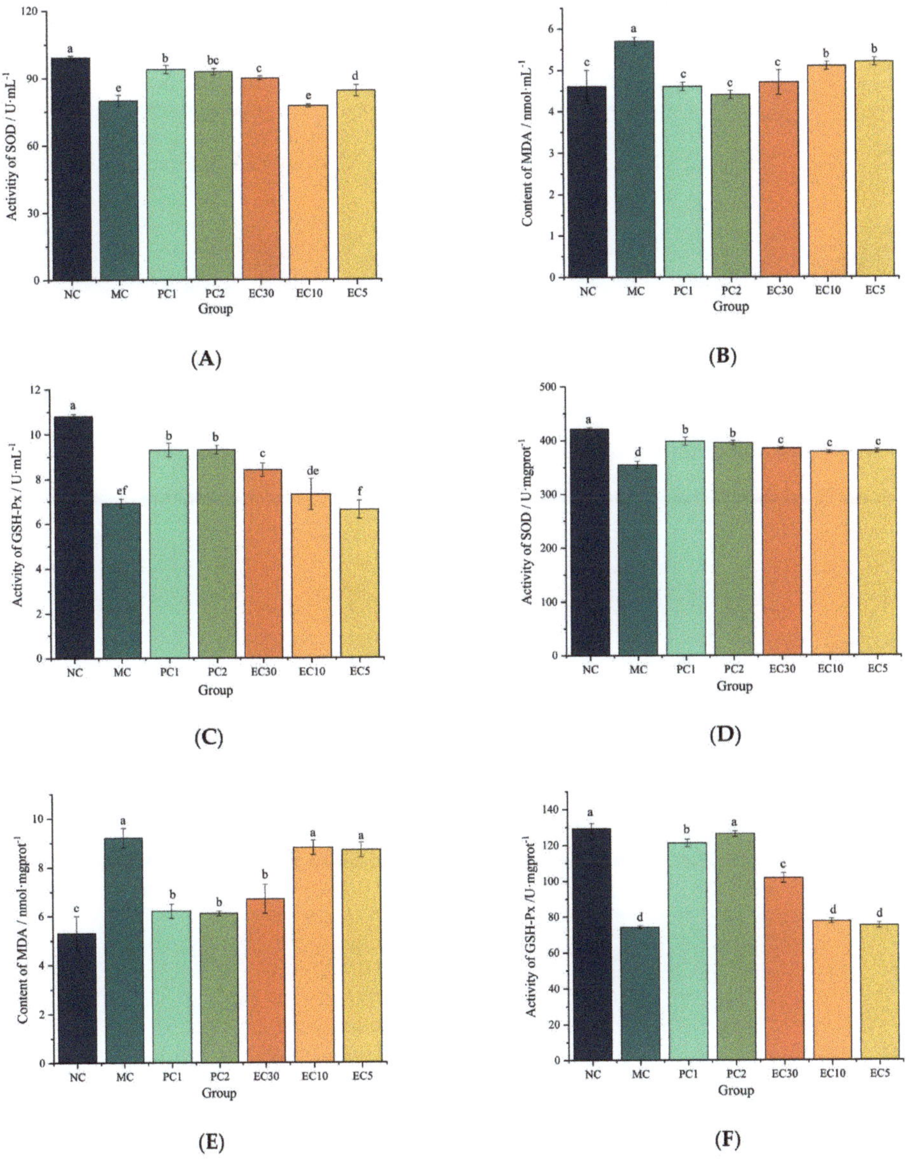

Figure 4. Effects of epimedin C on oxidative stress. (**A**) Superoxide dismutase (SOD) activity in serum; (**B**) malondialdehyde (MDA) content in serum; (**C**) glutathione peroxidase (GSH-Px) activity in serum; (**D**), SOD activity in livers; (**E**) MDA content in livers; (**F**) GSH-Px activity in livers. Different letters on the top of columns indicate significant differences between groups ($p < 0.05$).

3.1.3. Effects of Epimedin C on Blood Lipid Levels

As shown in Figure 5, the HDL-C contents in the serum and livers of mice in the group were significantly lower than those in the NC group ($p < 0.05$), while the LD contents were significantly higher than those in the NC group ($p < 0.05$). The HD contents in the PC1 and EC30 groups were notably higher than that in the MC g ($p < 0.05$), while the LDL-C contents in the PC1 and EC30 groups were markedly lo than that in the MC group ($p < 0.05$) (Figure 5). There were no significant differenc the HDL-C and LDL-C contents between the EC30 and NC groups in the serums ($p >$ (Similarly, the difference in the LDL-C content between the EC30 and NC groups in the li was not significant ($p > 0.05$). These results indicated that epimedin C dose-depende elevated HDL-C contents and reduced LDL-C contents in livers and serums.

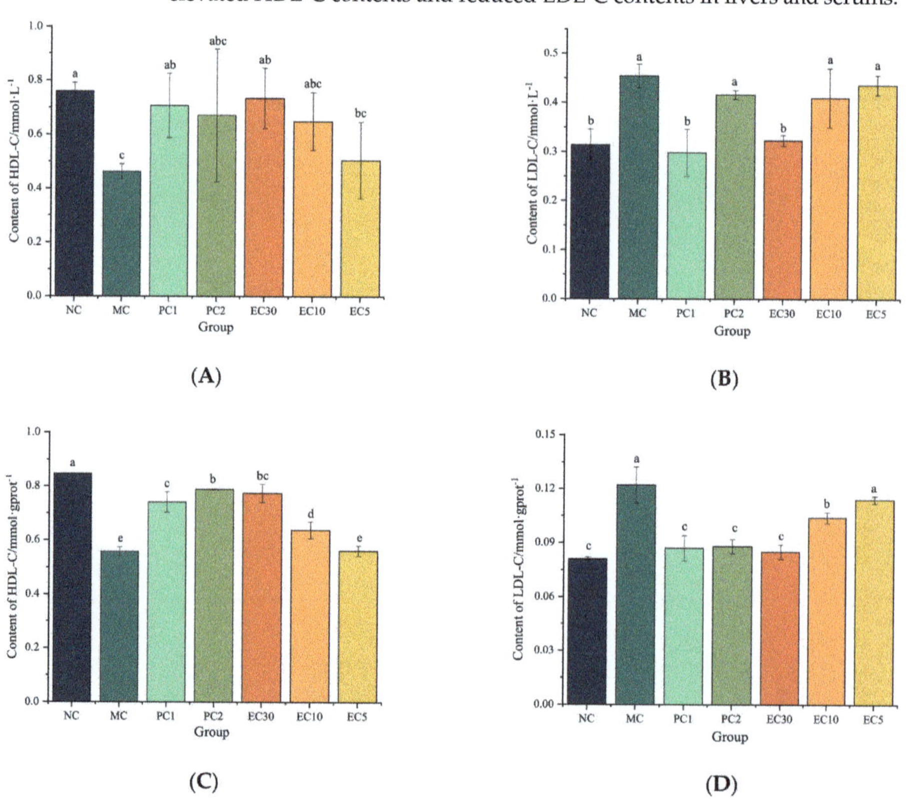

Figure 5. Effects of epimedin C on blood lipid levels. (**A**) High-density lipoprotein cholest (HDL-C) content in serums; (**B**) low-density lipoprotein cholesterol (LDL-C) content in ser (**C**) HDL-C content in livers; (**D**), LDL-C content in livers. Different letters on the top of colu indicate significant differences between groups ($p < 0.05$).

3.2. *Effects of Epimedin C on Liver Histopathological Changes*

Figure 6 illustrates the impacts of epimedin C on liver tissues and cells. In the group, the liver tissues had a complete structure with an orderly arrangement of hep lobules, hepatic cords, and hepatic sinusoids. In addition, the cytoplasm of hepatocytes homogeneous, and no cytoplasmic granules, oil droplets, or fibrous tissues were obser Conversely, the structures of liver cells and tissues were greatly destroyed in the group, and vacuolar degeneration was found in hepatocytes. Some hepatocytes beca oedematous, and some hepatocytes were surrounded with lipid droplets. In particular, outlines of some hepatocytes became blurred, and the arrangement of hepatic cords sinusoids became irregular (Figure 6B). Compared with the MC group, the structures of liver cells and tissues in each drug intervention group were basically complete with less vacuolar degeneration of hepatocytes and a more regular arrangement of hepatocyte cords and sinusoids (Figure 6C–G). In the EC30 group, the structures of liver cells and tissues were similar to those in the NC group (Figure 6G), which demonstrated the therapeutic effect of epimedin C on liver damage induced with T2DM.

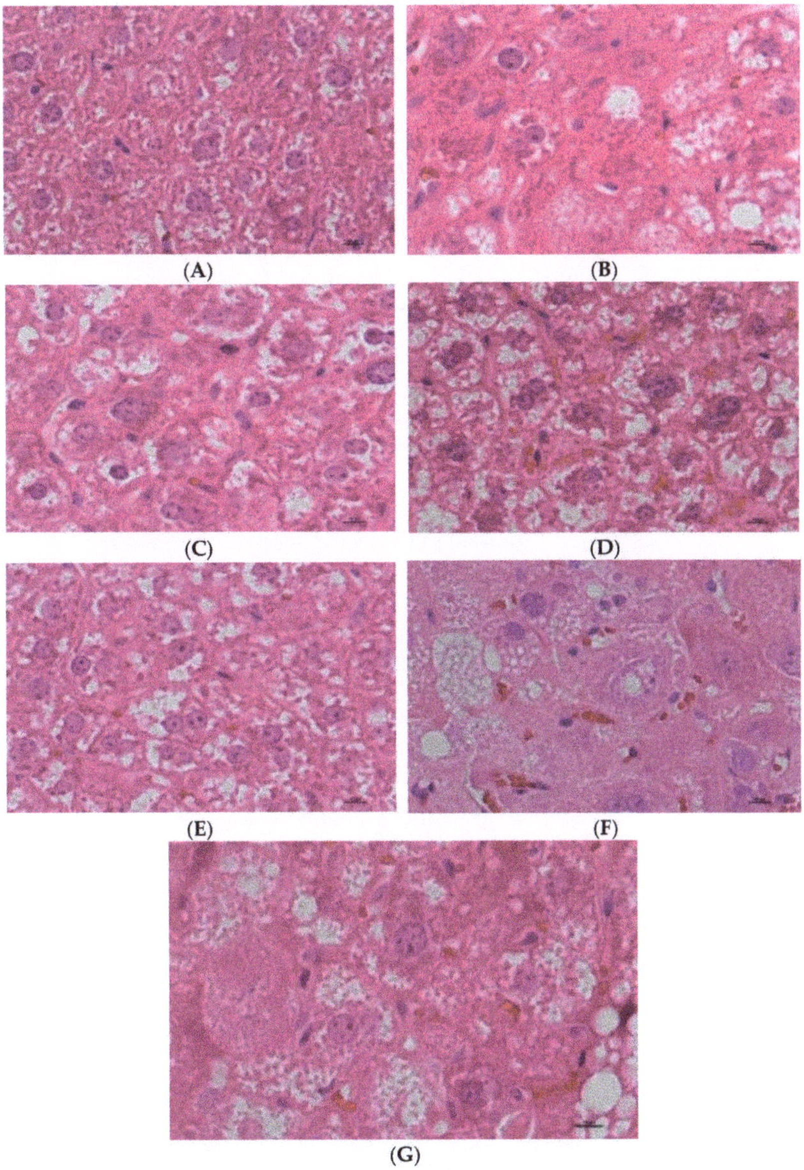

Figure 6. Effects of epimedin C on liver histopathological changes. Structures of liver cells and tissues in the NC group (**A**), the MC group (**B**), the PC1 group (**C**), the PC2 group (**D**), the EC30 group (**E**), the EC10 group (**F**), and the EC5 group (**G**).

3.3. Proteomic Analysis

3.3.1. Identification of DEPs

According to the results of the animal experiments, the levels of FBG, HOMA-IR, a oral glucose tolerance, as well as the MDA and LDL-C contents, in the EC30 group w significantly lower than those in the MC group, while the hepatic glycogen, insulin, a HDL-C contents, as well as the activities of SOD and GSH-Px in the EC30 group, w notably higher than those in MC group. In order to further ascertain the hypoglycen effect of epimedin C on diabetic mice and to reveal the mechanism of action of epimec C against T2DM, liver tissues of mice in the NC, MC, and EC30 groups were subjected proteomic analysis. Uniprot-Swissprot was used for a database search and value analy and 46,168 matched spectra were obtained, which were assigned to various peptid proteins, and/or enzymes. In the NC group vs. the MC group, a total of 324 DEPs w found, which included 161 down-regulated proteins and 163 up-regulated proteins. the EC30 group vs. the MC group, 331 DEPs, including 148 down-regulated protei and 183 up-regulated proteins were identified (Figures 7A, S2 and S3). A Venn diagr was plotted, and 139 common DEPs were identified between the NC group and the M group and the EC30 group and the MC group (Figure 7B). These 139 DEPs were used the hierarchical clustering analysis (Figure 7C). According to the heatmap, the EC30 a NC groups belonged to the same cluster, while the MC group was separated from th (Figure 7C), illustrating that the expression profile of DEPs in the EC30 group was simi to that in the NC group. The remarkable similarity in the expression profile of DEPs the EC30 and NC groups implied that epimedin C had a therapeutic effect on T2DM m In the boxed parts of the heatmap, there were 92 DEPs (54 up-regulated proteins and down-regulated proteins) in the EC30 group vs. the MC group (Figure 7C), which might the key proteins during T2DM treatment with epimedin C.

3.3.2. GO and KEGG Pathway Analyses

The GO and KEGG pathway assays involving 92 key DEPs are presented as histogra (Figure 8). According to the GO analysis, several proteins were intimately linked to G functions such as CC, MF, and BP (Table S4). These functional proteins included acetyl-C carboxylase 2 (Acacb), diazepam-binding inhibitor (Dbi), cytochrome P450 4A14 (Cyp4a threonine synthase-like 2 (Thnsl2), peroxisomal acyl-coenzyme A oxidase 2 (Acox2), glu mate oxaloacetate transaminase 1 (Got1), methionine adenosyltransferase 1 alpha (Mat cytosolic phosphoenolpyruvate carboxykinase (Pck1), etc. (Table S3 and Figure S4). garding CC, DEPs were mainly related to protein–lipid complexes, secondary lysoson high-density lipoprotein particles, and the perinuclear endoplasmic reticulum, most which contributed to lipid anabolism (Figure 8A). Regarding MF, DEPs were mainly c related with carboxylic acid binding, organic acid binding, fatty acid derivative bindi oxidoreductase activity, and carboxy-lyase activity. Regarding BP, DEPs took part in fa acid metabolism, carboxylic acid and organic acid anabolism or catabolism, and li catabolism (Figure 8A). According to the KEGG pathway analysis, 24 pathways were riched (Figure 8B). Among them, seven pathways were directly involved in the metabolis of fatty acids and their derivatives, namely, the peroxisome proliferator-activated recep (PPAR) signaling pathway, arachidonic acid metabolism, fatty acid metabolism, α-linole acid metabolism, unsaturated fatty acid biosynthesis, fatty acid degradation, and fat dig tion and absorption (Figure 8B). In contrast, one pathway (i.e., pyruvate metabolic pathw was indirectly related to fatty acid anabolism through the synthesis of intermediates (ace CoA). There were 10 DEPs annotated in KEGG pathways directly related to fatty a metabolism, including Dbi, Cyp4a14, Acox2, Pck1, fatty acid desaturase 1 (Fads1), fa acid desaturase 2 (Fads2), group XIIB secretory phospholipase A2-like protein (Pla2g1 apolipoprotein B-100 (Apob), prostaglandin E synthase 3 (Ptges3), and aldehyde dehyd genase family 3 member A2 (Aldh3a2). To sum up, a total of 14 proteins were identif with the GO and KEGG pathway analyses.

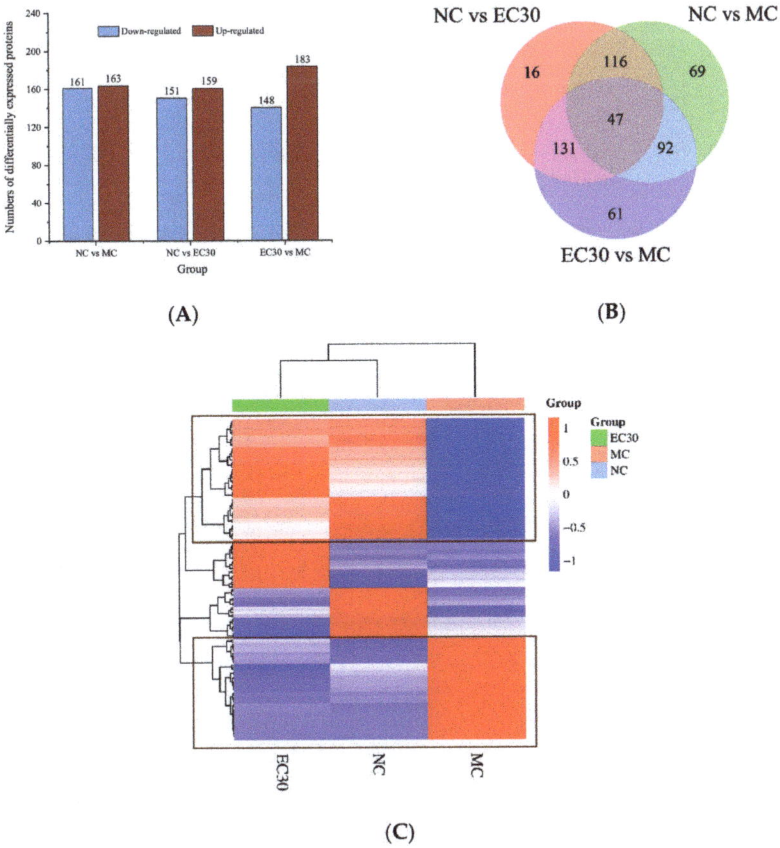

Figure 7. Differentially expressed protein (DEP) analyses in different groups. (**A**) The numbers of down-regulated and up-regulated DEPs in the NC group vs. the MC group, the NC group vs. the EC30 group, and the EC30 group vs. the MC group. (**B**) Venn diagram of DEPs in the NC group vs. the MC group, the NC group vs. the EC30 group and the EC30 group vs. the MC group. (**C**) Hierarchical clustering analysis of DEPs in the NC, MC, and EC30 groups.

3.3.3. DEPs Network Interaction Analysis

To understand the PPI networks of the key proteins, 92 DEPs were screened with the hierarchical clustering analysis and 14 DEPs characterized with GO and KEGG assays were analyzed using the SRTING tool. As shown in Figure 9A,B, the interaction score of DEPs varied greatly, and some proteins exhibited a high degree of protein–protein connectivity. For instance, Pla2g12b was closely associated with Fads2, Apob, Cyp4a14, and Pck1. In the EC30 group vs. the MC group, some proteins were up-regulated, such as Pla2g12b, Cyp4a14, and Apob. The expression levels of some proteins were down-regulated, such as Pck1 (Figure 9B). Pck1, Apob, Pla2g12b, and Cyp4a14 had a high degree of protein–protein connectivity in the PPI networks, which might be the key proteins involved in the PPAR signaling pathway, gluconeogenesis, fatty acid degradation, fat digestion and absorption, and arachidonic acid metabolism. According to the DEP network interaction analysis, epimedin C alleviated T2DM and regulated lipid metabolism partially by increasing the expression levels of Pla2g12b, Cyp4a14, and Apob and decreasing the expression level of Pck1 in the livers of T2DM mice.

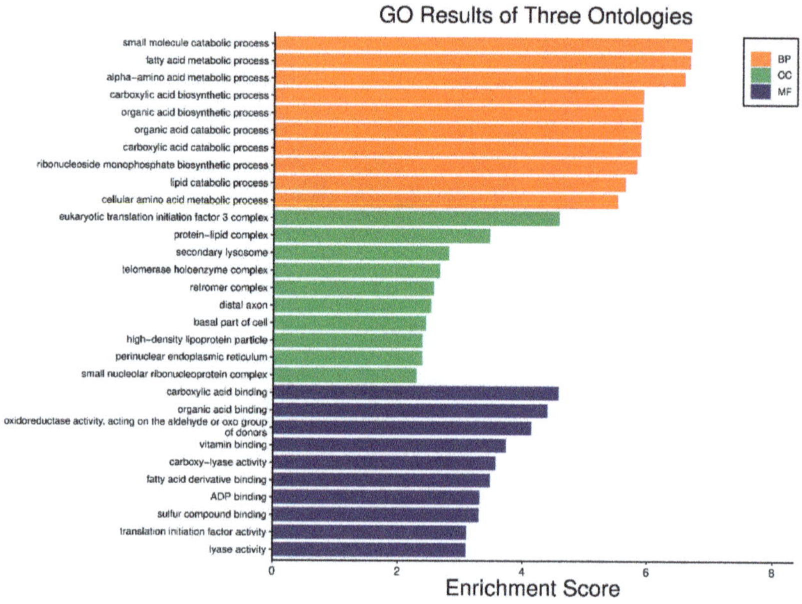

(A)

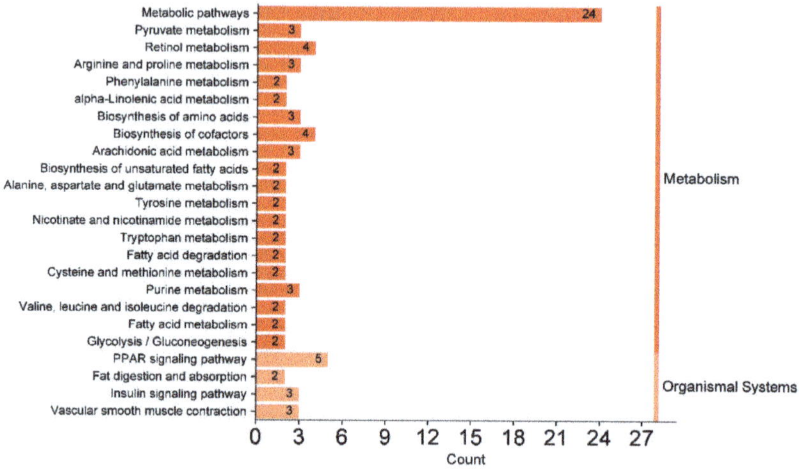

(B)

Figure 8. Gene ontology (GO) (**A**) and Kyoto Encyclopedia of Genes and Genomes (KEGG) pathw analyses (**B**) of DEPs.

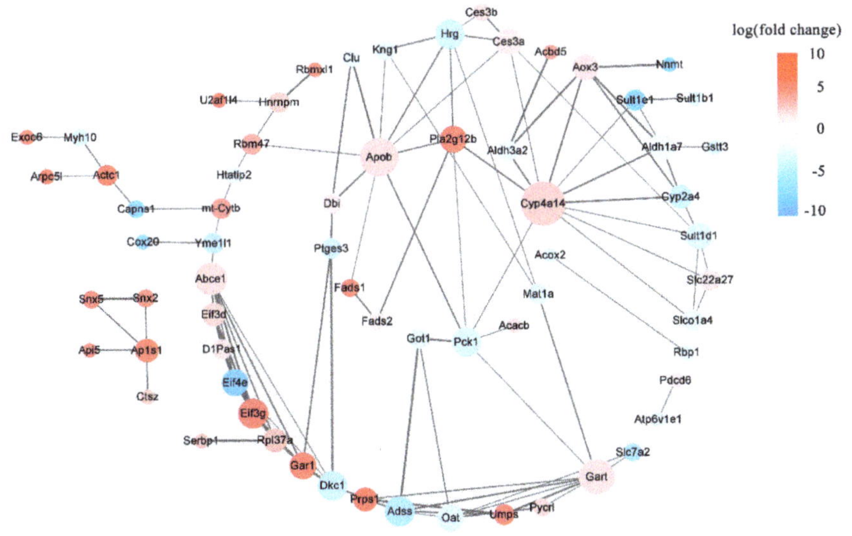

(A)

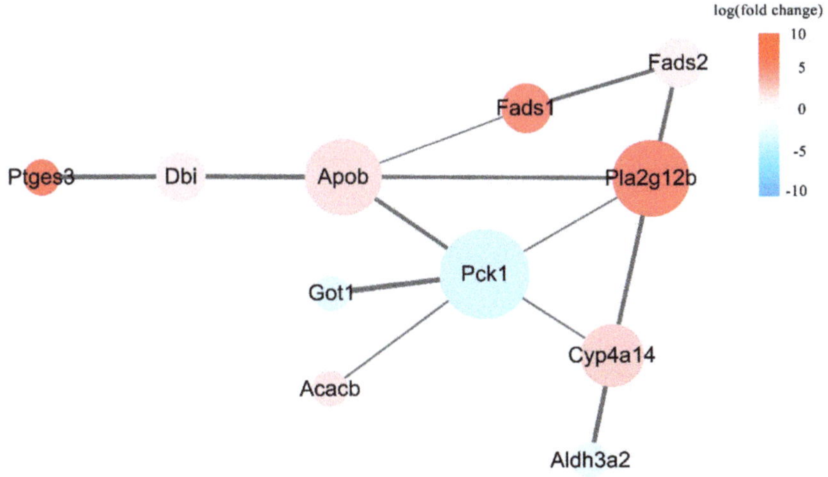

(B)

Figure 9. Protein–protein interaction (PPI) networks and degree of protein–protein connectivity of DEPs. (**A**) PPI network of 92 DEPs screened with a hierarchical clustering analysis. (**B**) PPI network of 14 DEPs characterized with GO and KEGG assays. Red nodes represent up-regulated DEPs, and blue nodes represent down-regulated DEPs. The size of each node reflects the degree of interaction. Line thickness indicates the interaction score between two DEPs, and only DEPs with an interaction score >0.4 are displayed.

4. Discussion

Normally, insulin secreted by pancreatic β-cells maintains FBG at a relatively sta level mainly by promoting hepatic glycogen synthesis and inhibiting gluconeogenesis [However, in diabetic mice, glucose metabolism in vivo is most probably disturbed partia due to pancreatic β-cells destruction and insulin secretion shortage. Thus, FBG mi reach a high level, hepatic gluconeogenesis might be hampered, and symptoms like insu resistance and oral glucose tolerance might appear [25]. Once insulin resistance takes pla not only does insulin no longer inhibit gluconeogenesis, but it also stimulates hepatic synthesis, both of which exacerbate insulin resistance and other symptoms [26]. A T2DM mice were treated with epimedin C in the present work, FBG, HOMA-IR, and o glucose tolerance significantly decreased, while the hepatic glycogen and insulin conte remarkably increased. It is suggested that epimedin C improves the secretion function pancreatic β-cells, attenuates insulin resistance, increases serum insulin contents, promo hepatic glycogen synthesis, and keeps FBG nearly constant in T2DM mice.

It was reported that oxidative stress in vivo is closely related to the occurrences T2DM and its complications [27]. T2DM patients with chronic hyperglycemia are prone accumulate advanced glycosylation end products (AGEs), which enhance the activity nicotinamide adenine dinucleotide phosphate (NADPH) oxidase and increase the amo of reactive oxygen species (ROS) [28,29]. Excessive ROS might attack the cytoplasn membrane, trigger cellular damage and dysfunction of organelles such as mitochond and endoplasmic reticulum, and exacerbate lipid peroxidation [30], which was regarded the major cause of NAFLD [31]. In this study, the SOD and GSH-Px activity in the liv and serums of T2DM mice treated with epimedin C in the EC30 group were significan higher than those in the MC group ($p < 0.05$) but lower than those in the NC gro Moreover, the MDA content in the EC30 group was remarkably lower than that in the M group ($p < 0.05$) but did not differ significantly from that in the NC group ($p > 0.05$). I known that SOD and GSH-Px have evident antioxidant activity, which can maintain homeostasis of in vivo antioxidant system by scavenging ROS. And MDA is the product intracellular lipid peroxidation. Numerous reports indicated that phenolic hydroxyl grou in the chemical structures of flavonoids could react with free radicals to reduce ROS in organism and to mitigate damage to tissues and cells resulting from peroxidation [32]. T effects of epimedin C on SOD and GSH-Px activities and MDA contents demonstrated protective ability against oxidant stress related to T2DM, which was possibly attributed its phenolic hydroxyl groups.

Because the occurrence and development of T2DM are usually accompanied metabolic disorders of lipid, intervention of lipid metabolism plays an important r in the prevention and treatment of T2DM. Cholesterol is mainly transported in the blo using LDL-C, which enters cells after specific binding with low-density lipoprotein recep (LDLR) on peripheral cell membranes and implements its physiological function af hydrolysis by lysosomes [33]. Excessive LDL-C will be phagocytosed by macropha on vascular walls and then be hydrolyzed to free cholesterol, which may bind with hi density lipoprotein (HDL) to form HDL-C. The resulting HDL-C may be ingested by li cells via scavenger receptor class B type 1 (SR-B1) [34]. Normally, cholesterol ingested by liver can be excreted through bile or converted to very low-density lipoprotein choleste (VLDL-C), which subsequently re-enters blood circulation [35]. However, T2DM patie capacities to transport HDL-C are frequently low. When cholesterol level exceeds the tients' capacities, HDL-C levels may decrease while LDL-C levels increase [36], which fo an important pathological basis for vascular atherosclerosis complicated by diabetes [In this study, the HDL-C contents in the livers and serums of mice in the EC30 group w significantly higher than those in the MC group, and the LDL-C contents in the EC30 gro were notably lower than those in the MC group. These results demonstrated the regulat effects of epimedin C on lipid metabolism disorders.

The pathogenesis of T2DM is very complex and involves numerous proteins a pathways [38]. Consequently, the label-free proteomic technique was used to clarify key pathways and identify the important proteins that were differentially expressed among various groups in the presence or absence of epimedin C in the present study. Notably, epimedin C was found to alleviate the abnormalities of lipid metabolism in T2DM mice mainly by regulating the expression levels of Pck1, Pla2g12b, Apob, and Cyp4a14. Pck1 is a rate-limiting enzyme in gluconeogenesis, which is sometimes regarded as a candidate target of anti-diabetic agents [39]. It was reported that liver-specific overexpression of Pck1 in mice resulted in FBG elevation, lipid deposition, and insulin resistance [40]. In this work, the expression level of Pck1 in the EC30 group was close to that in the NC group. Nevertheless, its expression level in the EC30 group was significantly lower than that in the MC group. Based on the KEGG pathway analysis in the EC30 group vs. the MC group, Pck1 was observed to take part in pyruvic acid metabolism, glycolysis, gluconeogenesis, and insulin signaling pathways (Figure S5). These results implied that epimedin C treated T2DM partly by inhibiting Pck1 expression, regulating glycolysis, promoting hepatic glycogen synthesis, and activating insulin signaling pathways. Apob and Pla2g12b play important roles in the assembly and secretion of hepatic low-density lipoproteins (VLDL), which are transcriptionally regulated by hepatocyte nuclear factor-4α (HNF-4α) [41]. Mice with knockouts of Apob and Pla2g12b genes secreted less VLDL but accumulated more triglycerides, cholesterol, and fatty acids in their livers, which frequently led to NAFLD [42,43]. In this study, the expression levels of Pla2g12b and Apob in the MC group were significantly lower than those in the NC group, while the expression levels of these two proteins in the EC30 group were close to those in the NC group. The GO and KEGG pathway assays indicated that Pla2g12b and Apob were co-enriched in lipolytic metabolism as well as fat digestion and absorption pathways in the EC30 group vs. the MC groups (Figures 8 and S5 and Table S4). These results suggested that epimedin C could increase the expression levels of Pla2g12b and Apob, promote the secretion of liver VLDL, facilitate lipolysis as well as fat digestion and absorption, reduce the deposition of lipid in livers, and thus alleviate the symptoms of T2DM. Cytochrome P450 (Cyp) epoxygenases can catalyze the conversion of arachidonic acid into other bioactive eicosanoids. Cyp4a14 is one of the important Cyp epoxygenases, which catalyze ω-hydroxylation of long-chain and medium-chain fatty acids such as arachidonic acid [44]. It was reported that the expression level of Cyp4a14 in obese mice was significantly lower than that in normal mice [45]. After treating obese mice with the lipid-lowering drug WY-14643, Cyp4a14 expression notably up-regulated, and lipid deposition remarkably reduced [46]. In this research, the expression level of Cyp4a14 in the MC group was significantly lower than that in the NC group, while its expression level in the EC30 group was close to that in the NC group. The GO and KEGG pathway assays also revealed that Cyp4a14 was enriched in fatty acid degradation pathways, especially arachidonic acid catabolism (Figures 8 and S5). These results suggested that epimedin C could increase the expression level of Cyp4a14, inhibit excessive lipid deposition, and protect the liver from lipotoxicity, all of which might contribute to the prevention and treatment of T2DM.

5. Conclusions

In this study, the therapeutic potential of epimedin C against T2DM was confirmed using an animal model, and the mechanism underlying the hypoglycemic activity of epimedin C was determined using the label-free proteomic technique for the first time. Notably, intervention with 30 mg·kg^{-1} epimedin C decreased levels of FBG, HOMA-IR, and oral glucose tolerance, as well as the contents of MDA and LDL-C in T2DM mice, while it increased the hepatic glycogen, insulin, and HDL-C contents, as well as the activities of SOD and GSH-Px. In addition, epimedin C effectively protected the structures of liver cells and tissues. There were 92 key DEPs enriched in the GO and KEGG pathways. In the EC30 group vs. the MC group, the expression level of Pck1 was down-regulated, while expression levels of Pla2g12b, Apob, and Cyp4a14 were up-regulated. According to the KEGG pathway assay, Pck1 participated in gluconeogenesis and the insulin signaling pathways, and Pla2g12b, Apob, and Cyp4a14 were the key proteins in fat digestion and the fatty acid degradation pathways. Pck1, Pla2g12b, Apob, and Cyp4a14 seemed to p important roles in the prevention and treatment of T2DM. In summary, epimedin C co inhibit Pck1 expression to maintain FBG at a relatively stable level, promote Pla2g1 Apob, and Cyp4a14 expressions to alleviate liver lipotoxicity, and protect liver tissues a cells from oxidant stress possibly by its phenolic hydroxyl groups.

Supplementary Materials: The following supporting information can be downloaded at: https://w mdpi.com/article/10.3390/nu16010025/s1, Table S1. Experimental groups and treatments (n = 1(each group); Table S2. Effects of epimedin C on fasting blood glucose (FBG) at 0 and 28 d; Table Effects of epimedin C on insulin content in serums and homeostasis model assessment of insu resistance (HOMA-IR); Table S4. Gene ontology (GO) assay of key differentially expressed prote (DEPs); Figure S1. Chemical structure of epimedin C; Figure S2. Volcano plots of differentially pressed proteins (DEPs); Figure S3. Hierarchical clustering analysis of DEPs; Figure S4. Gene-conc network (CNET) assay of gene ontology (GO) functions; Figure S5. Typical Kyoto Encyclopedi Genes and Genomes (KEGG) pathways.

Author Contributions: Investigation, X.Z.; writing—original draft, X.Z. and Z.L.; visualization, Z writing—review and editing, Z.L., M.S.G. and H.Z.; conceptualization, X.Y. and H.Z.; supervisi X.Y.; validation, Z.L., T.Y., J.F. and L.H. All authors have read and agreed to the published versio the manuscript.

Funding: This research was partially funded by the National Program of Belt and Road Innovat Talents Exchange of the Chinese Ministry of Science and Technology (DL2023173001L), the K Research and Development Program of Shaanxi Province and Ankang City (Key Laboratory of enriched Products Development and Quality Control of Ministry of Agriculture and Rural Affairs a National-Local Joint Engineering Laboratory of Se-enriched Food Development) in China (S2024- ZDCXL-ZDLSF-0183/Se-2023C04/2022ZDLSF05-11), and the Program of Academician and Exp Workstation in Puer City of Yunnan Province in China (202305AF150118).

Institutional Review Board Statement: All animal experiments were approved by the Advis Ethics Committee of the International Joint Research Center of Shaanxi Province for Food and Hea Sciences, Shaanxi Normal University (IJRC-A2020-001, approve date: 1 March 2020).

Informed Consent Statement: Not applicable.

Data Availability Statement: All data generated or analyzed during this study are included in t publication and the Supplementary Material.

Conflicts of Interest: The authors declare no conflict of interest.

References

1. Magliano, D.J.; Boyko, E.J.; IDF Diabetes Atlas 10th Edition Scientific Committee. *IDF Diabetes Atlas*, 10th ed.; Internatio Diabetes Federation: Brussels, Belgium, 2021. Available online: http://www.ncbi.nlm.nih.gov/books/NBK581934/ (accessed 23 September 2022).
2. Chen, P.L.; Yang, W.S. Human Genetics of Diabetes Mellitus in Taiwan. *Front. Biosci.* **2009**, *14*, 4535–4545. [CrossRef] [PubM
3. Artasensi, A.; Pedretti, A.; Vistoli, G.; Fumagalli, L. Type 2 Diabetes Mellitus: A Review of Multi-Target Drugs. *Molecules* **2** *25*, 1987. [CrossRef] [PubMed]
4. Kitagawa, N.; Hashimoto, Y.; Hamaguchi, M.; Osaka, T.; Fukuda, T.; Yamazaki, M.; Fukui, M. Liver Stiffness Is Associated w Progression of Albuminuria in Adults with Type 2 Diabetes: Nonalcoholic Fatty Disease Cohort Study. *Can. J. Diabetes* **2020**, 428–433. [CrossRef] [PubMed]
5. Ceriello, A. Postprandial Hyperglycemia and Diabetes Complications: Is it Time to Treat? *Diabetes* **2005**, *54*, 1–7. [CrossR [PubMed]
6. Spínola, V.; Llorent-Martínez, E.J.; Castilho, P.C. Polyphenols of *Myrica faya* Inhibit Key Enzymes Linked to Type II Diabetes a Obesity and Formation of Advanced Glycation End-Products (In Vitro): Potential Role in the Prevention of Diabetic Complicati *Food Res. Int.* **2019**, *116*, 1229–1238. [CrossRef] [PubMed]
7. Gomez-Delgado, F.; Katsiki, N.; Lopez-Miranda, J.; Perez-Martinez, P. Dietary Habits, Lipoprotein Metabolism and Cardiovasc Disease: From Individual Foods to Dietary Patterns. *Crit. Rev. Food Sci. Nutr.* **2021**, *61*, 1651–1669. [CrossRef] [PubMed]
8. Etxeberria, U.; de la Garza, A.L.; Campión, J.; Martínez, J.A.; Milagro, F.I. Antidiabetic Effects of Natural Plant Extracts Inhibition of Carbohydrate Hydrolysis Enzymes with Emphasis on Pancreatic Alpha Amylase. *Expert Opin. Ther. Targets* **2012** 269–297. [CrossRef]

Hu, Y.; Chen, X.; Hu, M.; Zhang, D.; Yuan, S.; Li, P.; Feng, L. Medicinal and Edible Plants in the Treatment of Dyslipidemia: Advances and Prospects. *Chin. Med.* **2022**, *17*, 113. [CrossRef]
Zhang, H.F.; Yang, X.H. Asian Medicine: Protect Rare Plants. *Nature* **2012**, *482*, 35. [CrossRef]
The State Pharmacopoeia Commission of the People's Republic of China. *Pharmacopoeia of the People's Republic of China*; Chinese Medical Science Press: Beijing, China, 2020; Volume I.
Zhang, H.F.; Yang, X.H.; Zhao, L.D.; Wang, Y. Ultrasonic-Assisted Extraction of Epimedin C from Fresh Leaves of *Epimedium* and Extraction Mechanism. *Innov. Food Sci. Emerg. Technol.* **2009**, *10*, 54–60. [CrossRef]
Xu, A.; Yao, M. The Mechanism of Improving Insulin Resistance of *Epimedium* in Type 2 Diabetic Rats. *Chin. J. Clin. Res.* **2012**, *25*, 731–733.
Song, L.; Wang, L.-W.; Zhou, Y.-T.; Liu, Y.; Bi, Y.-N.; Zhou, K. The Mechanism of Anti-Osteoporosis Effect of Epimedin C In Vivo. *Lishizhen Med. Mater. Med. Res.* **2021**, *32*, 555–558. [CrossRef]
Han, D.; Zhuang, Y.; Meng, J. Research Progress on the Role of SGK in the Development of Diabetes Mellitus. *Chin. J. Diabetes* **2023**, *31*, 394–397. [CrossRef]
Liu, Z.; Liu, M.; Fan, M.; Pan, S.; Li, S.; Chen, M.; Wang, H. Metabolomic-Proteomic Combination Analysis Reveals the Targets and Molecular Pathways Associated with Hydrogen Sulfide Alleviating NAFLD. *Life Sci.* **2021**, *264*, 118629. [CrossRef]
Rahman, M.O.; Ahmed, S.; Mazumder, T.; Salam, A.; Baral, P.K.; Rana, F.; Mitra, S.; Hossain, S.; Rahman, R.; Hussain, S. A Comparative Evaluation of Cardiac and Neurological Safety Status of Two Commonly Used Oral Hypoglycaemic Agents in T2-DM Swiss Albino Mice Model. *Metab. Open* **2022**, *14*, 100191. [CrossRef]
Liu, W.; Lv, X.; Huang, W.; Yao, W.; Gao, X. Characterization and Hypoglycemic Effect of a Neutral Polysaccharide Extracted from the Residue of *Codonopsis pilosula*. *Carbohydr. Polym.* **2018**, *197*, 215–226. [CrossRef]
Geidenstam, N.; Danielsson, A.P.H.; Spégel, P.; Ridderstråle, M. Changes in Glucose-Elicited Blood Metabolite Responses Following Weight Loss and Long Term Weight Maintenance in Obese Individuals with Impaired Glucose Tolerance. *Diabetes Res. Clin. Pract.* **2016**, *113*, 187–197. [CrossRef]
Jack, B.U.; Ramharack, P.; Malherbe, C.; Gabuza, K.; Joubert, E.; Pheiffer, C. *Cyclopia intermedia* (Honeybush) Induces Uncoupling Protein 1 and Peroxisome Proliferator-Activated Receptor Alpha Expression in Obese Diabetic Female Db/Db Mice. *Int. J. Mol. Sci.* **2023**, *24*, 3868. [CrossRef]
Yang, X.; Zhang, H.; Li, L.; Zhou, X.; Liu, Y.; Lai, J. Proteomic Analysis of Protective Effects of *Epimedium* Flavonoids against Ethanol-Induced Toxicity in Retinoic Acid-Treated SH-SY5Y Cells. *Molecules* **2022**, *27*, 1026. [CrossRef]
Liu, S.; Wang, F.; Chen, H.; Yang, Z.; Ning, Y.; Chang, C.; Yang, D. New Insights into Radio-Resistance Mechanism Revealed by (Phospho)Proteome Analysis of *Deinococcus radiodurans* after Heavy Ion Irradiation. *Int. J. Mol. Sci.* **2023**, *24*, 14817. [CrossRef]
Li, Y.X.; Lu, Y.P.; Tang, D.; Hu, B.; Zhang, Z.Y.; Wu, H.W.; Fan, L.J.; Cai, K.W.; Tang, C.; Zhang, Y.Q.; et al. Anthocyanin Improves Kidney Function in Diabetic Kidney Disease by Regulating Amino Acid Metabolism. *J. Transl. Med.* **2022**, *20*, 510. [CrossRef] [PubMed]
Zhang, X.; Yang, G.; Zhu, Y.; Zhang, D.; Xia, J.; Li, W. Effects of NAFLD Act on Glucose, insulin Metabolism and Liver Glycogen Synthesis in Rats. *J. Southeast Univ. Med. Sci. Ed.* **2020**, *39*, 420–424. [CrossRef]
Winzell, M.S.; Ahrén, B. The High-Fat Diet-Fed Mouse: A Model for Studying Mechanisms and Treatment of Impaired Glucose Tolerance and Type 2 Diabetes. *Diabetes* **2004**, *53*, S215–S219. [CrossRef] [PubMed]
McGarry, J.D. Banting Lecture 2001: Dysregulation of Fatty Acid Metabolism in the Etiology of Type 2 Diabetes. *Diabetes* **2002**, *51*, 7–18. [CrossRef] [PubMed]
Duan, W.X.; Yang, X.H.; Zhang, H.F.; Feng, J.; Zhang, M.Y. Chemical Structure, Hypoglycemic Activity, and Mechanism of Action of Selenium Polysaccharides. *Biol. Trace Elem. Res.* **2022**, *200*, 4404–4418. [CrossRef] [PubMed]
San Martin, A.; Foncea, R.; Laurindo, F.R.; Ebensperger, R.; Griendling, K.K.; Leighton, F. Nox1-Based NADPH Oxidase-Derived Superoxide Is Required for VSMC Activation by Advanced Glycation End-Products. *Free Radic. Biol. Med.* **2007**, *42*, 1671–1679. [CrossRef] [PubMed]
Yamagishi, S.; Maeda, S.; Matsui, T.; Ueda, S.; Fukami, K.; Okuda, S. Role of Advanced Glycation End Products (AGEs) and Oxidative Stress in Vascular Complications in Diabetes. *Biochim. Biophys. Acta* **2012**, *1820*, 663–671. [CrossRef]
Stefan, N.; Cusi, K. A Global View of the Interplay between Non-Alcoholic Fatty Liver Disease and Diabetes. *Lancet Diabetes Endocrinol.* **2022**, *10*, 284–296. [CrossRef]
Kuzmenko, D.I.; Udintsev, S.N.; Klimentyeva, T.K.; Serebrov, V.Y. Oxidative Stress in Adipose Tissue as a Primary Link in Pathogenesis of Insulin Resistance. *Biomed. Khim.* **2016**, *62*, 14–21. [CrossRef]
Dias, M.C.; Pinto, D.C.G.A.; Silva, A.M.S. Plant Flavonoids: Chemical Characteristics and Biological Activity. *Molecules* **2021**, *26*, 5377. [CrossRef]
Wang, L.; Song, B. The Research Perspective of Intestinal Cholesterol Absorption. *Chem. Life* **2014**, *34*, 318–328. [CrossRef]
Subramaniam, K.; Babu, L.A.; Shah, N. A Case of Premature and Recurrent Myocardial Infarction Associated with ABCA.1 Gene Mutation. *J. Postgrad. Med.* **2021**, *67*, 29–32. [CrossRef] [PubMed]
Zhang, Q.; Xu, C.; Tang, C. Regulation Mechanism of Intracellular Cholesterol Level. *Prog. Biochem. Biophys.* **2022**, *49*, 292–302. [CrossRef]
Yang, M.; Sui, D.; Chen, W.; Yang, M.; Yu, D. Study on Hypoglycemic Effect and Mechanism of Total Flavonoids of Propolis in

37. Yuan, Q.; Yu, Y.; Wang, Y.; Xu, C. The Study of Effection and Mechanism of Tangkening Granule on Rats Blood Glucose and Blc Fat. *Chin. J. Mod. Appl. Pharm.* **2009**, *26*, 184–187. [CrossRef]
38. Xiao, C.; Wu, Q.; Xie, Y.; Tan, J.; Ding, Y.; Bai, L. Hypoglycemic Mechanisms of *Ganoderma lucidum* Polysaccharides F31 in Db/ Mice via RNA-Seq and iTRAQ. *Food Funct.* **2018**, *9*, 6495–6507. [CrossRef]
39. Beale, E.G.; Hammer, R.E.; Antoine, B.; Forest, C. Disregulated Glyceroneogenesis: PCK1 as a Candidate Diabetes and Obes Gene. *Trends Endocrinol. Metab.* **2004**, *15*, 129–135. [CrossRef]
40. Ye, Q.; Liu, Y.; Zhang, G.; Deng, H.; Wang, X.; Tuo, L.; Chen, C.; Pan, X.; Wu, K.; Fan, J.; et al. Deficiency of Gluconeoge Enzyme PCK1 Promotes Metabolic-Associated Fatty Liver Disease through PI3K/AKT/PDGF Axis Activation in Male Mice. *N Commun.* **2023**, *14*, 1402. [CrossRef]
41. Guan, M.; Qu, L.; Tan, W.; Chen, L.; Wong, C.W. Hepatocyte Nuclear Factor-4 Alpha Regulates Liver Triglyceride Metabolism Part through Secreted Phospholipase A_2 GXIIB. *Hepatology* **2011**, *53*, 458–466. [CrossRef]
42. Fisher, E.; Lake, E.; McLeod, R.S. Apolipoprotein B100 Quality Control and the Regulation of Hepatic Very Low Dens Lipoprotein Secretion. *J. Biomed. Res.* **2014**, *28*, 178–193. [CrossRef]
43. Liu, Q.; Yang, M.; Fu, X.; Liu, R.; Sun, C.; Pan, H.; Wong, C.W.; Guan, M. Activation of Farnesoid X Receptor Promo Triglycerides Lowering by Suppressing Phospholipase A2 G12B Expression. *Mol. Cell Endocrinol.* **2016**, *436*, 93–101. [CrossR [PubMed]
44. Xu, X.; Zhang, X.A.; Wang, D.W. The Roles of CYP450 Epoxygenases and Metabolites, Epoxyeicosatrienoic Acids, in Cardiovas lar and Malignant Diseases. *Adv. Drug Deliv. Rev.* **2011**, *63*, 597–609. [CrossRef] [PubMed]
45. Zhang, X.; Li, S.; Zhou, Y.; Su, W.; Ruan, X.; Wang, B.; Zheng, F.; Warner, M.; Gustafsson, J.-Å.; Guan, Y. Ablation of Cytochro P450 Omega-Hydroxylase 4A14 Gene Attenuates Hepatic Steatosis and Fibrosis. *Proc. Natl. Acad. Sci. USA* **2017**, *114*, 3181–3 [CrossRef] [PubMed]
46. Xu, Z.; Shi, L.; Li, D.; Wu, Q.; Zhang, Y.; Gao, M.; Ji, A.; Jiang, Q.; Chen, R.; Zhang, R.; et al. Real Ambient Particulate Mat Induced Lipid Metabolism Disorder: Roles of Peroxisome Proliferators-Activated Receptor Alpha. *Ecotoxicol. Environ. Saf.* **2C** *231*, 113173. [CrossRef]

nutrients

MDPI

icle

ntihypertensive Effects of an Optimized Aged Garlic Extract Subjects with Grade I Hypertension and Antihypertensive rug Therapy: A Randomized, Triple-Blind Controlled Trial

é C. E. Serrano [1,*], Eva Castro-Boqué [2], Alicia García-Carrasco [2], María Inés Morán-Valero [3], niel González-Hedström [3], Marcelino Bermúdez-López [2], José Manuel Valdivielso [2], Alberto E. Espinel [3] and nuel Portero-Otín [1]

1 NUTREN-Nutrigenomics, Department of Experimental Medicine, University of Lleida, 25198 Lleida, Spain; manuel.portero@udl.cat
2 Vascular and Renal Translational Research Group, Institut de Recerca Biomèdica de Lleida, IRBLleida, 25198 Lleida, Spain; ecastro.lleida.ics@gencat.cat (E.C.-B.); agarcia@irblleida.cat (A.G.-C.); mbermudez@irblleida.cat (M.B.-L.); valdivielso@irblleida.cat (J.M.V.)
3 Pharmactive Biotech S.L.U., 28049 Madrid, Spain; ines.moran@pharmactive.eu (M.I.M.-V.); dgonzalez@pharmactive.eu (D.G.-H.); aespinel@pharmactive.eu (A.E.E.)
* Correspondence: josecarlos.serrano@udl.cat; Tel.: +34-973702408

Abstract: The use of garlic (*Allium sativum*) for treating arterial hypertension has been recognized as effective for several decades. However, tolerance to treatment is low, and several technological modifications have been developed to improve its tolerability, such as the aging process at controlled temperature and humidity. This study aims to validate the antihypertensive effects of an optimized extract of aged black garlic with low doses of s-allyl-cysteine (SAC) in a Grade I hypertensive population with drug treatment. A randomized, triple-blind, placebo-controlled parallel trial was developed, where a daily supplementation with 0.25 mg/day of SAC for 12 weeks was performed. A reduction in systolic and diastolic blood pressure of 1.8 mmHg (0.7 to 4.1 95% CI) and 1.5 mmHg (0.3 to 3.0 95% CI), respectively, was observed. Similarly, an increase in blood nitric oxide (10.3·µM, 1.1 to 19.5 95% CI) and antioxidant capacity (7×10^{-3} µM TE/min, (1.2 to 13×10^{-3} 95% CI) and a reduction in uric acid levels (−0.3 mg/dL, −0.5 to −0.001 95% CI) and ACE activity (−9.3 U/L; −18.4 to −0.4 95% CI) were observed. No changes in endothelial function and inflammatory cytokines were observed. It was concluded that low-dose SAC supplementation in an optimized black-garlic extract allows for an extra-significant reduction in blood pressure in a Grade I hypertensive population receiving drug treatment.

Keywords: hypertension; aged black garlic; blood pressure; s-allyl-cysteine

ion: Serrano, J.C.E.;
o-Boqué, E.; García-Carrasco, A.;
n-Valero, M.I.;
ález-Hedström, D.;
údez-López, M.; Valdivielso,
Espinel, A.E.; Portero-Otín, M.
ypertensive Effects of an
nized Aged Garlic Extract in
cts with Grade I Hypertension
Antihypertensive Drug Therapy:
ndomized, Triple-Blind
olled Trial. *Nutrients* **2023**, *15*,
https://doi.org/10.3390/
173691

emic Editors: Francesca
pieri and Lanfranco D'Elia

ved: 28 June 2023
ed: 28 July 2023
pted: 18 August 2023
shed: 23 August 2023

1. Introduction

A wealth of evidence exists to demonstrate that lowering blood pressure can substantially reduce premature morbidity and mortality [1]. However, despite the advances in understanding the pathology and the development of drugs to treat hypertension, it is estimated that only a small percentage of the population worldwide (around 32%) can control hypertension [2]. Indeed, resistant hypertension (subjects receiving three or more drugs of different classes at maximally tolerated doses but still have uncontrolled blood pressure) is found in a substantial part of the adult population [3]. The use of food-bioactive compounds and a healthy lifestyle are suitable options for non-pharmacological approaches to improving vascular health [4]. Among them, organosulfur compounds, a class of organic compounds containing sulfur as a critical component of their molecular structure, have been shown to benefit blood pressure [5]. They are found in various foods, such as garlic, onions, and cruciferous vegetables [6].

Several studies have shown that garlic can help lower blood pressure in indivi als with hypertension [7]. A mean reduction of around 9 and 6 mmHg in systolic a diastolic blood pressure could be expected due to garlic supplementation in hypertensi subjects (>140/90 mmHg). This effect is thought to be due to the presence of vario compounds, such as allicin, alliin, and S-allyl-L-cysteine (SAC), among others, which ha potent anti-inflammatory, antioxidant, and vasodilatory properties [8–10]. Neverthele although garlic is generally well-tolerated in moderate amounts as part of a balanced di some individuals may experience adverse effects when consuming large quantities of ra and cooked garlic or taking high doses of garlic supplements. Therefore, technologi modifications of raw garlic in different forms, such as smoked garlic, garlic oil macera garlic juice, garlic powder, and fermented garlic, have been developed to alleviate th discomforts and increase dietary tolerance.

Aged black garlic (ABG) is a type of modified garlic that contains a unique compo tion of organosulfur compounds and polyphenols compared to fresh garlic [11]. Duri aging, the allicin and other sulfur compounds undergo chemical changes that form n compounds, including SAC and melanoidin [12]. These compounds are responsible ABG's distinctive taste, aroma, and color. Regarding hypertension, ABG has been found have potential benefits due to the high levels of SAC, which have been shown to relax blo vessels, improve blood flow, and reduce oxidative stress [13], with a similar magnitude effects in cardiovascular physiology compared to raw garlic [14].

In this sense, this study aimed to evaluate the antihypertensive effects of a daily inta of a low-dose SAC-optimized ABG extract for 12 weeks in individuals with Grade I hyp tension following blood-pressure-reducing drug therapy. For that purpose, a randomiz triple-blind, placebo-controlled parallel trial was developed, where the primary outco was the reduction in office and home blood pressure. Secondary outcomes such as endot lial function, nitric oxide, inflammatory cytokines, angiotensin-converting enzyme activ and antioxidant capacity levels in blood were determined to explain the antihypertens effect of ABG.

2. Materials and Methods

2.1. Experimental Design

The study was designed as a triple-blind, placebo-controlled, randomized paral study to determine whether an optimized aged black garlic (ABG) extract could redu systolic and diastolic blood pressure in subjects with Grade I hypertension. The stu consisted of two periods. An initial period called the "Run-in phase", two weeks duration, involved all volunteers being treated with a placebo, and blood pressure w monitored daily and served as the basis for the evaluation of future changes in blo pressure. At least five consecutive days of home blood pressure monitoring are requi for a reliable diagnosis of home hypertension [15]. Therefore, two weeks of monitori is adequate to establish volunteers' basal blood pressure. Similarly, it helped determ volunteers that were non-adherent to the study protocol before the experimental phase. T second period was called the "Experimental phase". During this phase, the volunteers w randomly divided into groups—Group A and Group L—and were treated for 12 we with a placebo or active ingredient, respectively. Treatment time was determined based previous studies reported in different meta-analyses on garlic and hypertension [16,17

Volunteers were randomly assigned to each group by sequential number assignm and they were divided by sex after their acceptance to participate in the study. T study was carried out under the ethical principles established in the latest version of Declaration of Helsinki and Tokyo for human investigation and following Regulation (E 2016/679 of the European Parliament of the Council of 27 April 2016 on Data Protecti (RGPD); Law 14/2007 on biomedical research; and Organic Law 3/2018 on the Protect of Personal Data and Guarantee of Digital Rights. Participants provided informed cons before they participated in the study. The study protocol was approved by Hosp Universitari Arnau de Vilanova (CEIC-2435), Lleida, Spain, and the trial is registe at ClinicalTrials.gov NCT04915053. No changes to methods after trial commencement were introduced.

2.2. Intervention

The intervention consisted of a daily intake of one tablet after wake-up and before breakfast. Group L received an optimized ABG10+ -extract tablet manufactured by Pharmactive Biotech Products, S.L.U. (Madrid, Spain), according to a proprietary process. The extract was from garlic harvested in the area of Las Pedroñeras (Castilla la Mancha, Spain), and the tablets were manufactured at Instant Procès (La Roca del Vallès, Spain). Each ABG tablet contained 250 mg of ABG extract, which provided 0.25 mg of SAC per tablet plus 300 mg of the excipients shown in Table 1. Group A received a placebo tablet where the ABG extract was replaced by maltodextrin. Both pills were similar in appearance and smell and coded as A or L by the manufacturer. To guarantee blinding, none of the researchers involved in the study knew the composition of each tablet. Participants were instructed to maintain their dietary and physical habits and medical treatments. Changes in antihypertensive drug treatment were asked about at the end of the experimental phase. None of the volunteers reported changes in antihypertensive drug treatment. Volunteer recruitment and trial beginning started in September 2021 and ended in April 2022.

Table 1. Detailed composition of tablets employed in the study.

	Group A Placebo (mg per Tablet)	Group L ABG Extract (mg per Tablet)
ABG extract	-	250
Maltodextrin	250	-
Microcrystalline cellulose	90	90
Dicalcic phosphate	157	157
Sodium croscarmellose	10	10
Magnesium stearate	7	7
Sodium Alginate	3.06	3.06
Stearic acid	0.03	0.03
Oleic acid	1.54	1.54
Medium-chain triglycerides	2.80	2.80
Ethylcellulose	13.17	13.17
Hydroxypropyl methylcellulose	4.86	4.86
Hydroxypropyl cellulose	4.86	4.86
Talc	2.88	2.88
Titanium oxide	1.80	1.80
Garlic flavor	1.00	1.00

2.3. Study Population

Volunteers were recruited in collaboration with the Atherothrombotic Disease Detection and Treatment Unit (UDETMA) of the Hospital Universitari Arnau de Vilanova (Lleida, Spain) from the database of patients with Grade I hypertension. The inclusion criteria included: men and women over 18 years of age with diagnosed hypertension that were under treatment with one or more antihypertensive drugs, excluding ACE inhibitors. Exclusion criteria included suffering from morbid obesity (body mass index > 35 kg/m^2), serum LDL-cholesterol levels < 115 mg/dL (set point of low or moderate SCORE risk of cardiovascular events), fasting serum glucose levels > 126 mg/dL (cut-off parameter among the criteria of diabetes), anemia (hemoglobin < 13 g/dL in mean and <12 g/dL in women), chronic gastrointestinal disease, being pregnant or intending to become pregnant, lactation, intending to participate or have participated in a clinical trial or nutritional intervention study in the last 30 days before study enrolment, and allergy to garlic.

2.4. Sample Size Determination

The sample size was calculated based on the study published by Rothwell et al. [1 in which it was estimated that the within-individual visit-to-visit variability in SBP w between 11.4 and 14.9 mmHg (mean, standard deviation of 13.1 mmHg). The expect outcome after ABG-extract supplementation was a reduction of 9 mmHg in systolic blo pressure, following the results of a recent meta-analysis of garlic-extract intervention blood pressure management [7]. The number of volunteers needed in each interventi group for a two-sided test, a 0.05 Type I error rate, and an estimated power of 0.90 was

2.5. Outcomes

2.5.1. Blood Pressure

Office and home blood pressure were determined by the same oscillometric automa sphygmomanometers (Medisana BU 512, Medisana GmbH, Neuss, Germany), with a ma mum error tolerance for static pressure of ± 3 mmHg. Volunteers were seated comfortal in a quiet environment for 5 min before blood pressure measurements. Office and hor blood pressure were determined by the average of three recordings of systolic and diastc blood pressure, obtained at 3 min intervals after subjects had been seated on a chair wi their feet on the floor and arms supported at the heart level. For home blood press measurements, volunteers were instructed to take measurements in the morning, 15 n after getting up and urinating (if applicable); to sit comfortably; and to take blood pressu three times at 5 min intervals. They were provided with a notebook where they had record the measurements daily, including the exact time. Additionally, at the time of t appointments, the measurements provided by the volunteer were verified by consulti the memory of the sphygmomanometer and, if necessary, corrected in their notebook bas on the time and date of measurement.

2.5.2. Endothelial Function

The EndoPAT™ parameters reactive hyperemia index (RHI) and augmentation ind (AI_{75}) were assessed to evaluate endothelial function during the experimental peri (basal and 12 weeks after the intervention). Briefly, the thermoneutral room temperatu was always maintained (24 °C) during the analysis. Any restrictive clothing, watches, rings that could interfere with blood flow to the arms and fingers were removed from volunteer. Further, the upper-arm blood pressure cuff was applied snuggly. The volunt was comfortably lying in the study room and relaxed for at least 15 min before the analy The measurement protocol consisted of 1 min of reading, where the correct placement the sensors and the perceived signal were verified, followed by 5 min of basal conditio 5 min of occlusion, and 5 min of post-occlusion recording. Occlusion was performed increasing the blood pressure cuff to a supra-systolic level of at least 60 mmHg abo systolic blood pressure and no less than 200 mmHg. A total absence of signal from occluded hand verified blood-flow cessation. When the appearance of any movement v noted, cuff pressure was increased by an additional 50 mmHg, up to 300 mmHg. RHI a AI_{75} values were automatically calculated by EndoPAT™ 2000 software. Occlusion bord were verified in all measurements, and manual corrections were performed when necess

2.5.3. Nitric Oxide

Nitric oxide was determined by nitrate/nitrite concentration using a fluoromet assay kit (Abcam ab65327, Cambridge, UK) following manufacturer instructions. Ser samples were filtered with a 10 kDa Spin column (Abcam ab93349, Cambridge, UK) deproteinize the samples before the analysis. The analysis consisted of a two-step proc first, nitrate is converted to nitrite by nitrate reductase, and then nitrite reacts with fluorescent probe 2,3 diaminonaphthalene. A nitrite standard calibration curve was us to compare the fluorescence intensity of samples. The analysis was performed in triplic in each sample.

2.5.4. Antioxidant Capacity

Antioxidant capacity was measured by the FRAP (ferric reducing antioxidant power) assay. Briefly, 900 μL of the FRAP reagent containing 2,4,6-tri(2-pyridyl)-s-triazine, $FeCl_3$, and acetate buffer (300 mM, pH 3.6) was mixed with 90 μL of distilled water and 30 μL of plasma, or sodium phosphate buffer for the blank. Maximum absorbance values of this solution at 595 nm were taken every minute for 30 min at 37 °C, using a Beckman DU640 spectrophotometer (Beckman Instruments Inc., Fullerton, CA, USA). Antioxidant capacity was referenced to standards containing known concentrations of 6-hydroxy-2,5,7,8-tetramethyl chroman-2-carboxylic acid (Trolox) and expressed as μM Trolox Equivalents (TE). Total antioxidant capacity was defined as the slope of the increase in absorbance.

2.5.5. Inflammatory Cytokines

The simultaneous assessment of serum concentrations of IL-2, IL-4, IL-5, IL-8, INF-γ, inducible protein 10 (IP-10), and soluble CD40 ligand (sCD40L) was performed using commercially available multiplex bead-based immunoassay kits (Milliplex MAP Human Cytokine/Chemokine Magnetic Bead Panel, HCYTOMAG-60K, Millipore, Billerica, MA, USA) following manufacturer's instructions. Briefly, plasma samples (25 μL/well) or standards (25 μL well) were incubated with 25 μL of the pre-mixed bead sets in pre-wetted 96-well microtiter plates at 4 °C overnight. After washing, 25 μL of the fluorescent detection antibody mixture was added for 30 min, and 25 μL of streptavidin-phycoerythrin was added to each well for 30 min at room temperature. A range of 3.2–10,000 pg/mL recombinant cytokines was used to establish standard curves and maximize the assay's sensitivity and dynamic range. Cytokine levels were determined using a Luminex IS 100 (Luminex, Austin, TX, USA).

2.5.6. Angiotensin-Converting Enzyme (ACE) Activity

ACE activity was determined following the manufacturer's instruction of Spinreact's (41205, Sant Esteve de Bas, Spain) quantitative determination kit. Briefly, in this method, the direct substrate *N*-[3-(2-furyl)-acryloyl]-Lphenylalanylglyclglycine (FAPGG) is hydrolyzed to FAP and glycylglycine. The hydrolysis of FAPGG by volunteer serum ACE results in a decrease in absorbance at 340 nm. A standardized calibrator (Spinreact, 1002225, Sant Esteve de Bas, Spain) was employed to estimate ACE activity in samples.

2.5.7. Statistical Analysis

Statistical analysis was performed as per protocol. Primary and secondary parameters were measured at baseline and after 12 weeks of intervention. Data are presented as mean ± standard deviation of the mean. Two-way ANOVA and multiple comparisons, used to analyze the treatment effect, were performed using Sidak's test. A statistical analysis of initial differences between Groups A and L was performed by a two-tailed unpaired Student's t-test. A Chi-square test was performed for categorical variables. Correlation analysis between the change in blood pressure and confounding variables was performed by Pearson's correlation analysis. To determine the evolution in home blood pressure measurements, linear regressions were conducted between the days of treatment (0–84 days) and the mean of the blood pressure taken daily. Differences in slopes were monitored by a two-tailed t-test analysis. In all cases, the cut-off to determine significant differences between groups was set at a *p*-value below 0.05. Statistical analyses and graphs were achieved with GraphPad Prism v6.

3. Results

3.1. Study Design

The recruitment phase included creating a database of potential volunteers according to the inclusion and exclusion criteria. Of the 303 patients in the database, 89 volunteers showed interest in participating in the study and entered the run-in phase. Eight volunteers did not complete the run-in phase due to placebo intolerances, personal issues, and non-response to the next appointment. Eighty-one volunteers were randomized to particip in one of the two groups of the present study (Group A: Placebo and Group L: Acti compound) and started the experimental phase. Finally, 38 and 39 volunteers from Grou A and L completed the experimental phase. The leading causes of withdrawal from study were non-response to appointments in Groups A and L and intolerance to treatm in Group L, mainly due to gastroesophageal reflux. A flowchart of the study is depicted Figure 1. Table 2 describes the baseline characteristics of the participants in both grou where no differences between groups were found.

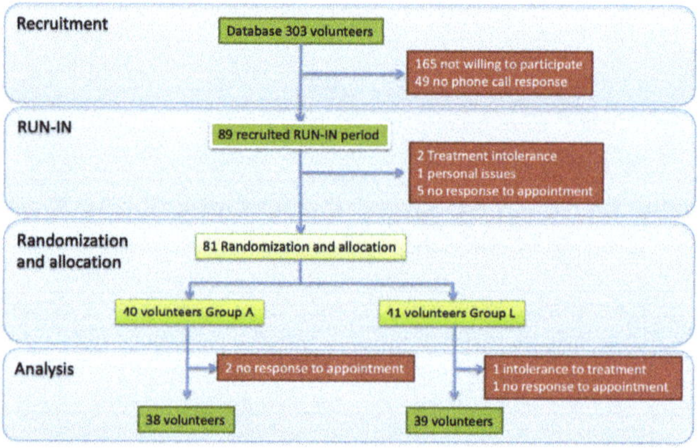

Figure 1. Study and participants flowchart.

3.2. *Compliance and Reported Adverse Events*

Treatment compliance was determined from the number of pills taken divided the total days of the treatment period. Some volunteers took extra treatment days d to volunteer availability adjustments in appointments. The mean treatment days in b groups were 87 ± 5 and 88 ± 5 for groups L and A, respectively. The mean compliance both groups was above 96%, where no difference in compliance was observed betwe groups (p = 0.4396). No data were excluded from the study.

Four volunteers reported treatment adverse events. Two volunteers from Grou (active compound) and one from Group A (placebo group) reported mild gastric discomf while one volunteer from Group A reported muscular cramps.

The treatment blinding of volunteers was executed by asking each participant in last appointment if they believed there where in the active or placebo group. Moreo 23% and 45% of volunteers in Groups L and A, respectively, thought they were in the act group. Therefore, it was determined that volunteer treatment blinding was appropria conducted. Regarding the researcher blinding, none of the researchers were aware of wh group the active compound belonged to until the finalization of the statistical analysis.

3.3. *Blood Pressure, Endothelial Function, and Related Variables*

Changes in blood pressure were determined through two methods: the meas ments in the office at baseline and 12 weeks and the daily measurement performed the volunteers at home. The results of the office pressure measurements are shown Table 3. At the time of the visit, no significant treatment effect was observed on syst and diastolic blood pressure based on two-way ANOVA analysis. However, post-h studies suggest that Group A (placebo) had significantly reduced systolic blood press levels. On the other hand, the evolution in the change in daily blood pressure was a analyzed (Figures 2 and 3 for systolic and diastolic blood pressure, respectively). Un these conditions, Group L's treatment showed a reduction in systolic and diastolic bl pressure levels of about 1.8 and 1.5 mmHg, respectively, after 12 weeks of treatment. 95% confidence intervals for these changes in blood pressure were from a maximum red tion of 4.1 and 3.0 mmHg and an increase of 0.7 and 0.3 mmHg in systolic and diastolic pressure, respectively.

Table 2. Baseline characteristics of participants by group allocation. Data are reported as mean and standard deviation. Statistical differences between groups were analyzed by two-tailed unpaired Student *t*-tests. *p*-values below 0.05 were considered as a statistical difference.

	Group A (Placebo)	Group L (ABG Extract)	*p*-Value
. of volunteers	38	39	
nder	Male: 20; Female: 18	Male: 20; Female 19	0.9057
e (years)	64.1 ± 5.9	63.7 ± 5.8	0.8018
stolic blood pressure (mmHg)	145.5 ± 17.2	148.8 ± 22.9	0.4876
astolic blood pressure (mmHg)	85.3 ± 10.8	85.7 ± 11.2	0.8765
lse pressure (mmHg)	60.0 ± 12.9	62.7 ± 15.7	0.4300
ars of diagnosis	15.4 ± 8.5	14.8 ± 11.2	0.8192
tihypertensive drug number			
1 type	13	13	0.6219
Two types	13	10	
Three or more types	12	16	
tihypertensive drug type			
Thiazides	25	27	0.9811
β-blockers	11	14	
Angiotensin II receptor blocker	15	16	
Calcium channel blockers	14	17	
ner types of drugs			
Statins	15	17	0.7826
Benzodiazepines	10	9	
Anticoagulants	8	10	
Proton pump inhibitors	12	8	
Jonsteroidal anti-inflammatory drug	14	10	
Antidiabetic	2	4	
atment adherence	96.7 ± 4.7	97.8 ± 6.8	0.4396
nimum	85%	72%	
ximum	108%	112%	
atment blinding			
/No	17/21	9/30	0.0559

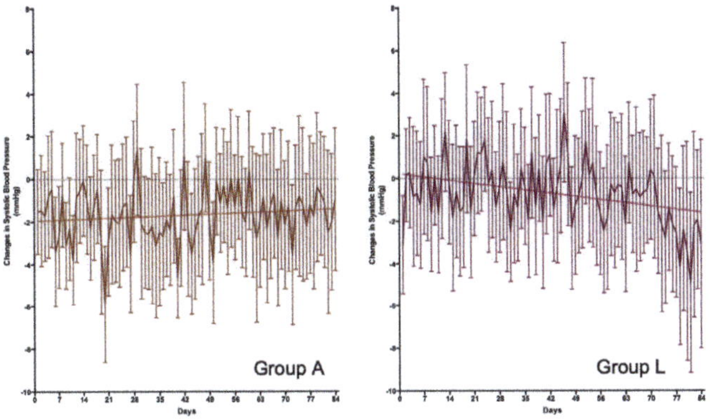

Figure 2. *Daily changes in systolic blood pressure.* Differences in the evolution of blood pressure were observed between Groups A and L ($p = 0.0033$ for the differences in slope). Group A ($n = 38$) did not show a linear regression ($p = 0.3044$), with a mean increase in systolic blood pressure of 6.7×10^{-3} mmHg/day after 12 weeks of treatment. Group L ($n = 39$) shows a good fit for a lineal regression ($p = 0.0021$), with a mean decrease in systolic blood pressure of 21.6×10^{-3} mmHg/day after 12 weeks of treatment.

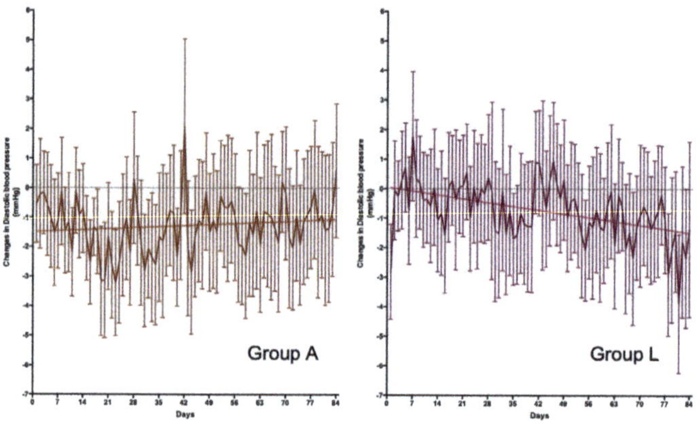

Figure 3. Daily changes in diastolic blood pressure. Differences in the evolution of blood press were observed between Groups A and L (p = 0.0002 for the differences in slope). Group A (n = did not show a linear regression (p = 0.2648), with a mean increase in diastolic blood pressure 5.1×10^{-3} mmHg/day after 12 weeks of treatment. Group L (n = 39) shows a good fit for a lin regression (p < 0.0001), with a mean decrease in diastolic blood pressure of 18.4×10^{-3} mmHg/c after 12 weeks of treatment.

Table 3. Changes in blood pressure, endothelial function, and related variables. Data are reported mean and standard deviation. Statistical differences between groups were analyzed by repeated m sures of two-way ANOVA (Basal-12 weeks) and Sidak's multiple comparison tests to determine ferences between time in both groups. p-values below 0.05 were considered as a statistical differen

	Group A (Placebo) (n = 38)			**Group L (ABG Extract) (n = 39)**			
	Basal	**12 Weeks**	**Sidak's p-Value**	**Basal**	**12 Weeks**	**Sidak's p-Value**	**Two-Wa ANOVA p-Valu**
Blood pressure (office)							
Systolic (mmHg)	144 ± 13	138 ± 14	0.0169	146 ± 21	143 ± 18	0.3011	0.3011
Diastolic (mmHg)	85 ± 7	83 ± 10	0.2497	85 ± 10	83 ± 9	0.2889	0.9219
Pulse pressure (mmHg)	59 ± 12	55 ± 14	0.0632	61 ± 16	60 ± 14	0.6126	0.2011
Endothelial function							
RHI	2.059 ± 0.414	2.033 ± 0.597	0.9583	2.016 ± 0.569	2.068 ± 0.728	0.8276	0.9753
AI_{75}	30.4 ± 14.9	26.7 ± 17.2	0.1767	25.0 ± 17.0	21.1 ± 15.1	0.1139	0.1185
Blood lipid profile							
Total cholesterol (mg/dL)	188 ± 43	188 ± 43	>0.9999	209 ± 40	209 ± 39	0.9984	0.9759
Triacylglycerides (mg/dL)	107 ± 53	107 ± 55	0.9986	122 ± 53	117 ± 43	0.8725	0.1394
HDL-cholesterol (mg/dL)	58 ± 15	58 ± 15	0.9939	59 ± 17	61 ± 21	0.8892	0.4172
Inflammatory cytokines							
IL-2 (pg/mL)	0.190 ± 0.657	0.223 ± 0.790	0.9021	0.894 ± 1.844	1.022 ± 2.043	0.2236	0.0282
IL-4 (pg/mL)	113 ± 541	121 ± 566	0.4704	213 ± 708	215 ± 705	0.9556	0.5079
IL-5 (pg/mL)	3.8 ± 10.5	3.6 ± 10.4	0.8234	6.4 ± 16.1	6.5 ± 16.0	0.9021	0.3859
IL-8 (pg/mL)	20.0 ± 29.8	21.6 ± 33.7	0.2054	25.0 ± 31.6	26.7 ± 33.0	0.2018	0.4948
IP-10 (pg/mL)	490 ± 252	446 ± 231	0.3753	450 ± 202	539 ± 332	0.0218	0.6238
sCD40L (pg/mL)	3856 ± 3927	2322 ± 2512	0.0543	3742 ± 3001	3211 ± 3337	0.6710	0.4978
Other variables							
Nitric oxide (μM)	88 ± 23	90 ± 27	0.9188	87 ± 19	97 ± 31	0.0254	0.5101
ACE activity (U/L)	78 ± 44	71 ± 41	0.1495	78 ± 40	69 ± 38	0.0398	0.9560
Uric acid (mg/dL)	5.6 ± 1.4	5.6 ± 1.3	0.9987	6.0 ± 1.4	5.7 ± 1.3	0.0483	0.3986
FRAP (μM TE/min)	0.111 ± 0.034	0.113 ± 0.029	0.6479	0.105 ± 0.036	0.113 ± 0.043	0.0161	0.6972

No changes in endothelial function parameters (RHI and AI_{75} values) were observed in either group. Similarly, no significant differences in the blood lipid profile were observed (Table 3).

Regarding the variables that could explain ABG-extract effects on blood pressure, except for IL-2 levels, there were no significant changes due to treatment in the variables included in Table 3. However, post-hoc analyses suggest that Group L increased nitric oxide, antioxidant capacity measured by the FRAP method, cytokine IP-10 levels, and reduced ACE activity and blood uric acid levels. On the other hand, Group A presented a reduction in the levels of the cytokine sCD40L.

3.4. *Confounding Variables Analysis*

No differences in response to the treatment were observed due to sex (*p*-value of 0.7544 and 0.8484 for systolic and diastolic blood pressure) or the intake of blood pressure-lowering drugs (*p*-value of 0.1183). Nevertheless, it was observed that the volunteers with lower basal values of RHI ($r = -0.311$ and *p*-value 0.0395) and higher basal levels of the cytokine IP10 ($r = 0.0077$ and *p*-value 0.0077) in Group L where the ones with a lower response to the treatment.

4. Discussion

This study aimed to validate whether daily supplementation with an optimized black-garlic extract could help to control blood pressure in medicated people with Grade I hypertension. The results of this study suggest a drop in systolic and diastolic blood pressure of 1.8 and 1.5 mmHg, respectively, at values measured at home, not observing significant changes in the measurements taken in the office.

Garlic has repeatedly shown a beneficial effect in lowering blood pressure [19]. The hypotensive response has been found in many products obtained from fresh, dried garlic and processed garlic through fermentation or aging. Some meta-analyses and reviews agree with such conclusions [17]. The calculated reduction in blood pressure is higher than the results shown in the current study. However, the conclusion presented in previous meta-analyses and reviews could be biased by notorious design flaws in the selected clinical trials. In some cases, the botanic material remained uncharacterized; in other cases, the active compound was not a defined traditional local product [20], or the study included additional interventions such as dieting.

On the other hand, it has also been shown that a daily ABG-extract supplementation, even at low doses of SAC (0.25 mg/day), induces significant effects on blood pressure control. This last point is critical to highlight since it demonstrates garlic extracts at low doses that are effective without inducing commonly reported side effects [21] or other adverse effects. For example, case reports have highlighted the possibility that garlic use at high doses may cause allergic reactions, the alteration of platelet function and coagulation [22], and gastrointestinal discomfort, among other things. Different sulfur-containing compounds may be responsible for the anticoagulant and antiplatelet effect [23]. Indeed, large quantities of garlic have been discouraged by reputed authorities before surgery or by persons under warfarin or other anticoagulant treatment because it may increase bleeding time. In this study, no changes in the levels of sCD40L have been observed, suggesting no potential effects of ABG consumption on coagulation [24]. Finally, the results of this study showed that only three volunteers reported mild discomfort, which did not interfere with treatment adherence.

The reduction in blood pressure measured at home is considered even more relevant than the measurement in the office since recent meta-analyses have indicated that home blood pressure monitoring better predicts cardiovascular morbidity and mortality than office blood pressure [25]. In this context, it is considered that for every 10 mmHg reduction in blood pressure, the risk of developing cardiovascular events and mortality is reduced by 13% [1]. The drop in blood pressure observed in this study did not reach 10 mmHg. However, it should be considered that even a modest blood pressure reduction can lead to meaningful gains in preventing incidents or recurrent cardiovascular disease [26]. Inde as observed in this study, even a 2 mmHg lower usual systolic blood pressure would me about 10% lower stroke mortality and about 7% lower mortality from ischaemic he disease or other vascular causes in middle age [27]. Similarly, it should be considered th the volunteers included in this study have a medical history of hypertension of an avera of 15 years, where despite the pharmacological treatment, they showed high blood pressu levels. Adding ABG extract to treat arterial hypertension effectively controls blood pressu in a population resistant to drug treatment without inducing side effects.

The changes in blood pressure observed in this study are consistent with previo observations of the effect of the same ABG extract on cardiovascular physiology. A clini trial in subjects with moderate hypercholesterolemia reported reduced diastolic blo pressure, particularly in men with diastolic blood pressure above 75 mmHg [28], witho changes in systolic blood pressure. However, it should be noted that the changes blood pressure were observed in subjects without hypertension diagnosis and medicati Similarly, in in-vitro studies with rat hearts subjected to ischemia and reperfusion using t Langendorff technique, it was observed that ABG extracts directly increased the relea of nitric oxide in aorta segments, inducing a relaxing effect on coronary arteries [11]. T previously described mechanisms of action suggest that the results may be related to t increased relaxation capacity of the endothelial wall. In this sense, it has been observ that ABG could induce its effects through an increase in nitric oxide levels; the antioxida capacity of the blood to preserve nitric oxide; the reduction of uric acid; and by reduci ACE activity. Other researchers have demonstrated the acute effects of other black-gar extracts on the relaxation capacity of the arterial wall, consistent with an increase in N levels [29]. On the other hand, in another study, it was not observed that vasorelaxati effects could be mediated by an increase in H_2S [30]. Nevertheless, no changes we observed in endothelial-relaxation capacity measured by RHI and AI_{75} indexes in th study. A similar finding is reported with other garlic extracts [31]. Interestingly, t changes in RHI were higher in subjects with higher initial values of RHI, suggesting th the volunteers that presented a reduced vasodilation capacity after hypoxia would r respond to ABG-extract treatment.

Regarding inflammation markers, previous studies suggest that components of A extract present anti-inflammatory properties in models of lipopolysaccharide toxin flammation induction [32]. In this study, the effects on inflammation markers were n significant, possibly due to the low inflammatory status of the volunteers and the l doses of ABG extract used. An increase in levels of IL-2 was observed at the end of t experimental period; however, this modification was observed in both groups, and it w considered a time-dependent effect. Similarly, there was an increase in the levels of IP-10 the ABG-extract treatment group, although the observed levels were in the range observ in healthy people (2205–4700 pg/mL) [33].

It is suggested that combination therapy is more effective than increasing dos of a single blood-pressure-reducing drug [34]. ABG-extract supplementation could involved in several controlling mechanisms. The reduction of oxidative stress and t antihypertensive effect via ACE inhibition have been proposed as well as H_2S-regulati and NO-stimulating properties [35]. The results from this study did not suggest a difference in blood pressure improvement in the volunteers with combination thera (two or more drugs) compared to monotherapy. However, due to the wide variability medications used by the volunteers, it is difficult to determine if there is an interacti between the ABG extract and the type and number of medications used. Similarly, ABG extract–dietary interaction was observed. No correlation was observed between intake of dietary raw garlic and onion and the change in blood pressure (with the char in systolic blood pressure $p = 0.6656$ and $p = 0.2638$ for garlic and onion, respective and for the difference in diastolic blood pressure $p = 0.6392$ and $p = 0.5370$ for garlic a onion, respectively).

This study has some limitations. Although a significant reduction in blood pressure was observed, the time of the day of administration can be optimized. Treatment-time differences affect the beneficial and adverse effects of different blood pressure-reducing drugs [36]. For example, α-adrenoceptor antagonists and β-blockers seem more effective upon awaking. In contrast, calcium-channel blockers, angiotensin-converting enzyme inhibitors, angiotensin II receptor blockers, and diuretics have been reported to be more effective at bedtime. How ABG extract improves blood pressure cannot be attributed to a unique mechanism of action, as described with drug therapy. Therefore, it is difficult to ascertain the ideal time and number of administration doses. For that purpose, ambulatory blood pressure monitoring should be employed in further studies to validate the ABG effect in a 24 h period and at different amounts and administration times. Similarly, this study lacks a dose-effect analysis. Previous studies have reported a dose-dependent effect, ranging between 240 and 960 mg of SAC/day [37]. Nevertheless, in this last study, it was observed that higher doses of an aged garlic extract induced a lower reduction in blood pressure. This could be linked to poorer compliance with and lesser tolerability of the extract to a different composition of aged garlic and ABG in bioactive compounds. In this sense, an analysis of the effectiveness of lower doses of ABG extract, its treatment optimization in the daytime, and the number of doses to be employed is desirable.

5. Conclusions

Based on the results obtained in this study, it can be concluded that daily supplementation with ABG extracts with low doses of SAC (0.25 mg/day) can induce significant reductions in blood pressure in individuals with Grade I hypertension and antihypertensive drug therapy.

Author Contributions: CRediT author statement: J.C.E.S.: Conceptualization, Methodology, Validation, Formal analysis, Writing, Review and Editing. E.C.-B.: Investigation. A.G.-C.: Investigation. M.I.M.-V.: Funding acquisition, Resources, Review and Editing. D.G.-H.: Funding acquisition, Resources, Review and Editing. M.B.-L.: Investigation. J.M.V.: Methodology, Resources. A.E.E.: Methodology, Resources, Funding Acquisition, Review and Editing. M.P.-O.: Conceptualization, Methodology, Validation, Writing, Review and Editing. All authors have read and agreed to the published version of the manuscript.

Funding: This research and APC was funded by Pharmactive Biotech S.L.U. grant number C21024.

Institutional Review Board Statement: The study was conducted by the Declaration of Helsinki and approved by the Institutional Review Board (or Ethics Committee) of Hospital Universitari Arnau de Vilanova (CEIC-2435, 25 March 2021) for studies involving humans.

Informed Consent Statement: Informed consent was obtained from all subjects involved in the study.

Data Availability Statement: The data presented in this study are available on request from the corresponding author.

Acknowledgments: We thank the personnel of the Biobank and Pharmacology units of the IRBLleida research institute for assistance with handling human samples.

Conflicts of Interest: A García-Carrasco, J.M. Valdivielso, and M. Portero-Otin do not declare any potential competing interests. Jose CE Serrano reports that Pharmactive Biotech Products S.L.U. provided financial support for study development. Maria Ines Moran Valero, Alberto E. Espinel, and Daniel Gonzalez Hedstrom report a relationship with Pharmactive Biotech Products SL that includes the employment of Pharmactive S.L., the company that produces and distributes the garlic extract used in this study.

References

1. Ettehad, D.; Emdin, C.A.; Kiran, A.; Anderson, S.G.; Callender, T.; Emberson, J.; Chalmers, J.; Rodgers, A.; Rahimi, K. Blo Pressure Lowering for Prevention of Cardiovascular Disease and Death: A Systematic Review and Meta-Analysis. *Lan* **2016**, *387*, 957–967. [CrossRef]
2. Chow, C.K.; Teo, K.K.; Rangarajan, S.; Islam, S.; Gupta, R.; Avezum, A.; Bahonar, A.; Chifamba, J.; Dagenais, G.; Diaz, R.; et Prevalence, Awareness, Treatment, and Control of Hypertension in Rural and Urban Communities in High-, Middle-, a Low-Income Countries. *JAMA* **2013**, *310*, 959–968. [CrossRef] [PubMed]
3. Lamirault, G.; Artifoni, M.; Daniel, M.; Barber-Chamoux, N.; Nantes University Hospital Working Group on Hypertensi Resistant Hypertension: Novel Insights. *Curr. Hypertens. Rev.* **2020**, *16*, 61–72. [CrossRef]
4. Borghi, C.; Cicero, A.F.G. Nutraceuticals with a Clinically Detectable Blood Pressure-Lowering Effect: A Review of Availa Randomized Clinical Trials and Their Meta-Analyses. *Br. J. Clin. Pharmacol.* **2017**, *83*, 163–171. [CrossRef] [PubMed]
5. Vazquez-Prieto, M.A.; Miatello, R.M. Organosulfur Compounds and Cardiovascular Disease. *Mol. Aspects Med.* **2010**, *31*, 540–5 [CrossRef]
6. Kris-Etherton, P.M.; Hecker, K.D.; Bonanome, A.; Coval, S.M.; Binkoski, A.E.; Hilpert, K.F.; Griel, A.E.; Etherton, T.D. Bioacti Compounds in Foods: Their Role in Preventing Cardiovascular Disease and Cancer. *Am. J. Med.* **2002**, *113*, 71–88. [CrossRef]
7. Ried, K. Garlic Lowers Blood Pressure in Hypertensive Individuals, Regulates Serum Cholesterol, and Stimulates Immunity: Updated Meta-Analysis and Review. *J. Nutr.* **2016**, *146*, 389S–396S. [CrossRef]
8. Piragine, E.; Citi, V.; Lawson, K.; Calderone, V.; Martelli, A. Regulation of Blood Pressure by Natural Sulfur Compounds: Fo on Their Mechanisms of Action. *Biochem. Pharmacol.* **2022**, *206*, 115302. [CrossRef]
9. Recinella, L.; Chiavaroli, A.; Masciulli, F.; Fraschetti, C.; Filippi, A.; Cesa, S.; Cairone, F.; Gorica, E.; De Leo, M.; Braca, et al. Protective Effects Induced by a Hydroalcoholic *Allium sativum* Extract in Isolated Mouse Heart. *Nutrients* **2021**, *13*, 23 [CrossRef]
10. Recinella, L.; Gorica, E.; Chiavaroli, A.; Fraschetti, C.; Filippi, A.; Cesa, S.; Cairone, F.; Martelli, A.; Calderone, V.; Veschi, et al. Anti-Inflammatory and Antioxidant Effects Induced by *Allium sativum* L. Extracts on an Ex Vivo Experimental Mode Ulcerative Colitis. *Foods* **2022**, *11*, 3559. [CrossRef] [PubMed]
11. García-Villalón, A.L.; Amor, S.; Monge, L.; Fernández, N.; Prodanov, M.; Muñoz, M.; Inarejos-García, A.M.; Granado, M. In Vi Studies of an Aged Black Garlic Extract Enriched in S-Allylcysteine and Polyphenols with Cardioprotective Effects. *J. Funct. Fo* **2016**, *27*, 189–200. [CrossRef]
12. Lu, X.; Li, N.; Qiao, X.; Qiu, Z.; Liu, P. Composition Analysis and Antioxidant Properties of Black Garlic Extract. *J. Food Drug A* **2017**, *25*, 340–349. [CrossRef]
13. Liu, J.; Zhang, G.; Cong, X.; Wen, C. Black Garlic Improves Heart Function in Patients With Coronary Heart Disease by Improv Circulating Antioxidant Levels. *Front. Physiol.* **2018**, *9*, 1435. [CrossRef]
14. Czompa, A.; Szoke, K.; Prokisch, J.; Gyongyosi, A.; Bak, I.; Balla, G.; Tosaki, A.; Lekli, I. Aged (Black) versus Raw Garlic agai Ischemia/Reperfusion-Induced Cardiac Complications. *Int. J. Mol. Sci.* **2018**, *19*, 1017. [CrossRef] [PubMed]
15. Groenland, E.H.; Bots, M.L.; Visseren, F.L.J.; McManus, R.J.; Spiering, W. Number of Measurement Days Needed for Obtainin Reliable Estimate of Home Blood Pressure and Hypertension Status. *Blood Press.* **2022**, *31*, 100–108. [CrossRef]
16. Ried, K. Garlic Lowers Blood Pressure in Hypertensive Subjects, Improves Arterial Stiffness and Gut Microbiota: A Review a Meta-Analysis. *Exp. Ther. Med.* **2020**, *19*, 1472–1478. [CrossRef]
17. Wang, H.P.; Yang, J.; Qin, L.Q.; Yang, X.J. Effect of Garlic on Blood Pressure: A Meta-Analysis. *J. Clin. Hypertens.* **2015**, *17*, 223– [CrossRef] [PubMed]
18. Rothwell, P.M.; Howard, S.C.; Dolan, E.; O'Brien, E.; Dobson, J.E.; Dahlöf, B.; Sever, P.S.; Poulter, N.R. Prognostic Significanc Visit-to-Visit Variability, Maximum Systolic Blood Pressure, and Episodic Hypertension. *Lancet* **2010**, *375*, 895–905. [CrossR [PubMed]
19. Xiong, X.J.; Wang, P.Q.; Li, S.J.; Li, X.K.; Zhang, Y.Q.; Wang, J. Garlic for Hypertension: A Systematic Review and Meta-Anal of Randomized Controlled Trials. *Phytomedicine* **2015**, *22*, 352–361. [CrossRef]
20. Nakasone, Y.; Nakamura, Y.; Yamamoto, T.; Yamaguchi, H. Effect of a Traditional Japanese Garlic Preparation on Blood Press in Prehypertensive and Mildly Hypertensive Adults. *Exp. Ther. Med.* **2013**, *5*, 399–405. [CrossRef]
21. Borrelli, F.; Capasso, R.; Izzo, A.A. Garlic (*Allium sativum* L.): Adverse Effects and Drug Interactions in Humans. *Mol. Nutr. F Res.* **2007**, *51*, 1386–1397. [CrossRef] [PubMed]
22. Vaes, L.P.J.; Chyka, P.A. Interactions of Warfarin with Garlic, Ginger, Ginkgo, or Ginseng: Nature of the Evidence. *A Pharmacother.* **2000**, *34*, 1478–1482. [CrossRef] [PubMed]
23. Ryu, J.H.; Kang, D. Physicochemical Properties, Biological Activity, Health Benefits, and General Limitations of Aged Bl Garlic: A Review. *Molecules* **2017**, *22*, 919. [CrossRef]
24. Pamukcu, B.; Lip, G.Y.H.; Snezhitskiy, V.; Shantsila, E. The CD40-CD40L System in Cardiovascular Disease. *Ann. Med.* **2011**, 331–340. [CrossRef]
25. Bliziotis, I.A.; Destounis, A.; Stergiou, G.S. Home versus Ambulatory and Office Blood Pressure in Predicting Target Or Damage in Hypertension: A Systematic Review and Meta-Analysis. *J. Hypertens.* **2012**, *30*, 1289–1299. [CrossRef]

Canoy, D.; Nazarzadeh, M.; Copland, E.; Bidel, Z.; Rao, S.; Li, Y.; Rahimi, K. How Much Lowering of Blood Pressure Is Required to Prevent Cardiovascular Disease in Patients With and Without Previous Cardiovascular Disease? *Curr. Cardiol. Rep.* **2022**, *24*, 851. [CrossRef] [PubMed]
Lewington, S.; Clarke, R.; Qizilbash, N.; Peto, R.; Collins, R. Age-Specific Relevance of Usual Blood Pressure to Vascular Mortality: A Meta-Analysis of Individual Data for One Million Adults in 61 Prospective Studies. *Lancet* **2002**, *360*, 1903–1913. [CrossRef]
Valls, R.M.; Companys, J.; Calderón-Pérez, L.; Salamanca, P.; Pla-Pagà, L.; Sandoval-Ramírez, B.A.; Bueno, A.; Puzo, J.; Crescenti, A.; Del Bas, J.M.; et al. Effects of an Optimized Aged Garlic Extract on Cardiovascular Disease Risk Factors in Moderate Hypercholesterolemic Subjects: A Randomized, Crossover, Double-Blind, Sustainedand Controlled Study. *Nutrients* **2022**, *14*, 405. [CrossRef]
Baik, J.S.; Min, J.H.; Ju, S.M.; Ahn, J.H.; Ko, S.H.; Chon, H.S.; Kim, M.S.; Shin, Y. Il Effects of Fermented Garlic Extract Containing Nitric Oxide Metabolites on Blood Flow in Healthy Participants: A Randomized Controlled Trial. *Nutrients* **2022**, *14*, 5238. [CrossRef]
Leitão, R.; de Oliveira, G.V.; Rezende, C.; Volino-Souza, M.; Mesquita, J.; de Carvalho, L.L.; Alvares, T.S. Improved Microvascular Reactivity after Aged Garlic Extract Intake Is Not Mediated by Hydrogen Sulfide in Older Adults at Risk for Cardiovascular Disease: A Randomized Clinical Trial. *Eur. J. Nutr.* **2022**, *61*, 3357–3366. [CrossRef]
Gruenwald, J.; Bongartz, U.; Bothe, G.; Uebelhack, R. Effects of Aged Garlic Extract on Arterial Elasticity in a Placebo-Controlled Clinical Trial Using EndoPAT^{TM} Technology. *Exp. Ther. Med.* **2020**, *19*, 1490–1499. [CrossRef]
Kim, D.G.; Kang, M.J.; Hong, S.S.; Choi, Y.H.; Shin, J.H. Antiinflammatory Effects of Functionally Active Compounds Isolated from Aged Black Garlic. *Phytother. Res.* **2017**, *31*, 53–61. [CrossRef]
Kim, H.O.; Kim, H.S.; Youn, J.C.; Shin, E.C.; Park, S. Serum Cytokine Profiles in Healthy Young and Elderly Population Assessed Using Multiplexed Bead-Based Immunoassays. *J. Transl. Med.* **2011**, *9*, 113. [CrossRef] [PubMed]
Paz, M.A.; De-La-Sierra, A.; Sáez, M.; Barceló, M.A.; Rodríguez, J.J.; Castro, S.; Lagarón, C.; Garrido, J.M.; Vera, P.; Coll-De-Tuero, G. Treatment Efficacy of Anti-Hypertensive Drugs in Monotherapy or Combination: ATOM Systematic Review and Meta-Analysis of Randomized Clinical Trials According to PRISMA Statement. *Medicine* **2016**, *95*, e4071. [CrossRef]
Shouk, R.; Abdou, A.; Shetty, K.; Sarkar, D.; Eid, A.H. Mechanisms Underlying the Antihypertensive Effects of Garlic Bioactives. *Nutr. Res.* **2014**, *34*, 106–115. [CrossRef] [PubMed]
Smolensky, M.H.; Hermida, R.C.; Ayala, D.E.; Tiseo, R.; Portaluppi, F. Administration-Time-Dependent Effects of Blood Pressure-Lowering Medications: Basis for the Chronotherapy of Hypertension. *Blood Press. Monit.* **2010**, *15*, 173–180. [CrossRef] [PubMed]
Ried, K.; Frank, O.R.; Stocks, N.P. Aged Garlic Extract Reduces Blood Pressure in Hypertensives: A Dose-Response Trial. *Eur. J. Clin. Nutr.* **2013**, *67*, 64–70. [CrossRef]

nutrients

MDPI

Article

Totum-070, a Polyphenol-Rich Plant Extract, Prevents Hypercholesterolemia in High-Fat Diet-Fed Hamsters by Inhibiting Intestinal Cholesterol Absorption

Cédric Langhi [1,*], Marie Vallier [1], Yolanda F. Otero [1], Maheva Maura [2], Florian Le Joubioux [2], Hugo Groult [3], Oussama Achour [4], Ratna Budhi Pebriana [5], Martin Giera [5], Bruno Guigas [6], Thierry Maugard [3], Benoit Chassaing [7], Sébastien Peltier [8], Jean-Marie Bard [9] and Pascal Sirvent [1]

1 R&D Riom Center, Valbiotis, 20-22 rue Henri et Gilberte Goudier, 63200 Riom, France
2 R&D Center, Valbiotis, 23 Avenue Albert Einstein, 17000 La Rochelle, France
3 Equipe BCBS (Biotechnologies et Chimie des Bioressources pour la Santé), UMR (Unité Mixte de Recherc CNRS (Centre National de la Recherche Scientifique) 7266 LIENSs (LIttoral ENvironnement Et Sociétés), Rochelle Université, 17042 La Rochelle, France
4 BioAqtiv, Equipe BCBS (Biotechnologies et Chimie des Bioressources pour la Santé), LIENSs (LIttoral ENvironnement Et Sociétés), UMR (Unité Mixte de Recherche) 7266 CNRS (Centre National de la Recher Scientifique), La Rochelle Université, 17042 La Rochelle, France
5 Center for Proteomics and Metabolomics, Leiden University Medical Center, Albi-nusdreef 2, 2333 ZA Lei The Netherlands
6 Department of Parasitology, Leiden University Medical Center, Albinusdreef 2, 2333 ZA Leiden, The Netherlands
7 Team "Mucosal Microbiota in Chronic Inflammatory Diseases", Institut Cochin, INSERM (Institut National la Santé et de la Recherche Médicale) U1016, CNRS UMR 8104, Université Paris Cité, 75014 Paris, France
8 R&D Périgny Center, Valbiotis, 12F rue Paul Vatine, 17180 Périgny, France
9 Laboratoire de Biochimie Générale et Appliquée, UFR (Unité de Formation et de Recherche) de Pharmac ISOMer-UE 2160, IUML-Institut Universitaire Mer et Littoral-FR3473 CNRS, Université de Nantes, 44035 Nantes, France
* Correspondence: cedric.langhi@valbiotis.com; Tel.: +33-4-63-54-50-04

Citation: Langhi, C.; Vallier, M.; Otero, Y.F.; Maura, M.; Le Joubioux, F.; Groult, H.; Achour, O.; Pebriana, R.B.; Giera, M.; Guigas, B.; et al. Totum-070, a Polyphenol-Rich Plant Extract, Prevents Hypercholesterolemia in High-Fat Diet-Fed Hamsters by Inhibiting Intestinal Cholesterol Absorption. *Nutrients* **2023**, *15*, 5056. https://doi.org/10.3390/nu15245056

Academic Editors: Sonia Ramos Rivero and María Angeles Martín

Received: 3 November 2023
Revised: 6 December 2023
Accepted: 7 December 2023
Published: 9 December 2023

Abstract: Atherosclerotic cardiovascular disease is the leading cause of mortality worldwide, a hypercholesterolemia is a central risk factor for atherosclerosis. This study evaluated the effects Totum-070, a plant-based polyphenol-rich supplement, in hamsters with high-fat diet (HFD)-indu dyslipidemia. The molecular mechanisms of action were explored using human Caco2 enterocy Totum-070 supplementation reduced the total cholesterol (−41%), non-HDL cholesterol (−47 and triglycerides (−46%) in a dose-dependent manner, compared with HFD. HFD-induced hep steatosis was also significantly decreased by Totum-070, an effect associated with the reduction various lipid and inflammatory gene expression. Upon challenging with olive oil gavage, the p prandial triglyceride levels were strongly reduced. The sterol excretion in the feces was increased the HFD-Totum-070 groups compared with the HFD group and associated with reduction of intesti cholesterol absorption. These effects were confirmed in the Caco2 cells, where incubation w Totum-070 inhibited cholesterol uptake and apolipoprotein B secretion. Furthermore, a microbi composition analysis revealed a strong effect of Totum-070 on the alpha and beta diversity of bacte species and a significant decrease in the Firmicutes to Bacteroidetes ratio. Altogether, our findi indicate that Totum-070 lowers hypercholesterolemia by reducing intestinal cholesterol absorpti suggesting that its use as dietary supplement may be explored as a new preventive strategy cardiovascular diseases.

Keywords: triglycerides; lipoproteins; liver; dyslipidemias; enterocytes; polyphenols; nutraceuti

1. Introduction

Atherosclerotic cardiovascular disease (ASCVD) is the main cause of mortality worldwide [1,2]. It is well-established that an increased plasma low-density lipoprotein cholesterol (LDL-C) level is the primary risk factor for the development of coronary heart disease and atherosclerosis [3,4]. Therefore, lifestyle modifications, particularly a healthy diet, are important steps in ASCVD primary prevention. However, lifestyle changes are not easily achieved and maintained. In people with moderate cholesterol levels, dietary supplements may contribute to improving the health status in addition to a balanced diet. In recent years, many animal studies and clinical trials have shown the beneficial effects of polyphenolic compounds as cholesterol-lowering agents [5–8]. Many plants have been explored to reduce blood cholesterol, such as artichokes [9,10], olive leaves [11,12], and goji berries [13]. The hypolipidemic effect of artichoke leaf extracts has been well described in various animal models, such as high-fat-fed [14,15] and alloxan-induced diabetic rats receiving a daily gavage of artichoke leaf extract [16]. Furthermore, Qiang et al. demonstrated that Golden Syrian hamsters fed a cholesterol-enriched diet supplemented with artichoke leaf extract prevented hypercholesterolemia [17]. The effect of olive leaf extract on lipid metabolism has been described in HFD-fed and ovariectomized rats supplemented with olive leaf extract [18,19]. Wang et al. also showed that New Zealand rabbits fed an HFD for 6 weeks with olive leaf extract exhibited a reduction in hyperlipidemia [20]. The hypolipidemic effect of Goji fruit extract was demonstrated in rats fed an HFD for 8 weeks [21]. This effect was reproducible in other studies with rats fed a cholesterol-enriched diet for 45 days [22] or for 60 days [23]. Luo et al. demonstrated in alloxan-induced diabetic rabbits that 10 days of oral administration of Goji fruit extract reduced hypercholesterolemia [24]. Finally, the hypocholesterolemic effect of black pepper has been reported in various animal models [25,26].

Totum-070 is a polyphenol-rich formulation based on five plant extracts (olive leaves (*Olea europaea* L.), artichoke leaves (*Cynara scolymus* L.), chrysanthellum, aerial part, (*Chrysanthellum indicum* subsp. *afroamericanum* B.L. *Turner*), goji fruits (*Lycium barbarum*), and black pepper fruits (*Piper nigrum* L.)) to lower cholesterol in individuals with mild to moderate hypercholesterolemia (French priority patent number: FR1559965). The aim of this study was to evaluate the effects of a 12-week supplementation with Totum-070 on hypercholesterolemia in hamsters fed a high-fat diet (HFD) and to decipher the mechanisms of Totum-070's effect on lipoprotein metabolism.

2. Materials and Methods

2.1. Totum-070 (Lipidrive®) Characterization

The batch used in this study was chemically characterized using the Folin–Ciocalteu method (total polyphenol estimation), the Dubois method (total sugar estimation), the phospho–vanillin method (total lipid estimation) [27,28], the O-phthalaldehyde method (total protein estimation) [29], and HPLC-UV/visible/MS (Agilent Technologies Infinity series 1200 and 1260, Santa Clara, CA, USA) with a C18 column (250 × 4.6 mm, 5 μm, Phenomenex, Torrance, CA, USA) and a HILIC Silica column (150 × 4.6 mm, 5 μm, Waters, Etten-Leur, The Netherlands) (quantification of potential compounds of interest). The results of this analysis are presented in Table 1. Totum-070 is available from Valbiotis (Perigny, France) under the brand name Lipidrive®.

Table 1. Chemical characterization of Totum-070.

Compound Type (Sorted by Families)	Totum-070 Content (*w*/*w*)
Total sugars	41.35
Total lipids	7.30
Total proteins	2.05
Betaine	0.23
Total phenolic compounds	15.42
Mono-caffeoylquinic acids	
Chlorogenic acid	0.81
Cryptochlorogenic acid	0.01
Other mono-caffeoylquinic acids	0.25
Dicaffeoylquinic acids	
Cynarin	0.16
3-5 dicaffeoylquinic acid	0.18
4-5 dicaffeoylquinic acid	0.14
Other dicaffeoylquinic acids	0.11
Caffeic acid	0.01
Oleuropein	4.6
Oleuropein isomers	0.47
Hydroxytyrosol	0.08
Luteolin	0.01
Luteolin-7-O-glucoside	0.27
Luteolin-7-O-glucoside isomer	0.02
Luteolin-4-O-glucoside	0.06
Luteolin-7-O-glucuronide	0.29
Apigenin-7-O-glucoside	0.03
Apigenin-7-O-glucuronide	0.17
Apigenin-6-*C*-glucoside-8-*C*-arabinoside (Shaftoside)	0.01
Apigenin-6,8-*C*-diglucoside (Vicenin 2)	0.02
Apigenin-7-O-rutinoside	0.01
Eriodictyol-7-O-glucoside	0.04
Flavanomarein	0.14
Marein	0.05
Maritimein	0.05
Rutin	0.02
Verbascoside	0.08
Terpenes and terpenoids	
Oleanolic acid	1.64
Saponins	
Chrysanthellin A	0.28
Chrysanthellin B	0.27
Iridoids	
Oleoside	0.01
Cynaropicrin	0.07
Alkaloids	
Piperine	0.06

2.2. *Animals*

All the animal procedures were approved by the local ethics committee (C2E
Auvergne, France, under the number 18588-2019012216136952V5 dated 2 October 20
and comply with the ARRIVE guidelines. Male golden Syrian hamsters were purchas
from Janvier Labs (Le Genest-Saint-Isle, France). All the hamsters were housed at the anir
facility center (Valbiotis R&D Center, Riom, France) at 22 °C under a standard 12 h light-
h dark cycle, 2 or 3 hamsters per cage. Upon arrival, the hamsters were fed a normal d
(ND, D18102403, Research Diets, New Brunswick, NJ, USA) for 2 weeks of acclimatizat
until the age of 8 weeks. Thereafter (i.e., study start), the animals were randomly assigr
to five groups with similar mean body weight and fat mass: (i) ND (n = 15); (ii) H
(45 kcal% fat, mainly hydrogenated coconut oil, 17 kcal% fructose, 1.4 gm% choleste
D99122211, Research Diets, New Brunswick, NJ, USA) (n = 16); (iii) HFD supplement

with 3% (*wt*/*wt*) of Totum-070 (HFD-T070 3%; $n = 16$); (iv) HFD supplemented with 4.25% (*wt*/*wt*) of Totum-070 (HFD-T070 4.25%; $n = 16$); and (v) HFD supplemented with 5% (*wt*/*wt*) of Totum-070 (HFD-T070 5%; $n = 16$). Food and water were supplied ad libitum. Supplementation was for 12 weeks. The hamsters were weighed every week. Fresh food was distributed every 2 to 3 days and food intake was recorded. The ND and HFD energy densities were 3.85 kcal·g^{-1} and 4.5 kcal·g^{-1}, respectively. The lipid parameters were monitored in the blood before the study's start (week 0), after 6 weeks of supplementation (week 6), and at the study's end (week 12). The hamsters were fasted for 6 h, and blood was collected from the gingival vein. Then, the blood was incubated at room temperature for 30 min and centrifuged at 2000× *g* for 10 min. The serum was harvested and stored at −80 °C until analysis. The body composition was assessed with an echo MRI system (Zynsser Analytic) at week 0, week 6, and week 12. Feces were collected from each cage during 3 full days (from day 75 to day 78). At the study's end, the animals were fasted for 6 h, and then anaesthetized with isoflurane. Blood was collected by cardiac puncture, and then the hamsters were sacrificed by cervical dislocation.

2.3. Serum Lipid Parameters

High-density lipoprotein (HDL)-enriched serum fractions were isolated by incubating the serum samples with the LDL/VLDL precipitation buffer (Abcam, Boston, MA, USA). The cholesterol levels in the serum (total cholesterol) and in the purified HDL fraction (HDL-C) were quantified enzymatically using the CHOD-PAP colorimetric assay kit (Biolabo SAS, Maizy, France). The non-HDL-C fraction of each sample was calculated by subtracting the HDL-C concentration from the total cholesterol concentration. The serum triglyceride concentration was measured with a colorimetric assay kit (Cayman chemical, Ann Arbor, MI, USA).

2.4. Liver Triglycerides

The liver triglyceride level was determined with the triglyceride enzymatic assay kit by Cayman Chemical. The hamster livers (100 mg) were homogenized in 1 mL of a solution containing 5% NP40 in water, followed by centrifugation (10,000× *g* at 4 °C for 10 min) to recover the supernatant. The triglyceride concentration was normalized to the liver sample weight and expressed as mg/mg liver weight.

2.5. Histology

At the study's end, the liver tissue was fixed in paraformaldehyde at 4 °C for 48 h and kept in 70% ethanol for 2 to 3 h before automated tissue processing. The samples were embedded in paraffin, and 4 µm sections were cut with a microtome. The sections were dewaxed and rehydrated in successive xylene and ethanol baths. Then, they were stained in Gill II hematoxylin for 4 min, and after the removal of excess stain in an alcohol acid bath, they were transferred to a 0.25% eosin Y bath for 45 s. Finally, the sections were dehydrated with ethanol and xylene, and slides were mounted with Eukitt medium. Other parts of the livers were frozen in OCT medium, and 10 µm tissue sections were cut with a cryostat and fixed in cold 10% formalin for 5 min. The slides were stained with Oil Red O at 60 °C in an oven for 8 min. The Oil Red O areas were quantified using ImageJ software (version: v1.54g). Stained areas of at least 10 µm^2 and with a circularity between 0.1 and 1 were selected for quantifying the Oil Red O surface as a percentage of the total surface. At least three images per liver were analyzed.

2.6. Fecal Neutral Sterols

The fecal neutral sterols were quantified in the feces collected from 5 HFD cages and 5 HFD–T070 5% cages (n = 2 or 3 hamsters per cage). The feces were dried at 60 °C for 72 h and weighted. The sterols were analyzed as previously published with minor modifications in the selected ion monitoring mode [30]. Briefly, 0.5 mg of fecal homogenate in 90% *v*/*v* isopropanol was hydrolyzed using 1 M NaOH in 80% *v*/*v* of ethanol. Then, the sterols were extracted twice with methyl tert-butyl ether and derivatized with a mixture of *N*-methyl-(trimethylsilyl)trifluoroacetamide, trimethylsilyl chloride, and *N*-trimethylsilyl-imidaz Cholestane, cholesterol-d7, desmosterol-d6, and 25-hydroxy cholesterol-d6 were used internal standards, and the sterols were quantified against external calibration lines. T sterol concentrations were expressed as µg/mg of dry feces. The calculated fecal neut sterol excretion was expressed per cage and as µg/day.

2.7. Serum Lipopolysaccharide (LPS) Load Quantification

The LPS concentration was determined in the serum obtained from cardiac punct at the study's end, as described elsewhere [31]. Hek-blue mTLR4 cells (from InVivog San Diego, CA, USA) were plated in 96-well plates (25,000 cells/well) in 180 µL of detecti medium/well. After 24 h, 20 µL of LPS standard or diluted serum sample was added 180 µL of detection medium to each well in duplicate. After 24 h of incubation at 37 with 5% CO_2, the absorbance was measured at 630 nm using a Spark microplate reac (Tecan, Männedorf, Switzerland).

2.8. Intestinal Cholesterol Absorption Test

The intestinal cholesterol absorption test was performed according to the dual-isoto plasma ratio method described by Turley et al. with modifications [32,33]. Briefly, at t study's end, the non-fasted hamsters were anesthetized with isoflurane. Then, 2.5 µ of [1,2-^{3}H(N)]-cholesterol (Perkin Elmer, Every, France), completely dissolved in 200 of intralipid (I141, Sigma-Aldrich, Saint-Louis, MO, USA), was injected directly in t gingival vein. Immediately afterwards, 1 µCi [4-^{14}C]-cholesterol (Perkin Elmer, Eve France) dissolved in 600 µL olive oil was administered by gavage. The hamsters w returned to their cage, where they received their assigned diet for 72 h. Then, the hamst were anesthetized with isoflurane and were bled from the heart to collect blood, which w then incubated at room temperature for 30 min and centrifuged at 2000× *g* for 10 min. determine the proportion of [^{14}C]- and [^{3}H]-cholesterol still in circulation after 72 h, 300 of each serum sample and the original mixtures were added to 3 mL Ultima Gold liqu scintillation cocktails (Perkin Elmer, Every, France). The vials were shaken vigorou and the radioactivity was counted using a TRI-CARB 4910TR Liquid Scintillation Coun (Perkin Elmer, Every, France).

2.9. Oral Fat Tolerance Test

This test was performed at week 11. After overnight fasting, the hamsters receiv 300 µL of olive oil by gavage, and the triglyceride concentration in the serum was quantifi before (0), 150, 330, 510, and 720 min after gavage.

2.10. Hepatic Triglyceride Production Test

After 12 weeks of supplementation, the triglyceride secretion rate in the blood w quantified using Poloxamer 407. This nonionic detergent inhibits the tissue lipoprot lipase and, thus, the removal of triglycerides from the circulation [34]. A 100 g/L Poloxan 407 solution in an injection-grade saline solution was injected intraperitoneally in overnig fasted hamsters (1000 mg/kg body weight). Blood samples were collected immediately and at 45, 90, and 120 min after injection to quantify the triglyceride concentration. T liver triglyceride secretion rate was calculated by multiplying the slope of the triglyceri concentration increase over time by the intravascular distribution volume (estimated 3.8 mL/100 g body weight) [35,36].

2.11. Lipoprotein Lipase Activity Assay

The post-heparin lipoprotein lipase activity was measured using a previously scribed method [37,38]. After 12 weeks, the hamsters were fasted for 4 h and then ar thetized with isoflurane. Then, 100 µL (40 U) of sodium heparin (Sanofi, Paris, Fran was injected into the gingival vein. After 5 min (to allow lipoprotein lipase release in plasma), the animals were sacrificed, then blood was collected by cardiac puncture and separated by centrifugation at 2000× *g* at 4 °C for 10 min. The total lipase activity was determined using the LPL assay kit (Abcam, Boston, MA, USA). The hepatic lipase activity was then determined by incubating the reaction mixture in the presence of 1 M NaCl, which specifically blocks LPL activity. The lipoprotein lipase activity was calculated as the difference between the total lipase activity and hepatic lipase activity [38,39].

2.12. Quantitative Reverse Transcription-PCR (RT-qPCR)

The total mRNA was extracted from snap-frozen tissues using TRIzol® (Invitrogen (Waltham, MA, USA), Life Technologies (Carlsbad, CA, USA)). cDNA was synthesized from 2 μg RNA with a high-capacity cDNA transcription kit (Applied Biosystems (Waltham, MA, USA), Life Technologies). PCR amplification was carried out using the CFX system (Bio-Rad, Marnes-la-Coquette, France) with SybrGreen primers (Eurofins, Luxembourg). The ΔΔCt method was used to quantify the mRNA levels. Gene expression was normalized to the housekeeping gene *Polr2a*. The data are presented using the Rq, normalized to the control group, with the following equation:

$$Rq = 2^{-\Delta\Delta Ct}\ [\Delta Ct = Ct\ (target) - Ct\ (\mathit{Polr2a});\ \Delta\Delta Ct = \Delta Ct\ (sample) - \Delta Ct\ (control)].$$

The primer list is in Table S4.

2.13. Protein Extraction and Western Blot Analysis

The liver samples were homogenized in 20 μL/mg RIPA buffer (50 mM Tris-HCl, 150 mM NaCl, 1% NP40, 0.5% sodium deoxycholate, 0.1% SDS) on ice using a glass potter. Just before use, protease (P8340, Sigma, Saint-Louis, MO, USA) and phosphatase (88667, Thermo Fisher Scientific, Waltham, MA, USA) inhibitors were added to the buffer. The homogenized samples were centrifuged at 14,000× *g* at 4 °C for 10 min, and the supernatant was collected. The membrane proteins were extracted using the Mem-PER Plus Membrane Protein Extraction Kit (Thermo Fisher Scientific, Waltham, MA, USA), following the manufacturer's protocol. The protein content was determined using a commercial Lowry-based DC protein assay (Bio-Rad, Marnes-la-Coquette, France). All the samples were adjusted to a standard concentration (15 μg) and then diluted with 2× Laemmli buffer. Western blotting was performed as described by Chavanelle et al. [40] with the following specificities: the membranes were incubated at 4 °C with primary antibodies against an LDL receptor (Biovision, Milpitas, CA, USA), MTTP (Abcam, Boston, MA, USA), ACAT2 (Cell signaling, Saint-Cyr-L'Ecole, France), and NPC1L1 (Cell signaling, Saint-Cyr-L'Ecole, France), all at 1:1000. After overnight incubation, the membranes were washed with Tris-buffered saline/0..% Tween-20 and incubated with anti-rabbit horseradish peroxidase-conjugated secondary antibodies (1:2000) at room temperature for 1 h. Then, the membranes were washed three times in Tris-buffered saline/0.5%Tween-20 before the addition of an enhanced chemiluminescent solution (Clarity Western ECL; Bio-Rad) for 1 min. Band images were acquired with the Bio-Rad ChemiDoc system, and the band density was determined using Image Lab V6.0 (Bio-Rad). Stain-free blot images were used as the total protein loading control for data normalization [41].

2.14. Cell Cultures

Human colon colorectal adenocarcinoma Caco-2 cells (Sigma, Saint-Louis, MO, USA) were maintained in DMEM 4.5 g/L glucose supplemented with 2 mM glutamine, 1% nonessential amino acids, 20% fetal bovine serum, and 1% penicillin/streptomycin. For the experiments, Caco-2 cells were seeded on 1.0 μm transparent polyethylene terephthalate membrane inserts and allowed to differentiate for 21 days. Radioactive micelles were prepared freshly according to [42]. Briefly, lipids were added to the cells in the apical pole in the DMEM/serum-free medium as a complex lipid emulsion containing 0.4 mM oleic acid, 0.2 mM L-α-lysophosphatidylcholine, 0.05 mM cholesterol, 0.2 mM 2-oleoylglycerol, and 2 mM taurocholic acid and supplemented with [1,2-^{3}H(N)]-cholesterol (2 μCi/mL).

For the [^{3}H]-cholesterol uptake assay, Caco-2 cells were incubated with radioactive micel and with/without Totum-070 for 1 h. Then, the cell layers were washed, and the c lysates were collected for protein determination. The cell lysates were mixed directly the liquid scintillators (Ultima Gold, Perkin Elmer, Every, France). The radioactivity w counted with a TRI-CARB 4910TR Liquid Scintillation Counter and normalized to the c protein content. In some conditions, ezetimibe (Sigma, Saint-Louis, MO, USA) was add directly in the apical medium as a positive control of [^{3}H]-cholesterol uptake inhibiti For the [^{3}H]-cholesterol secretion assay, the Caco-2 cells were first loaded with radioacti micelles in the upper compartment for 24 h. Then, the cells were washed, and the api medium was changed to DMEM containing Totum-070 at the indicated concentrati After overnight incubation, the basolateral medium was collected and the lipoprote secreted in the basolateral medium were separated from the free tracers with a PD- Sephadex-G25 desalting column (Sigma, Saint-Louis, MO, USA) [43]. The purified fracti was suspended in a liquid scintillator (Ultima Gold, Perkin Elmer, Every, France), a the radioactivity was counted with a TRI-CARB 4910TR Liquid Scintillation Counter. T radioactivity in each vial was normalized to the cell protein content. For the apolipoprot B (ApoB) secretion measurement, a protocol similar to the one for cholesterol secreti quantification was used, but without radiolabeled cholesterol during the cell loading w micelles. The ApoB secretion into the basolateral medium after overnight incubation w quantified with the Human Apolipoprotein B ELISA Kit (Abcam, Boston, MA, USA). MT and ACAT2 were detected by Western blotting using cells in the same culture conditi described above. For NPC1L1 detection by Western blotting in whole-cell and in the c membrane lysates, a protocol similar to the one for the cholesterol uptake assay was us but with non-radioactive micelles.

2.15. RNA Sequencing Analysis

The livers were collected from the euthanized hamsters after 12 weeks of supplem tation and immediately frozen in liquid nitrogen. The RNA purity and concentration w determined using a NanoDrop ND-2000 Spectrophotometer (Thermo Fisher Scientific) a a Qubit fluorometer (Thermo Scientific), respectively. The RNA integrity was assess on an RNA 6000 NanoChip using a 2100 Bioanalyzer (Agilent Technologies). The rR was depleted from 200 ng of the total RNA using a NEBNext rRNA Depletion Kit (H man/Mouse/Rat) (New England BioLabs, Ipswich, MA, USA) following the manufactu protocol. Fragmented double-strand cDNA was synthesized using a NEBNext Ultra Directional RNA Library Prep Kit (New England Biolabs) according to the manufacture instructions. The libraries were amplified for 9 cycles and purified using Ampure XP bea (Beckman Coulter, Brea, CA, USA), and their concentration was measured using a Qubit Fluorometer and a Qubit dsDNA HS Assay Kit (Thermo Fisher Scientific). Then, the libr quality was confirmed by a size analysis on a Bioanalyzer 2100 instrument and a DN 1000 Assay reagent kit (Agilent Technologies). These data were used for pooling librar at equimolar concentrations to normalize each library. Single-end, 75-cycle sequencing w performed on a NextSeq 500 Sequencing System using NextSeq 500/550 High Output kit (75 cycles) (Illumina, San Diego, CA, USA). FASTQ files were generated using bcl2fa (Illumina). The reads were trimmed with Trimmomatic (version 0.36) to filter poor qua reads and to cut poor quality bases and adapters. The filtered reads were mapped to reference genome (*Mesocricetus auratus*, assembly MesAur1) using STAR (version 2.7. The STAR software was also used to produce the count matrix to assign reads to featu using transcriptome annotations from the GTF file. The quality control statistics w summarized using MultiQC (version: v1.14). The descriptive analysis was performed w the DESeq2 R package. The differential analysis was performed with the edgeR R packa TMM normalization was applied. The glmQLFit and glmQLFTest functions were us to identify differentially expressed genes in the three groups (ND, HFD, and HFD-T 5%). The genes of interest were selected based on the significance (Benjamini–Hochb correction, corrected p-value < 0.05) and effect size (fold-change) for each comparison. T gene lists were annotated with the DAVID database (database for annotation, visualization, and integrated discovery, https://david.ncifcrf.gov/home.jsp, accessed on 1 February 2022) and *Cricetulus griseus* ortholog gene identifiers.

2.16. Microscale Thermophoresis (MST) Binding Affinity Assay

The binding affinity of Totum-070 for the *N*-terminal domain (22–284 amino acids) of the NPC1L1 protein (NPC1L1-NTD) was assessed using MST. Recombinant human NPC1L1-NTD (MBS7102155, MyBiosource, San Diego, CA, USA) was labeled with the fluorescent RED-NHS dye using a RED-NHS protein labeling kit (NanoTemper Technologies, Munich, Germany) according to the manufacturer's instructions. Briefly, the dye and NPC1L1-NTD were incubated, and the tagged protein was purified. The resultant tagged solution was then diluted in a binding buffer (50 mM Tris-HCl pH 7.5, 150 mM NaCl, 1 mM DTT, and 0.1% Fos-choline 13) to achieve a final NPC1L1-NTD concentration of 20 nM in the capillaries. Standard capillaries for red fluorescence emission recording were filled with a 1:1 (*v*/*v*) mixture of tagged NPC1L1-NTD and diluted ligand. The fluorescence was measured using a NanoTemper Monolith NT.115 instrument (NanoTemper Technologies, Munich, Germany) at 37 °C, with a fluorescence excitation power of 90% and the MST power set to high. The data were analyzed with NanoTemper analysis software(version: v2.3) (NanoTemper Technologies, Munich, Germany).

2.17. Microbiota Analysis

The cecum content was harvested from the euthanized hamsters after 12 weeks of supplementation and immediately frozen in liquid nitrogen. DNA was extracted from 50–100 mg of the sample using a FastDNA^{TM} Spin Kit for feces (MP Biomedicals, Illkirch, France) and FastPrep-24 5G (MP Biomedicals, Illkirch, France) following the manufacturer's instructions. The DNA was eluted in 50 μL of elution buffer, pre-heated at 55 °C, and normalized to 20 ng/μL. Microbial 16S libraries were prepared at the Plateforme Génome Transcriptome de Bordeaux, Bordeaux, France, by amplification and sequencing of the V3-V4 region of the 16S rRNA gene on an Illumina MiSeq machine, using a 2 × 250 base pair Illumina v2 kit. Quality filtering and processing were achieved with Cutadapt v4.0 and Python v3.9.9, USEARCH v11.0.667, and mothur v1.48.0 to identify 97% of the operational taxonomic units [44]. Abundance tables were imported and analyzed in R v4.1. The alpha and beta diversities were estimated using vegan 2.6-2 and compared using the Kruskal–Wallis and MANOVA tests, respectively. The indicator taxa were selected using the KW test and multiple testing correction with false discovery rate (FDR) correction. For the indicator taxa, an additional linear model was tested to determine the type of observed effect. Correlations between the relative abundance of the taxa and arbitrary values were tested using the Pearson correlation coefficient. Arbitrary units were allocated to the experimental groups, and six models were tested: (1) Totum-070 induces a return to normal that goes beyond the ND value (ND = 25, HFD = 75; HFD-T070 5% = 10); (2) Totum-070 induces a return to normal (ND = 25, HFD = 75; HFD-T070 5% = 25); (3) Totum-070 induces a partial return to normal (ND = 25, HFD = 75; HFD-T070 5% = 50); (4) Totum-070 does not rescue the changes induced by the HFD (ND = 25, HFD = 75; HFD-T070 5% = 75); (5) Totum-070 accentuates the HFD effect (ND = 25, HFD = 75; HFD-T070 5% = 100); and (6) Totum-070, but not the HFD, induces a change (ND = 100, HFD = 100; HFD-T070 5% = 10). The indicator taxa were identified by selecting taxa significantly associated with the supplementation group in the Kruskal–Wallis test, and for which at least one of the linear models was significant. If multiple linear models were significant, the model with the highest effect size was reported in Tables S2 and S3. Correlations between the indicator taxa and cholesterol levels were assessed using the Pearson correlation with FDR correction for multiple testing.

2.18. Statistical Analysis

Prism V.9.5.1 (GraphPad Software) was used for the statistic tests and figure drawi Specifically, the Shapiro–Wilk normality test was used to determine whether the da followed a Gaussian distribution. If the data were not normally distributed, the Krusk Wallis nonparametric test was used, followed by the Dunn test for post hoc comparis When the normal distribution was assumed, the data were compared with a one-way two-way ANOVA and Tukey's test for multiple comparisons. For measurements repeat over time, differences among the groups and time points were evaluated using a repeat measure two-way ANOVA, followed by the Tukey's post hoc test for multiple comparisc If missing data did not allow using the repeated-measures 2-way ANOVA, a mixed-effe analysis was used. The values are presented as the mean ± SEM, unless specified otherwi Differences were considered significant at $p < 0.05$.

3. Results

3.1. Supplementation with Totum-070 Reduces Hyperlipidemia and Improves Liver Steatosis a Inflammation in Hamsters without an Effect on Body Weight

To demonstrate Totum-070's efficacy in a diet-induced hyperlipidemia model, t body weight- and fat mass-matched Syrian hamsters were fed the ND, HFD, or H with Totum-070 mixed at different concentrations: 3.5% (HFD-T070 3.5%), 4.25% (HF T070 4.25%), and 5% (HFD-T070 5%). The total caloric intake was not different betwe the ND and HFD groups, which could be explained by the significantly reduced fo intake by the HFD animals compared with the ND animals (Table 2). Similarly, the bo weights at the study's end were similar between the ND and HFD groups (with/withc supplementation), although the fat mass tended to increase, and the lean mass tend to decrease in the HFD groups (with/without supplementation) compared with the N group (Table 2). Food intake was comparable in the three supplemented groups and t HFD group, demonstrating that Totum-070 was well tolerated at the tested concentratic

Table 2. Effect of the different diets on body weight, body composition, food intake, a caloric intake.

Diet **Supplementation**	**ND** **-**	**HFD** **-**	**HFD** **T070 3.5%**	**HFD** **T070 4.25%**	**HFD** **T070 5%**
Final body weight (g)	129.2 ± 2.6	124.4 ± 2.7	124.8 ± 2.6	126.5 ± 3.3	123.4 ± 1.2
Final fat mass (g)	18.12 ± 1.3	20.14 ± 0.6	22.79 ± 1	23.36 ± 1 *	22.26 ± 1
Final lean mass (g)	106.3 ± 2.2	98.61 ± 2.3	97.17 ± 1.9 *	98.74 ± 2.7	96.78 ± 1.8 *
Cumulative food intake (g)	436 ± 7.1	375 ± 11.5 **	386 ± 10.5	401.5 ± 8	397 ± 15.4
Cumulative caloric intake (Kcal)	1680 ± 27.3	1688 ± 52	1700 ± 46.3	1766 ± 35.2	1748 ± 68

Hamsters were fed a normal diet (ND, $n = 15$), a high-fat diet ($n = 16$), or a high-fat diet supplemented w Totum-070 at the indicated concentration (T070 3.5%, T070 4.25% and T070 5%, $n = 16$ per group) for 12 weeks. < 0.05 and ** $p < 0.01$ versus ND. Data are the mean ± SEM.

Next, to determine Totum-070's effect on the lipid profile, serum samples were c lected before the study (week 0), during the study (week 6), and at the study's end (we 12). As expected, dyslipidemia was observed in all the HFD groups, with a strong increa in the total cholesterol (Figure 1A), non-HDL-C (Figure 1B), and TG (Figure 1C), compa with the ND group at week 6 and week 12. However, their concentration was significan lower in the three Totum-070 groups (in a dose-dependent manner) compared with HFD group at week 6 and week 12. At week 12, the total cholesterol, non-HDL-C, and were reduced by 41% ($p < 0.001$), by 47% ($p < 0.001$), and by 46% ($p < 0.05$) in the HFD-T 5% group compared with the HFD group. A slight increase in HDL-C was observed the HFD groups compared with the ND group throughout the study; however, Totum- did not have any effect on the HDL-C level compared with the HFD group (Figure 1 Altogether, these results demonstrate that Totum-070 has both hypocholesterolemic a hypotriglyceridemic properties.

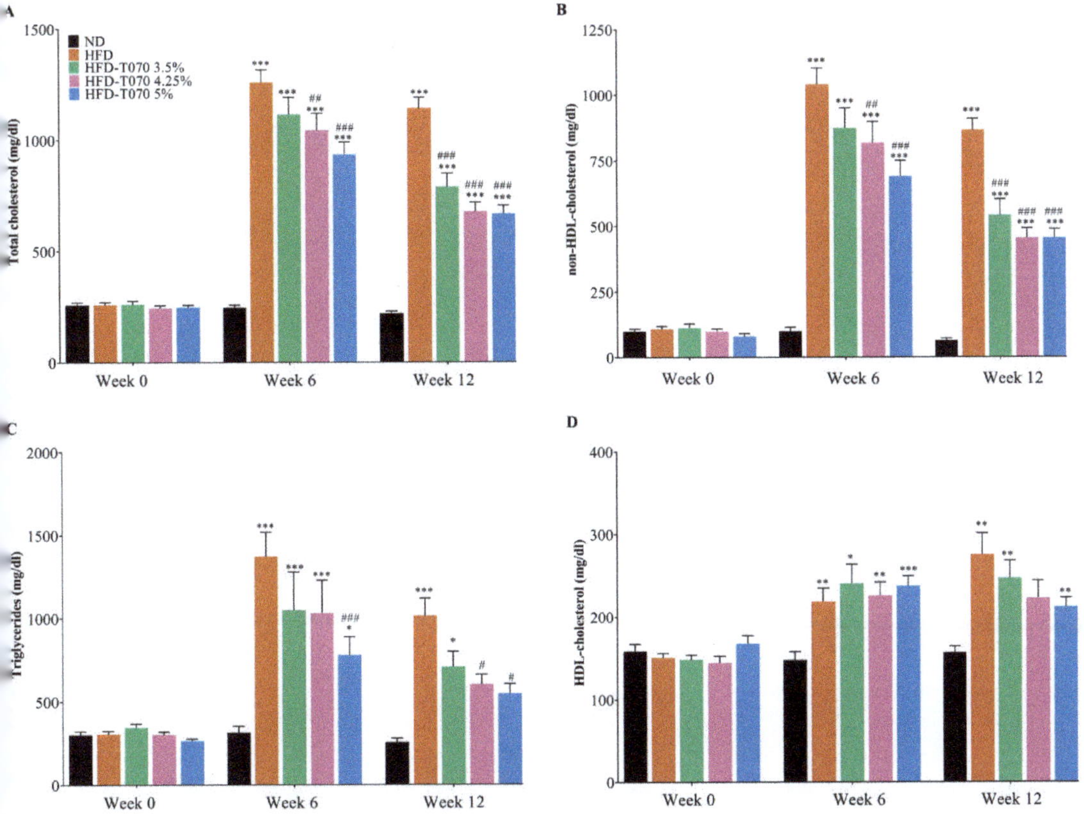

Figure 1. Supplementation with Totum-070 prevents hypercholesterolemia and hypertriglyceridemia in hamsters fed a high-fat diet. Male hamsters (n = 15/16 per group) were fed a normal diet (ND), high-fat diet (HFD), or HFD supplemented with Totum-070 at the following concentrations: 3.5% (HFD-T070 3.5%), 4.25% (HFD-T070 4,25%), and 5% (HFD-T070 5%). Blood samples were collected before the study's start (week 0), during the study (week 6), and at the study's end (week 12). (**A**) Concentration of total cholesterol; (**B**) non-HDL-cholesterol; (**C**) triglycerides; and (**D**) HDL-cholesterol in serum. * $p < 0.05$, ** $p < 0.01$, and *** $p < 0.001$ versus ND. # $p < 0.05$, ## $p < 0.01$, and ### $p < 0.001$ Totum-070 groups versus HFD. Data are the mean ± SEM.

Then, to assess Totum-070's effect on liver lipid metabolism (Figure 2), liver tissue samples were stained with hematoxylin–eosin and Oil Red O (Figure 2A). After 12 weeks, the livers from the HFD groups displayed micro-vesicular steatosis of moderate to severe intensity and increased neutral lipid accumulation. These changes were reduced in the liver samples from the HFD-T070 3.5%, 4.25%, and 5% groups, as indicated by Oil Red O staining quantification (Figure 2B). Enzymatic assays in the liver extracts confirmed the decrease in hepatic TG content in the hamsters supplemented with Totum-070 compared with the HFD group (Figure 2C). An independent experiment was performed to assess the effect of Totum-070 on the hepatic TG output compared with HFD alone at week 12. Monitoring of the TG concentration in serum for 120 min following Poloxamer 407 injection, a nonionic detergent that inhibits lipoprotein lipase [34], showed that the liver TG secretion rate was similar in the HFD and HFD-T070 5% groups (Figure 2D).

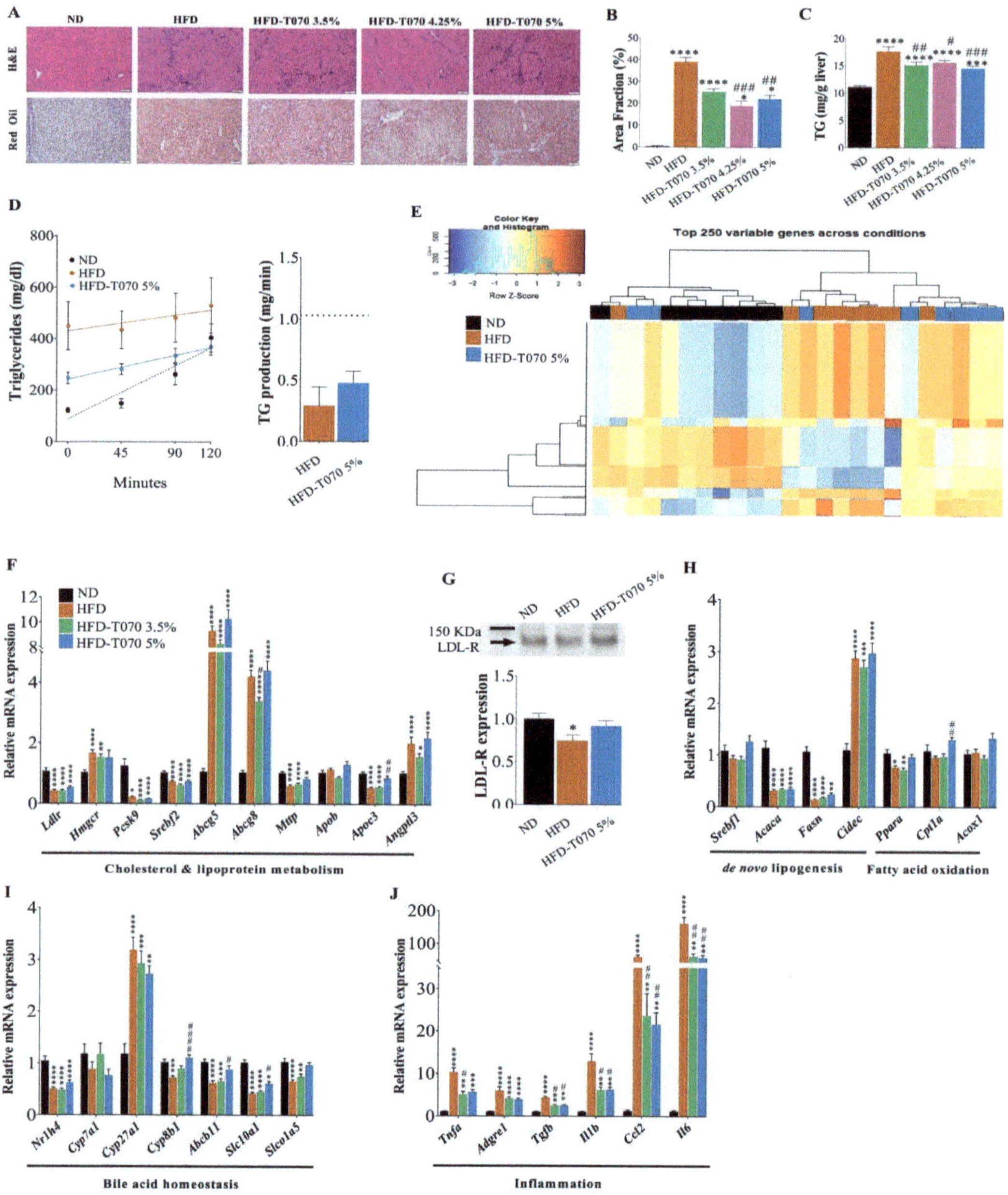

Figure 2. Effect of Totum-070 on liver lipid metabolism. (**A**) Paraffin-embedded sections of hams livers stained with hematoxylin and eosin (H&E). Frozen sections of hamster livers stained w Oil Red O and counterstained with hematoxylin. Scale bars, 100 µm. (**B**) Quantification of Oil I O in liver sections (at least three images per liver) shows reduction in neutral lipid accumulat in hamsters supplemented with Totum-070. (**C**) Triglyceride (TG) content in liver. (**D**) Change triglyceride concentration in serum samples collected at the indicated time points after injectior

Poloxamer 407, and calculation of hepatic triglyceride (TG) production in HFD (n = 11) and HFD-T070 5% (n = 14) hamsters. An age-matched control group of hamsters (n = 6) was fed the ND. **(E)** Heatmap showing the top 250 genes differentially expressed (RNA-sequencing data) in hamster livers in the ND, HFD, and HFD-T070 5% groups (n = 8 per group) **(F)** Relative expression in liver of selected genes encoding proteins implicated in cholesterol and lipoprotein metabolism (n = 15/16 per group). **(G)** Immunoblot of LDL receptor expression in hamster liver samples from the indicated groups and quantification. **(H)** Relative expression in liver of selected genes encoding proteins implicated in lipid metabolism. **(I)** Relative expression in liver of selected genes encoding proteins implicated in bile acid metabolism. **(J)** Relative expression in liver of selected genes encoding proteins implicated in inflammation. * $p < 0.05$, ** $p < 0.01$, *** $p < 0.001$, and **** $p < 0.00001$ versus ND. # $p < 0.05$, ## $p < 0.01$, ### $p < 0.001$, and #### $p < 0.00001$ Totum-070 groups versus HFD. Data are the mean ± SEM.

To determine the liver metabolic adaption induced by Totum-070, the transcriptomes of livers from the ND, HFD, and HFD-T070 5% groups were analyzed by RNA sequencing (n = 8 per group). A data analysis revealed that 3434 genes were downregulated in the HFD group compared with the ND group and that 3567 genes were upregulated in the HFD group compared with the ND group. It should be noted that differentially expressed genes represented more than 50% of all the genes tested when comparing the two groups. In addition, 325 genes were downregulated and 186 genes were upregulated in the HFD-T070 5% group compared with the HFD group. The top 250 most variable genes in the three groups are shown in the heatmap in Figure 2E. This heatmap indicates that the ND and HFD groups were the most transcriptionally distinct, whereas the HFD-T070 5% group exhibited an intermediate profile. Among the top canonical pathways differentially regulated between the HFD and HFD-T070 5% groups (Table S1), the most important was the canonical metabolic pathway, and the implicated genes are listed in Table 3.

A detailed analysis of the expression of key regulators of hepatic lipid metabolism performed by RT-qPCR analysis (n = 15/16 per group) showed that the expression of *Ldlr* (encoding LDL receptor, a major actor of LDL removal from circulation) was drastically reduced in the livers from the HFD groups compared with the ND group, although a reduction was not observed in the HFD-T070 5% group (Figure 2F). This is consistent with the Western blot analysis (Figure 2G) showing that the LDL receptor was decreased by 26% in the HFD compared with the ND group ($p < 0.05$), and that its expression was nearly restored to normal level in the HFD-T070 5% group ($p = 0.085$ versus HFD). Other genes were also strongly affected by the HFD. Thus, *Abcg5* and *Abcg8* were upregulated in the HFD livers as expected because they are induced upon cholesterol accumulation in hepatocytes. Conversely, *Pcsk9* and *Srebf2* were downregulated in the livers from the HFD hamsters, in accordance with their repression upon the intracellular cholesterol level increase. However, no difference between the 3.5% and 5% Totum-070 groups and the HFD group was observed. *Fasn* and *Acaca* (genes implicated in de novo lipogenesis) were downregulated in the HFD group (Figure 2H), but this effect was slightly counteracted for *Fasn* in the 5% Totum-070 group ($p = 0.11$ compared to HFD). Similar results were obtained for *Mttp* and *Apoc3* (lipoprotein metabolism regulation), and for *Ppara*, *Cpt1a*, and *Acox1* (fatty acid oxidation) (Figure 2H). An analysis of the genes implicated in bile acid metabolism showed that the expression of *Cyp7a1*, which encodes the rate-limiting enzyme to convert cholesterol into bile acids in the liver, was similarly expressed in all the groups (Figure 2I). Conversely, secondary enzymes implicated in bile acid production pathways exhibited different expression patterns. Indeed, *Cyp27a1* was induced by the HFD, and Totum-070 did not have any effect ($p > 0.99$), while *Cyp8b1* was downregulated by the HFD, and its expression was restored by Totum-070 supplementation. Similarly, *Abcb11*, which encodes a bile acid transporter that promotes bile acid excretion into the bile duct, and *Slc10a1* and *Slco1a5*, transporters implicated in hepatocyte uptake of bile acids, were downregulated by the HFD, but their expression was restored by Totum-070 supplementation (Figure 2I). To further characterize Totum-070's effects in the liver, the expression of inflammation markers (*Tnfα*, *Adgre1*, *Tgfβ*, *Il1β*, *Ccl2*, *Il6*) was investigated. Compared with the ND group,

Totum-070 attenuated this effect. Collectively, these results indicate that Totum-070 ten to normalize the expression of liver genes altered by the HFD.

Table 3. Genes that belong to the canonical metabolic pathway and that are differentially express in HFD hamster livers compared with HFD-T070 5% hamster livers ($n = 8$ per group).

Log Fold-Change	p-Values HFD	Golden Hamster Gene ID	Gene Name	Protein
2.43	0.001509	ENSMAUG00000000351	*Lipg*	lipase G, endothelial type
2.39	0.017545	ENSMAUG00000000465	*Acly*	ATP-citrate synthase
2.24	0.000089	ENSMAUG00000001919	*Atp6v0d2*	V-type proton ATPase subunit d 2
2.08	0.000001	ENSMAUG00000008803	*Alpl*	alkaline phosphatase
2.06	0.000125	ENSMAUG00000004155	*Asns*	asparagine synthetase
1.51	0.000422	ENSMAUG00000009435	*Bst1*	ADP-ribosyl cyclase 2
1.47	0.001238	ENSMAUG00000005505	*Mboat2*	lysophospholipid acyltransferase 2
1.35	0.002228	ENSMAUG00000018983	*Alox5*	polyunsaturated fatty acid 5-lipoxygenase
1.21	0.000716	ENSMAUG00000001135	*B3galnt1*	UDP-GalNAc:beta-1,3-*N*-acetylgalactosaminyltransferase 1
1.18	0.000011	ENSMAUG00000000598	N/A	N/A
1.14	0.000212	ENSMAUG00000009261	*Mthfd2*	NAD-dependent methylene tetrahydrofolate dehydrogenase cyclohydrolase
1.14	0.000757	ENSMAUG00000017913	*Mboat1*	lysophospholipid acyltransferase 1
1.07	0.003228	ENSMAUG00000020920	*Hk2*	hexokinase-2
1.07	0.001393	ENSMAUG00000007951	*Elovl7*	elongation of very long chain fatty acids protein 7
1.02	0.000411	ENSMAUG00000016079	*Acss1*	acetyl-coenzyme A synthetase 2-like mitochondrial
1.01	0.002529	ENSMAUG00000004073	*Sqle*	squalene monooxygenase
1.00	0.001829	ENSMAUG00000019624	*Pla2g7*	platelet-activating factor acetylhydrolase
−1.02	0.005813	ENSMAUG00000003438	*Gck*	hexokinase-4

3.2. *Totum-070 Modulates Intestinal Cholesterol Metabolism in Hamsters*

To better understand Totum-070's effect on cholesterol metabolism, neutral ste excretion was quantified in fecal samples from the HFD and HFD-T070 5% hamst collected from day 75 to day 78. The total fecal neutral sterol excretion was slightly high (+17%; $p = 0.22$) in the HFD-T070 5% group than in the HFD group (Figure 3A), suggesti a decrease in intestinal cholesterol absorption during the study. To test this hypothesis, intestinal cholesterol absorption was quantified in the HFD and HFD-T070 5% groups usi a dual cholesterol isotope procedure [32,33]. A comparison of the [^{14}C]/[^{3}H] choleste ratio in the blood suggested a slight intestinal cholesterol absorption reduction (by 12%) the HFD-T070 5% group ($p = 0.25$) (Figure 3B). The serum TG concentration was significan higher in the HFD group than the HFD-T070 5% group at all time points (Figure 3C). each group, the TG values were corrected to the baseline concentration (T0), and the area under the curve (AUC) was calculated. The net AUC of serum TG was significan reduced, by 39%, in the HFD-T070 5% group compared with the HFD group ($p < 0.0$ Conversely, the ex vivo post-heparin lipoprotein lipase activity assay results were different between the HFD and HFD-T070 5% groups (Figure 3D), excluding any Totu 070 effect on TG hydrolysis in circulation. Altogether, these data strongly suggest th

Totum-070's TG-lowering action could be mediated through the inhibition of intestinal lipid absorption.

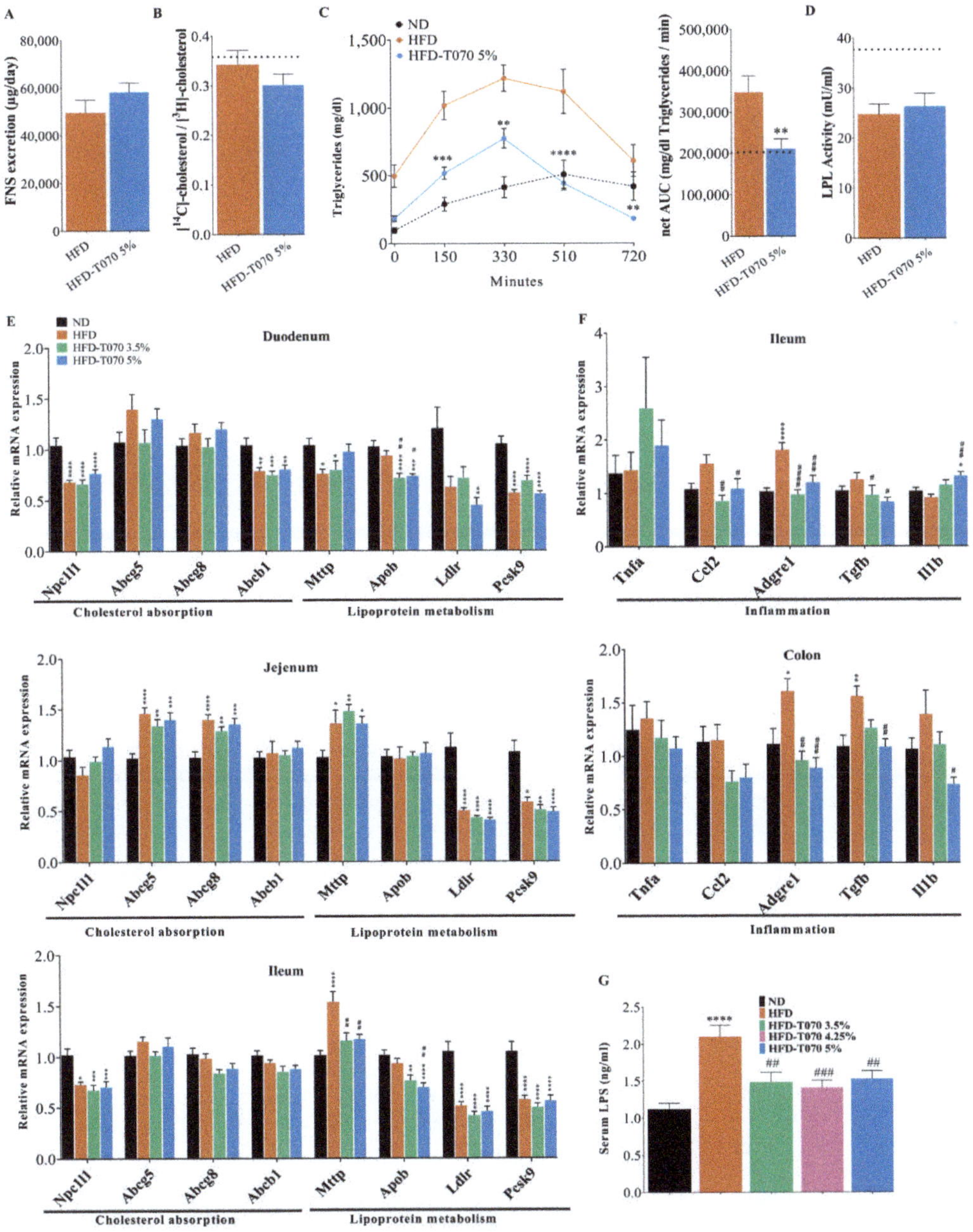

Figure 3. Effect of Totum-070 on intestinal lipid metabolism. (**A**) Excretion of fecal neutral sterols (FNS) quantified in fecal samples collected from day 75 to day 78. For each group (HFD and HFD-T070 5), pooled feces from 5 cages ($n = 2/3$ hamsters per cage) were used. (**B**) Results of the intestinal cholesterol absorption test performed after 12 weeks of supplementation in hamsters fed an HFD

(n = 14) or HFD supplemented with T070 5% (n = 14). An age-matched control group of hamst (n = 6) was fed the ND. (**C**) Changes in triglyceride concentration in serum after olive oil gava and calculation of the net area under the curve (AUC) in HFD (n = 13) and HFD-T070 5% (n = hamsters after the oral fat tolerance test. An age-matched control group of hamsters (n = 6) was the ND. (**D**) Measurement of lipoprotein lipase (LPL) activity after heparin injection in HFD (n = and HFD-T070 5% (n = 14) hamsters. An age-matched control group of hamsters (n = 6) was fed ND. (**E**) Expression of genes encoding proteins implicated in lipid and lipoprotein metabolism duodenum, jejunum and ileum. (**F**) Expression of genes encoding inflammation markers in ile and colon. (**G**) Concentration of lipopolysaccharides (LPS) in the serum after 12 weeks. * p < 0 ** p < 0.01, *** p < 0.001, and **** p < 0.00001 versus ND. # p < 0.05, ## p < 0.01, ### p < 0.001, and # p < 0.0001 (Totum-070 groups versus HFD). Data are the mean ± SEM.

Next, in order to thoroughly investigate Totum-070's effects in the intestine, ge expression was analyzed in the different parts of the intestine. Totum-070 exposure 12 weeks did not change the expression of the main players implicated in choleste absorption in the duodenum, jejunum, and ileum (Figure 3E). In the ileum, *Slc10a2*, coding the ileal bile acid transporter protein (IBAT), was increased in the HFD group < 0.001) (Figure S1A), whereas its induction was slightly blunted in the HFD-T070 3. and HFD-T070 5% groups, although not significantly (p = 0.06 and p = 0.09 versus H respectively). An analysis of the genes encoding various pro-inflammatory markers (C *Adgre1*, *Tgfb*) in the ileum and colon indicated a reduction of inflammation in the HFD-T 3.5% and HFD-T070 5% groups (Figure 3F). This observation was supported by the dos of circulating LPS in the serum samples at the study's end (Figure 3G). The serum L concentration was increased by 87.3% in the HFD (p < 0.0001) group compared with ND group. Compared with the HFD group, the serum LPS concentration was significan reduced by 29% (p < 0.01), 33% (p < 0.001), and 27% (p < 0.01) in the HFD-T070 3. HFD-T070 4.25%, and HFD-T070 5% groups, respectively.

3.3. Totum-070 Inhibits Cholesterol Uptake and Secretion in Enterocytes

To determine Totum-070's effect on cholesterol intestinal homeostasis, differentia Caco-2 cells were incubated with [^{3}H]-cholesterol with or without the presence of Tot 070 for 1 h, and then the [^{3}H]-cholesterol uptake by the cells was measured. Ezetim was used as control because it blocks the NPC1L1, the cholesterol transporter protein. 150 μM, ezetimibe reduced the [^{3}H]-cholesterol uptake by 58% (p < 0.0001) compared w the vehicle (Figure 4A). This effect was not increased by using higher concentrations. Tot 070 significantly decreased the cholesterol uptake at 1 g/L (p < 0.01) (Figure 4A). As NPC inhibition might explain the reduction of [^{3}H]-cholesterol uptake by Totum-070, the c were incubated with [^{3}H]-cholesterol, 150 μM ezetimibe, and/or 1 g/L Totum-070. their own, Totum-070 (1 g/L) and ezetimibe (150 μM) decreased the cholesterol uptake 34.5% (p < 0.001) and by 62.5%, respectively, but their combination did not further decre the cholesterol uptake (Figure 4B). The combination of 75 μM ezetimibe, which reduced cholesterol uptake by 49.4% on its own (p < 0.001), and 1 g/L Totum-070 led to a decre in the cholesterol uptake by 61% compared with the control (p < 0.001) (Figure 4B). Th findings suggest that NPC1L1 activity is required for Totum-070 inhibition of choleste uptake. A Western blot analysis of whole Caco-2 cell lysates (Figure 4C) and the membr extracts (Figure 4D) of cultures treated for the cholesterol uptake analysis showed effect of 1 g/L Totum-070 on NPC1L1 protein expression. The *N*-terminal domain the NPC1L1 protein binds cholesterol and plays essential roles in cholesterol uptake [Following the evidence that Totum-070 inhibited the NPC1L1 activity without changing expression, we examined whether Totum-070 could interact with the *N*-terminus par NPC1L1 through a direct binding event. Human recombinant NPC1L1-NTD was incuba with Totum-070 at concentrations ranging from 0.015 to 500 μg/mL, and the interacti were measured using microscale thermophoresis (MST). As detailed by Jerabek-Willem et al. [46], MST is a fluorescence-based technique to quantify an analysis of interacti between biomolecules, and its interest for characterizing the binding interaction betw proteins and plant extracts has been validated previously [47]. Here, we demonstrate that NPC1L1-NTD interacted with Totum-070 with a dissociation constant (K_d) of 5.934 µg/mL (Figure 4E).

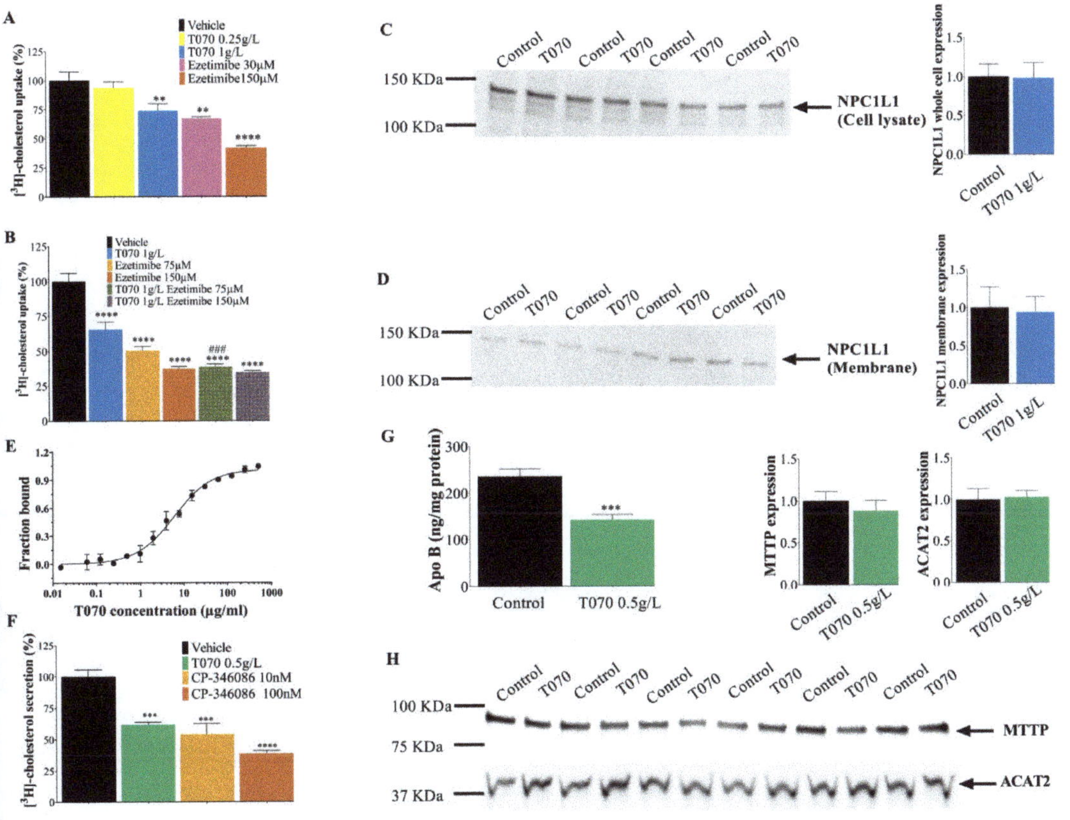

Figure 4. Totum-070 supplementation inhibits cholesterol uptake and secretion by differentiated human Caco-2 enterocytes. (**A**) Caco-2 cells differentiated on inserts were incubated (1 h) with [^{3}H]-cholesterol-micelles at the apical pole with or without the presence of Totum-070 in the culture medium in the upper compartment. Ezetimibe was used as a positive control of [^{3}H]-cholesterol uptake inhibition. Data are expressed as the percentage of control (vehicle). (**B**) Similar protocol as in (**A**) but combining Totum-070 and ezetimibe. The results suggest that Totum-070 may inhibit [^{3}H]-cholesterol uptake in the same manner as ezetimibe. Data are expressed as the percentage of control (vehicle). (**C**,**D**) Immunoblot and quantification of the NPC1L1 transporter expression in whole-cell lysates (**C**) and membrane protein extracts (**D**) of Caco-2 cells. (**E**) Fraction of Totum-070 bound to NPC1L1-NTD, measured by MST, $n = 2$ for all points. Raw fluorescence (in counts) was normalized by expressing them as the fraction bound to the target. (**F**) Caco-2 cells differentiated on inserts were incubated with [^{3}H]-cholesterol-micelles in the apical pole for 24 h before overnight incubation with or without the presence of Totum-070 in the culture medium in the upper compartment. CP-346086 was used as positive control of [^{3}H]-cholesterol secretion inhibition. Data are expressed as the percentage of [^{3}H]-cholesterol secretion in the basolateral medium in control cells (vehicle). (**G**) ApoB secretion in the basolateral medium from differentiated Caco-2 cells. (**H**) Immunoblot and quantification of MTTP and ACAT2 expression in differentiated Caco-2 cells. ** $p < 0.01$, *** $p < 0.001$, and **** $p < 0.00001$ versus vehicle. ### $p < 0.001$, Totum-070 1 g/L versus Totum-070 1 g/L + Ezetimibe 75 µM. Data are the mean ± SEM.

Re-packaging of intracellular esterified cholesterol and secretion into chylomicro is another mechanism that influences intestinal cholesterol absorption. To assess Totu 070's effect on cholesterol secretion, differentiated Caco-2 cells were incubated with [^{3}I cholesterol for 24 h prior to overnight incubation with Totum-070, followed by [^{3}I cholesterol quantification in the basolateral medium. Incubation with 0.5 g/L Totum-C led to a reduction of 38.4% of cholesterol secretion compared with control ($p < 0.001$) (F ure 4F). Incubation with CP-346086, an inhibitor of MTTP enzymatic activity used a positive control, decreased cholesterol secretion by 46% at 10 nM ($p < 0.001$) and by 6 at 100 nM ($p < 0.0001$) (Figure 4F). Consistent with the inhibition of cholesterol secreti ApoB secretion in the basolateral medium was also reduced by 0.5 g/L Totum-070 (−39. compared with control, $p < 0.001$) (Figure 4G). Of note, a Western blot analysis of t Caco-2 cells after overnight incubation with 0.5 g/L Totum-070 showed no effect on t expression level of MTTP and ACAT2, two major enzymes controlling the secretion esterified cholesterol (Figure 4H). Altogether, these results demonstrate that Totum-C inhibits intestinal cholesterol absorption, including cholesterol uptake and chylomicr secretion.

3.4. *Totum-070 Influences Gut Microbiota in Hamsters*

Whole 16S rRNA gene sequencing of the cecum content was used to assess the imp of the ND, HFD, and HFD supplemented with 5% Totum-070 on the intestinal microbi composition at the study's end. At the phylum level, evenness was significantly increas in the HFD ($p < 0.01$) and HFD-T070 5% ($p < 0.001$) groups compared with the ND gro (Figure 5A), without any difference between HFD and HFD-T0710 5%. At the genus le species richness was similar in the three groups (Figure 5B). Conversely, evenness w increased in the HFD-T070 5% group compared with both the HFD ($p < 0.001$) and N groups ($p < 0.001$) (Figure 5C). A principal coordinate analysis of the relative abundance intestinal microbial communities at the phylum level using the Bray–Curtis dissimilar index followed by PERMANOVA showed that the microbial composition differed sign cantly among the three groups (Bray–Curtis, adonis: $p < 0.0001$ for all the groups, R_{NDvsH} = 0.24, $R_{NDvsHFD\text{-}T0705\%}$ = 0.60, $R_{HFDvsHFD\text{-}T0705\%}$ = 0.31) (Figure 5D). The same analysis the genus level confirmed that the HFD strongly affected the microbial composition in cecum compared with the ND (Bray–Curtis, adonis: $p < 0.0001$, $R_{NDvsHFD}$ = 0.26 (Figure and Jaccard, adonis: $p < 0.0001$, $R_{NDvsHFD}$ = 0.12 (Figure 5F)). The microbial composition the HFD-T070 5% group differed from that of the ND group (Bray–Curtis, adonis: $p < 0.0$ $R_{NDvsHFD\text{-}T0705\%}$ = 0.36 (Figure 5E) and Jaccard, adonis: $p < 0.0001$, $R_{NDvsHFD\text{-}T0705\%}$ = 0 (Figure 5F)), as well as the HFD group (Bray–Curtis, adonis: $p < 0.0001$, $R_{HFDvsHFD\text{-}T07}$ = 0.21 (Figure 5E) and Jaccard, adonis: $p < 0.0001$, $R_{HFDvsHFD\text{-}T0705\%}$ = 0.17 (Figure 5 This highlighted a stronger impact of the HFD with Totum-070 supplementation than H alone at the genus level.

Given the strong effect on the alpha and beta diversity measures in the HFD-T070 group, and the visible differences in the average composition at the phylum and ger levels (Tables S2 and S3, respectively), an indicator species analysis was performed identify the taxa specifically affected by the diet. At the phylum level, HFD significantly creased the abundance of Firmicutes (−15%, $p < 0.01$) and Actinobacteria (−57%, $p < 0.$ compared with ND (Table S2) and increased the abundance of Campylobacterota, ferribacteres, and Proteobacteria (+211%, $p < 0.05$; +123%, $p < 0.05$; and +44%, $p < 0$ vs. ND, respectively). The Firmicutes/Bacteroidetes ratio tended to decrease in the H group (−15%; not significant) compared with the ND group (Table S2). Supplementati with 5% Totum-070 accentuated the reduction in Firmicutes abundance compared w the ND group (−34%, $p < 0.0001$) and the HFD group (−23%, $p < 0.05$). Conversely, abundance of Bacteroidetes was higher in the HFD-T070 5% group than the HFD gro (+23%, $p < 0.001$). Therefore, the Firmicutes/Bacteroidetes ratio further decreased in HFD-T070 5% group compared with the ND (−47%, $p < 0.001$) and HFD (−37%, $p < 0.0$ groups (Table S2). At the genus level (Table S3), 39 genera were significantly differe between at least two groups. An analysis of the correlation of the abundance of these genera with the total cholesterol level in each animal at the study's end showed that 14 of the genera influenced by Totum-070 were correlated with the total cholesterol level (Table S3). Specifically, the abundance of uncl. *Muribaculaceae*, *Duncaniella*, and uncl. *Desulfovibrionaceae* was inversely correlated with the cholesterol level. These genera were decreased in the HFD group compared with the ND group (−37%, $p < 0.01$; −37%, ns and −32%, ns, respectively). Supplementation with Totum-070 significantly increased their abundance compared with the HFD group (+52%, $p < 0.01$; +45%, $p < 0.05$; and +54%, $p < 0.05$, respectively) and restored their abundance to the level observed in the ND group. Among the genera the abundance of which was positively correlated with the cholesterol level, *Lactococcus* and *Flintibacter* were higher in the HFD group than the ND group (+120%, ns and +59%, $p < 0.05$, respectively). In the HFD-T070 5% group, their abundance was reduced compared with HFD (−55%, $p < 0.05$ and −29%, $p < 0.05$) to levels comparable to those in the ND group. Altogether, HFD-T070 5% strongly affected the gut microbial composition and influenced the abundance of several cholesterol-related genera in correlation with its hypocholesterolemic action.

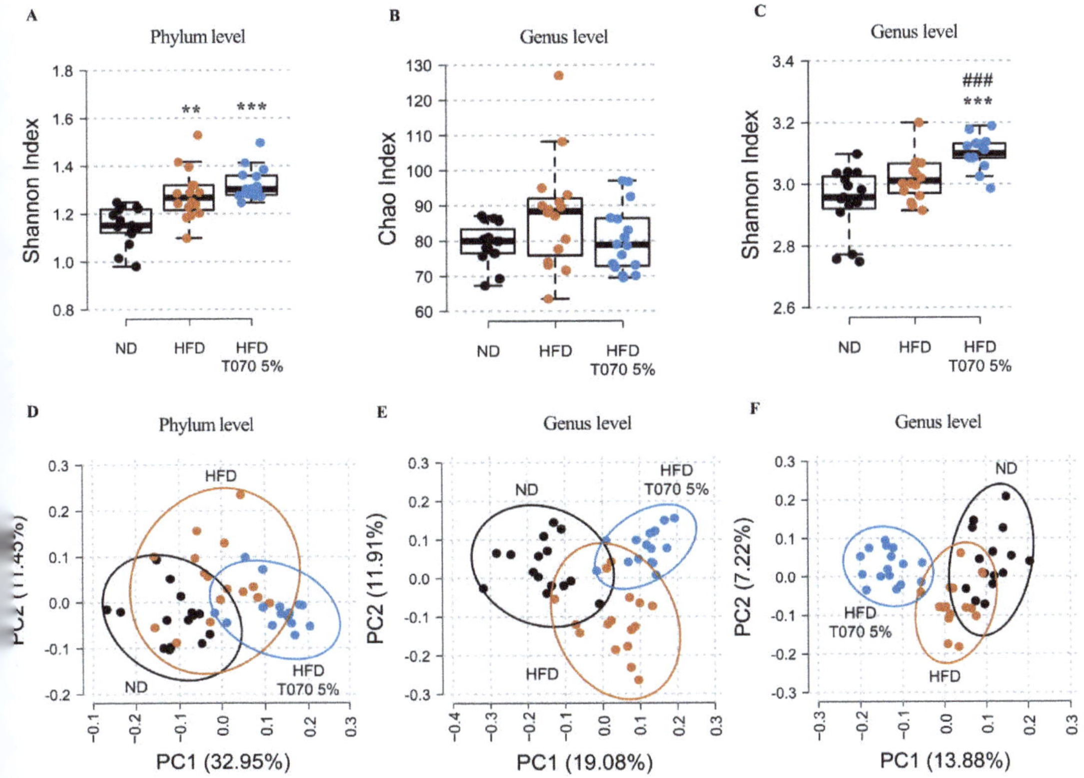

Figure 5. Effect of Totum-070 on cecal microbiota richness, diversity, and composition. (**A**) Shannon diversity index at the phylum level. (**B**) Rarefied Chao index (index of richness) at the genus level. (**C**) Shannon diversity index at the genus level. (**D**,**E**) Principal coordinate analyses based on the Bray–Curtis dissimilarity index at the phylum (**D**) and genus (**E**) levels, respectively, with 95% confidence ellipses. (**F**) Principal coordinate analysis based on the Jaccard dissimilarity index at the genus levels, with 95% confidence ellipses. Boxplots show the median, 25th, and 75th percentiles and adjacent values (n = 15/16 per group). ** $p < 0.01$ and *** $p < 0.001$ versus ND. ### $p < 0.001$ Totum-070 5% group versus HFD.

4. Discussion

This study demonstrated the efficacy of Totum-070, a polyphenol-rich mixture, reduce total cholesterol, non-HDL cholesterol, and triglyceride levels in a hamster moc of diet-induced dyslipidemia and identified in vivo and in vitro pathways through whi Totum-070 prevents hyperlipidemia. This dose-dependent cholesterol reduction was served from week 6 of supplementation until the study's end (week 12). Importantly, m of the total cholesterol lost was non-HDL-C (Figure 1D), a particularly interesting res given the link between elevated plasma non-HDL-C and ASCVD [48,49]. The hypocho terolemic effect of Totum-070 is independent of the modulation of caloric intake and bo weight (Table 2). Besides its beneficial effect on serum cholesterol and triglycerides, sup mentation with Totum-070 reduced triglyceride accumulation in liver (Figure 2C). Signs hepatic steatosis were clearly identified in the HFD group but were reduced in the grou receiving HFD with Totum-070. Moreover, Totum-070 had a systemic anti-inflammat effect by preventing elevation of serum LPS (Figure 3G) and counteracting the inducti of pro-inflammatory genes in the liver (Figure 2J) and intestine (Figure 3F). These resu strengthened Totum-070's role in the prevention of cardiometabolic disorders commo associated with hyperlipidemia [50,51].

Totum-070 combines five plant extracts selected to act on different organs involv in the regulation of cholesterol homeostasis. The effect on dyslipidemia of several these plant extracts and compounds present in Totum-070 has already been studied animal models, such as the hypolipidemic effect of artichoke leaf extracts in rats fed HFD [14,52,53], the beneficial action of olive leaf extracts in HFD-mice [54] and rats [18, and hyperlipemia reduction by Goji berries [21,24] and black pepper extracts [25] in r and mice. The choice of animal model in the present work is also fundamental. Inde Syrian hamsters better mimic cholesterol metabolism and the enzymatic pathways involv in human-like lipoprotein disorders, compared with other rodent models that are natura resistant to diet-induced hypercholesterolemia. Their receptor-mediated uptake of LDL cholesteryl ester transfer protein activity, and hepatic ApoB-100 and intestinal ApoB secretion, comparable to those described in humans [56,57], allow for studying the eff of a food supplement on HFD-induced dyslipidemia. Plasma cholesterol levels depe on several parameters, including endogenous cholesterol synthesis, cholesterol remo from circulation, dietary cholesterol absorption, and cholesterol reabsorption from b The intestine and liver play central roles in cholesterol homeostasis and, consequen in hypercholesterolemia and cardiovascular risk. In this study, an HFD's effects on li and lipoprotein metabolism in the liver and intestine were investigated to understa Totum-070's mechanisms of action. A gene expression analysis of the main regulat for cholesterol (Figure 2F) and fatty acid (Figure 2H) metabolism in the liver showed to scarce effects of Totum-070 compared to HFD-fed hamsters. An investigation of b acid homeostasis depicted a normalization of the transcripts altered upon HFD with 5% Totum-070 supplementation. LDL-C is removed from plasma predominantly by LDL receptor, a transmembrane protein expressed at the plasma membrane of liver ce and which is responsible for LDL particle internalization via receptor-mediated endo tosis [58]. Therefore, regulation of the liver LDL receptor is a key mechanism by wh therapeutic agents can decrease cholesterol levels. Previous reports documented L receptor upregulation in cultured hepatocytes when cells were incubated with piperine [or hydroxytyrosol [60], two biomolecules present in Totum-070 (Table 1). In our study, li LDL receptor expression was reduced by 26% in the HFD group, but in the HFD-T 5% group, its expression was restored to the level observed in the ND group (Figure 2 This observation may partly explain the reduction of non-HDL-C in HFD-fed hamst supplemented with Totum-070. In addition, LDL receptor implication in the productior newly synthetized triglyceride-rich particles in the liver is a mechanism that regulates hepatic triglyceride output [61,62]. Modulation of the liver's secretion of triglyceride-r particles might partly explain Totum-070's hypotriglyceridemic effect. However, the liv triglyceride production was similar in the Totum-070-supplemented groups and in

HFD group (Figure 2D). The activity of lipoprotein lipase, which mediates the hydrolysis of circulating triglycerides, was also comparable between the groups (Figure 3D). Conversely, the oral fat tolerance test showed that the amplitude and duration of postprandial hypertriglyceridemia were significantly reduced in the Totum-070 group compared with the HFD group (Figure 3C). This is an important result because postprandial hyperlipidemia is a cardiovascular risk factor, especially in patients with metabolic syndrome [63]. As no difference was observed in the lipoprotein lipase activity, Totum-070 might lower postprandial hypertriglyceridemia by decreasing the release of intestinal triglyceride-rich lipoproteins. Several biomolecules present in Totum-070 (i.e., piperine [64] and luteolin [65,66]) have been previously reported to modulate intestinal lipid metabolism, notably through the inhibition of NPC1L1 expression and trafficking to the cellular membrane. This transmembrane protein localized at the apical membrane of enterocytes is involved in cholesterol uptake by enterocytes. As suggested by the effect of ezetimibe, an NPC1L1 inhibitor on postprandial hypertriglyceridemia [67], NPC1L1 inhibition may also lead to the reduction in intestinal triglyceride production [67] Totum-070 reduced cholesterol uptake in vitro, and this effect was blunted by co-incubation with ezetimibe (Figure 4B). This suggests that NPC1L1 activity is required for a Totum-070-mediated reduction of cholesterol uptake. Totum-070's effect was not due to the regulation of NPC1L1 expression or location at the cell membrane (Figure 4C,D), but possibly through the inhibition of NPC1L1 activity. The structure of NPC1L1 or the uptake route may have been altered because cholesterol uptake by Caco-2 cells was inhibited by the Totum-070 bioactive molecules within 1 h of incubation. For example, ezetimibe binding to the middle extracellular domain of NPC1L1 causes conformational changes in the NPC1L1 protein to disturb NPC1L1 and cholesterol interactions [68]. The *N*-terminus part of NPC1L1 contains the sterol-binding domain that is essential for extracellular cholesterol binding to NPC1L1 and uptake into the cell [45]. Using MST, we demonstrated direct binding of Totum-070 to the *N*-terminal domain of the NPC1L1 protein (Figure 4E). Thus, we could speculate that Totum-070's interaction with NPC1L1-NTD would suppress cholesterol binding to NPC1L1, resulting in the inhibition of NPC1L1 transport. Totum-070 also interfered with cholesterol secretion into newly synthetized particles in vitro (Figure 4F). The exact molecular mechanisms need to be elucidated, because Totum-070 did not have any effect on MTTP or ACAT2 expression in the Caco-2 cells (Figure 4H). This reduction in cholesterol uptake in the intestine was confirmed in vivo, as fecal neutral sterols were increased by 17% in hamsters supplemented with Totum-070 compared with the HFD group (Figure 3A). In line with the increased cholesterol excretion in the feces, dietary cholesterol absorption in the intestine was decreased in the Totum-070-supplemented groups (Figure 3B). Altogether, these results demonstrate that intestinal cholesterol absorption was reduced by Totum-070 supplementation, suggesting that chronic supplementation with Totum-070 (12 weeks) might contribute to decrease cholesterol in this model. At the end, our data show that Totum-070 supplementation is an effective strategy to reduce intestinal cholesterol absorption.

The gut microbiota has emerged as a crucial factor that influences cholesterol metabolism [69,70], and many studies have already linked the imbalance or maladaptation of the microbial community, defined as "dysbiosis", to host lipid metabolism and metabolic disorders [71,72]. A meta-analysis found that the Firmicutes to Bacteroidetes ratio is increased in HFD-fed mice compared with ND-fed mice [73]. Importantly, an increase in the Firmicutes to Bacteroidetes ratio has been associated with cardiovascular risk factors, such as obesity [71] and hypertension [72]. Moreover, this ratio has been significantly correlated with the circulating total cholesterol concentration in mice [74] and hamsters [75]. In our study, the Firmicutes to Bacteroidetes ratio did not change significantly between the ND group and HFD group but decreased significantly in the Totum-070-supplemented groups compared with the HFD group (Table S2). Moreover, uncl. *Muribaculaceae*, *Duncaniella*, and uncl. *Desulfovibrionaceae*, which were negatively correlated with the cholesterol level, decreased in the HFD group and were restored to normal levels upon Totum-070 supplementation (Table S3). The abundance of *Duncaniella*, which has anti-inflammatory properties [76], reduced in western diet-fed mice [74], and negatively correlated with LDL-cholesterol [77], as observed in the present study. Simila the *Muribaculaceae* community was significantly reduced in the intestinal flora of HI mice [78], whereas its increase might contribute to the serum cholesterol-lowering eff of plant proteins in dyslipidemic hamsters [75]. *Muribaculaceae* may also be a potent biomarker of cholesterol reduction [79]. Moreover, our study shows that the abundan of *Lactococcus* and *Flintibacter* positively correlated with total cholesterol level (Table S *Lactococcus* and *Flintibacter* were more abundant in the HFD group compared with t ND group, but supplementation with Totum-070 reduced their abundances to the lev observed in the ND group. Consistent with our observation, previous reports describ increased *Lactococcus* [80] and *Flintibacter* [81] in animal models fed an HFD. Altogeth the analysis of the cecal microbiota showed alterations of the microbial balance in the H group compared with the ND group. Our results suggest that the Totum-070-supplement group displayed an alternative profile compared with the HFD and ND groups. So specific taxa were influenced by Totum-070 in correlation with its beneficial effect on cho terol lowering. However, it remains to be clarified whether these changes are a cause o consequence of the hypocholesterolemic effect.

This study has potential limitations. It has been shown that the lipid profile c fers by sex, with estrogen providing protection in non-menopausal women [82,83]. this study, we exclusively used male hamsters; therefore, future studies could explo the effect of Totum-070 on females. Regarding NPC1L1, this work explored its role Totum-070's intestinal effects. However, new studies in genetically modified mice lacki NPC1L1 [84,85] would complement the understanding of Totum-070's cholesterol-loweri mechanism. Additionally, the actions of Totum-070 were limited to lipid homeostasis, a the potential effects of Totum-070 in preventing or delaying the atherosclerotic process w not covered in this study. Although hamsters exhibit metabolic similarities more compa ble to humans than other rodents, they do not develop aortic atherosclerotic lesions [8 Therefore, the use of a suitable preclinical model prone to atherosclerosis developme such as LDL receptor- or ApoE-deficient mice [86], will be required to evaluate the be ficial effects of Totum-070 on atherosclerosis. Finally, no adverse effects from Totum-0 supplementation were observed in our animal studies. To our knowledge, only mi to no gastrointestinal effects have been reported in trials using plant extracts comprisi Totum-070, such as olive leaf extracts [87] or artichoke leaf extracts [88]. Future high-qua clinical studies will need to confirm the hypocholesterolemic potential of this formulati in humans and document its safety and tolerance.

5. Conclusions

In conclusion, we demonstrated that Totum-070 can reduce hypercholesterolem in a hamster model of diet-induced dyslipidemia. Furthermore, the HFD-induced pertriglyceridemia and hepatic triglyceride accumulation were reduced by Totum-0 supplementation. In vivo and in vitro analyses in hamsters and human enterocytes, spectively, demonstrated that Totum-070 directly inhibits intestinal cholesterol absorpti potentially through its interaction with the *N*-terminal domain of NPC1L1. The additio beneficial effects observed on the gut microbiota composition also might contribute Totum-070's mode of action, making Totum-070 an innovative natural ingredient to prev blood cholesterol elevation, a major risk factor for ASCVD.

Supplementary Materials: The following supporting information can be downloaded at: ht //www.mdpi.com/article/10.3390/nu15245056/s1, Figure S1: expression of genes encoding prote implicated in bile acid homeostasis in ileum; Table S1: list annotation of differentially expres hepatic genes within KEGG pathways between HFD and HFD T070 5% groups using the *Cricet griseus'* ortholog gene identifiers; Table S2: cecal microbiota: indicator phyla. Indicator taxa w identified as those the abundance of which differed significantly in ≥2 groups; Table S3: cecal mi biota: indicator genera. Indicator taxa were identified as those whose abundance differ significa between two groups or more.; Table S4: List of hamster primers used for qPCR.

Author Contributions: Conceptualization, C.L., M.V.,M.M., Y.F.O., F.L.J., H.G., O.A., R.B.P., M.G., B.G., T.M., B.C., S.P., J.-M.B. and P.S.; methodology, M.V., M.M., F.L.J., H.G., O.A., R.B.P., M.G., B.G., T.M., B.C. and.; software, M.V.; validation, Y.F.O.; formal analysis, C.L., M.V. and M.M.,.; investigation, C.L.; data curation, C.L., M.V. and F.L.J.; writing—original draft preparation, C.L.; writing—review and editing, Y.F.O., B.G., T.M., B.C., J.-M.B. and P.S.; visualization, C.L.; supervision, S.P., J.-M.B. and P.S.; funding acquisition, S.P. All authors have read and agreed to the published version of the manuscript.

Funding: This work was supported by a FEDER (Fonds Européen de Développement Régional) grant number 2015RPC-BAFE-114.

Institutional Review Board Statement: The study was conducted according to the guidelines of the Declaration of Helsinki, and approved by the local Ethics Committee C2E2A, Auvergne, France, under the number 18588-2019012216136952V5 dated 2 October 2020.

Informed Consent Statement: Not applicable.

Data Availability Statement: The raw RNA sequencing data from the hamster livers has been deposited in GEO under BioProject PRJNA945002 with the accession number GSE227411 (BioSamples: GSM7100119 to GSM7100142). The raw 16S rRNA gene sequencing data from hamster cecal microbiota has been deposited in the Sequence Read Archive under BioProject PRJNA549583 with the following accession numbers: SRR22824159 to SRR22824205 (BioSamples: SAMN32320356 to SAMN32320402). All the remaining data described in the manuscript are shown in the manuscript and Supplementary Data.

Acknowledgments: We would like to thank HELIXIO for performing the RNA sequencing and analysis. We thank Elisabetta Andermarcher for reviewing and correcting the manuscript. Part of the experiments (16S rRNA gene sequencing) were performed at the Genome Transcriptome Facility of Bordeaux.

Conflicts of Interest: Cédric Langhi, Marie Vallier, Yolanda F. Otero, Maheva Maura, Florian Le Joubioux, and Pascal Sirvent are employees of Valbiotis. Sébastien Peltier is CEO of Valbiotis. Jean-Marie Bard is a member of the scientific committee and a stock shareholder of Valbiotis. Bruno Guigas is a member of the scientific committee of Valbiotis. Benoit Chassaing reports honorarium and consulting fees from Nestlé, Procter and Nobles, and Qiagen.

Abbreviations

ACAT2	acetyl-CoA acetyltransferase 2
ASCVD	Atherosclerotic cardiovascular disease
AUC	Area under the curve
FNS	Fecal neutral sterols
HDL-C	High-density lipoprotein cholesterol
HFD	High-fat diet
HTPT	Hepatic triglycerides production test
IBAT	Ileal bile acid transporter
LDL-C	Low-density lipoprotein cholesterol
LPL	Lipoprotein lipase
LPS	Lipopolysaccharides
MTTP	Microsomal triglyceride transfer protein
MST	Microscale thermophoresis
ND	Normal diet
NTD	N-terminal domain
NPC1L1	Niemann-pick C1-like 1
OFTT	Oral fat tolerance test
TG	Triglycerides
VLDL	Very low-density lipoprotein

References

1. Dagenais, G.R.; Leong, D.P.; Rangarajan, S.; Lanas, F.; Lopez-Jaramillo, P.; Gupta, R.; Diaz, R.; Avezum, A.; Oliveira, G.B Wielgosz, A.; et al. Variations in common diseases, hospital admissions, and deaths in middle-aged adults in 21 countries fr five continents (PURE): A prospective cohort study. *Lancet* **2020**, *395*, 785–794. [CrossRef]
2. Farzadfar, F. Cardiovascular disease risk prediction models: Challenges and perspectives. *Lancet Glob. Health* **2019**, *7*, e1288–e1 [CrossRef]
3. Boren, J.; Chapman, M.J.; Krauss, R.M.; Packard, C.J.; Bentzon, J.F.; Binder, C.J.; Daemen, M.J.; Demer, L.L.; Hegele, R.A.; Nich S.J.; et al. Low-density lipoproteins cause atherosclerotic cardiovascular disease: Pathophysiological, genetic, and therape insights: A consensus statement from the European Atherosclerosis Society Consensus Panel. *Eur. Heart J.* **2020**, *41*, 2313–2 [CrossRef]
4. Prospective Studies, C.; Lewington, S.; Whitlock, G.; Clarke, R.; Sherliker, P.; Emberson, J.; Halsey, J.; Qizilbash, N.; Peto, Collins, R. Blood cholesterol and vascular mortality by age, sex, and blood pressure: A meta-analysis of individual data from prospective studies with 55,000 vascular deaths. *Lancet* **2007**, *370*, 1829–1839. [CrossRef]
5. Zanotti, I.; Dall'Asta, M.; Mena, P.; Mele, L.; Bruni, R.; Ray, S.; Del Rio, D. Atheroprotective effects of (poly)phenols: A focus cell cholesterol metabolism. *Food Funct.* **2015**, *6*, 13–31. [CrossRef] [PubMed]
6. Morze, J.; Osadnik, T.; Osadnik, K.; Lejawa, M.; Jakubiak, G.; Pawlas, N.; Gasior, M.; Schwingshackl, L.; Banach, M. Comparat effect of nutraceuticals on lipid profile: A protocol for systematic review and network meta-analysis. *BMJ Open* **2020**, *10*, e032 [CrossRef] [PubMed]
7. Osadnik, T.; Golawski, M.; Lewandowski, P.; Morze, J.; Osadnik, K.; Pawlas, N.; Lejawa, M.; Jakubiak, G.K.; Mazur, Schwingschackl, L.; et al. A network meta-analysis on the comparative effect of nutraceuticals on lipid profile in adults. *Pharma Res.* **2022**, *183*, 106402. [CrossRef]
8. Cicero, A.F.G.; Colletti, A.; Bajraktari, G.; Descamps, O.; Djuric, D.M.; Ezhov, M.; Fras, Z.; Katsiki, N.; Langlois, M.; Latkovskis, et al. Lipid-lowering nutraceuticals in clinical practice: Position paper from an International Lipid Expert Panel. *Nutr. Rev.* **2** *75*, 731–767. [CrossRef]
9. Wider, B.; Pittler, M.H.; Thompson-Coon, J.; Ernst, E. Artichoke leaf extract for treating hypercholesterolaemia. *Cochrane Data Syst. Rev.* **2013**, *3*, CD003335. [CrossRef]
10. Sahebkar, A.; Pirro, M.; Banach, M.; Mikhailidis, D.P.; Atkin, S.L.; Cicero, A.F.G. Lipid-lowering activity of artichoke extracts systematic review and meta-analysis. *Crit. Rev. Food Sci. Nutr.* **2018**, *58*, 2549–2556. [CrossRef]
11. Acar-Tek, N.; Agagunduz, D. Olive Leaf (*Olea europaea* L. folium): Potential Effects on Glycemia and Lipidemia. *Ann. Nutr. Me* **2020**, *76*, 10–15. [CrossRef]
12. Vogel, P.; Kasper Machado, I.; Garavaglia, J.; Zani, V.T.; de Souza, D.; Morelo Dal Bosco, S. Polyphenols benefits of olive leaf (C *europaea* L) to human health. *Nutr. Hosp.* **2014**, *31*, 1427–1433. [CrossRef]
13. Kulczyński, B.; Gramza-Michałowska, A. Goji Berry (*Lycium barbarum*): Composition and Health Effects—a Review. *Pol. J. F Nutr. Sci.* **2016**, *66*, 67–75. [CrossRef]
14. Ben Salem, M.; Affes, H.; Dhouibi, R.; Charfi, S.; Turki, M.; Hammami, S.; Ayedi, F.; Sahnoun, Z.; Zeghal, K.M.; Ksouda, K. Ef of Artichoke (*Cynara scolymus*) on cardiac markers, lipid profile and antioxidants levels in tissue of HFD-induced obesity. *A Physiol. Biochem.* **2022**, *128*, 184–194. [CrossRef] [PubMed]
15. Ben Salem, M.; Ksouda, K.; Dhouibi, R.; Charfi, S.; Turki, M.; Hammami, S.; Ayedi, F.; Sahnoun, Z.; Zeghal, K.M.; Affes, H. MS/MS Analysis and Hepatoprotective Activity of Artichoke (*Cynara scolymus* L.) Leaves Extract against High Fat Diet-Indu Obesity in Rats. *Biomed. Res. Int.* **2019**, *2019*, 4851279. [CrossRef] [PubMed]
16. Ben Salem, M.; Ben Abdallah Kolsi, R.; Dhouibi, R.; Ksouda, K.; Charfi, S.; Yaich, M.; Hammami, S.; Sahnoun, Z.; Zeg K.M.; Jamoussi, K.; et al. Protective effects of Cynara scolymus leaves extract on metabolic disorders and oxidative stres alloxan-diabetic rats. *BMC Complement. Altern. Med.* **2017**, *17*, 328. [CrossRef] [PubMed]
17. Qiang, Z.; Lee, S.O.; Ye, Z.; Wu, X.; Hendrich, S. Artichoke extract lowered plasma cholesterol and increased fecal bile acid Golden Syrian hamsters. *Phytother. Res.* **2012**, *26*, 1048–1052. [CrossRef] [PubMed]
18. Poudyal, H.; Campbell, F.; Brown, L. Olive leaf extract attenuates cardiac, hepatic, and metabolic changes in high carbohydra high fat-fed rats. *J. Nutr.* **2010**, *140*, 946–953. [CrossRef] [PubMed]
19. Yoon, L.; Liu, Y.N.; Park, H.; Kim, H.S. Olive Leaf Extract Elevates Hepatic PPAR alpha mRNA Expression and Improves Ser Lipid Profiles in Ovariectomized Rats. *J. Med. Food* **2015**, *18*, 738–744. [CrossRef]
20. Wang, L.; Geng, C.; Jiang, L.; Gong, D.; Liu, D.; Yoshimura, H.; Zhong, L. The anti-atherosclerotic effect of olive leaf ext is related to suppressed inflammatory response in rabbits with experimental atherosclerosis. *Eur. J. Nutr.* **2008**, *47*, 235– [CrossRef]
21. Cui, B.; Liu, S.; Lin, X.; Wang, J.; Li, S.; Wang, Q.; Li, S. Effects of Lycium barbarum aqueous and ethanol extracts on high-fat- induced oxidative stress in rat liver tissue. *Molecules* **2011**, *16*, 9116–9128. [CrossRef]
22. Pai, P.G.; Habeeba, P.U.; Ullal, S.D.; Shoeb, P.A.; Pradeepti, M.S.; Ramya, K. Evaluation of Hypolipidemic Effects of Lycium barum (goji Berry) in a Murine Model. *J. Nat. Remedies* **2013**, *13*, 4–8.
23. Crepaldi, L.D.; Mariano, I.R.; Trondoli, A.J.P.C.; Moreno, F.N.; Piovan, S.; Formigoni, M.; Salgueiro-Pagadigorria, C.L.; Go V.A.F.; Brito, M.d.N.; Garcia, R.F. Goji Berry (*Lycium barbarum*) Extract Improves Biometric, Plasmatic and Hepatic Parameter Rats Fed a High-Carbohydrate Diet. *J. Pharm. Pharmacol.* **2018**, *6*, 877–889.

Luo, Q.; Cai, Y.; Yan, J.; Sun, M.; Corke, H. Hypoglycemic and hypolipidemic effects and antioxidant activity of fruit extracts from Lycium barbarum. *Life Sci.* **2004**, *76*, 137–149. [CrossRef] [PubMed]
Parim, B.; Harishankar, N.; Balaji, M.; Pothana, S.; Sajjalaguddam, R.R. Effects of *Piper nigrum* extracts: Restorative perspectives of high-fat diet-induced changes on lipid profile, body composition, and hormones in Sprague-Dawley rats. *Pharm. Biol.* **2015**, *53*, 1318–1328. [CrossRef]
Vijayakumar, R.S.; Surya, D.; Senthilkumar, R.; Nalini, N. Hypolipidemic Effect of Black Pepper (*Piper nigrum* Linn.) in Rats Fed High Fat Diet. *J. Clin. Biochem. Nutr.* **2002**, *32*, 31–42. [CrossRef]
Cottet, J.; Etienne, J. Determination of serum lipids by the sulfo-phospho-vanillic method of E. Chabrol and R. Charonnat. *Bull. Acad. Natl. Med.* **1965**, *149*, 331–338. [PubMed]
Cheng, Y.S.; Zheng, Y.; VanderGheynst, J.S. Rapid quantitative analysis of lipids using a colorimetric method in a microplate format. *Lipids* **2011**, *46*, 95–103. [CrossRef]
Benson, J.R.; Hare, P.E. O-phthalaldehyde: Fluorogenic detection of primary amines in the picomole range. Comparison with fluorescamine and ninhydrin. *Proc. Natl. Acad. Sci. USA* **1975**, *72*, 619–622. [CrossRef] [PubMed]
Giera, M.; Plossl, F.; Bracher, F. Fast and easy in vitro screening assay for cholesterol biosynthesis inhibitors in the post-squalene pathway. *Steroids* **2007**, *72*, 633–642. [CrossRef]
Maillard, F.; Vazeille, E.; Sauvanet, P.; Sirvent, P.; Bonnet, R.; Combaret, L.; Chausse, P.; Chevarin, C.; Otero, Y.F.; Delcros, G.; et al. Preventive Effect of Spontaneous Physical Activity on the Gut-Adipose Tissue in a Mouse Model That Mimics Crohn's Disease Susceptibility. *Cells* **2019**, *8*, 33. [CrossRef]
Turley, S.D.; Daggy, B.P.; Dietschy, J.M. Psyllium augments the cholesterol-lowering action of cholestyramine in hamsters by enhancing sterol loss from the liver. *Gastroenterology* **1994**, *107*, 444–452. [CrossRef]
Turley, S.D.; Herndon, M.W.; Dietschy, J.M. Reevaluation and application of the dual-isotope plasma ratio method for the measurement of intestinal cholesterol absorption in the hamster. *J. Lipid Res.* **1994**, *35*, 328–339. [CrossRef] [PubMed]
Ono-Moore, K.D.; Ferguson, M.; Blackburn, M.L.; Issafras, H.; Adams, S.H. Application of an In Vivo Hepatic Triacylglycerol Production Method in the Setting of a High-Fat Diet in Mice. *Nutrients* **2016**, *9*, 16. [CrossRef] [PubMed]
Lewis, G.F.; Uffelman, K.; Naples, M.; Szeto, L.; Haidari, M.; Adeli, K. Intestinal lipoprotein overproduction, a newly recognized component of insulin resistance, is ameliorated by the insulin sensitizer rosiglitazone: Studies in the fructose-fed Syrian golden hamster. *Endocrinology* **2005**, *146*, 247–255. [CrossRef] [PubMed]
Taghibiglou, C.; Carpentier, A.; Van Iderstine, S.C.; Chen, B.; Rudy, D.; Aiton, A.; Lewis, G.F.; Adeli, K. Mechanisms of hepatic very low density lipoprotein overproduction in insulin resistance. Evidence for enhanced lipoprotein assembly, reduced intracellular ApoB degradation, and increased microsomal triglyceride transfer protein in a fructose-fed hamster model. *J. Biol. Chem.* **2000**, *275*, 8416–8425. [CrossRef] [PubMed]
Cheema, S.K.; Cornish, M.L. Bio F1B hamster: A unique animal model with reduced lipoprotein lipase activity to investigate nutrient mediated regulation of lipoprotein metabolism. *Nutr. Metab.* **2007**, *4*, 27. [CrossRef] [PubMed]
Lewis, G.F.; Murdoch, S.; Uffelman, K.; Naples, M.; Szeto, L.; Albers, A.; Adeli, K.; Brunzell, J.D. Hepatic lipase mRNA, protein, and plasma enzyme activity is increased in the insulin-resistant, fructose-fed Syrian golden hamster and is partially normalized by the insulin sensitizer rosiglitazone. *Diabetes* **2004**, *53*, 2893–2900. [CrossRef]
Yamazaki, T.; Kishimoto, K.; Ezaki, O. The ddY mouse: A model of postprandial hypertriglyceridemia in response to dietary fat. *J. Lipid Res.* **2012**, *53*, 2024–2037. [CrossRef]
Chavanelle, V.; Otero, Y.F.; Le Joubioux, F.; Ripoche, D.; Bargetto, M.; Vluggens, A.; Montaurier, C.; Pickering, G.; Ducheix, G.; Dubray, C.; et al. Effects of Totum-63 on glucose homeostasis and postprandial glycemia: A translational study. *Am. J. Physiol. Endocrinol. Metab.* **2021**, *320*, E1119–E1137. [CrossRef]
Gilda, J.E.; Gomes, A.V. Western blotting using in-gel protein labeling as a normalization control: Stain-free technology. *Methods Mol. Biol.* **2015**, *1295*, 381–391. [CrossRef]
Chateau, D.; Pauquai, T.; Delers, F.; Rousset, M.; Chambaz, J.; Demignot, S. Lipid micelles stimulate the secretion of triglyceride-enriched apolipoprotein B48-containing lipoproteins by Caco-2 cells. *J. Cell Physiol.* **2005**, *202*, 767–776. [CrossRef]
Nakano, T.; Inoue, I.; Katayama, S.; Seo, M.; Takahashi, S.; Hokari, S.; Shinozaki, R.; Hatayama, K.; Komoda, T. Lysophosphatidylcholine for efficient intestinal lipid absorption and lipoprotein secretion in caco-2 cells. *J. Clin. Biochem. Nutr.* **2009**, *45*, 227–234. [CrossRef]
Vallier, M.; Suwandi, A.; Ehrhardt, K.; Belheouane, M.; Berry, D.; Cepic, A.; Galeev, A.; Johnsen, J.M.; Grassl, G.A.; Baines, J.F. Pathometagenomics reveals susceptibility to intestinal infection by Morganella to be mediated by the blood group-related B4galnt2 gene in wild mice. *Gut Microbes* **2023**, *15*, 2164448. [CrossRef] [PubMed]
Betters, J.L.; Yu, L. NPC1L1 and cholesterol transport. *FEBS Lett.* **2010**, *584*, 2740–2747. [CrossRef] [PubMed]
Jerabek-Willemsen, M.; André, T.; Wanner, R.; Roth, H.M.; Duhr, S.; Baaske, P.; Breitsprecher, D. MicroScale Thermophoresis: Interaction analysis and beyond. *J. Mol. Struct.* **2014**, *1077*, 101–113. [CrossRef]
Haguet, Q.; Le Joubioux, F.; Chavanelle, V.; Groult, H.; Schoonjans, N.; Langhi, C.; Michaux, A.; Otero, Y.F.; Boisseau, N.; Peltier, S.L.; et al. Inhibitory Potential of alpha-Amylase, alpha-Glucosidase, and Pancreatic Lipase by a Formulation of Five Plant Extracts: TOTUM-63. *Int. J. Mol. Sci.* **2023**, *24*, 3652. [CrossRef]
Verbeek, R.; Hovingh, G.K.; Boekholdt, S.M. Non-high-density lipoprotein cholesterol: Current status as cardiovascular marker. *Curr. Opin. Lipidol.* **2015**, *26*, 502–510. [CrossRef] [PubMed]

49. Arsenault, B.J.; Rana, J.S.; Stroes, E.S.; Despres, J.P.; Shah, P.K.; Kastelein, J.J.; Wareham, N.J.; Boekholdt, S.M.; Khaw, K.T. Beyo low-density lipoprotein cholesterol: Respective contributions of non-high-density lipoprotein cholesterol levels, triglycerid and the total cholesterol/high-density lipoprotein cholesterol ratio to coronary heart disease risk in apparently healthy men a women. *J. Am. Coll. Cardiol.* **2009**, *55*, 35–41. [CrossRef]
50. Eckel, R.H.; Grundy, S.M.; Zimmet, P.Z. The metabolic syndrome. *Lancet* **2005**, *365*, 1415–1428. [CrossRef]
51. Kolovou, G.D.; Anagnostopoulou, K.K.; Cokkinos, D.V. Pathophysiology of dyslipidaemia in the metabolic syndrome. *Postgr Med. J.* **2005**, *81*, 358–366. [CrossRef]
52. Kusku-Kiraz, Z.; Mehmetcik, G.; Dogru-Abbasoglu, S.; Uysal, M. Artichoke leaf extract reduces oxidative stress and lipoprot dyshomeostasis in rats fed on high cholesterol diet. *Phytother. Res.* **2010**, *24*, 565–570. [CrossRef]
53. Kucukgergin, C.; Aydin, A.F.; Ozdemirler-Erata, G.; Mehmetcik, G.; Kocak-Toker, N.; Uysal, M. Effect of artichoke leaf extract hepatic and cardiac oxidative stress in rats fed on high cholesterol diet. *Biol. Trace Elem. Res.* **2010**, *135*, 264–274. [CrossRef]
54. Cheurfa, M.; Abdallah, H.H.; Allem, R.; Noui, A.; Picot-Allain, C.M.N.; Mahomoodally, F. Hypocholesterolaemic and antioxid properties of Olea europaea L. leaves from Chlef province, Algeria using in vitro, in vivo and in silico approaches. *Food Ch Toxicol.* **2019**, *123*, 98–105. [CrossRef] [PubMed]
55. Olmez, E.; Vural, K.; Gok, S.; Ozturk, Z.; Kayalar, H.; Ayhan, S.; Var, A. Olive Leaf Extract Improves the Atherogenic Lipid Pro in Rats Fed a High Cholesterol Diet. *Phytother. Res.* **2015**, *29*, 1652–1657. [CrossRef] [PubMed]
56. Dillard, A.; Matthan, N.R.; Lichtenstein, A.H. Use of hamster as a model to study diet-induced atherosclerosis. *Nutr Metab (Lo* **2010**, *7*, 89. [CrossRef] [PubMed]
57. Bhatia, G.; Rizvi, F.; Saxena, R.; Puri, A.; Khanna, A.K.; Chander, R.; Wulff, E.M.; Rastogi, A.K. In vivo model for dyslipider with diabetes mellitus in hamster. *Indian. J. Exp. Biol.* **2003**, *41*, 1456–1459. [PubMed]
58. Blacklow, S.C. Versatility in ligand recognition by LDL receptor family proteins: Advances and frontiers. *Curr. Opin. Struct. B* **2007**, *17*, 419–426. [CrossRef] [PubMed]
59. Ochiai, A.; Miyata, S.; Shimizu, M.; Inoue, J.; Sato, R. Piperine Induces Hepatic Low-Density Lipoprotein Receptor Expressi through Proteolytic Activation of Sterol Regulatory Element-Binding Proteins. *PLoS ONE* **2015**, *10*, e0139799. [CrossRef] [PubM
60. Zhang, X.; Qin, Y.; Wan, X.; Liu, H.; Iv, C.; Ruan, W.; Lu, L.; He, L.; Guo, X. Hydroxytyrosol Plays Antiatherosclerotic Effe through Regulating Lipid Metabolism via Inhibiting the p38 Signal Pathway. *Biomed. Res. Int.* **2020**, *2020*, 5036572. [CrossRef
61. Twisk, J.; Gillian-Daniel, D.L.; Tebon, A.; Wang, L.; Barrett, P.H.; Attie, A.D. The role of the LDL receptor in apolipoprotei secretion. *J. Clin. Investig.* **2000**, *105*, 521–532. [CrossRef]
62. Blasiole, D.A.; Oler, A.T.; Attie, A.D. Regulation of ApoB secretion by the low density lipoprotein receptor requires exit from endoplasmic reticulum and interaction with ApoE or ApoB. *J. Biol. Chem.* **2008**, *283*, 11374–11381. [CrossRef]
63. Nakamura, K.; Miyoshi, T.; Yunoki, K.; Ito, H. Postprandial hyperlipidemia as a potential residual risk factor. *J. Cardiol.* **2016**, 335–339. [CrossRef]
64. Duangjai, A.; Ingkaninan, K.; Praputbut, S.; Limpeanchob, N. Black pepper and piperine reduce cholesterol uptake and enha translocation of cholesterol transporter proteins. *J. Nat. Med.* **2013**, *67*, 303–310. [CrossRef] [PubMed]
65. Kobayashi, S. The Effect of Polyphenols on Hypercholesterolemia through Inhibiting the Transport and Expression of Niema Pick C1-Like 1. *Int. J. Mol. Sci.* **2019**, *20*, 4939. [CrossRef] [PubMed]
66. Nekohashi, M.; Ogawa, M.; Ogihara, T.; Nakazawa, K.; Kato, H.; Misaka, T.; Abe, K.; Kobayashi, S. Luteolin and quercetin af the cholesterol absorption mediated by epithelial cholesterol transporter niemann-pick c1-like 1 in caco-2 cells and rats. *P ONE* **2014**, *9*, e97901. [CrossRef]
67. Pirillo, A.; Catapano, A.L.; Norata, G.D. Niemann-Pick C1-Like 1 (NPC1L1) Inhibition and Cardiovascular Diseases. *Curr. M Chem.* **2016**, *23*, 983–999. [CrossRef]
68. Huang, C.S.; Yu, X.; Fordstrom, P.; Choi, K.; Chung, B.C.; Roh, S.H.; Chiu, W.; Zhou, M.; Min, X.; Wang, Z. Cryo-EM structure NPC1L1 reveal mechanisms of cholesterol transport and ezetimibe inhibition. *Sci. Adv.* **2020**, *6*, eabb1989. [CrossRef]
69. Ghazalpour, A.; Cespedes, I.; Bennett, B.J.; Allayee, H. Expanding role of gut microbiota in lipid metabolism. *Curr. Opin. Lipi* **2016**, *27*, 141–147. [CrossRef] [PubMed]
70. Kriaa, A.; Bourgin, M.; Potiron, A.; Mkaouar, H.; Jablaoui, A.; Gerard, P.; Maguin, E.; Rhimi, M. Microbial impact on choleste and bile acid metabolism: Current status and future prospects. *J. Lipid Res.* **2019**, *60*, 323–332. [CrossRef]
71. Yu, Y.; Raka, F.; Adeli, K. The Role of the Gut Microbiota in Lipid and Lipoprotein Metabolism. *J. Clin. Med.* **2019**, *8*, 22 [CrossRef] [PubMed]
72. Zhang, X.; Gerard, P. Diet-gut microbiota interactions on cardiovascular disease. *Comput. Struct. Biotechnol. J.* **2022**, *20*, 1528–1 [CrossRef] [PubMed]
73. Bisanz, J.E.; Upadhyay, V.; Turnbaugh, J.A.; Ly, K.; Turnbaugh, P.J. Meta-Analysis Reveals Reproducible Gut Microbio Alterations in Response to a High-Fat Diet. *Cell Host Microbe* **2019**, *26*, 265–272.e264. [CrossRef] [PubMed]
74. Kunstner, A.; Aherrahrou, R.; Hirose, M.; Bruse, P.; Ibrahim, S.M.; Busch, H.; Erdmann, J.; Aherrahrou, Z. Effect of Difference the Microbiome of Cyp17a1-Deficient Mice on Atherosclerotic Background. *Cells* **2021**, *10*, 1292. [CrossRef] [PubMed]
75. Tong, L.T.; Xiao, T.; Wang, L.; Lu, C.; Liu, L.; Zhou, X.; Wang, A.; Qin, W.; Wang, F. Plant protein reduces serum cholesterol lev in hypercholesterolemia hamsters by modulating the compositions of gut microbiota and metabolites. *iScience* **2021**, *24*, 1034 [CrossRef] [PubMed]

Feng, P.; Li, Q.; Liu, L.; Wang, S.; Wu, Z.; Tao, Y.; Huang, P.; Wang, P. Crocetin Prolongs Recovery Period of DSS-Induced Colitis via Altering Intestinal Microbiome and Increasing Intestinal Permeability. *Int. J. Mol. Sci.* **2022**, *23*, 3832. [CrossRef] [PubMed]
Zhang, J.; Yi, C.; Han, J.; Ming, T.; Zhou, J.; Lu, C.; Li, Y.; Su, X. Gut microbiome and metabolome analyses reveal the protective effect of special high-docosahexaenoic acid tuna oil on d-galactose-induced aging in mice. *Food Sci. Nutr.* **2022**, *10*, 3814–3827. [CrossRef]
Liu, H.; Zhu, H.; Xia, H.; Yang, X.; Yang, L.; Wang, S.; Wen, J.; Sun, G. Different effects of high-fat diets rich in different oils on lipids metabolism, oxidative stress and gut microbiota. *Food Res. Int.* **2021**, *141*, 110078. [CrossRef]
Li, Y.; Chen, M.; Ma, Y.; Yang, Y.; Cheng, Y.; Ma, H.; Ren, D.; Chen, P. Regulation of viable/inactivated/lysed probiotic Lactobacillus plantarum H6 on intestinal microbiota and metabolites in hypercholesterolemic mice. *NPJ Sci. Food* **2022**, *6*, 50. [CrossRef]
Lang, J.M.; Sedgeman, L.R.; Cai, L.; Layne, J.D.; Wang, Z.; Pan, C.; Lee, R.; Temel, R.E.; Lusis, A.J. Dietary and Pharmacologic Manipulations of Host Lipids and Their Interaction With the Gut Microbiome in Non-human Primates. *Front. Med.* **2021**, *8*, 646710. [CrossRef]
Just, S.; Mondot, S.; Ecker, J.; Wegner, K.; Rath, E.; Gau, L.; Streidl, T.; Hery-Arnaud, G.; Schmidt, S.; Lesker, T.R.; et al. The gut microbiota drives the impact of bile acids and fat source in diet on mouse metabolism. *Microbiome* **2018**, *6*, 134. [CrossRef]
Palmisano, B.T.; Zhu, L.; Eckel, R.H.; Stafford, J.M. Sex differences in lipid and lipoprotein metabolism. *Mol. Metab.* **2018**, *15*, 45–55. [CrossRef] [PubMed]
Russo, G.T.; Giandalia, A.; Romeo, E.L.; Cucinotta, D. Gender differences in lipoprotein metabolism. *Ital. J. Gend.-Specif. Med.* **2015**, *1*, 58–65. [CrossRef]
Davies, J.P.; Scott, C.; Oishi, K.; Liapis, A.; Ioannou, Y.A. Inactivation of NPC1L1 causes multiple lipid transport defects and protects against diet-induced hypercholesterolemia. *J. Biol. Chem.* **2005**, *280*, 12710–12720. [CrossRef] [PubMed]
Davis, H.R., Jr.; Zhu, L.J.; Hoos, L.M.; Tetzloff, G.; Maguire, M.; Liu, J.; Yao, X.; Iyer, S.P.; Lam, M.H.; Lund, E.G.; et al. Niemann-Pick C1 Like 1 (NPC1L1) is the intestinal phytosterol and cholesterol transporter and a key modulator of whole-body cholesterol homeostasis. *J. Biol. Chem.* **2004**, *279*, 33586–33592. [CrossRef]
Kamato, D.; Ilyas, I.; Xu, S.; Little, P.J. Non-Mouse Models of Atherosclerosis: Approaches to Exploring the Translational Potential of New Therapies. *Int. J. Mol. Sci.* **2022**, *23*, 12964. [CrossRef]
Hermans, M.P.; Lempereur, P.; Salembier, J.P.; Maes, N.; Albert, A.; Jansen, O.; Pincemail, J. Supplementation Effect of a Combination of Olive (*Olea europea* L.) Leaf and Fruit Extracts in the Clinical Management of Hypertension and Metabolic Syndrome. *Antioxidants* **2020**, *9*, 872. [CrossRef]
Rondanelli, M.; Riva, A.; Petrangolini, G.; Allegrini, P.; Bernardinelli, L.; Fazia, T.; Peroni, G.; Gasparri, C.; Nichetti, M.; Faliva, M.A.; et al. The Metabolic Effects of Cynara Supplementation in Overweight and Obese Class I Subjects with Newly Detected Impaired Fasting Glycemia: A Double-Blind, Placebo-Controlled, Randomized Clinical Trial. *Nutrients* **2020**, *12*, 3298. [CrossRef]

nutrients

Article

Nutritional and Physiological Properties of *Thymbra spicata*: I Vitro Study Using Fecal Fermentation and Intestinal Integrity Models

Mohamad Khalil [1], Francesca Piccapane [2], Mirco Vacca [3], Giuseppe Celano [3], Laura Mahdi [1], Valeria Perniola [1], Carmen Aurora Apa [3], Alessandro Annunziato [3], Ilaria Iacobellis [3], Giuseppe Procino [2], Maria Calasso [3], Maria De Angelis [3], Rosa Caroppo [2] and Piero Portincasa [1,]*

[1] Clinica Medica "A. Murri", Department of Precision and Regenerative Medicine and Ionian Area (DiMePr University of Bari Medical School, 70124 Bari, Italy; mohamad.khalil@uniba.it (M.K.); lauramahdi5@gmail.com (L.M.); valeriaperniola17@gmail.com (V.P.)

[2] Department of Bioscience, Biotechnology and Environment, University of Bari, 70125 Bari, Italy; francesca.piccapane@uniba.it (F.P.); giuseppe.procino@uniba.it (G.P.); rosa.caroppo@uniba.it (R.C.)

[3] Department of Soil, Plant and Food Sciences, University of Bari Aldo Moro, Via Amendola 165/a, 70126 Bari, Italy; mirco.vacca@uniba.it (M.V.); giuseppe.celano@uniba.it (G.C.); carmen.apa@uniba.it (C.A. alessandro.annunziato@uniba.it (A.A.); ilaria.iacobellis@uniba.it (I.I.); maria.calasso@uniba.it (M.C.); maria.deangelis@uniba.it (M.D.A.)

* Correspondence: piero.portincasa@uniba.it

Abstract: (Poly)phenolic-rich Mediterranean plants such as *Thymbra spicata* have been associa with several health-promoting effects. The nutritional value, as well as physiological interactior *T. spicata* with the gastrointestinal tract, has not been investigated before. The nutritional composit of *T. spicata* leaves was here characterized by standard analytical methods. *T. spicata* leaves w subjected to ethanolic extraction, simulated gastrointestinal digestion, and anaerobic microl gut fermentation. Phenols/flavonoid contents and radical scavenging activity were assessed colorimetric methods. The volatile organic compounds (VOCs) were detected by gas chromatogra coupled with mass spectrometry. The effect on intestinal integrity was evaluated using a Cac monolayers mounted in a Ussing chamber. *T. spicata* contains a high amount of fiber (12.3%) a unsaturated fatty acids (76% of total fat). A positive change in VOCs including short-chain fatty ac was observed without significant change in viable microbe. *T. spicata* and carvacrol (main phen compound) enhanced ionic currents in a concentration-dependent manner without compromis the Caco-2 monolayer's integrity. These effects were partially lost upon simulated digestion a completely abolished after colonic fermentation in line with polyphenols and carvacrol cont Conclusion: *T. spicata* represents a promising nutrient for the modulation of gut microbiota and gut barrier. Further studies must better define its mechanisms of action.

Keywords: *Thymbra spicata*; carvacrol; gut microbiota; intestinal permeability; polyphenols

Citation: Khalil, M.; Piccapane, F.; Vacca, M.; Celano, G.; Mahdi, L.; Perniola, V.; Apa, C.A.; Annunziato, A.; Iacobellis, I.; Procino, G.; et al. Nutritional and Physiological Properties of *Thymbra spicata*: In Vitro Study Using Fecal Fermentation and Intestinal Integrity Models. *Nutrients* **2024**, *16*, 588. https://doi.org/10.3390/nu16050588

Academic Editors: Sonia Ramos Rivero and María Angeles Martín

Received: 22 December 2023
Revised: 9 February 2024
Accepted: 11 February 2024
Published: 21 February 2024

1. Introduction

Dietary components and eating patterns can affect important health outcomes. E dence points to a close link between dietary pattern and decreased mortality [1]. Spec factors that can influence the choices include personal preference, gender, social and c tural aspects, as well as economic considerations. As underscored by several meta-analy and societies, including the WHO [2], the American Institute for Cancer Research and the US Department of Agriculture [4], a high-quality and healthy diet rely on limited consumption of saturated and industrial trans fats, red and processed meats, su sodium, and alcohol. By contrast, the consumption of fruits, vegetables, legumes, nuts, a whole grains is greatly encouraged. The Mediterranean diet (MD) mirrors several of above-mentioned beneficial features, since it is based on the consumption of vegetab fruits, nuts, olive oil, and many herbs. Besides the richness of MD in essential nutrients, vegetables and herbs contain a high amount of polyphenols [5]. Polyphenols are a diverse group of naturally occurring compounds linked to various health-promoting effects. The gut–polyphenols interaction occurs after the ingestion of food, where polyphenols are modified by digestive enzymes [6]. Small amounts of polyphenols are absorbed in the small intestine, while the greatest amount remains unabsorbed and accumulates in the colon, where it is extensively metabolized by gut microbiota prior to reaching the circulation [7–9]. Information about polyphenol biotransformation, bioavailability, and in vivo physiological effects is still yet to be delineated.

Since the gut microbiota play a pivotal role in host homeostasis and energy balance, changes in their composition can be associated with disease states through the promotion of immune-mediated inflammatory disorders and increasing the intestinal permeability, ultimately leading to the impairment of intestinal barrier function. Evidence from pre-clinical and clinical studies suggests that polyphenols or polyphenol-rich food can positively influence gut health by promoting the growth of beneficial gut bacteria and suppressing the growth of harmful bacteria. This can contribute to a balanced gut microbiome and gut barrier [7].

Za'atar is one of the most popular plant-based foods in the Eastern Mediterranean region. Za'atar is a mixture of different thyme-like plant leaves including *Origanum syriacum* and *Thymbra Spicata*, fruits of *Rhus coriaria* (Sumac), and sesame seeds [10]. The mixture contains hundreds of antioxidant compounds, especially polyphenols. Za'atar compounds have been studied from a gastrointestinal perspective, concerning both gut microbiota and gastrointestinal diseases, suggesting that Za'atar polyphenols may provide beneficial effects in the complex interplay between the diet, GM, and intestinal permeability [11].

In Lebanon, *Thymbra spicata* is widely diffused. It belongs to the *Lamiaceae* family and it is rich in polyphenols such as rosmarinic acid, carvacrol (CVL) and different flavonoids [10]. *T. spicata* leaves and flowers are used as food ingredients in salad and in tea infusions. Different biological and bio-functional effects have been reported for *T. spicata* including anti-oxidant and antisteatotic [12], antihypercholesterolaemic [13,14], antitumoral [15], antiviral [16], and antimicrobial [17–19] activities.

CVL is a monoterpenoid phenolic compound found in the essential oils of various plants, particularly in the *Lamiaceae* family. *T. spicata* contains up to 80% CVL in its essential oil [20]. CVL has gained attention among the scientific community due to its diverse biological activities and its potential health benefits [21–23]. CVL exerts antimicrobial [24], antioxidant [25], anti-inflammatory [26], and antisteatotic effects [27], and is being explored for various pharmacological and nutraceutical applications.

To our knowledge, no studies have reported the effects of *T. spicata* on the gastrointestinal tract. Based on pre-clinical studies concerning compounds found in *T. spicata*, we have theorized that *T. spicata* may have potential nutraceutical applicability for disorders involving the gastrointestinal tract and gut–liver axis. The translational value of such studies would be of major interest. The objective of this research is to investigate the nutritional profile, physiological implications, and bioactive properties of *T. spicata*, particularly emphasizing the interest in simulating in vitro gastrointestinal digestion and microbial fermentation. This approach allows for a comprehensive exploration of how *T. spicata* affects gut microbiota composition and metabolites. Additionally, the study aims to evaluate the impact of *T. spicata* on the intestinal epithelial function, including permeability and electrophysiological characteristics.

2. Methods

2.1. Plant Collection and Extraction

Leaves of *Thymbra spicata* L. were collected from flowering plants growing in "Maarakeh", South Lebanon, during June 2022. The voucher specimen (L1.125/1) was authenticated by Prof. G. Tohme (CNRS, Beirut, Lebanon). The solvent extraction method was used for ethanolic extraction, as described previously by Khalil et al. [15].

2.2. *Nutritional Value of Thymbra spicata*

The nutritional value of *T. spicata* leaves was determined by Food Safety LAB (Corato, BA, Italy) accredited and certified by international standards and requiremer The energy value (Kcal and kJ) was determined according to Regulation (EU) 1169/2011. Sodium chloride was measured according to the L-MI068 method. Carb drates were calculated as total sugar content (DLgs n° 77 16/02/1993 GU n° 69 24/03/1 dietary fiber (according to AOAC 985.29), and simple sugars (L-MI014 rev.0 2016). T fat was calculated according to DM 23-07-1994 SO n°4 G.U. n°186 del 10/08/1994 single fatty acids were calculated by the AOCS Ce 2-66:2009 + AOCS Ce 1e-91:2001 met The total proteins content was calculated by UNI CEN ISO/TS 16634-2:2016. Ashes evaluated via ISO 2171:2007, and humidity was calculated using UNI EN ISO 712:201

2.3. *Simulated Digestion and Colonic Fermentation*

2.3.1. In Vitro Simulation of Gastrointestinal Digestion

Simulated *T. spicata* digestion was carried out in vitro using different solutions n icking the oral, gastric, and intestinal fluids [28]. Briefly, 5 g of dried *T. spicata* leaves distilled water (25 mL) were homogenized for 3 min with a lab stomacher (Bag M Interscience International; Roubaix, France). The oral phase accounted for the addi of 10 mg of α-amylase dissolved in 3.125 mL of 1 mM $CaCl_2$ solution, and samples w incubated for 30 min at 37 °C under stirring conditions (50 rpm). To simulate gastric fl 1.35 mg of pepsin was dissolved in 12.5 mL of 0.1 M HCl (pH 2.0, adjusted with a solution to 6 M) and, after its addition, samples underwent the second incubation step 3 h at 37 °C under stirring conditions (150 rpm). Finally, the intestinal fluid was simula by adding 280 mg of pancreatin and 1.75 mg of bile salts dissolved in 62.5 mL of 0. $NaHCO_3$ (pH 7.0, adjusted with a NaOH solution to 6 M) to the samples. At the en the third incubation (3 h at 37 °C under stirring conditions (150 rpm)), the enzymatic digested samples were used as substrates for the fecal microbiota inoculum.

2.3.2. In Vitro Colonic Fermentation of Digested *T. spicata*

As detailed by Vacca et al., the digested samples constituted a part of the fecal me To obtain fecal media, a fecal pool containing equal stool samples from 3 healthy donors added, in a ratio of 1:5 (*w*/*v*), to distilled water. The suspension, homogenized for 3 with a lab stomacher in bags with a filter (250 μm), was centrifuged at 10,000× *g* for 10 to recover supernatants. The previously digested *T. spicata* samples were added in a r of 1:4 (*v*/*v*) to supernatants and supplemented with K_2HPO_4•2 g/L, $C_2H_3NaO_2$•5 g $C_6H_{17}N_3O_7$•2 g/L, $MgSO_4$•0.2 g/L, $MnSO_4$•0.05 g/L, glucose•2 g/L, inulin•4 g/L, fru oligosaccharides•4 g/L and Tween 80 polysorbate•1 mL/L, then sterilized at 121 for 20 min. A cold-sterilized cysteine HCl (0.5 g/L), haemin (0.02 g/L), and vitamin (10 μL/L) solution was finally added to form three different fecal media. Each fecal medi was inoculated with a fecal slurry (32% *w*/*v*) obtained by processing fresh feces from healthy donor collected within 1 h. The simulated colonic fermentation in fecal batc (i.e., fecal medium with fecal slurry inoculum) accounted for a single-step anaerobic 4 incubation at 37 °C under stirring conditions (150 rpm). Three final samples (namely, f media and *T. spicata* digest (FMTD); fecal slurry (FS); and fecal slurry and *T. spicata* dig (FSTD)) were obtained as depicted in Figure 1.

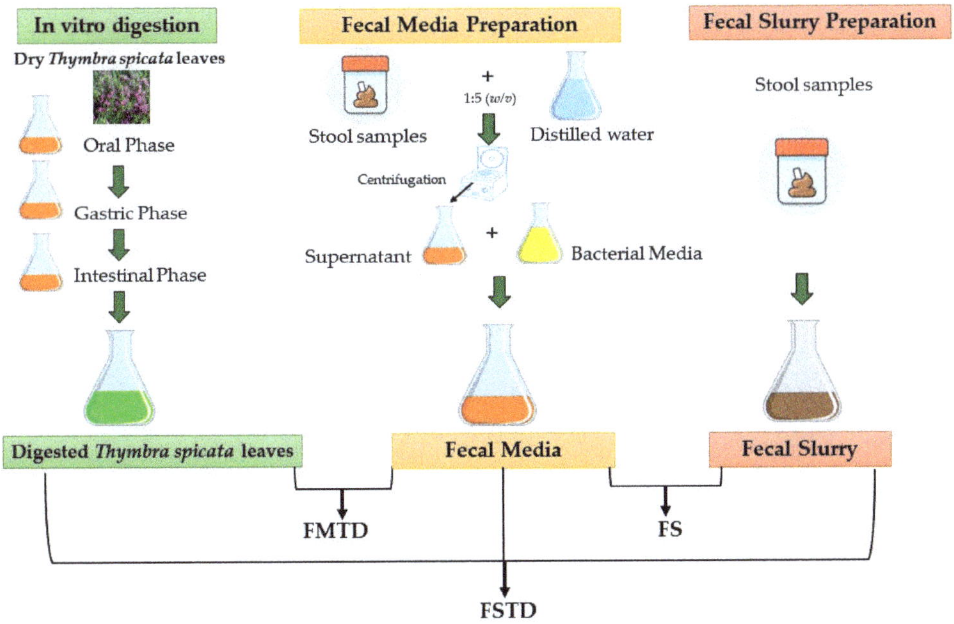

Figure 1. Representative scheme for the three samples obtained by in vitro simulation of gastrointestinal digestion and in vitro colonic fermentation. FMTD, fecal media and *T. spicata* digest; FS, fecal slurry; FSTD, fecal slurry and *T. spicata* digest.

2.4. Fecal Microbiota Characterization

2.4.1. Enumeration of Cultivable Microorganisms

Aliquots (5 g) from each fecal batch were added to 45 mL of sterile NaCl (0.9% *w*/*v*) solution and a serial 10-fold dilution was carried out subsequently. To inspect the viable microbiota, Plate Count Agar (*PCA*), Wilkins–Chalgren anaerobe agar (*WCAn*), de Man, Rogosa and Sharpe (*MRS*) agar, M17 agar, Violet Red Bile Glucose Agar (*VRBGA*) and modified Bifidobacterium agar (*mBifA*) were used as culture media for counts of total aerobes microbial (TAMC), total anaerobes microbial (TANMC), lactic acid bacteria (LAB), lactococci/streptococci, total coliforms and fecal *Bifidobacterium*, respectively. An exception was made for *mBifA*, which was purchased from Becton Dickinson GmbH (Heidelberg; Germany); other than this, all other media were purchased from Oxoid Ltd. (Basingstoke, Hampshire, England). *WCAn* and *mBifA* were anaerobically incubated, while the other media were incubated aerobically. The time and temperature of incubation followed those defined by the related manufacturer.

2.4.2. DNA Extraction

Total DNA was extracted from stool samples. An aliquot (500 µL) of each sample was used for total DNA extraction. Samples were diluted in 1 mL of PBS-EDTA (phosphate buffer 0.01 M, pH 7.2, 0.01 M EDTA) and centrifuged (14,000× *g* at 4 °C for 5 min). Each pellet was washed twice. The extraction was performed using the FastDNA® Pro Soil-Direct Kit (MP Biomedicals, Irvine, CA, USA). The quality check of the final DNA was carried out by spectrophotometric measurement at 260, 280, and 230 nm using a NanoDrop® ND-1000 Spectrophotometer (ThermoFisher Scientific Inc., Milan, Italy).

2.4.3. Real Time PCR

Fecal bacteria were quantified by Real Time PCR (qPCR). The primers used for microbial investigation are listed and described in Table 1. qPCR reactions were carried out on an

Applied Biosystems 7300 Real-Time PCR System (Waltham, MA, USA). The total reacti mix (25 µL) contained 12.5 µL of SYBR Green Mix (# 1725271, Bio-Rad Laboratories S. Milano, Italy), 0.1 µL of 0.2 µM of primer, 11.4 µL of DNase and RNase-free water, a 1 µL of template. Each reaction was performed in triplicate. The amplification progr consisted of 1 cycle of 95 °C for 3 min, followed by 40 cycles of 95 °C for 5 s, an appropri annealing temperature (Table 1) for 30 s and 72 °C for 35 s. The PCR amplicon melti curve analysis started at a temperature of 60 °C and increased at 1 °C/5 s until reachi the final temperature of 95 °C.

Table 1. List of primers and related features.

Target	Primers	Primer Sequence (5′–3′)	Gene	Product Size (bp)	T Annealing (°C)	Referen
All Bacteria	Uni331F	TCCTACGGGAGGCAGCAGT	16S rRNA	466	58	[29]
	Uni797R	GGACTACCAGGGTATCTATCCTGTT				
Lactobacillus genus	Lp-F	AAAATCATGCGTGCGGGTAC	16S rRNA	341	55	[29]
	Lp-R	ATGTTGCGTTGGCTTCGTCT				
Bifidobacterium genus	Bifid-F	CTCCTGGAAACGGGTGG	16S rRNA	550	55	[29]
	Bifid-R	GGTGTTCTTCCCGATATCTACA				
Escherichia coli subgroup (*E. coli*, *Hafnia alvei*, *Shigella* spp.)	Eco-F	GTTAATACCTTTGCTCATTGA	V3-V4 hyper-variable regions	340	60	[30]
	Eco-R	ACCAGGGTATCTAATCCTGTT				
Clostridium coccoides group	Ccoc-F Ccoc-R	AAATGACGGTACCTGACTAA CTTTGAGTTTCATTCTTGCGAA	16S rRNA	440	50	[29]

Preliminary qPCR results have been reported in Log (Copy Number). The Co Numbers (CN) and the logarithms were calculated based on the DNA concentration a amplicon length.

2.4.4. Fecal Batch Volatile Organic Compound (VOC) Analysis

For the analysis of volatile organic compounds (VOCs), 2 mL each of the fecal medi and *T. spicata* digest, fecal slurry and fecal slurry, and *T. spicata* digest samples was taken a placed inside 20 mL vials that were closed with a magnetic cap with polytetrafluoroethyle septa (PTFE)/silicone. As internal standard, 10 µL of 4-methyl-2-pentanol was added Clarus 680 gas chromatograph (Perkin Elmer, Beaconsfield, UK) was used for GC-I analysis. For the extraction of volatile compounds, a CombiPAL injector sample syst (CTC Analytucs, Zwingen, Switzerland) and a 50/30 µm DVB/CAR/PDMS (Supel Bellefonte, PA, USA) were used. Helium was used as a carrier gas with a flow rate 1 mL/min. Firstly, the samples were equilibrated for 15 min at 45 °C. To extract the vola compounds, the fiber was exposed to 45 °C for 45 min in the headspace of the vials. T injection took place by heating the fiber inside the heated injection port to 220 °C in splitl mode. An Rtx-WAX column 30 m × 0.25 mm × 0.25 µm (Restek, Bellafonte, PA, USA) a coupled Clarus SQ8MS (Perkin Elmer) were used for VOCs separation with the source at a temperature of 250 °C and the mass transfer line MS at a temperature of 230 °C. T oven temperature was initially 35 °C for 8 min and then increased to 60 °C with an incre of 4 °C/min, 160 °C with an increase of 6 °C/min and finally a temperature of 200 with an increase of 20 °C/min, at which it was maintained for 15 min. Electron ionizat masses were recorded at 70 eV in the mass-to-charge ratio range from m/Z 34 to 3 For peak identification, chromatograms were analyzed using the NIST/EPA/NIH M Spectral Library with Search Program—Data ver. 2.3. Gaithersburg, MD, USA: Natio Institute of Standards and Technology (NIST); 2017) library. A peak area threshold >1,000,000 and a probability of match of 85% or greater was used to identify the compour The relative concentration (expressed in µg/g of 4-methyl-2-pentanol) was estimated the interpolation of the relative area with the air of the internal standard. The absol concentrations of short chain fatty acids (SCFAs) and CVL (main phenolic and volatile compounds in *T. spicata*) in the samples were detected using standard curves.

2.5. Total Phenol Quantification

The total phenol content (TPC) was determined for each extract using the Folin–Ciocalteu method. Briefly, 100 µL aliquots of each sample were incubated with 0.5 mL of 10% (*w*/*v*) Folin–Ciocalteau reagent. After 5 min, 1.5 mL of Na_2CO_3 (2% *w*/*v*) was added and incubated in the dark at room temperature for 30 min. The absorbance was measured at 760 nm using a UV–Vis spectrophotometer Ultrospec 300 against a blank. The results were derived from a calibration curve of gallic acid (0–250 µg/mL) prepared from a stock solution and expressed in Gallic Acid Equivalents (GAE) per mL of extract.

2.6. Total Flavonoid Quantification

Total flavonoid content (TFC) was determined for each extract using aluminum chloride ($AlCl_3$) colorimetric method. Briefly, 1 mL aliquot of each sample was mixed with 0.2 mL of 10% (*w*/*v*) methanolic $AlCl_3$ solution, 0.2 mL (1 M) potassium acetate and 5.6 mL distilled water. After = incubation at room temperature in the dark for 30 min, the absorbance was measured at 415 nm using a UV–Vis spectrophotometer Ultrospec 300. The results were derived from a calibration curve of quercetin (0–200 µg/mL) prepared from a stock solution. The results have been expressed in mg of Quercetin Equivalent (QE) per mL of extract.

2.7. DPPH Radical Scavenging Assay

The radical scavenging activity of each extract was measured using the 1,1-diphenyl-2-picrylhydrazyl (DPPH) method. Briefly, from each sample, a 100 µL serial dilution was prepared to the following concentrations: 100%, 50%, 25%, 12.5%, and 6.25%. Then, in a 96-multiwell plate, 50 µL of each sample were added to 200 µL of DPPH (0.1 mM in methanol) solution. The mixtures were incubated at 37 °C in darkness for 20 min, and then the absorbance was measured at 490 nm using a microplate reader UV–Vis spectrophotometer Ultrospec 300 against an equal amount of DPPH solution as a blank. The percentage of DPPH scavenging was estimated using the equation:

$$\% \text{ scavenging activity} = [(\text{Abs control} - \text{Abs sample})]/(\text{Abs control})] \times 100$$

2.8. Cell Culture

Caco-2 cells were grown and maintained in DMEM GlutaMAX supplemented with 10% fetal bovine serum, 100 i.u./mL penicillin, 100 µg/mL streptomycin and 1% non-essential amino acids, at 37 °C in a humidified atmosphere containing 5% CO_2.

For the Ussing chamber experiments, Caco-2 cells were grown on Snapwell™ (Corning Incorporated, Corning, NY, USA) supports with a 1.12 cm^2 cell growth area (#3801); the cells were seeded at a density of 2.2×10^5 cells/cm^2 on filters.

2.9. Ussing Chamber Studies

Transepithelial measurements were performed on Caco-2 cells grown as monolayers for 20 days. Inserts were mounted into Easy Mount Ussing chambers (Physiologic Instruments Inc., Reno, NV, USA).

Apical and basolateral hemi-chambers were filled with 5 mL of Ringer's solutions containing (in mM): 115 NaCl, 25 $NaHCO_3$, 0.4 KH_2PO_4, 2.4 K_2HPO_4, 1.2 $MgCl_2$, 1.2 $CaCl_2$, and 10 glucose (pH 7.4, 300 mOsm). The solution was gassed with a 95% O_2 and 5% CO_2 mixture and constantly recirculated by gas bubble lifting.

The short circuit current (Isc) and the transepithelial resistance (Rt) were measured as previously described [31].

2.10. Statistical Analyses

As appropriate, data were expressed as means with standard error of the mean (SEM For continuous variables, comparisons were undertaken using the two-tailed t-test, Tuke test, or Mann–Whitney U test corrected for multiple comparisons (Dunn's statistics). T analysis of variance was carried out by Two-way ANOVA testing. For all compariso differences were considered as significant when reaching a p-value (p) < 0.05. Subsequen for normalization (Z-score), multivariate analyses were carried out by running a princip component analysis (PCA) or hierarchical clustering (based on Ward's metrics and E clidean distance) in the R-environment by running the "FactoMineR" package (Multivari Exploratory Data Analysis and Data Mining) version 2.4, available in the CRAN reposito A volcano plot combining results from the Fold Change (FC) analysis and nonparamet Wilcoxon rank-sum test was used to intuitively select significant VOCs based on eith biological significance, statistical significance, or both. Statistical analyses and a graph vi alization of the electrophysiological data were performed using GraphPad Prism versi 8.0 (GraphPad Software, San Diego, CA, USA).

3. Results

3.1. Chemical and Nutritional Value of *Thymbra spicata*

Leaves of *T. spicata* were fully characterized, as depicted in Table 2. The data refer macronutrients, minerals (ash), and energy content.

Table 2. Analytical values of chemical composition and nutritional values of *Thymbra spicata* leav

Compounds	Quantity	Compounds	Quanti
Total fat (g/100 g)	4.3	Total carbohydrates (g/100 g)	52.7
Saturated fatty acids (g/100 g)	0.9	Total dietary fiber (g/100 g)	12.3
Fatty acid methyl esters (% of fat content)		Simple sugars (g/100 g)	
C12:0—Lauric acid	<0.1	Fructose	0.03
C14:0—Myristic acid	<0.1	Galactose	n.d
C16:0—Palmitic acid	17.5	Glucose	1.69
C16:1—Palmitoleic acid	<0.1	Lactose	n.d
C17:0—Heptadecanoic acid	<0.1	Maltose	n.d
C17:1—Heptadecenoic acid	<0.1	Sucrose	0.02
C18:0—Stearic acid	2.6	Sum sugars	1.74
C18:1—Oleic acid	40.6	Nitrogen (g/100 g)	0.99
C18:2—Linoleic acid	35.5	Total protein (g/100 g)	6.19
C20:0—Arachic acid	2.6	Moisture (g/100 g)	17.0
C18:3—Linolenic acid	0.5	Salt (g/100 g)	0.009
C20:1—Eicosenoco acid	<0.1	Ash (g/100 g)	9.11
C22:0—Beenic acid	<0.1	Energy value (Kcal/100 g)	324
C22:1—Erucic acid	<0.1	Energy value (KJ/100 g)	1367
C24:0—Lignoceric acid	<0.1		

Cx:y, number of carbon atoms:number of double covalent bonds; n.d, not determined.

More than 50% of plant materials were carbohydrates (52.7 g per 100 g of the *T. spicata* leaves). Total dietary fiber was measured at 12.3 g/100 g, and the sum of the simple sugars (listed by specific sugar types, along with their values in g per 100 g) was 1.74 g. Total protein and fat content were 6.19 and 4.3 g/100 g, respectively. The contents of saturated fatty acids were low (0.9 g), while unsaturated fatty acids were more represented. Oleic acid (C18:1) and linoleic acid (C18:2) accounted for 40.6% and 30.5% of fat content, respectively. Other constituents, such as moisture and ash, were present in moderate amounts (17.0 and 9.11 g/100 g, respectively). Finally, the energy value of *T. spicata* leaves was 324 kcal/100 g, which is equal to 1367 kJ/100 g.

3.2. Effect of *T. spicata* on Fecal Microbiota after Simulated Digestion and Colonic Fermentation

In vitro simulated digestion was performed to inspect the physiological effects of *T. spicata* on the viable cell density of the main cultivable bacterial groups belonging to the intestinal microbiota. The differences observed were determined through the simulated digestion of *T. spicata*, comparing it against the same medium and culturing conditions without the presence of the plant's macro and micronutrients. While a similar trend characterized the experiments, indicating a decreasing density from FS to FSTD, no significant differences were detected based on the presence of *T. spicata* in the medium. In fact, the densities of both total aerobes (~9 logs CFU/mL) and anaerobes (~8 logs) were comparable between FS and FSTD samples (Table 3). Consistently with these findings, even *Lactococci* and *Bifidobacterial* cells did not exhibit substantial variations. Although not statistically significant ($p > 0.05$), lactobacilli and *Enterobacteriaceae* showed a ~0.5 log higher density in FS compared to FSTD. The most notable difference was observed in the density of *Enterocci*, reaching approximately 5 logs in FS, while reporting mean values of 3.5 logs in FSTD. *Clostridi*, on the other hand, demonstrated the lowest *p*-value ($p = 0.09$) among the sample comparisons, with a higher density in FS (~8.8 logs) compared to FSTD (~8.3 logs).

Table 3. Viable microbes of fecal slurry (FS) and fecal slurry and *T. spicata* digest (FSTD).

Viale Microbes (Log UFC/mL)	FS (Control)	FSTD	*p*-Value
Total aerobes	9.3 ± 0.3	9.0 ± 0.4	0.61
Total anaerobes	8.0 ± 0.5	8.0 ± 0.4	0.99
Lactobacilli	5.0 ± 1.0	4.4 ± 1.0	0.69
Lactococci	7.8 ± 0.4	7.6 ± 0.5	0.81
Bifiobacteria	7.1 ± 0.7	6.9 ± 0.5	0.86
Enterobacteriaceae	8.8 ± 0.2	8.2 ± 0.6	0.39
Enterococci	4.9 ± 0.1	3.5 ± 1.4	0.46
Clostridi	8.8 ± 0.1	8.3 ± 0.2	0.09

Data are expressed as Mean ± SEM; *p*-value obtained by Student's *t*-test.

In addition, to better inspect the effects of *T. spicata* on microbiota composition in the fecal batch, a panel of microbial genera and species was investigated by qPCR analysis (Figure 2). In line with the viable microbial cells count, the Log (Copy Number) values for all tested bacteria, which were calculated based on the DNA concentration and amplicon length, were comparable between FS and FSTD.

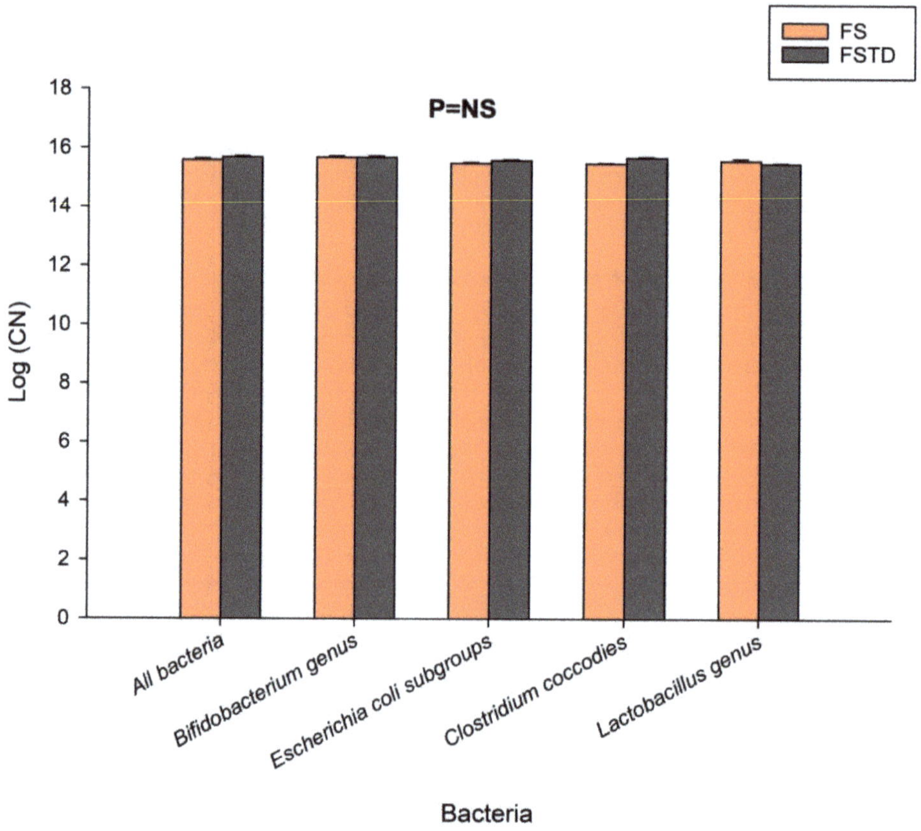

Figure 2. Log (Copy Number) based on DNA concentration and amplicon length after a qPCR extracted DNA derived from the batch fermentation of FS and FSTD. Tested bacteria were: all bacte *Bifidobacterium genus*, *Escherichia coli* subgroups, *Clostridium coccodies*, and *Lactobacillus genus*. *p*-va obtained by Student's *t*-test.

3.3. Effects of *T. spicata* on Fecal Volatile Organic Compounds (VOCs)

With the aim of ascertaining the impact of *T. spicata* on the metabolic activities of main cultivable bacterial group from the gut microbiota, we performed HS-SPME GC-l analyses to evaluate the VOCs. For this purpose, the volatile profiles were evaluat in the fecal medium and *T. spicata* digest (FMTD), fecal slurry (FS) and fecal slurry a *T. spicata* digest (FSTD) groups (Supplementary Table S1). In total, 94 VOCs were identif and divided into the following chemical classes: alcohols (n = 8), carboxylic acids (n = hydrocarbons (n = 23), indoles (n = 3), phenols (n = 5) and terpenes (n = 18), plus anot 28 compounds not belonging to any of the classes previously mentioned. In addition, results obtained were analyzed by multivariate analysis of the principal component (PC The unsupervised PCA, which explained 68.3% of the variance, showed that there wa difference distribution among samples containing fecal slurry and samples with only fe medium (Figure 3).

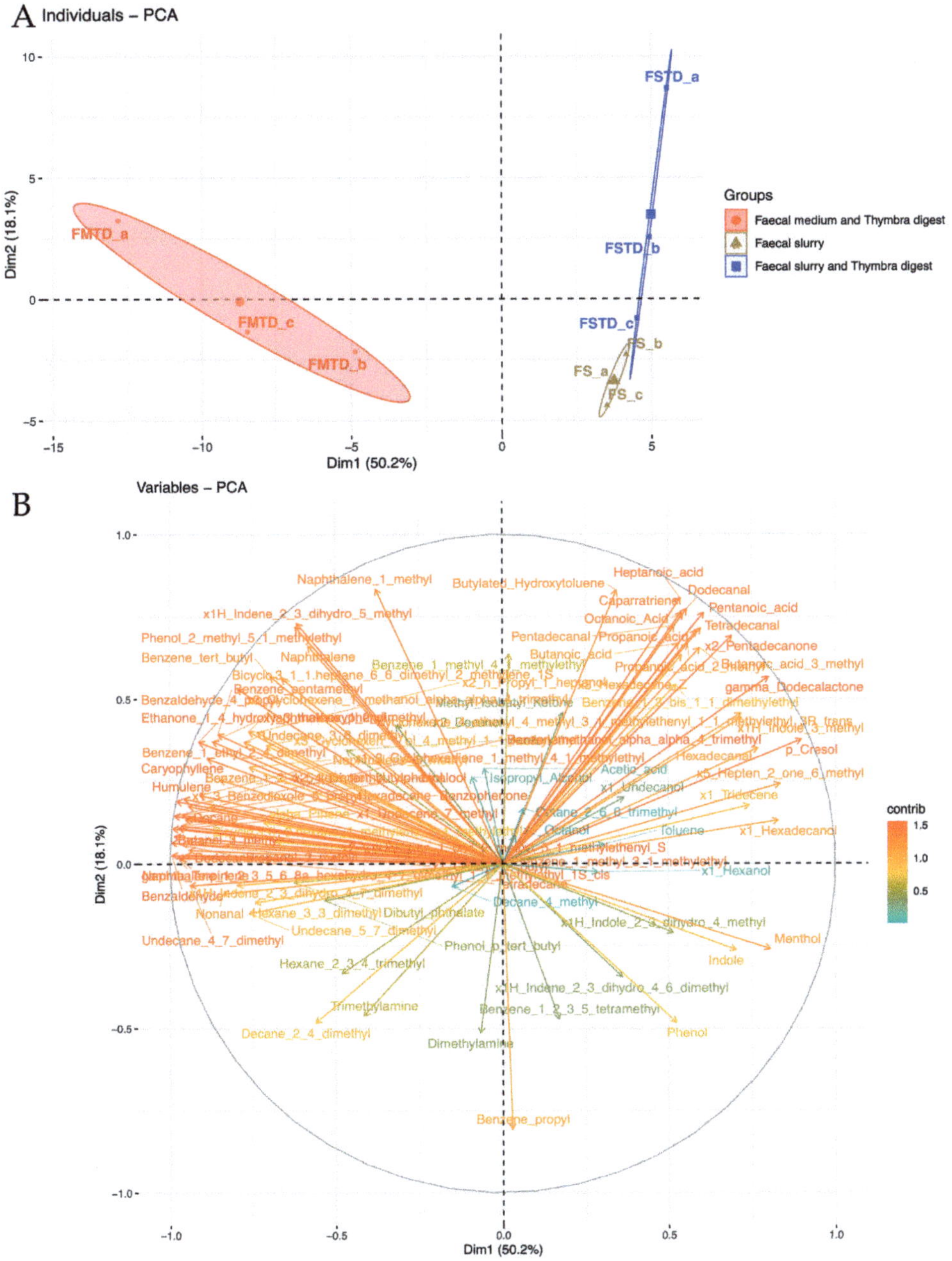

Figure 3. Principal component analysis (PCA) of volatile organic compounds (VOCs) in fecal medium and *T. spicata* digest (FMTD), fecal slurry (FS) and fecal slurry and *T. spicata* digest (FSTD) samples. The distributions of sample replicates (a,b,c) were plotted according to the first two components Dim 1 (50.2%) and Dim 2 (18.1.%) (**A**), and the contributions of each variable are shown in the loading plot (**B**).

The analysis has shown that the FMTD samples were placed in the first and four quadrants, while samples of fecal slurry with and without the addition of *T. spicata* dig were placed in the second and third quadrants. To evaluate the statistical differences of t VOCs between those of FS and FSTD, a paired comparison was undertaken through t Wilcoxon rank–sum test and a fold change analysis (log2FC > 2) (Figure 4).

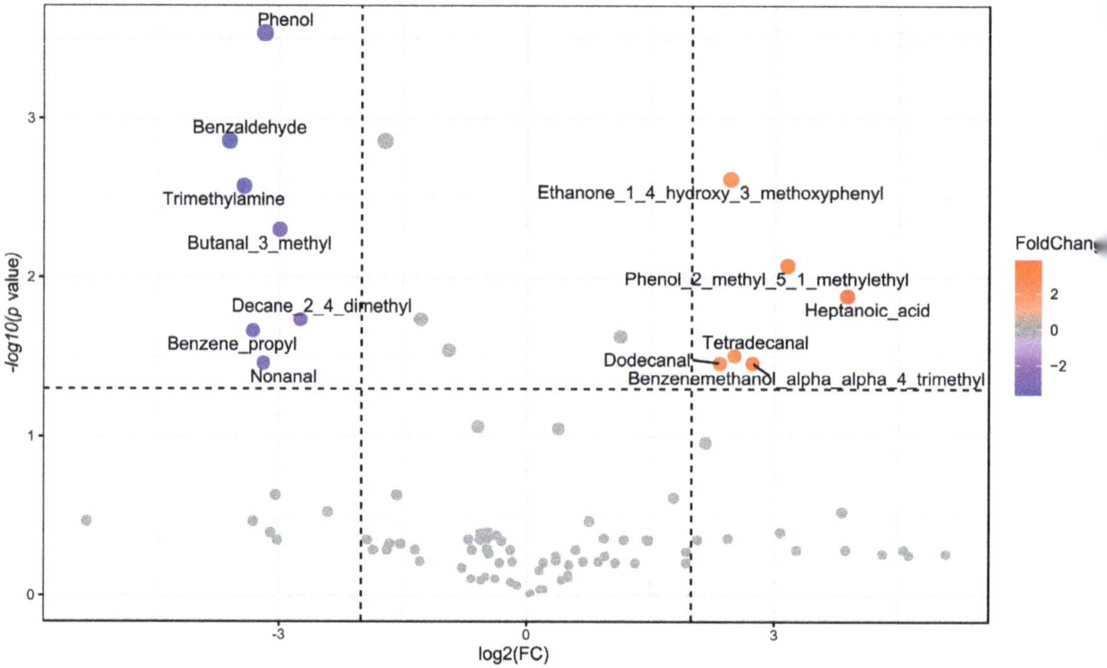

Figure 4. Volcano plot of statistically significant VOCs in fecal slurry. Increases (red) and decrea (blue) in VOC concentration in FSTD compared to FS. Non-statistically significant differences VOCs are reported as grey dots. Wilcoxon rank-sum tests ($p < 0.05$) combined with a fold char analysis (log2FC > 2) were used.

As shown in the Volcano plot, the addition of *T. spicata* led to increases in tetra canal; dodecanal; heptanoic acid; ethanone, 1-(3-hydroxy-4-methoxyphenyl)-; pher 2-methyl-5-(1-methylethyl)- and benzenemethanol, alpha,alpha,4-trimethyl-, and to creases in phenol; nonanal; decane,2,4-dimethyl-; benzene, propyl-; butanal, 3-meth trimethylamine and benzaldehyde.

The concentrations of SCFAs and CVL (the main volatile phenolic compound *T. spicata*) were also evaluated. As depicted in Table 4, higher concentrations of propan acid, butanoic acid, and isovaleric acid were observed in FSTD compared to FS, while the FMTD, as expected, SCFAs were not detected.

Table 4. Concentrations (ppm) of short-chain fatty acids in headspace of fecal medium and *T. spicata* digest (FMTD), fecal slurry (FS) and fecal slurry and *T. spicata* digest (FSTD) samples.

SCFAs (ppm)	FMTD	FS	FSTD
Acetic acid	n.d.	10.7 ± 2.3	10.1 ± 1.2
Propanoic acid	n.d.	4.8 ± 2.1 *	8.5 ± 2.2 *
Isobutyric acid	n.d.	3.2 ± 0.9	1.8 ± 1.2
Butanoic acid	n.d.	3.6 ± 1.2	4.9 ± 1.1
Isovaleric acid	n.d.	1.5 ± 1.30	2.7 ± 0.5

Data are expressed as mean ± SEM. Asterisk (*) indicates statistically significant differences (p-value < 0.05) evaluated by One-way ANOVA followed by Tukey's test. n.d., not detected.

3.4. *Effects of Microbial Fermentation on T. spicata Phenolic Content and Radical Scavenging Activity*

The possible modification of the phenolic content, flavonoid content, and antioxidant capacity of *T. spicata* after its interaction with human intestinal microbiota was investigated by colorimetric assays on FS, FMTD, and FSTD. As shown in Figure 5A, the TPC and TFC in FS were lower (32.1 ± 5.4 and 16.5 ± 3.7, respectively) than in FMDT and FSDT (138.9 ± 22.5 vs. 139.4 ± 27.3 μgGAE/mL for TPC and 43.3 ± 5.8 vs. 48.6 ± 8.3 μgQE/mL for TFC, respectively).

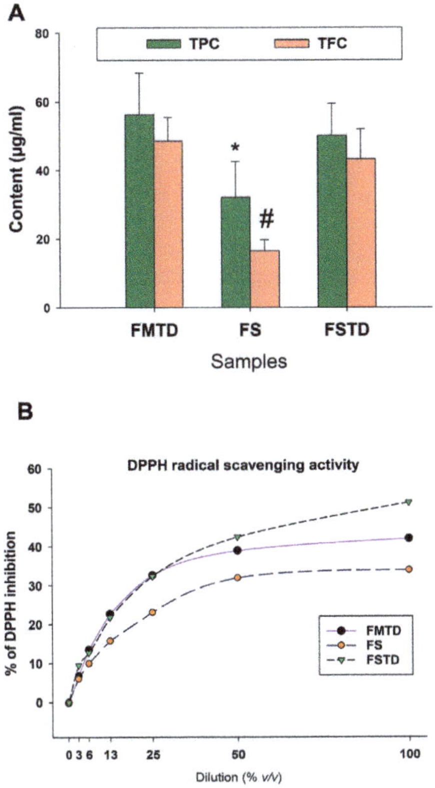

Figure 5. (**A**) Total phenol (TPC) and flavonoid (TFC) contents measured in FS, FMTD and FSTD. TFC is expressed as μg Gallic Acid Equivalent GAE per mL of sample. TFC is expressed as μg Quercetin

Equivalent QE per mL of sample. * indicates difference compared to another sample ($p < 0.05$) TPC; # is difference compared to another sample ($p < 0.05$) for TFC. (**B**) Dose-dependent DPPH radi scavenging activity expressed as a percentage of radical inhibition.

The antioxidant capacity was tested by a free-radical scavenging ability (DPPH) ass The dose-dependent DPPH inhibition (%) is represented in Figure 5B. In line with TPC a TFC, the maximum percentage of DPPH inhibition was observed in TE, and the low percentage was seen in FS. FMTD and FSTD showed similar trends of radical scavengi with slightly higher activity observed in FSTD compared to FMTD at higher concentratic (5 mg/mL).

The content of CVL was calculated in all samples (Figure 6). The highest CVL co centration was found in FMTD (54.96 mg/kg). The CVL concentration was dramatica decreased to 5 mg/kg in FSTD, while the CVL concentration in FS was 0.43 mg/kg.

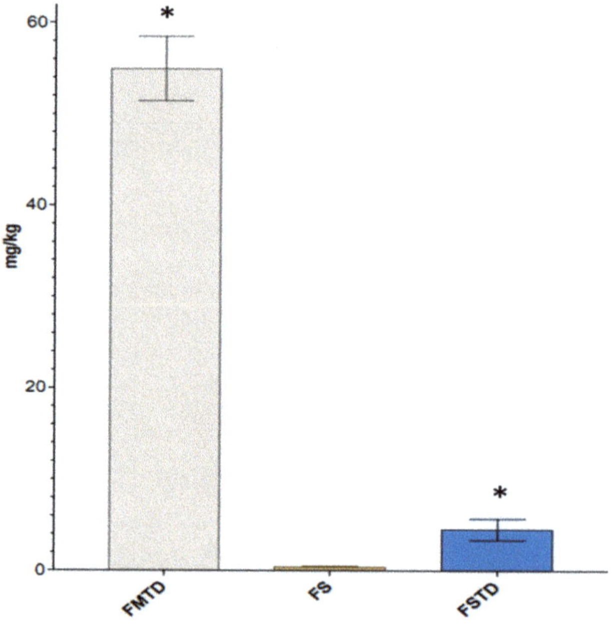

Figure 6. Concentration (mg/kg) of carvacrol determined by GC-MS in fecal medium and *T. spic* digest (FMTD), fecal slurry (FS) and fecal slurry and *T. spicata* digest (FSTD) samples. Statisti differences were evaluated by One-way ANOVA followed by Tukey's test. Asterisk (*) indica statistically significant differences (p-value < 0.05).

3.5. *Effects of T. spicata Ethanolic Extract and Carvacrol on the Intestinal Barrier Function*

To evaluate the effect of *T. spicata* ethanolic extract (TE) on the intestinal barr function, the electrophysiological properties of Caco-2 monolayers were investigated us the Ussing chamber technique. This technique monitors concurrent changes in short-circ current (Isc), a measure of net electrogenic transepithelial ion flux, as well as transepithe resistance (Rt), as a measure of barrier integrity.

Following an equilibration period of 15 min, the baseline values of short-circuit curr (Isc) and transepithelial resistance (Rt) were 2.94 ± 0.23 μA/cm^2 and 517 ± 31.8 Ω/c respectively (n = 37). The addition of increasing concentrations of TE at the apical s of Caco-2 monolayers increased the Isc when applied at 25 and 50 μg/mL (+24.8% a +82.7% compared to baseline, respectively), without affecting the Rt (Figure 7B–D).

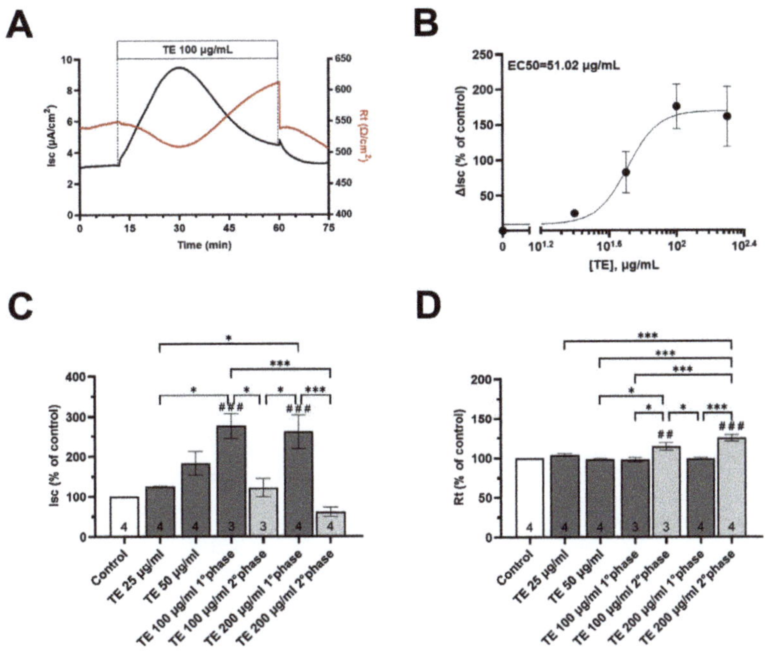

Figure 7. Effect of apical addition of *T. spicata* ethanolic extract on Caco-2 monolayers. (**A**) Representative traces of short circuit current (Isc) and transepithelial resistance (Rt) after *T. spicata* ethanolic extract (TE) addition (100 µg/mL). (**B**) Dose–response curve of *T. spicata* ethanolic extract-evoked Isc changes. The plots show the means (represented by black dots) ± the SEM of Isc measured at each concentration. The half-maximal inhibition constant (EC_{50}) for TE was 51.02 µg/m. (**C**,**D**) Changes in the electrophysiological parameters induced by apical TE addition. Results are expressed as percentages of the Isc and Rt basal values. Sample size is represented by the number at the bottom of each bar. Values are reported as mean values ± SEM, with the control represented by a white bar and TE addition at concentrations of 25, 50, 100, and 200 µg/ml depicted by light gray and dark gray bars for the first and second phases, respectively. ## $p < 0.01$, ### $p < 0.001$ vs. basal condition calculated by one-way analysis of variance (ANOVA) with Dunnett's multiple comparison test; * $p < 0.05$, *** $p < 0.001$ values calculated by one-way analysis of variance (ANOVA) with Sidak's multiple comparison test, indicating how much the results differ from those of other treatments.

Of note, higher concentrations of TE (100 and 200 µg/mL) evoked biphasic responses, with a significant and transient increase in the Isc (+175.6% and 161.1%) not associated with changes in the Rt (−2.21 and −0.72%), followed by a slight increase in the Isc at 100 µg/mL (+22.1%) and a decrease at 200 µg/mL (−38.7%) accompanied by a significant increase in the Rt (+14.5% and +25.1%) (Figure 7A,C,D).

The dose–response curve of TE was characterized by measuring the percentage change in short-circuit current with increasing concentrations of TE. The EC_{50} value, representing the concentration at which Isc exhibits a 50% change from baseline, was determined to be 51.02 µg/mL (Figure 7B).

Given that CVL is the major component of TE (constituting about 25% of TE [15]), our specific aim was to determine whether the response observed after the addition of TE could be attributed to the presence of CVL. To address this, we investigated the effect of CVL by using 6.25, 12.5, and 25 and 50 µg/mL, which correspond to the amounts of CVL contained in the different doses of the ethanolic extract tested. As illustrated in Figure 8B,C, the application of increasing concentrations of CVL at the apical side resulted in an increase in Isc (+39% and +52,4% for 6.25 and 12.5 µg/mL, respectively) without any change in the Rt. However, when a higher concentration of CVL (25 and 50 μg/mL) was appli a biphasic response similar to that evoked by 100 and 200 μg/mL TE was observed. fact, the presence of CVL elicited a significant increase in the Isc (+66.9% and 106.3%) w no change in the Rt, followed by an Isc reduction (−50.8% and −97.1%) with a significa increase in Rt (+21% and +42.5%). The similarities in the TE and CVL responses are ve intriguing since we could speculate that CVL plays a key role in modulating the electri properties of intestinal cells.

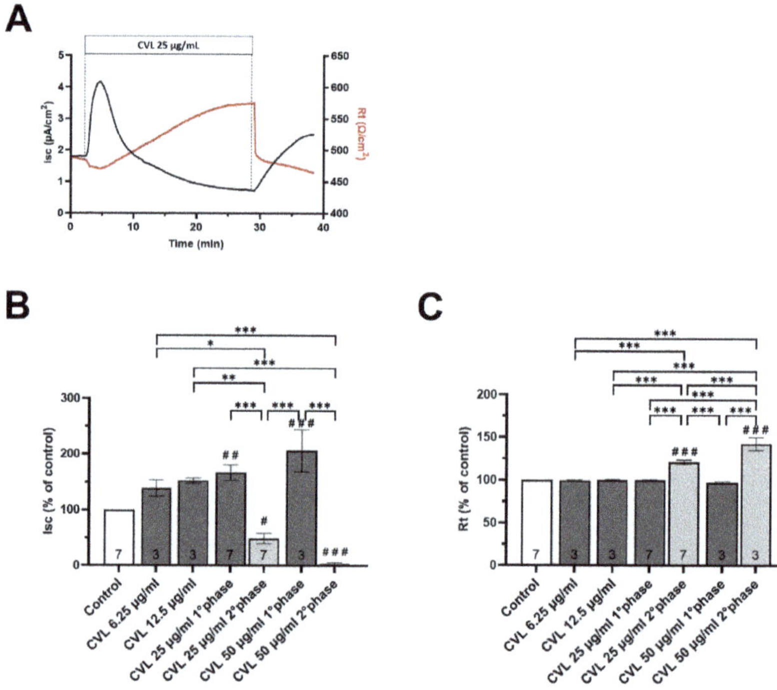

Figure 8. Effect of apical addition of Carvacrol on Caco-2 monolayers. (**A**) Representative traces short circuit current (Isc) and transepithelial resistance (Rt) after carvacrol (CVL) addition (25 μg/m (**B**,**C**) Changes in the electrophysiological parameters induced by apical CVL addition. Results expressed as percentages of the Isc and Rt basal values. Sample size is represented by the numbe the bottom of each bar. Values are reported as mean values ± SEM, with the control represented a white bar and TE addition at concentrations of 25, 50, 100, and 200 μg/ml depicted by light g and dark gray bars for the first and second phases, respectively. # $p < 0.05$, ## $p < 0.01$, ### $p < 0.0$ vs. basal condition calculated by one-way analysis of variance (ANOVA) with Dunnett's multi comparison test; * $p < 0.05$, ** $p < 0.01$, *** $p < 0.001$ values calculated by one-way analysis of varia (ANOVA) with Sidak's multiple comparisons test, indicating how much the results differ from th of other treatments.

3.6. *In Vitro Digestion and Microbial Fermentation Influences the Effect of T. spicata on the Intestinal Barrier*

To study the role of the gut microbiome on the effect of *T. spicata* on Caco-2 monolay we examined the effects of FMTD, FS, and FSTD. These series of experiments were p formed using a concentration of 100 μg/mL, which, as described previously, induced maximal cell response. As shown in Figure 9A,B, the addition of FMTD elicited a respo characterized by two phases. Initially there was a substantial and statistically significa increase in the Isc (+49%) and no change in the Rt. Subsequently, in the second pha the I_{SC} decreased and the Rt increased (−31.2% and +30% compared to the basal val respectively). Interestingly, the apical application of FS or FSTD did not affect the electrical properties of the Caco-2 monolayers since neither the Isc nor the Rt were significantly altered. Moreover, the Isc responses evoked by FMTD were significantly lower than those observed for TE, but did not differ from the response to CVL.

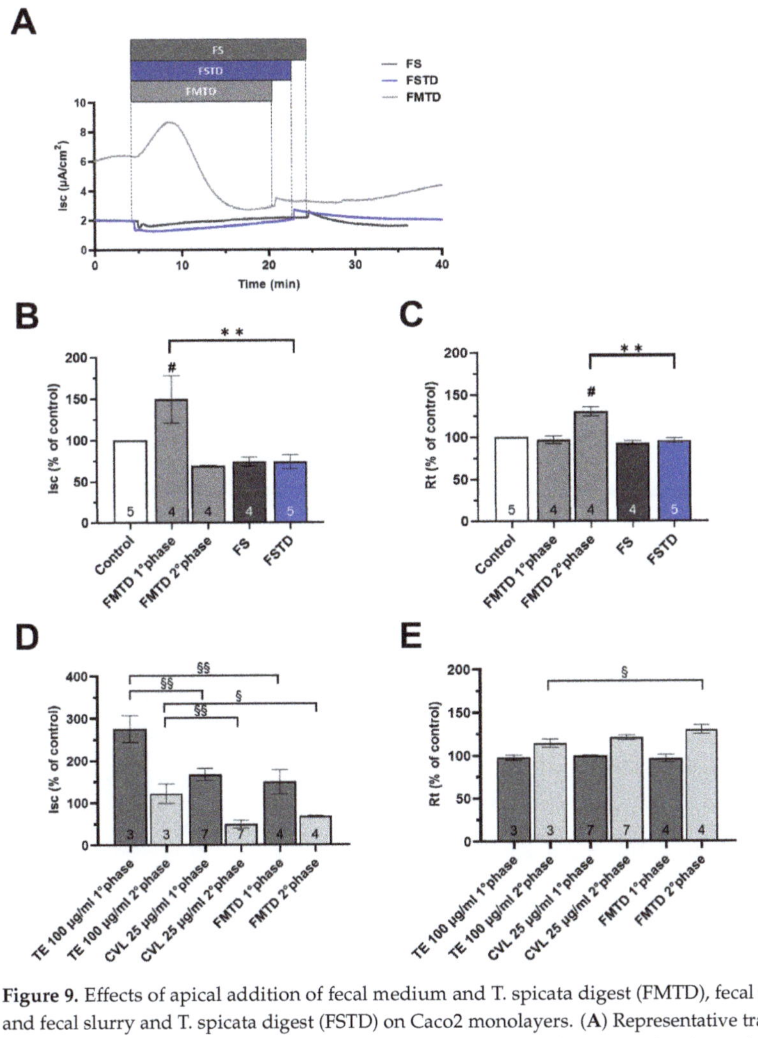

Figure 9. Effects of apical addition of fecal medium and T. spicata digest (FMTD), fecal slurry (FS), and fecal slurry and T. spicata digest (FSTD) on Caco2 monolayers. (**A**) Representative trace of short circuit current (Isc) after FMTD, FS and FSTD addition. (**B**,**C**) Changes in the electrophysiological parameters induced by apical FMTD, FS and FSTD addition. (**D**,**E**) Changes in the electrophysiological parameters induced by apical *T. spicata* ethanolic extract, carvacrol and FMTD addition. Results are expressed as percentages of the Isc and Rt basal values. Sample size is represented by the number at the bottom of each bar. Values are reported as mean values ± SEM, with the control represented by a white bar and TE addition at concentrations of 25, 50, 100, and 200 µg/ml depicted by light gray and dark gray bars for the first and second phases, respectively. # $p < 0.05$ vs. basal condition calculated by one-way analysis of variance (ANOVA) with Dunnett's multiple comparison test; ** $p < 0.001$ vs. FSTD condition calculated by Bonferroni's multiple comparisons test; § < 0.05, §§ $p < 0.01$ values calculated by one-way analysis of variance (ANOVA) with Bonferroni's multiple comparisons test, indicating how much the results differ from those of other treatments.

4. Discussion

The MD incorporates a wide variety of plant-based foods, including fruits, vegetab whole grains, nuts, seeds, and herbs [32]. Besides its richness in essential nutrients, fib and vitamins, the components of the MD are rich in polyphenols. The potential effe of these substances are mainly dependent on their interaction with the gut barrier, wh polyphenols may contribute to the modulation of gut microbiota by increasing the be ficial bacteria and decreasing the overgrowth of pathogenic bacteria. On the other ha the gut microbiota may bio-transform polyphenols to increase their bioavailability [Understanding the physiological interactions of components/substances in the MD w gut microbiota and gut barrier is of key importance and has attracted the attention researchers worldwide.

Thymbra spicata is rich in polyphenols and essential oils with a long-standing l tory and tradition of use in the Eastern Mediterranean region for medical and foc purposes. *Thymbra spicata* is a thyme-like plant belonging to the *Lamiaceae* family [Recently, *T. spicata* has gained much popularity as a remedy to combat different metabc disorders [34]. A protective role of *T. spicata* was also reported as it was used as an a genotoxic [35] and anti-coagulant agent [36]. However, aspects including the nutritio value and the physiological interaction between the ingested *T. spicata* and the gut have s not been identified. In this preliminary study, we assessed, for the first time, the nutritio and caloric profiles of *T. spicata* leaves. These aspects may give important insight into nutritional and bio-functional characteristics of *T. spicata*, helping us to better understa its physiological interaction with gut microbiota and the intestinal barrier, which w assessed by well-established in vitro models. *T. spicata* contains a considerable amoun fiber and unsaturated fatty acids, as well as CVL (polyphenols). In line with its high fi content, we found an increase in the production of SCFAs after microbial fermentation.

A growing body of scientific research supports the potential health benefits of diet polyphenols on the gastrointestinal tract and the gut microbiota. Plant-based polyphen undergo metabolization and biotransformation during gastrointestinal digestion, a through their interaction with the gut microbiota. Polyphenols and their metabolites th exert several effects on the human gastrointestinal tract, including the modulation of microbiota, the improvement of the intestinal barrier, and a reduction in inflammation a oxidative stress [37].

We previously reported the possible antisteatotic and antioxidant effects of *T. spicata* the in vitro hepatic steatosis model, whereby we proved that *T. spicata* ameliorated hep steatosis by reducing lipid droplet numbers and sizes, intracellular ROS production, a lipid peroxidation [12]. CVL, a monoterpene phenolic compound, is the main compou of *T. spicata* essential oils and alcoholic extracts. Previously, we reported an antisteatc effect, as well as the albumin-binding properties in a high-fat environment, of CVL us in vitro models [27]. In addition, we have recently reported possible patterns of phenc compounds in *T. spicata* and their bioactivity during gastrointestinal transit via a simula in vitro gastrointestinal (GI) digestion of *T. spicata* [28]. The simulated GI digestion redu the phytochemical contents in *T. spicata* extracts but improved their antioxidant potent

Given their beneficial effects, *T. spicata* and CVL were additionally investigated this study in term of their physiological interaction with gut barriers. We incuba the digested *T. spicata* with a fecal microbiota fermentation model to understand possible effects of *T. spicata* on the growth of human microbiota. *T. spicata* tended reduce bacterial growth. Due to its richness in essential oils, especially CVL, *T. spic* might exert an antimicrobial effect [17–19,38]. This could be useful for the managemen pathogenic bacterial overgrowth during dysbiosis, which may occur under different dise conditions. Several studies using cellular or animal models have described the modulat effects of CVL and CVL-rich essential oils (EOs) on gut microbiota [11]. EOs rich in C increased the relative abundance of some beneficial species such as *Bacilli*, *Lactobacilla* and *Veillonellaceae* in weaned piglets [39], and improved microbiota composition, as well piglet health and performance [40]. Additionally, CVL increased the relative abundanc

Lactobacillus spp. in the chicken gut [41]. CVL supplementation also improved gut dysbiosis by enhancing the abundance of beneficial bacteria such as *Firmicutes*, and reducing the abundance of harmful bacteria such as *Proteobacteria* [42]. The treatment of germ-free zebrafish colonized by microbiota with CVL-rich EOs exerted an anti-inflammatory effect by reducing serum amyloid and interleukin 1β, 8, and upregulated the expression of tight junction proteins such as claudin-1 and occludin-2 [43]. In intestinal cells, i.e., Caco-2 [44], Vero cells [45], and IPEC-J2 [46], CVL treatment resulted in a significant reduction of bacterial pathogenicity and toxicity in vitro. One additional aspect to consider is the possibility that the changes to the gut microbiota induced by *T. spicata* can have beneficial effects on the gut bile acid pool and their overall effects as anti-inflammatory agents, and can act agonists of ileal membrane-associated and nuclear receptors.

The effects of *T. spicata* on microbiota are supported by metabolomic analyses of VOCs. This analysis has allowed us to highlight the differences in microbiota metabolism, and facilitated the detection and quantification of metabolites upon *T. spicata* fermentation. The results indicate that *T. spicata* stimulated specific microbial pathways, resulting in the increased production of hepatonic acid, medium-chain fatty acids (MCFAs), and SCFAs (mainly propanoic acid). Beneficial effects of SCFAs are expected on the gut barrier, as well as at a systemic (metabolic) level. The increase in MCFAs and SCFAs can be beneficial in patients with irritable bowel syndrome, where MCFAs and SCFAs are reduced [47]. Besides this, the partial increase in SCFAs could be attributed to the high content of fiber in *T. spicata* leaves. The increase in other metabolites such as ethanone, 1-(3-hydroxy-4-methoxyphenyl)-, phenol, 2-methyl-5-(1-methylethyl)-, benzenemethanol and alpha,alpha,4-trimethyl- point to the modulation of microbial activities associated with phenolic metabolism and biotransformation. Notably, a decrease in trimethylamine is observed in the *T. spicata* metabolome, which could be of great relevance to cardiovascular health, as trimethylamine is a precursor of trimethylamine N-oxide (TMAO), which has been linked to cardiovascular risk [48].

Regarding the effects of *T. spicata* on the intestinal barrier function, ethanolic extract (TE) and CVL, as well as digested *T. spicata* with or without fecal fermentation samples, were tested. We found that the apical addition of the TE (100 and 200 μg/mL) and CVL (50 μg/mL) transiently increased the ionic permeability without affecting the integrity of Caco-2 monolayers, as suggested by the biphasic increase in short circuit current (Isc) and absence of changes in transepithelial resistance (Rt). Interestingly, both TE and CVL exhibited a biphasic response with a significant increase in intestinal transepithelial resistance observed in the second phase. The biphasic response observed in Caco-2 monolayers exposed to 25 μg/mL of CVL might be attributed to various molecular and cellular processes. The initial increase in short-circuit current could be linked to the activation of ion channels or transporters on the apical membrane of Caco-2 cells.

CVL, known for its modulatory effects on ion channels, may initially enhance the movement of ions across the cell membrane, resulting in an increased Isc. The subsequent reduction in Isc might be associated with cellular adaptation or regulatory mechanisms. It is possible that prolonged exposure to CVL triggers feedback mechanisms that downregulate ion transport, leading to a decrease in Isc. The observed biphasic response evoked by CVL shares similarities with the response induced by 100 μg/mL of *T. spicata* ethanolic extract. This is not surprising, since CVL is the most abundant component of the ethanolic extract, and therefore it is conceivable that it could play a key role in determining the electrical changes recorded in our studies.

Both TE and CVL initially increased Isc, suggesting a commonality in their early effects on ion transport. The subsequent reduction in Isc in the second phase might involve shared or convergent signaling pathways.

The ability of the *T. spicata* to increase the ionic permeability and the transepithelial resistance was maintained after simulated gastrointestinal digestion (FMDT sample), although with lower efficacy due to the partial loss of phytochemical compounds during digestion, as reported by our previous study [28], and CVL, which was lower in the digested sample than the ethanolic extract. However, when pre-digested *T. spicata* were subjected human microbial fermentation, mimicking the physiological interaction between digest food and healthy human gut microbiota, an effect on Caco-2 monolayers was not observ This could be also explained by the dramatic reduction in CVL in the fermented samp The results of this series of experiments demonstrate that although the fecal slurry (us as negative control) does not have any significant influence on intestinal function, it h a strong impact on the ability of *T. spicata* to modulate the permeability of the intesti barrier, and its function, since the biological effects observed in non-fermented *T. spicata* (and FMDT) on the Isc and Rt were completely abolished. Taken together, we can observ correlation between CVL and the effects on Caco-2 monolayers, and therefore, we cou speculate that CVL plays a key role in determining the biological effects of *T. spicata*.

Although no studies have elucidated the effects of *T. spicata* on Caco-2 cells, a few stu ies have reported the beneficial effects of CVL. For example, CVL at lower concentrations erted a cytoprotective and DNA-protection effect on free radical-injured Caco-2 cells [49, The dual-phase response in Caco-2 monolayers exposed to FMTD at 100 μg/mL lik involves intricate interactions between *T. spicata* components and the gut, in terms polyphenolic metabolites and digestion, leading to distinct cellular responses. This respo differs from the effects seen with *T. spicata* ethanolic extract or CVL, indicating uniq microbial–plant–metabolite interactions, supported by the different levels of polyphen and polyphenol metabolites.

The gut microbiome is known for its dynamic metabolic activities. Metabolites p duced during fecal medium preparation may contain active metabolites (i.e., enzymes) th could interact with *T. spicata*, and which may play a role in modulating cellular respons Furthermore, microbial signals could potentially interact with the cellular pathways volved in ion transport. To exclude such effects, we also tested the effects of fecal slu (FS) on Caco-2 monolayers, and found no significant effect.

Although no significant changes in the amount of polyphenols (TPC) or the radi scavenging activity (DPPH) were observed between FMTD and FSTD, the CVL cont in FMTD was ten times higher than in FSTD. This crosstalk and the content of CVL m contribute to the changes observed in Isc. Unravelling these mechanisms will enhar our understanding of how *T. spicata*'s different interactions shape cellular responses in t gut epithelium.

T. spicata manifested an increase in intestinal barrier resistance under physiologi conditions. This could be relevant to the further evaluation of the beneficial effects *T. spicata* on the modulation of gut barrier integrity during conditions of "leaky gut", wh the intestine's epithelial resistance is altered due to different states of epithelial damag

5. Conclusions

Research has shown that dietary polyphenols can positively influence the gastrointe nal tract and gut microbiota. *Thymbra spicata*, a Mediterranean plant rich in polyphen may play an important role in gut physiology. Here, we found that *T. spicata*'s nutritio and phytochemical composition, as well as its antioxidant potential, elevate its potential u as a valuable component of the diet. Furthermore, its interaction with the gut microbi demonstrates its ability to help maintain a balanced microbial environment and ben cial metabolites, and it can potentially address pathogenic bacterial overgrowth duri dysbiosis, a condition associated with various diseases. The ethanolic extract of *T. spic* has shown promise for use in enhancing intestinal barrier resistance, which is crucial preserving gut integrity and preventing conditions such as leaky gut syndrome.

In addition, understanding the alterations in microbial populations and their potent functional implications in response to *T. spicata* is a critical aspect that warrants furt investigation. Further studies must better define the mechanisms underlying the benefic effects of *T. spicata* on microbial growth and the ionic permeability and barrier functio the intestinal tissue.

Supplementary Materials: The following supporting information can be downloaded at: https://www.mdpi.com/article/10.3390/nu16050588/s1, Table S1. Concentration of volatile organic compounds (VOCs), expressed as ppm, of fecal medium and Thymbra digest (FMTD), fecal slurry (FS) and fecal slurry and Thymbra digest (FSTD) samples. Data are expressed as average ± standard deviation.

Author Contributions: Conceptualization, M.K. and P.P.; methodology, M.K., F.P., C.A.A., A.A. and I.I.; software, L.M. and V.P.; formal analysis, M.V. and G.C.; investigation, M.D.A.; writing—original draft preparation, M.K.; writing—review and editing, P.P., R.C. and G.P.; supervision, M.C.; funding acquisition, P.P. and M.D.A. All authors have read and agreed to the published version of the manuscript.

Funding: P.P. is a grant recipient and MK is an assistant researcher (RTDA) in a project funded under the National Recovery and Resilience Plan (NRRP), Mission 4 Component 2 Investment 1.3—Call for tender No. 341 of 15/03/2022 of Italian Ministry of University and Research funded by the European Union—Next Generation EU Award Number: Project code PE0000003, Concession Decree No. 1550 of 11/10/2022 adopted by the Italian Ministry of University and Research, CUP D93C22000890001, Project title "Research and innovation network on food and nutrition Sustainability, Safety and Security—Working ON Foods" (ONFoods). P.P. is the coordinator of the B4HT project "Box for Health by Tradition & Innovation: promoting sustainable mediterranean diet by Healthy Foods" funded by the PRIMA project, Section 2—Multi-topic 2022. Project partners: University of Bari Aldo Moro (Italy), University of Genoa (Italy), Lebanese University (Lebanon), and University of Monastir (Tunisia). PP is recipient of HORIZON-HLTH-2022-STAYHLTH-01-05-two-stage Project 101080329—PAS GRAS with the following partners: Universidade De Coimbra, Portugal; Uppsala Universitet, Sweden; Universidade Nova De Lisboa, Portugal; Fundacio Eurecat (Eurecat), Barcelona, Spain; Consiglio Nazionale Delle Ricerche (Cnr), Roma Italy; Instituto Politecnico De Viana De Castelo, Viana Do Castelo 4900-347, Portugal; Technische Universitaet Muenchen (Tum), Muenchen, Germany; Instytut Biologii Doswiadczalnej Im. M. Nenckiego Polskiej Akademii Nauk (Nencki), Warszawa, Poland; Instituto Pedro Nunes Associacao Para A Inovacao E Desenvolvimento Em Ciencia E Tecnologia (Ipn), Coimbra, Portugal; The European Society For Clinical Investigation (Esci), Utrecht, Netherlands; Mediagnost Gesellschaft Fur Forschung Und Herstellung Von Diagnostika Gmbh (Mediagnost), Reutlingen, Germany; Martin-Luther-Universitat Halle-Wittenberg (Mlu), Halle, Germany; Associacao Protectora Dos Diabeticos De Portugal (Apdp), Lisboa 1250-203, Portugal; Agdcentro Associacao De Ginastica Do Centro (Agcentro), Coimbra, Portugal.

Institutional Review Board Statement: Not applicable.

Informed Consent Statement: Not applicable.

Data Availability Statement: The data presented in this study are available on request from the corresponding author.

Acknowledgments: In memory of Lucantonio Debellis for his profound influence on electrophysiology research and for his enthusiastic support over the years. In memory of David Q.-H. Wang for his longstanding fruitful collaboration in the field of pathophysiology related to gastrointestinal diseases. We dedicate this work to their enduring legacy.

Conflicts of Interest: The authors declare no conflict of interest.

erences

Sotos-Prieto, M.; Bhupathiraju, S.N.; Mattei, J.; Fung, T.T.; Li, Y.; Pan, A.; Willett, W.C.; Rimm, E.B.; Hu, F.B. Association of Changes in Diet Quality with Total and Cause-Specific Mortality. *N. Engl. J. Med.* **2017**, *377*, 143–153. [CrossRef]
World Health Organization. Healthy Diet. 2023. Available online: https://www.who.int/en/news-room/fact-sheets/detail/healthy-diet (accessed on 15 December 2023).
World Cancer Research Fund/American Institute for Cancer Research. Cancer Prevention Recommendations. 2023. Available online: https://www.wcrf.org/dietandcancer/cancer-prevention-recommendations (accessed on 15 December 2023).
US Department of Agriculture. Dietary Guidelines for Americans: 2020–2025. US Department of Health and Human Services 2020. 2023. Available online: https://www.dietaryguidelines.gov/sites/default/files/2020-12/Dietary_Guidelines_for_Americans_2020-2025.pdf (accessed on 15 December 2023).
Di Salvo, E.; Gangemi, S.; Genovese, C.; Cicero, N.; Casciaro, M. Polyphenols from Mediterranean Plants: Biological Activities for Skin Photoprotection in Atopic Dermatitis, Psoriasis, and Chronic Urticaria. *Plants* **2023**, *12*, 3579. [CrossRef]
Eran Nagar, E.; Okun, Z.; Shpigelman, A. Digestive fate of polyphenols: Updated view of the influence of chemical structure and the presence of cell wall material. *Curr. Opin. Food Sci.* **2020**, *31*, 38–46. [CrossRef]

7. Wan, M.L.Y.; Co, V.A.; El-Nezami, H. Dietary polyphenol impact on gut health and microbiota. *Crit. Rev. Food Sci. Nutr.* **2021**, 690–711. [CrossRef]
8. Scott, M.B.; Styring, A.K.; McCullagh, J.S.O. Polyphenols: Bioavailability, Microbiome Interactions and Cellular Effects on Hea in Humans and Animals. *Pathogens* **2022**, *11*, 770. [CrossRef]
9. Cheng, H.; Zhang, D.; Wu, J.; Liu, J.; Zhou, Y.; Tan, Y.; Feng, W.; Peng, C. Interactions between gut microbiota and polyphenols mechanistic and metabolomic review. *Phytomed. Int. J. Phytother. Phytopharm.* **2023**, *119*, 154979. [CrossRef]
10. Khalil, M.; Rita Caponio, G.; Diab, F.; Shanmugam, H.; Di Ciaula, A.; Khalifeh, H.; Vergani, L.; Calasso, M.; De Angelis, Portincasa, P. Unraveling the beneficial effects of herbal Lebanese mixture "Za'atar". History, studies, and properties of a poten healthy food ingredient. *J. Funct. Foods* **2022**, *90*, 104993. [CrossRef]
11. Khalil, M.; Abdallah, H.; Razuka-Ebela, D.; Calasso, M.; De Angelis, M.; Portincasa, P. The Impact of Za'atar Antioxid Compounds on the Gut Microbiota and Gastrointestinal Disorders: Insights for Future Clinical Applications. *Antioxidants* **2** *12*, 426. [CrossRef]
12. Khalil, M.; Khalifeh, H.; Baldini, F.; Salis, A.; Damonte, G.; Daher, A.; Voci, A.; Vergani, L. Antisteatotic and antioxidant activitie *Thymbra spicata* L. extracts in hepatic and endothelial cells as in vitro models of non-alcoholic fatty liver disease. *J. Ethnopharma* **2019**, *239*, 111919. [CrossRef] [PubMed]
13. Akkol, E.K.; Avci, G.; Küçükkurt, I.; Keleş, H.; Tamer, U.; Ince, S.; Yesilada, E. Cholesterol-reducer, antioxidant and liver protect effects of *Thymbra spicata* L. var. spicata. *J. Ethnopharmacol.* **2009**, *126*, 314–319. [CrossRef] [PubMed]
14. Avci, G.; Kupeli, E.; Eryavuz, A.; Yesilada, E.; Kucukkurt, I. Antihypercholesterolaemic and antioxidant activity assessmen some plants used as remedy in Turkish folk medicine. *J. Ethnopharmacol.* **2006**, *107*, 418–423. [CrossRef]
15. Khalil, M.; Khalifeh, H.; Baldini, F.; Serale, N.; Parodi, A.; Voci, A.; Vergani, L.; Daher, A. Antitumor Activity of Ethanolic Ext from *Thymbra spicata* L. aerial Parts: Effects on Cell Viability and Proliferation, Apoptosis Induction, STAT3, and NF-kB Signal *Nutr. Cancer* **2021**, *73*, 1193–1206. [CrossRef] [PubMed]
16. Toker, E.B.; Yeşilbağ, K. In vitro antiviral activity of *Thymbra spicata* L. extract on bovine respiratory viruses (BCoV, BPIV-3, BR BVDV and BoHV-1). *J. Appl. Microbiol.* **2022**, *132*, 2625–2632. [CrossRef] [PubMed]
17. Gedikoğlu, A.; Sökmen, M.; Çivit, A. Evaluation of Thymus vulgaris and *Thymbra spicata* essential oils and plant extracts chemical composition, antioxidant, and antimicrobial properties. *Food Sci. Nutr.* **2019**, *7*, 1704–1714. [CrossRef] [PubMed]
18. Haroun, M.F.; Al-Kayali, R.S. Synergistic effect of *Thymbra spicata* L. extracts with antibiotics against multidrug-resist *Staphylococcus aureus* and *Klebsiella pneumoniae* strains. *Iran. J. Basic Med. Sci.* **2016**, *19*, 1193–1200. [PubMed]
19. Kerem, S.; Koşar, N.; Tekin, F.; Güreser, A.S.; Özbek, Ö. Investigation of antimicrobial activities and molecular characterizatio the species belong to *Origanum*, *Thymus* and *Thymbra genera* by ISSR. *Mol. Biol. Rep.* **2023**, *50*, 289–298. [CrossRef] [PubMed]
20. Soylu, E.M.; Soylu, S.; Kurt, S. Antimicrobial activities of the essential oils of various plants against tomato late blight dise agent *Phytophthora infestans*. *Mycopathologia* **2006**, *161*, 119–128. [CrossRef] [PubMed]
21. Hoca, M.; Becer, E.; Vatansever, H.S. Carvacrol is potential molecule for diabetes treatment. *Arch. Physiol. Biochem.* **2023**, [CrossRef] [PubMed]
22. Mączka, W.; Twardawska, M.; Grabarczyk, M.; Wińska, K. Carvacrol-A Natural Phenolic Compound with Antimicrol Properties. *Antibiotics* **2023**, *12*, 824. [CrossRef]
23. Imran, M.; Aslam, M.; Alsagaby, S.A.; Saeed, F.; Ahmad, I.; Afzaal, M.; Arshad, M.U.; Abdelgawad, M.A.; El-Ghorab, A Khames, A.; et al. Therapeutic application of carvacrol: A comprehensive review. *Food Sci. Nutr.* **2022**, *10*, 3544–3561. [CrossR
24. Al-Tawalbeh, D.; Alkhawaldeh, Y.; Abu Sawan, H.; Al-Mamoori, F.; Al-Samydai, A.; Mayyas, A. Assessment of carvacrol-antibi combinations' antimicrobial activity against methicillin-resistant *Staphylococcus aureus*. *Front. Microbiol.* **2023**, *14*, 1349 [CrossRef] [PubMed]
25. Şimşek, H.; Gür, C.; Küçükler, S.; İleritürk, M.; Akaras, N.; Öz, M.; Kandemir, F.M. Carvacrol Reduces Mercuric Chloride-Indu Testicular Toxicity by Regulating Oxidative Stress, Inflammation, Apoptosis, Autophagy, and Histopathological Changes. *B Trace Elem. Res.* **2023**. [CrossRef]
26. Zhao, W.; Deng, C.; Han, Q.; Xu, H.; Chen, Y. [Corrigendum] Carvacrol may alleviate vascular inflammation in diabetic db/ mice. *Int. J. Mol. Med.* **2023**, *52*, 90. [CrossRef] [PubMed]
27. Khalil, M.; Serale, N.; Diab, F.; Baldini, F.; Portincasa, P.; Lupidi, G.; Vergani, L. Beneficial Effects of Carvacrol on In Vitro Moc of Metabolically-Associated Liver Steatosis and Endothelial Dysfunction: A Role for Fatty Acids in Interfering with Carvac Binding to Serum Albumin. *Curr. Med. Chem.* **2022**, *29*, 5113–5129. [CrossRef]
28. Diab, F.; Khalil, M.; Lupidi, G.; Zbeeb, H.; Salis, A.; Damonte, G.; Bramucci, M.; Portincasa, P.; Vergani, L. Influence of Simula In Vitro Gastrointestinal Digestion on the Phenolic Profile, Antioxidant, and Biological Activity of *Thymbra spicata* L. Extra *Antioxidants* **2022**, *11*, 1778. [CrossRef] [PubMed]
29. Kwok, L.Y.; Zhang, J.; Guo, Z.; Gesudu, Q.; Zheng, Y.; Qiao, J.; Huo, D.; Zhang, H. Characterization of fecal microbiota acr seven Chinese ethnic groups by quantitative polymerase chain reaction. *PLoS ONE* **2014**, *9*, e93631. [CrossRef]
30. Jian, C.; Luukkonen, P.; Yki-Järvinen, H.; Salonen, A.; Korpela, K. Quantitative PCR provides a simple and accessible method quantitative microbiota profiling. *PLoS ONE* **2020**, *15*, e0227285. [CrossRef]
31. Piccapane, F.; Gerbino, A.; Carmosino, M.; Milano, S.; Arduini, A.; Debellis, L.; Svelto, M.; Caroppo, R.; Procino, G. Aquapori Facilitates Transmesothelial Water Permeability: In Vitro and Ex Vivo Evidence and Possible Implications in Peritoneal Dial *Int. J. Mol. Sci.* **2021**, *22*, 12535. [CrossRef]

Farias-Pereira, R.; Zuk, J.B.; Khavaran, H. Plant bioactive compounds from Mediterranean diet improve risk factors for metabolic syndrome. *Int. J. Food Sci. Nutr.* **2023**, *74*, 403–423. [CrossRef]
Corrêa, T.A.F.; Rogero, M.M.; Hassimotto, N.M.A.; Lajolo, F.M. The Two-Way Polyphenols-Microbiota Interactions and Their Effects on Obesity and Related Metabolic Diseases. *Front. Nutr.* **2019**, *6*, 188. [CrossRef]
Diab, F.; Zbeeb, H.; Baldini, F.; Portincasa, P.; Khalil, M.; Vergani, L. The Potential of Lamiaceae Herbs for Mitigation of Overweight, Obesity, and Fatty Liver: Studies and Perspectives. *Molecules* **2022**, *27*, 5043. [CrossRef] [PubMed]
Dirican, E.; Turkez, H.; Toğar, B. Modulatory effects of *Thymbra spicata* L. different extracts against the mercury induced genotoxicity in human lymphocytes in vitro. *Cytotechnology* **2012**, *64*, 181–186. [CrossRef]
Omar, G.; Abdallah, L.; Barakat, A.; Othman, R.; Bourinee, H. In vitro haemostatic efficacy of aqueous, methanol and ethanol plant extracts of three medicinal plant species in Palestine. *Braz. J. Biol.* **2020**, *80*, 763–768. [CrossRef] [PubMed]
Lan, H.; Wang, H.; Chen, C.; Hu, W.; Ai, C.; Chen, L.; Teng, H. Flavonoids and gastrointestinal health: Single molecule for multiple roles. *Crit. Rev. Food Sci. Nutr.* **2023**, 1–19. [CrossRef]
Kiliç, T. Analysis of essential oil composition of *Thymbra spicata* var. spicata: Antifungal, antibacterial and antimycobacterial activities. *Z. Naturforschung. C J. Biosci.* **2006**, *61*, 324–328. [CrossRef] [PubMed]
Li, Y.; Fu, X.; Ma, X.; Geng, S.; Jiang, X.; Huang, Q.; Hu, C.; Han, X. Intestinal Microbiome-Metabolome Responses to Essential Oils in Piglets. *Front. Microbiol.* **2018**, *9*, 1988. [CrossRef]
Hall, H.N.; Wilkinson, D.J.; Le Bon, M. Oregano essential oil improves piglet health and performance through maternal feeding and is associated with changes in the gut microbiota. *Anim. Microbiome* **2021**, *3*, 2. [CrossRef]
Kelly, C.; Gundogdu, O.; Pircalabioru, G.; Cean, A.; Scates, P.; Linton, M.; Pinkerton, L.; Magowan, E.; Stef, L.; Simiz, E.; et al. The In Vitro and In Vivo Effect of Carvacrol in Preventing Campylobacter Infection, Colonization and in Improving Productivity of Chicken Broilers. *Foodborne Pathog. Dis.* **2017**, *14*, 341–349. [CrossRef]
Mooyottu, S.; Flock, G.; Upadhyay, A.; Upadhyaya, I.; Maas, K.; Venkitanarayanan, K. Protective Effect of Carvacrol against Gut Dysbiosis and Clostridium difficile Associated Disease in a Mouse Model. *Front. Microbiol.* **2017**, *8*, 625. [CrossRef]
Ran, C.; Hu, J.; Liu, W.; Liu, Z.; He, S.; Dan, B.C.; Diem, N.N.; Ooi, E.L.; Zhou, Z. Thymol and Carvacrol Affect Hybrid Tilapia through the Combination of Direct Stimulation and an Intestinal Microbiota-Mediated Effect: Insights from a Germ-Free Zebrafish Model. *J. Nutr.* **2016**, *146*, 1132–1140. [CrossRef]
Upadhyay, A.; Arsi, K.; Wagle, B.R.; Upadhyaya, I.; Shrestha, S.; Donoghue, A.M.; Donoghue, D.J. Trans-Cinnamaldehyde, Carvacrol, and Eugenol Reduce Campylobacter jejuni Colonization Factors and Expression of Virulence Genes In Vitro. *Front. Microbiol.* **2017**, *8*, 713. [CrossRef]
Mooyottu, S.; Kollanoor-Johny, A.; Flock, G.; Bouillaut, L.; Upadhyay, A.; Sonenshein, A.L.; Venkitanarayanan, K. Carvacrol and trans-cinnamaldehyde reduce *Clostridium difficile* toxin production and cytotoxicity in vitro. *Int. J. Mol. Sci.* **2014**, *15*, 4415–4430. [CrossRef]
Burt, S.A.; Adolfse, S.J.; Ahad, D.S.; Tersteeg-Zijderveld, M.H.; Jongerius-Gortemaker, B.G.; Post, J.A.; Brüggemann, H.; Santos, R.R. Cinnamaldehyde, Carvacrol and Organic Acids Affect Gene Expression of Selected Oxidative Stress and Inflammation Markers in IPEC-J2 Cells Exposed to *Salmonella typhimurium*. *Phytother. Res.* **2016**, *30*, 1988–2000. [CrossRef] [PubMed]
Lavelle, A.; Sokol, H. Gut microbiota-derived metabolites as key actors in inflammatory bowel disease. *Nat. Rev. Gastroenterol. Hepatol.* **2020**, *17*, 223–237. [CrossRef] [PubMed]
Wang, Z.; Klipfell, E.; Bennett, B.J.; Koeth, R.; Levison, B.S.; Dugar, B.; Feldstein, A.E.; Britt, E.B.; Fu, X.; Chung, Y.M.; et al. Gut flora metabolism of phosphatidylcholine promotes cardiovascular disease. *Nature* **2011**, *472*, 57–63. [CrossRef] [PubMed]
Horváthová, E.; Sramková, M.; Lábaj, J.; Slamenová, D. Study of cytotoxic, genotoxic and DNA-protective effects of selected plant essential oils on human cells cultured in vitro. *Neuro Endocrinol. Lett.* **2006**, *27* (Suppl. S2), 44–47. [PubMed]
Slamenová, D.; Horváthová, E.; Sramková, M.; Marsálková, L. DNA-protective effects of two components of essential plant oils carvacrol and thymol on mammalian cells cultured in vitro. *Neoplasma* **2007**, *54*, 108–112. [PubMed]

nutrients

MDPI

Article

Bile Acids and Short-Chain Fatty Acids Are Modulated after Onion and Apple Consumption in Obese Zucker Rats

Claudia Balderas, Begoña de Ancos and Concepción Sánchez-Moreno *

Institute of Food Science, Technology and Nutrition (ICTAN), Spanish National Research Council (CSIC), ES-28040 Madrid, Spain; c.balderas@ictan.csic.es (C.B.); ancos@ictan.csic.es (B.d.A.)
* Correspondence: csanchezm@ictan.csic.es

Abstract: Gut microorganisms are involved in the development and severity of different cardiovascu diseases, and increasing evidence has indicated that dietary fibre and polyphenols can interact w the intestinal microbiota. The study objective was to investigate the effect of onion and apple int on the major types of microbial-derived molecules, such as short-chain fatty acids (SCFAs) and b acids (BAs). Obese Zucker rats were randomly assigned (n = eight rats/group) to a standard d (OC), a standard diet/10% onion (OO), or a standard diet/10% apple (OA). Lean Zucker rats fe standard diet served as a lean control (LC) group. Faecal samples were collected at baseline, an weeks later, the composition of the microbial community was measured, and BA and SCFA lev were determined using high-performance liquid chromatography–mass spectrometry (HPLC-MS) a gas chromatography–mass spectrometry (GC-MS), respectively. Rats fed onion- and apple-enrich diets had increased abundance of beneficial bacteria, such as *Bifidobacterium* spp. and *Lactobaci* spp., enhanced SCFAs (acetic, propionic, isobutyric, and valeric acids), decreased excretion of so BAs, mainly of the primary (CA, α-MCA, and β-MCA) and secondary type (ω-MCA, HDCA, N DCA, and LCA), and increased amount of taurine- and glycine-conjugated BAs compared to the group. The contribution of specific bioactive compounds and their metabolites in the regulation the microbiome and the pathways linked to SCFA and BA formation and their relationship with so diseases needs further research.

Keywords: functional foods and ingredients; prebiotic effect; microbiome; short-chain fatty ac primary and secondary bile acids; conjugated bile acids; glycine- and taurine-conjugated bile aci

Citation: Balderas, C.; de Ancos, B.; Sánchez-Moreno, C. Bile Acids and Short-Chain Fatty Acids Are Modulated after Onion and Apple Consumption in Obese Zucker Rats. *Nutrients* **2023**, *15*, 3035. https://doi.org/10.3390/nu15133035

Academic Editor: Gary David Lopaschuk

Received: 11 June 2023
Revised: 26 June 2023
Accepted: 28 June 2023
Published: 5 July 2023

1. Introduction

Bile acids (BAs) play a critical role in maintaining overall health, as they are essent endogenous compounds synthesised from cholesterol in the liver and metabolised the intestine by the gut microbiota. The circulation of BAs is very efficient and pla an important physiological role in lipid absorption and the regulation of choleste homeostasis [1,2]. Humans and rodents exhibit substantial differences in respect of b acid composition. The BAs that are newly synthesised from cholesterol and formed the hepatocytes are termed primary BAs. Cholesterol is converted into BAs by pa ways that involve 17 different enzymes, many of which are preferentially expressed the liver [3]. Under normal physiological conditions, the primary BAs are synthesis via two major pathways, the classical (neutral) and the alternative (acidic) pathways The primary BAs produced in humans are cholic acid (CA) and chenodeoxycholic a (CDCA), while rodents produce CA, CDCA, α-muricholic acid (α-MCA), and β-murich acid (β-MCA). After biosynthesis from cholesterol and before excretion from the hepa cyte, bile acid molecules are conjugated with glycine or taurine, which converts a we acid to a strong acid. In contrast, the BAs generated by subsequent modifications carri out by the gut microbiota are termed secondary BAs, including deoxycholic acid (DC lithocholic acid (LCA), ursodeoxycholic acid (UDCA), ω-muricholic acid (ω-MCA), a their conjugated counterparts [4–6]. Secondary BAs are hypothesised to be interkingdom signalling molecules. By understanding the dynamic intricacies in the gut microbiota–bile acid–host axis, insight into a variety of disease states could be elucidated [7]. Obesity is associated with an altered BA metabolism, and these products of microbial metabolism, along with short-chain fatty acids (SCFAs), among others, have been associated with an increased risk of cardiovascular disease, inflammation, and metabolic disorders [8,9]. The relationship between the BAs and microbiota appears to be bidirectional; gut microbes influence BA composition, and, in turn, BAs modulate the gut microbial community [10,11]. An imbalance in gut microbiota is considered to be an important factor in chronic inflammation diseases, such as obesity and autoimmune and metabolic diseases [12,13]. Most diseases only lead to an increase or decrease in the concentration of one or several BAs but not total BAs. The separation of various BAs and the detection of their concentration changes in samples are conducive to scientific research.

Diet, especially fibre intake, serves as a potent modulator of the gut microbiome [14]. Various studies have shown that different types of dietary fibres or fibres with different physiological and chemical properties are related to changes in the bile acid profile of rats [15]. Higher fibre diets are associated with various health benefits, in part due to their ability to increase microbial diversity and promote the growth of beneficial intestinal bacteria, such as *Bifidobacteria*, *Lactobacillus*, and *Eubacterium*, which produce SCFAs [16].

SCFAs are bioactive metabolites excreted by anaerobic bacteria in the colon after the degradation of dietary fibre and indigestible carbohydrates. The SCFA concentration in the gut depends on the microbiota composition, gut transit time, the regulation of host energy metabolism, and the fibre content of the host diet. Acetate, propionate, and butyrate are the most abundant, representing 90–95% of the total SCFAs present in the colon [17]. Dietary fibre and its associated compounds digested by the microbiota and their resulting metabolites, especially SCFAs, are significantly associated with health beneficial effects, such as via proposed anti-inflammatory mechanisms [18]. Thus, microbial metabolites including SCFAs may be suitable biomarkers in clinical practice, as their detection in blood plasma and in faeces is usually affordable. Emerging evidence based on numerous animal studies has shown that gut microbiota and its metabolites, particularly SCFAs, play an important role in hypertension, inflammatory bowel disease, diabetes, and obesity, among other diseases [19–25].

Onions and apples are both considered sources of dietary fibre and other phytochemicals, such as polyphenols, although their fibre content and compositions may differ. Onions are rich mainly in flavonoids, highlighting the presence of quercetin derivatives, such as quercetin 3,4′-*O*-diglucoside and quercetin 4′-*O*-glucoside, in addition to other glycosides of quercetin, kaempferol and isorhamnetin, and anthocyanins [26]. In apples, quercetin (mainly in the skin) is found as a mixture of the derivatives of 3-*O*-galactoside, 3-*O*-glucoside, 3-*O*-rhamnoside, and 3-*O*-rutinoside, in addition to other phenolic compounds, such as the derivatives of hydroxycinnamic acid (chlorogenic acid and caffeic acid) and flavanols, such as (+)-catechin, (−)-epicatechin, and the chalcones phloridzin and phloretin [27]. Animal studies suggest that polyphenol or polyphenol rich foods induce changes within gut microbiota and its metabolic output of trimethylamine *N*-oxide, SCFAs, BAs, and small phenolic acids [28]. Therefore, including these plant-based foods in the diet can provide essential nutrients and contribute to overall gut health. Onions, particularly the outer layers, are rich sources of both soluble and insoluble fibres. Soluble fibre, in the form of fructooligosaccharides (FOSs) and inulin, where the FOSs are one of the most acceptable prebiotic substances together with galactooligosacharides and lactulose [29], while insoluble fibre, such as cellulose and hemicellulose, provide bulk to the stool and help maintain regular bowel movements [30]. Apples, on other hand, are known for their high content of soluble fibre, specifically pectin. Pectin is a complex polysaccharide that has prebiotic properties, selectively stimulating the growth of health-promoting gut bacteria [31]. The fermentation of pectin by gut microbiota results in the production of SCFAs. Additionally, pectin has been shown to bind to BAs in the intestine, reducing their reabsorptions and promoting the synthesis of new BAs from cholesterol, which can help lower circulati cholesterol levels [31,32].

Previously, it has been shown that oxidative stress biomarkers (protein carbon 8-hydroxy-20-deoxyguanosine, and 8-epi-prostaglandin $F_{2\alpha}$), inflammatory and vascu injury biomarkers (PAI-1, TIMP-1, VEGF, sICAM-1, sE-Selectin, and MCP-1), and lep were lower in rats fed the onion and apple diets compared with rats fed the stand diet. In addition, endothelial impairment was partially reversed, and the superox content and gene expression of NLRP3, NFKβ1, and COX2 decreased [33]. González-Pe et al. (2017) [34] reported significant changes in specific primary and secondary BAs both the unconjugated and conjugated forms in rats fed a high-cholesterol diet enrich with onions. In addition, Fernández-Jalao et al. (2021) [35] reported phenolic metabol and SCFA production as well as the modulation of certain bacterial groups, exerted an onion ingredient treated using high pressure processing and subjected to a dynar gastrointestinal digestion and colon fermentation simulator. Based on these results a the extended literature findings regarding the relationships among dietary patterns, g microbiota, BA and SCFA metabolism, and metabolic health, the purpose of the pres study was to investigate the effect of the onion and apple intake in an obesity animal mo on the major types of microbial-derived molecules, such as BAs and SCFAs, linked to development of inflammation and metabolic disturbances to understand the interacti between diet, gut microbiota, and BA and SCFA metabolism.

2. Materials and Methods

2.1. Onion and Apple Powder Preparation

Onion and apple powder preparation is displayed in Supplementary Materials a Methods. Supplementary Table S1 shows the nutritional composition, phytochemi compounds, and antioxidant activity of the onion and apple powders. Analyses w carried out using the methods previously cited in Balderas et al. (2022) [33].

2.2. Animal Model and Experimental Design

The present study was approved by the Spanish Ministry of Economy, Indus and Competitiveness Advisory Committee (project AGL2016-76817-R) and by the Eth Committee of the Complutense University of Madrid (Spain) (reference: PROEX 133/ All experiments were performed in compliance with Directive 2010/63/UE regarding protection of animals used for scientific purposes. All necessary steps were taken to prev any potential animal suffering. Animal model and experimental design are display in Supplementary Materials and Methods [36–40]. Supplementary Table S2 shows composition of the experimental diets [33].

2.3. Faeces Collection and Caecum Sampling

Once a week, the animals were housed in individual metabolic cages for 24 h for fae sample collection. Faeces were stored at −20 °C until analysis. For the microbial analy and SCFA and BA determination, the faeces corresponding to Week 0 (start or baseli and Week 8 (end of the experimental feeding trial) were used. At Week 8 of the feedi trial, in order to avoid interassay variations that could affect the comparison of data fr the different groups, animals in fasting conditions were euthanised by decapitation, tak randomly one animal at a time, from each one of four groups. The caecum was collect and weighed before being frozen in liquid nitrogen and stored at −80 °C until analysis

2.4. Caecum pH and Antioxidant Activity Determination

A portion of the caecal (0.5 g) content was diluted in water immediately after sampli the pH was measured using a microelectrode (Crison micropH2000). 2,20-azinobis ethylbenzothiazoline-6-sulfonic acid) radical cation ($ABTS^{\bullet+}$) scavenging capacity a ferric-reducing antioxidant power (FRAP) in caecum were measured using the methc described in Colina-Coca et al. (2017) [38].

2.5. *Microbial Analysis*

A sample of approximately 1 g of faeces was diluted in 9 mL peptone water (Cultimed-Panreac, San Fernando de Henares, Madrid, Spain) and homogenised with vortex (Vortex model LabDancer S40, VWR; Radnor, PA, USA); in addition, a series of 10-fold dilution (10^{-1}–10^{-11}) was prepared. A given amount of each dilution (1 mL) was plated onto a selective media. The methods for the determination of total aerobic bacteria count, *Enterobacter* spp., β-glucuronidase positive *Escherichia coli*, sulphite-reducing *Clostridium* spp., *Bifidobacterium* spp., and *Lactobacillus* spp. are described in Colina-Coca et al. (2017) [38]. Colony-forming units (CFUs) were defined as colonies measuring at least 1 mm in diameter. Microbial counts were expressed as log CFU/g fresh weight (fw).

2.6. *Short-Chain Fatty Acid Analysis*

2.6.1. Short-Chain Fatty Acid Extraction

Faecal samples were weighed and suspended in 1 mL of 0.5% phosphoric acid and frozen at −20 °C immediately after collection. Once thawed, the faecal suspensions were homogenised with a vortex for about 2 min and centrifuged for 10 min at 13,951× *g*. Supernatants were extracted with 1 mL of butanol and centrifuged for 10 min at 13,951× *g*. Prior to analysis, a 200 µL volume of the organic phase was transferred into a tube, and 100 µL of 4-methyl valeric acid was added as internal standard (394 µM final concentration). Samples were directly injected onto the GC-MS column.

2.6.2. Short-Chain Fatty Acid Identification and Quantification Using Gas Chromatography–Mass Spectrometry (GC–MS)

The GC-MS system consisted of an Agilent 7890A gas chromatograph (Agilent Technologies, Palo Alto, CA, USA) equipped with an automatic injector G2613A Series (Agilent) and coupled with a mass spectrometer (5975C). Column DB-WAXtr (60 m × 0.25 mm × 0.25 µm) and helium as carried gas (1.5 mL/min) were used for the separation. Samples (2 µL) were injected in splitless mode with an injector temperature of 250 °C. The column temperature was initially 50 °C; after 2 min, it was gradually increased 15 °C/min until 150 °C, then increased 5 °C/min until 200 °C. To reach the final temperature of 240 °C, it was increased 15 °C/min. The final temperature was maintained for 20 min (total run time 41.3 min).

Solvent delay was 3.5 min. The detector was operated in electron impact ionization mode (electron energy 70 eV), scanning the 30–250 m/z range. The temperature of the ion source, quadrupole, and interface were 230, 150, and 280 °C, respectively. All samples were analysed in randomised order.

The concentration (µM) of each SCFA was calculated using the linear regression equations ($R^2 \geq 0.99$) from the corresponding curves of standards obtained with six different concentrations. A characteristic single ion was selected for the quantification of each compound. Standards employed were acetic, propionic, isobutyric, butyric, isovaleric, valeric, and caproic acids (Sigma Chemical Co., St. Louis, MO, USA), employing the Enhanced ChemStation G1701DA software and the Nist 08 and HP-Wiley 138 libraries. Results were expressed as µmol/g fresh weight (fw).

2.7. *Bile Acid Analysis*

2.7.1. Bile Acid Extraction

A total of 100 mg of each pool sample was weighed into a Pyrex tube, and 1.5 mL of ethanol was added and mixed using vortex vigorously for 1 min. The samples were subjected to sonication for 2 min and then heated at 80 °C for 30 min. After cooling down to room temperature, the samples were heated at 100 °C for 3 min and centrifuged at 18,894× *g* for 10 min at 4 °C. The supernatants were carefully transferred to the new tubes with 500 µL of ethanol, followed by centrifugation at 18,894× *g* for 10 min at 4 °C. The collected supernatants were dried by evaporation in a SpeedVac concentrator. The dried extracts were resuspended in one millilitre of methanol and transferred into HPLC vi after being filtered through a 0.22 μm pore size filter.

2.7.2. Bile Acid Identification Using High-Performance Liquid Chromatography–Quadrupole Time-of-Flight Mass Spectrometry (HPLC–QTOF MS)

The chromatography was performed on an Agilent 1200 series HPLC (Agilent Te nologies, Waldbronn, Germany) using a C18 Hypersil ODS stainless steel column (250 m × 4.6 mm, 5 μm) (Teknokroma, Barcelona, Spain) at 60 °C. Mobile phases consisted of 5 m ammonium acetate in Milli Q-water (A) and methanol (B). Reverse phase was employed the separation using gradient elution mode, and the separation was carried out in 30 m under the following conditions: 8 min 65% B; 20 min 95% B; 25 min 5% B. The colur was re-equilibrated to starting conditions for 5 min prior to each analysis. Flow rate w 0.8 mL/min, and injection volume was 3 μL.

The system was coupled to hybrid quadrupole–time of flight mass spectrome with JetStream ESI source QTOF-MS (G6530, Agilent Technologies, Waldbronn, German operating in negative ionization mode, following mass spectrometer parameters: capilla voltage set at −3500 V, fragmentor voltage was 200 V, and skimmer voltage was 65 V. T drying gas temperature was set to 300 °C, the sheath gas temperature was set to 250 nebuliser gas flow rate was set to 12 L/min, and the nebuliser pressure was set to 45 ps Data acquisition was collected in centroid mode using a scan rate of 1.01 scan per seco operated in full scan mode from 50 to 1200 m/z. During the analysis, two reference mass 121.0508 m/z and 922.0097 m/z, were continuously measured. All samples were analys in randomised order.

MS/MS collision energy was set at 10, 20, and 40 V. Data acquisition and MS a MS/MS data were processed through MassHunter Workstation software (version B.08. Agilent Technologies, Waldbronn, Germany).

The raw data files from LC-QTOF-MS were visually inspected and processed using t molecular feature extraction (MFE) tool in the MassHunter™ Qualitative Analysis softw package, version B05.00 (Agilent Technologies Inc., Santa Clara, CA, USA). The M algorithm groups ions related by charge state, isotopic distribution, and/or the presence adducts and dimers using the accuracy of the mass measurements.

2.8. Statistical Analysis

The univariate statistical analysis was performed using IBM SPSS Statistics Version 2 (IBM Corp., Armonk, NY, USA). The results were expressed as mean ± standard deviati (SD). Data were analysed using one-way ANOVA. In order to verify the homogeneity the variances, a Levene's test was applied. Tamhane's T2 (equal variances not assumed) Bonferroni (equal variances assumed) post hoc tests were used to determine differen between groups ($p < 0.05$). Mean values within the same group (LC, OC, OO, OA) the start (baseline or Week 0) and the end of the experimental trial (Week 8) for so parameters were statistically tested using Student's t-test ($p < 0.05$).

The multivariate statistical analysis was performed using Unscrambler 11.0 a MetaboAnalyst 5.0 [41] software. In order to evaluate the distribution of samples and id tify the variance between the groups of study, classification using unsupervised princip component analysis (PCA), supervised analysis based on partial least-squares discrimin analysis (PLS-DA), the values of variable importance in the projection (VIP), and heatma were performed.

3. Results

3.1. Prebiotic Effect

Within the obese groups, there was not a significant change in the body weight ga The food efficiency ratio was significantly lower in the OO and OA groups compared w the OC group, which indicates the positive effect of dietary interventions [33]. There w no statistically significant differences when analysing the caecum pH and dry and fresh weight (Table 1).

Table 1. Prebiotic effect parameters in rats fed the lean control/obese control, obese onion 10%, and obese apple 10% diets for 8 weeks.

	LC	OC	OO	OA
ecum				
	7.16 ± 0.39 ^{a}	7.22 ± 0.4 ^{a}	7.34 ± 0.42 ^{a}	7.01 ± 0.29 ^{a}
ight (g fw)	0.55 ± 0.03 ^{a}	0.55 ± 0.03 ^{a}	0.54 ± 0.02 ^{a}	0.54 ± 0.03 ^{a}
TS$^{\bullet+}$ (μmol TE/100 g fw)	3.16 ± 0.06 ^{b}	2.45 ± 0.15 ^{a}	3.23 ± 0.13 ^{b}	3.00 ± 0.17 ^{b}
AP (μmol TE/100 g fw)	2.31 ± 0.06 ^{b}	1.89 ± 0.05 ^{a}	2.30 ± 0.05 ^{b}	2.31 ± 0.06 ^{b}
ces				
—Week 0	7.00 ± 0.57 ^{aA}	6.63 ± 0.39 ^{aA}	6.73 ± 0.41 ^{aA}	6.71 ± 0.28 ^{aA}
—Week 8	7.07 ± 0.39 ^{abA}	6.73 ± 0.28 ^{aA}	7.35 ± 0.43 ^{bB}	7.07 ± 0.37 ^{abB}
ight—Week 0 (g fw)	3.19 ± 0.24 ^{aA}	3.89 ± 0.30 ^{bB}	4.45 ± 0.24 ^{cB}	4.30 ± 0.51 ^{bcB}
ight—Week 8 (g fw)	3.02 ± 0.20 ^{aA}	3.36 ± 0.34 ^{aA}	3.79 ± 0.32 ^{bA}	3.35 ± 0.33 ^{aA}
al aerobic bacteria count (log CFU/g fw)—Week 0	8.76 ± 0.07 ^{aB}	8.92 ± 0.10 ^{bB}	8.97 ± 0.08 ^{bB}	8.98 ± 0.10 ^{bB}
al aerobic bacteria count (log CFU/g fw)—Week 8	6.91 ± 0.42 ^{aA}	8.79 ± 0.02 ^{cA}	8.12 ± 0.06 ^{bA}	7.88 ± 0.28 ^{bA}
erobacter spp. (log CFU/g fw)—Week 0	8.16 ± 0.01 ^{aB}	8.10 ± 0.14 ^{aA}	8.12 ± 0.68 ^{aA}	8.08 ± 0.78 ^{aB}
erobacter spp. (log CFU/g fw)—Week 8	7.48 ± 0.16 ^{abA}	8.06 ± 0.47 ^{cA}	7.84 ± 0.13 ^{bcA}	7.20 ± 0.60 ^{aA}
lucuronidase-positive *Escherichia coli* (log CFU/g fw)—Week 0	8.08 ± 0.65 ^{aB}	8.60 ± 0.41 ^{bB}	8.77 ± 0.10 ^{bB}	8.78 ± 0.18 ^{bA}
lucuronidase-positive *Escherichia coli* (log CFU/g fw)—Week 8	7.52 ± 0.20 ^{aA}	7.99 ± 0.47 ^{bA}	8.44 ± 0.18 ^{cA}	8.61 ± 0.15 ^{cA}
al coliform bacteria count (log CFU/g fw)—Week 0	8.08 ± 0.65 ^{aB}	8.60 ± 0.41 ^{bA}	8.77 ± 0.10 ^{bB}	8.78 ± 0.18 ^{bA}
al coliform bacteria count (log CFU/g fw)—Week 8	7.52 ± 0.20 ^{aA}	7.99 ± 0.47 ^{bA}	8.44 ± 0.18 ^{cA}	8.61 ± 0.15 ^{cA}
fite-reducing *Clostridium* spp. (log CFU/g fw)—Week 0	6.64 ± 0.14 ^{bB}	4.57 ± 0.39 ^{aA}	4.55 ± 0.54 ^{aA}	4.65 ± 0.53 ^{aA}
fite-reducing *Clostridium* spp. (log CFU/g fw)—Week 8	6.23 ± 0.02 ^{aA}	6.61 ± 0.05 ^{bB}	6.42 ± 0.09 ^{abB}	6.57 ± 0.33 ^{bB}
dobacterium spp. (log CFU/g fw)—Week 0	9.09 ± 0.03 ^{bB}	8.49 ± 0.13 ^{aB}	8.50 ± 0.40 ^{aB}	8.51 ± 0.35 ^{aB}
dobacterium spp. (log CFU/g fw)—Week 8	8.32 ± 0.11 ^{cA}	7.88 ± 0.04 ^{abA}	7.77 ± 0.14 ^{aA}	7.99 ± 0.03 ^{bA}
tobacillus spp. (log CFU/g fw)—Week 0	9.07 ± 0.09 ^{bB}	8.58 ± 0.13 ^{aB}	8.53 ± 0.24 ^{aB}	8.55 ± 0.06 ^{aB}
tobacillus spp. (log CFU/g fw)—Week 8	8.35 ± 0.10 ^{dA}	7.67 ± 0.02 ^{aA}	7.85 ± 0.08 ^{bA}	8.11 ± 0.02 ^{cA}

Data are presented as mean ± SD (n = 8/group). LC, lean control: lean Zucker rats fed a standard diet; OC, obese control: obese Zucker rats fed a standard diet; OO, obese onion 10%: obese Zucker rats fed a standard diet containing 10% onion; OA, obese apple 10%: obese Zucker rats fed a standard diet containing 10% apple. Mean values within a row with different superscript small letters were significantly different; $p < 0.05$ (one-way ANOVA and posterior Tamhane's T2 and Bonferroni post hoc tests were used as appropriate; superscript small letters indicate Tamhane's T2 post hoc test). Mean values within a column in the same group (LC, OC, OO, and OA) for the same parameter (in faeces) with different superscript capital letters were significantly different; $p < 0.05$ (Student's t-test). Superscript capital letters indicate that equal variances were not assumed.

At the beginning of the experiment, the pH of the faeces also did not show any significant differences among the four groups. However, the OO (7.35 ± 0.43) and OA (7.07 ± 0.37) groups showed a significant increased faecal pH at the end of experiment compared to the OC group (6.73 ± 0.28).

The antioxidant capacity using the ABTS method in the caecum was higher in the LC group (3.16 ± 0.06 μmol TE/100 g fw) and significantly increased after onion (OO: 3.23 ± 0.13 μmol TE/100 g fw) and apple consumption (OA: 3.00 ± 0.17 μmol TE/100 g fw) compared with the OC group (2.45 ± 0.15 μmol TE/100 g fw). The results reflected the same trend when the antioxidant capacity was measured using the FRAP method (Table 1).

The abundance of gut microbiota was analysed in faeces in order to investigate the changes in microbial species in response to diet. The results in all groups showed that the total aerobic bacteria population decreased at the end of the 8-week period. However, when the groups were analysed individually after 8 weeks of experimental feeding, the total aerobic bacteria count was significantly higher in the OC group (8.79 ± 0.02 log CFU/g fw) compared to the OO (8.12 ± 0.06 log CFU/g fw), OA (7.88 ± 0.28 log CFU/g fw), and LC (6.91 ± 0.42 log CFU/g fw) groups. A similar trend was shown in *Enterobacter* spp. after the 8-week experimental period, with the OC group (8.06 ± 0.47 log CFU/g fw) compared to OO (7.84 ± 0.13 log CFU/g fw), OA (7.20 ± 0.60 log CFU/g fw), and LC (7.48 ± 0.16 log CFU/g fw) groups. Within the obese groups, at the end of the 8-week period, the levels of *Escherichia coli* were significantly higher in the onion- (OO: 8.44 ± 0.18 log CFU/g fw) and apple- (OA: 8.61 ± 0.15 log CFU/g fw) supplemented groups compared with the OC group

(7.99 ± 0.47 log CFU/g fw). At Week 0, *Clostridium* spp. in the OC, OO, and OA grou (4.57 ± 0.39, 4.55 ± 0.54, 4.65 ± 0.53 log CFU/g fw, respectively) were significantly lov than in the LC group (6.64 ± 0.14 log CFU/g fw). At Week 8, *Clostridium* spp. increas in the obese groups (OC: 6.61 ± 0.05 log CFU/g fw, OO: 6.42 ± 0.09 log CFU/g fw, C 6.57 ± 0.33 log CFU/g fw), although no significant differences among the groups w observed. *Bifidobacterium* spp. and *Lactobacillus* spp. at Week 0 showed significantly hig levels in the LC group (9.09 ± 0.03 and 9.07 ± 0.09 log CFU/g fw, respectively) compa with the OC (8.49 ± 0.13 and 8.58 ± 0.13 log CFU/g fw, respectively), OO (8.50 ± 0.40 a 8.53 ± 0.24 log CFU/g fw, respectively), and OA (8.51 ± 0.35 and 8.55 ± 0.06 log CFU fw, respectively) groups. After 8 weeks of the experimental diet, *Bifidobacterium* spp. a *Lactobacillus* spp. followed the same trend as at the beginning of the experiment in the group (8.32 ± 0.11 and 8.35 ± 0.10 log CFU/g fw, respectively) compared with the (group (7.88 ± 0.04 and 7.67 ± 0.02 log CFU/g fw, respectively). On the other hand, general, the groups supplemented with onion and apple showed a higher concentration *Bifidobacterium* spp. and *Lactobacillus* spp. (OO: 7.77 ± 0.14 and 7.85 ± 0.08 log CFU/g respectively) (OA: 7.99 ± 0.03 and 8.11 ± 0.02 log CFU/g fw, respectively) compared w the OC group.

3.2. Faecal Short-Chain Fatty Acid Composition in Response to Dietary Onion and Apple Inta

The faecal SCFA contents are shown in Table 2. The two major SFCAs were acetic a propionic acids.

Table 2. Short-chain fatty acid concentration in faeces from rats fed lean control/obese control, ob onion 10%, and obese apple 10% diets for 8 weeks.

	LC	OC	OO	OA
Acetic acid (μmol/g fw)	46.32 ± 4.54 ^{b}	35.58 ± 1.45 ^{a}	55.01 ± 4.98 ^{c}	45.42 ± 3.92
Propionic acid (μmol/g fw)	5.47 ± 0.75 ^{c}	2.61 ± 0.03 ^{a}	4.40 ± 0.17 ^{b}	4.42 ± 1.02
Isobutyric acid (μmol/g fw)	0.46 ± 0.06 ^{c}	0.32 ± 0.01 ^{a}	0.39 ± 0.04 ^{b}	0.35 ± 0.02
Butyric acid (μmol/g fw)	0.27 ± 0.02 ^{c}	0.13 ± 0.02 ^{b}	0.10 ± 0.02 ^{a}	0.13 ± 0.01
Isovaleric acid (μmol/g fw)	0.31 ± 0.05 ^{c}	0.19 ± 0.05 ^{b}	0.19 ± 0.04 ^{b}	0.12 ± 0.02
Valeric acid (μmol/g fw)	0.11 ± 0.01 ^{a}	0.15 ± 0.02 ^{b}	0.24 ± 0.01 ^{c}	0.37 ± 0.02
Caproic acid (μmol/g fw)	0.13 ± 0.01 ^{b}	0.12 ± 0.004 ^{ab}	0.12 ± 0.01 ^{ab}	0.11 ± 0.01

Data are presented as mean ± SD (n = 8/group). LC, lean control: lean Zucker rats fed a standard diet; obese control: obese Zucker rats fed a standard diet; OO, obese onion 10%: obese Zucker rats fed a standard containing 10% onion; OA, obese apple 10%: obese Zucker rats fed a standard diet containing 10% apple. M values within a row with different superscript small letters were significantly different; $p < 0.05$ (one-way ANC and posterior Tamhane's T2 and Bonferroni post hoc tests were used as appropriate, and small letters indi Tamhane's T2 post hoc test).

After 8 weeks of experimental feeding, the LC group showed greater concentrati in most of the SFCAs in comparison to the OC group, particularly acetic (46.32 ± 4.54 35.58 ± 1.45 μmol/g fw), propionic (5.47 ± 0.75 vs. 2.61 ± 0.03 μmol/g fw), isobuty (0.46 ± 0.06 vs. 0.32 ± 0.01 μmol/g fw), butyric (0.27 ± 0.02 vs. 0.13 ± 0.02 μmol/g f and isovaleric (0.31 ± 0.05 vs. 0.19 ± 0.05 μmol/g fw) acids. There was no signific difference in caproic acid between the LC (0.13 ± 0.01 μmol/g fw) and OC (0.12 ± 0.(μmol/g fw) groups, while valeric acid was lower in the LC group (0.11 ± 0.01 μmol/g compared with the OC group (0.15 ± 0.02 μmol/g fw).

Onion and apple supplementation changed the concentration of SCFAs: ac (55.01 ± 4.98 and 45.42 ± 3.92 μmol/g fw, respectively), propionic (4.40 ± 0.17 a 4.42 ± 1.02 μmol/g fw, respectively), isobutyric (0.39 ± 0.04 and 0.35 ± 0.02 μmol/g respectively), and valeric (0.24 ± 0.01 and 0.37 ± 0.02 μmol/g fw, respectively) acids w significantly higher compared to those in the OC group. There were no significant dif ences in the caproic acid concentration between the OO (0.12 ± 0.01 μmol/g fw) and ((0.11 ± 0.01 μmol/g fw) groups with regard to the OC group (0.12 ± 0.004 μmol/g fw)

3.3. Bile Acid Content in Response to Dietary Onion and Apple Intake

The chromatograms for the analysis were obtained from 32 samples at 0 and 8 weeks of the experimental period and 14 quality controls (QCs). All chromatograms were aligned, and the data were filtered to remove the noise. The samples were categorised as coming from lean or obese at Week 0 and LC, OO, OC, and OA at Week 8. The robustness of the analytical procedure was evaluated using the principal component analysis (PCA) method and was proved by the clustering of QC samples. To evaluate the bile acid level differences between the groups, the data were processed with the QCs removed from the data matrix.

In total, 22 BAs were identified using the LC-MS negative mode and further evaluated. The results are shown in Table 3, with their identification confirmed through their characteristic fragments provided by the databases and the literature and by comparison with the MS/MS spectrum of the commercial analytical standard when it was available.

Table 3. Bile acid detection using HPLC-QTOF MS from faecal samples in rats fed the lean control/obese control, obese onion 10%, and obese apple 10% diets for 8 weeks.

le Acid (Abbreviation)	Formula	RT (min)	Monoisotopic Mass	(M-H)	Fragments
uro-α-muricholic acid (T-α-MCA)	$C_{26}H_{45}NO_7S$	2.88	515.2917	514.284398	514.2897; 496.3058; 479.3058; 358.7687; 80.9621; 65.6705
uroursodeoxycholic acid (TUDCA)	$C_{26}H_{45}NO_6S$	3.27	499.2968	498.289483	498.2880; 479.9107; 465.9335; 393.2521; 159.0796; 96.9606
uro-β-muricholic acid (T-β-MCA)	$C_{26}H_{45}NO_7S$	3.29	515.2917	514.284398	514.2897; 496.3058; 479.3058; 358.7687; 80.9621; 65.6705
muricholic acid (ω-MCA)	$C_{24}H_{40}O_5$	3.35	408.2876	407.280298	407.2799; 391.2347; 373.1968; 345.2663; 179.0411; 59.011
urocholic acid (TCA)	$C_{26}H_{45}NO_7S$	3.37	515.2917	514.284398	514.2897; 496.9256; 479.3105; 357.2196; 80.9621
ycocholic acid (GCA)	$C_{26}H_{43}NO_6$	3.40	465.3090	464.301762	464.3091; 447.1784; 405.2479; 379.2552; 357.1239
Muricholic acid (α-MCA)	$C_{24}H_{40}O_5$	4.00	408.2876	407.280298	407.2791; 389.2682; 373.2247; 345.2770; 60.0348
muricholic acid (β-MCA)	$C_{24}H_{40}O_5$	4.17	408.2876	407.280298	407.2805; 389.2735
urochenodeoxycholic acid (TCDA)	$C_{26}H_{45}NO_6S$	4.35	499.2968	498.289483	498.2881; 480.2670; 466.2540; 388.2578; 374.1424; 80.9626
ocholic acid (HCA)	$C_{24}H_{40}O_5$	4.70	408.2876	407.280298	407.2805; 389.2735
sodeoxycholic acid (UDCA)	$C_{24}H_{40}O_4$	4.85	392.2927	391.285384	391.2526; 373.9815; 357.7511; 329.7645; 221.0703; 59.0103
ycodeoxycholic acid (GDCA)	$C_{26}H_{43}NO_5$	4.93	449.3141	448.306847	448.3165; 433.2967; 407.2899; 389.2625; 329.0834; 74.0232
ycochenodeoxycholic acid (GCDCA)	$C_{26}H_{43}NO_5$	5.16	449.3141	448.306847	448.3165; 433.2967; 407.2899; 389.2625; 329.0834; 74.0232
urodeoxycholic acid (TDCA)	$C_{26}H_{45}NO_6S$	5.33	499.2968	498.289483	498.2881; 480.2670; 466.2540; 388.2578; 374.1424; 80.9626
olic acid (CA)	$C_{24}H_{40}O_5$	5.53	408.2876	407.280298	407.2808; 391.2378; 371.2674; 345.2730; 289.7773; 131.6716

Table 3. *Cont.*

Bile Acid (Abbreviation)	Formula	RT (min)	Monoisotopic Mass	(M-H)	Fragments
Hyodeoxycholic acid (HDCA)	$C_{24}H_{40}O_4$	5.65	392.2927	391.285384	391.2839; 373.2686 345.2772; 327.2668 284.2067; 59.0152
Nutriacholic acid (NCA)	$C_{24}H_{38}O_4$	6.65	390.2770	389.269734	
Taurolithocholic acid (TLCA)	$C_{26}H_{45}NO_5S$	6.75	483.3018	482.294569	482.2910; 464.4037 389.2734; 349.1847 79.9556
Deoxycholic acid (DCA)	$C_{24}H_{40}O_4$	7.66	392.2927	391.285384	391.2905; 355.2557 345.2856; 327.2615 140.6033; 57.0309
Chenodeoxycholic acid (CDCA)	$C_{24}H_{40}O_4$	9.22	392.2927	391.285384	391.2872; 373.9914 345.2780; 329.2756 140.6033; 59.0132
Lithocholic acid (LCA)	$C_{24}H_{40}O_3$	16.33	376.2977	375.290469	375.2906; 357.2796 329.3094; 191.4657 50.0429
Glycolithocholic acid (GLCA)	$C_{26}H_{43}NO_4$	22.69	433.3192	432.311933	433.3379; 405.2467 387.2367; 373.2224 359.1364; 59.0162

Regarding the multivariate data analysis and to discriminate between the grou principal component analysis (PCA) and partial least-squares discriminant analysis (P DA) models were built, using log transformation and autoscaling (mean-centered a divided by the standard deviation of each variable). The ellipse in all PCA score plots is t 95% confidence interval for each model.

During the first step, the bile acid profiles of the lean and obese groups at Week 0 w compared using PCA. For all animals, the first principal component (PC-1) explains 4 of the variance. The lean group is markedly separated from the obese group along PC (Figure 1).

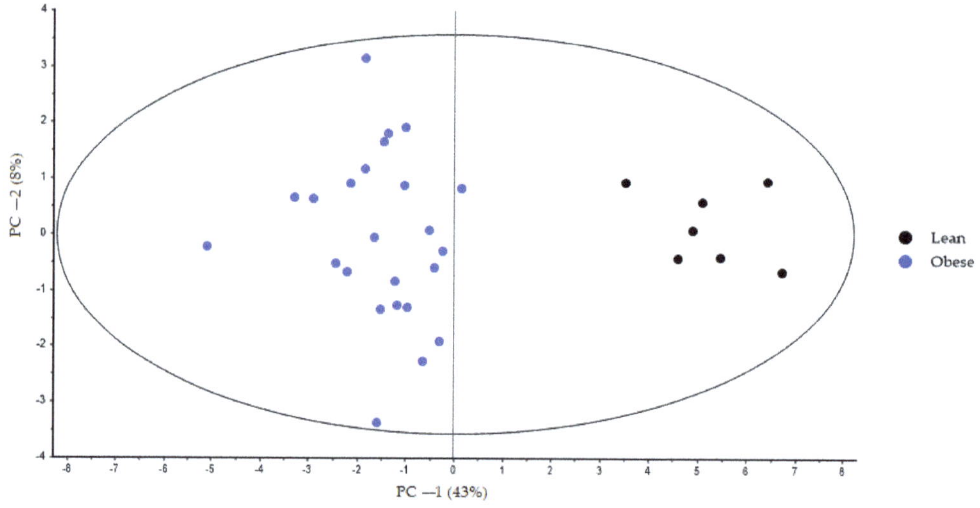

Figure 1. PCA score plot based on bile acid faecal samples at the start of the experimental trial (We 0) from lean (black dots) and obese (blue dots) groups generated using Unscrambler 11.0 softwa All samples were analysed in negative ionization mode. Both groups were fed a standard diet duri the adaptation period. PCA, principal component analysis.

At the end of the experiment, the PCA showed a clear separation between all the groups, and PC-1 explains 38% of the variance, where is possible to see the latent variable 1 separated the LC, OO, and OA groups with respect to the OC group (Figure 2). The second principal component (PC-2) explains 20% of the variance. Variations in these distributions can be explained, in part, by changes in the bile acid profile derived from the diet. These changes can be observed more clearly in (Figure 3), where PC-1 explains 42% of the variance, and the variable latent 1 completely separates the OO and OA groups in relation to the OC group, while the latent variable 2 further separates the two groups fed the supplemented diets along PC-2.

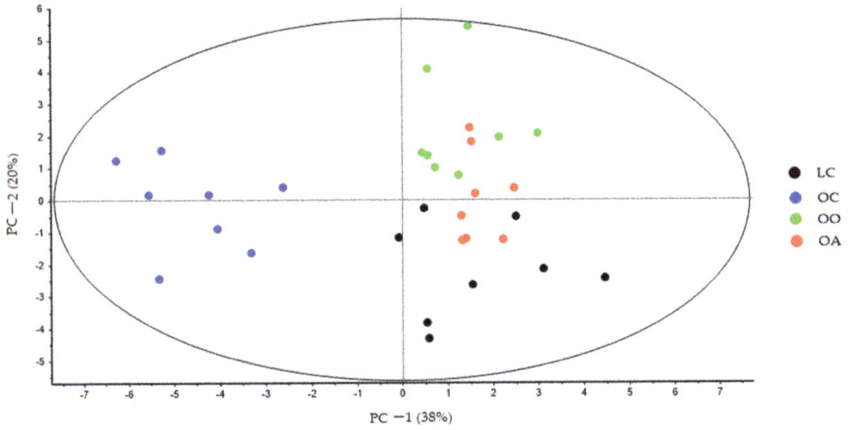

Figure 2. PCA score plot of bile acid faecal samples at the end of the experimental trial (Week 8) from LC (black dots), OC (blue dots), OO (green dots), and OA (red dots) groups generated using Unscrambler 11.0 software. LC, lean control: lean Zucker rats fed a standard diet; OC, obese control: obese Zucker rats fed a standard diet; OO, obese onion 10%: obese Zucker rats fed a standard diet containing 10% onion; OA, obese apple 10%: obese Zucker rats fed a standard diet containing 10% apple. PCA, principal component analysis.

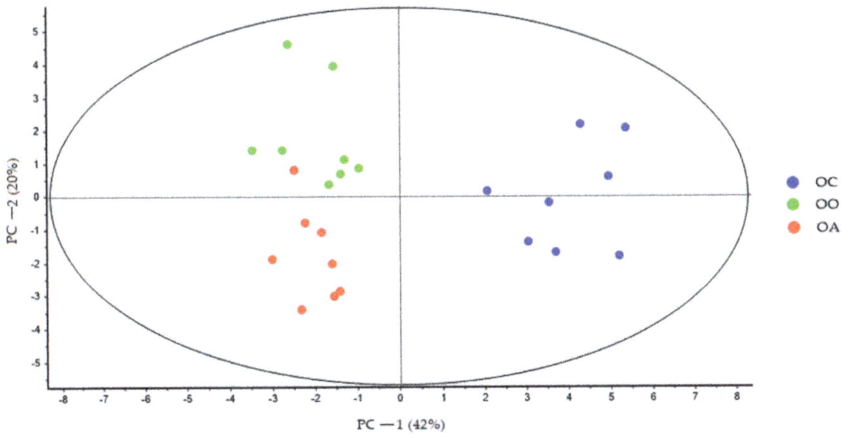

Figure 3. PCA score plot of bile acid faecal samples at the end of the experimental trial (Week 8) from OC (blue dots), OO (green dots), and OA (red dots) groups generated using Unscrambler 11.0 software. OC, obese control: obese Zucker rats fed a standard diet; OO, obese onion 10%: obese Zucker rats fed a standard diet containing 10% onion; OA, obese apple 10%: obese Zucker rats fed a standard diet containing 10% apple. PCA, principal component analysis.

To analyse the metabolites playing a more significant role, four different comparison LC vs. OC; OO vs. OC; OA vs. OC; and OO vs. OA (Figure 4A–D)—were evalua A PLS-DA of the bile acid profiles at Week 8 was carried out to better determine differences among the diet consumption. The variable importance in the projection (V scores obtained from the loading coefficients of the PLS-DA model indicated the metaboli that contributed to the observed group separation.

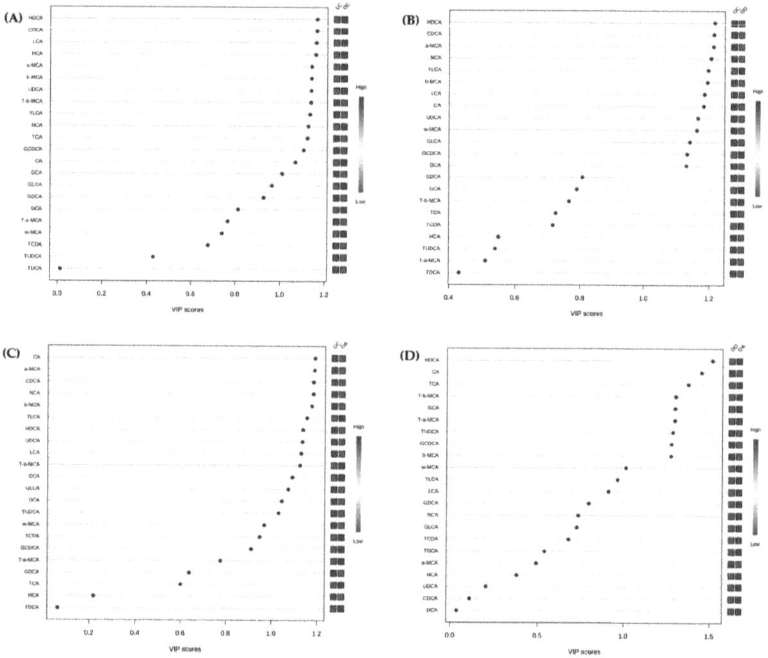

Figure 4. Variable importance in the projection (VIP) scores plots are shown for (**A**) LC and OC; (**B**) and OO; (**C**) OC and OA; and (**D**) OO and OA. VIP score ($p < 0.05$) of bile acids identified accordin PLS-DA for each diet at the end of the experimental trial (Week 8) generated using MetaboAnalyst software after data processing. The coloured boxes on the right indicate the relative concentration each bile acid in each group and are organised by descending order of VIP score. Red colour indica high, and blue colour indicates low. LC, lean control: lean Zucker rats fed a standard diet; OC, ob control: obese Zucker rats fed a standard diet; OO, obese onion 10%: obese Zucker rats fed a stand diet containing 10% onion; OA, obese apple 10%: obese Zucker rats fed a standard diet contain 10% apple. CA, cholic acid; CDCA, chenodeoxycholic acid; DCA, deoxycholic acid; GCA, glycoch acid; GCDCA, glycochenodeoxycholic acid; GDCA, glycodeoxycholic acid; GLCA, glycolithoch acid; HCA, hyocholic acid; HDCA, hyodeoxycholic acid; LCA, lithocholic acid; a-MCA, α-murich acid; b-MCA, β-muricholic acid; w-MCA, ω-muricholic acid; NCA, nutriacholic acid; TCA, tauroch acid; TCDA, taurochenodeoxycholic acid; TDCA, taurodeoxycholic acid; TLCA, taurolithocholic a T-a-MCA, tauro-α-muricholic acid; T-b-MCA, tauro-β-muricholic acid; TUDCA, tauroursodeoxych acid; UDCA, ursodeoxycholic acid.

The BAs of faecal samples showed significant changes, and the percentage of change each feature has been calculated by the difference between the abundances for both samp and then dividing this difference by the abundance for the control group and is express as a percentage. An upward arrow to indicate an increase in percentage and a downwa arrow to indicate a decrease in percentage were used (Table 4). In order to provide intuitive visualization and see how the 22 BAs changed across samples in the differ study groups, a hierarchical clustering heatmap was performed using standardised valu a Pearson distance measure, and a Ward clustering algorithm (Figure 5). The colour shows the relative concentration of each bile acid among the different samples. A red colour means that a sample, for a given bile acid, has a read count that is about 2 standard deviations greater than the mean count of that bile acid in all samples, while the blue colour represents the opposite.

Table 4. Bile acid differences from faecal samples in rats fed the lean control/obese control, obese onion 10%, and obese apple 10% diets for 8 weeks.

					Week 0	Week 8			
e	Abbreviation	Formula	RT (min)	Monoisotopic Mass	% Change Lean vs. Obese	% Change LC vs. OC	% Change OO vs. OC	% Change OA vs. OC	% Change OO vs. OA
nary	CA	$C_{24}H_{40}O_5$	5.53	408.2876	↑	↓	↓	↓	↑
nary	CDCA	$C_{24}H_{40}O_4$	9.22	392.2927	↓	↓	↓	↓	↓
nary (mouse)	α-MCA	$C_{24}H_{40}O_5$	4.00	408.2876	↓	↓	↓	↓	↑
nary (mouse)	β-MCA	$C_{24}H_{40}O_5$	4.17	408.2876	↓	↓	↓	↓	↑
ondary (mouse)	ω-MCA	$C_{24}H_{40}O_5$	3.35	408.2876	↑	↓	↓	↓	↓
ondary	HCA	$C_{24}H_{40}O_5$	4.70	408.2876	↓	↓	↓	↓	↓
ondary	UDCA	$C_{24}H_{40}O_4$	4.85	392.2927	↓	↓	↓	↓	↓
ondary	HDCA	$C_{24}H_{40}O_4$	5.65	392.2927	↓	↓	↓	↓	↓
ondary	NCA	$C_{24}H_{38}O_4$	6.65	390.2770	↓	↓	↓	↓	↓
ondary	DCA	$C_{24}H_{40}O_4$	7.66	392.2927	↓	↓	↓	↓	↓
ondary	LCA	$C_{24}H_{40}O_3$	16.33	376.2977	↓	↓	↓	↓	↓
coconjugated	GCA	$C_{26}H_{43}NO_6$	3.40	465.3090	↑	↑	↑	↑	↓
coconjugated	GDCA	$C_{26}H_{43}NO_5$	4.93	449.3141	↑	↑	↓	↓	↓
coconjugated	GCDCA	$C_{26}H_{43}NO_5$	5.16	449.3141	↑	↑	↑	↑	↑
coconjugated	GLCA	$C_{26}H_{43}NO_4$	22.69	433.3192	↓	↓	↓	↓	↓
roconjugated	TUDCA	$C_{26}H_{45}NO_6S$	3.27	499.2968	↑	↑	↓	↓	↑
roconjugated	TCA	$C_{26}H_{45}NO_7S$	3.37	515.2917	↑	↑	↓	↑	↓
roconjugated	TCDA	$C_{26}H_{45}NO_6S$	4.35	499.2968	↑	↑	↑	↑	↓
roconjugated	TDCA	$C_{26}H_{45}NO_6S$	5.33	499.2968	↑	↑	↓	↑	↓
roconjugated	TLCA	$C_{26}H_{45}NO_5S$	6.75	483.3018	↑	↑	↑	↑	↓
roconjugated (mouse)	T-α-MCA	$C_{26}H_{45}NO_7S$	2.88	515.2917	↑	↑	↓	↑	↓
roconjugated (mouse)	T-β-MCA	$C_{26}H_{45}NO_7S$	3.29	515.2917	↓	↑	↑	↑	↓

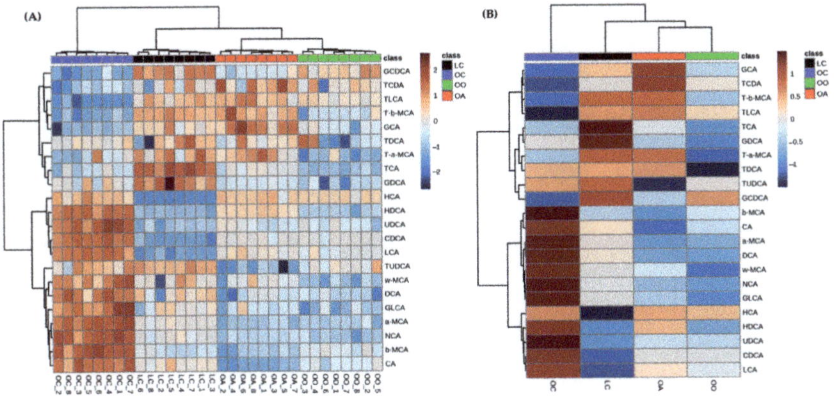

Figure 5. Heatmap of the individual (**A**) and average (**B**) faecal bile acid relative concentration in each group of the study at the end of the experimental trial (Week 8) generated using MetaboAnalyst 5.0 software after data processing. LC, lean control: lean Zucker rats fed a standard diet; OC, obese control: obese Zucker rats fed a standard diet; OO, obese onion 10%: obese Zucker rats fed a standard diet containing 10% onion; OA, obese apple 10%: obese Zucker rats fed a standard diet containing 10% apple. CA, cholic acid; CDCA, chenodeoxycholic acid; DCA, deoxycholic acid; GCA, glycocholic acid; GCDCA, glycochenodeoxycholic acid; GDCA, glycodeoxycholic acid; GLCA, glycolithocholic acid; HCA, hyocholic acid; HDCA, hyodeoxycholic acid; LCA, lithocholic acid; a-MCA, α-muricholic acid; b-MCA, β-muricholic acid; w-MCA, ω-muricholic acid; NCA, nutriacholic acid; TCA, taurocholic acid; TCDA, taurochenodeoxycholic acid; TDCA, taurodeoxycholic acid; TLCA, taurolithocholic acid; T-a-MCA, tauro-α-muricholic acid; T-b-MCA, tauro-β-muricholic acid; TUDCA, tauroursodeoxycholic acid; UDCA, ursodeoxycholic acid.

The sample clustering shows four clear clusters, corresponding to each group of d (LC, OC, OA, and OO). Regarding variable clustering, the primary and secondary BAs clearly influenced by the OC group, while the glycoconjugated and tauroconjugated B seemed to be more related to the supplemented diets (OO and OA).

Certain differences were observed, in the case of GCA, and some of the taurocon gated (TLCA, TCDA, TDCA, T-α-MCA, and T-β-MCA) have relatively high concentratio particularly in the OA group with respect to the OO group, where T-α-MCA, T-β-MC and mainly TDCA have a relatively low concentration in the OO group.

The rest of the BAs, in general, had a relatively low concentration in comparison w the obese control (OC) group. On the other hand, TUDCA exhibited a low concentration the OA group, while the opposite was observed in the groups that consumed the standa diet (LC and OC). HCA showed a relatively low concentration in the LC group compar to the rest of the groups.

4. Discussion

In the present study, as expected, obese Zucker rats were significantly hyperpha compared to the lean controls, resulting in a greater food intake and body weight [3 However, when the food efficiency ratio was calculated in the OO and OA groups, it w significantly lower in comparison to the OC group, suggesting a positive effect of oni and apple dietary interventions. Similar results have been reported by Cho et al. (2013) [and Yoshinari et al. (2012) [43], confirming the body-weight-reducing effect of apples a onions, respectively.

Regarding the prebiotic effect and the impact in the short-chain fatty acid compositi and bile acid content of the dietary interventions, obese rats showed disturbances in t gut microbiota and the levels of SFCAs and BAs. However, some of these changes ha been restored to normal levels after dietary supplementation with onion and apple. T benefits of onion and apple feeding on animals have been reported [33,34,44,45].

The link between the gut microbial composition and/or obesity-related patholog has been and will continue to be investigated by different research groups [46–48]. this context, it is common to use rodents as a model study; for example, the colonizati of germ-free mice with microbiota harvested from conventionally raised mice leads an increase in body fat content and insulin resistance [49]. Another study in genetica obese rats carrying a defective leptin receptor (Zucker fa/fa) suggested that *Bifidobacteri* spp. may have a role in defining an obese or lean phenotype [50]. In the present stu at the beginning of the experiment, *Enterobacter* spp. did not differ between the le and obese groups. The total aerobic bacteria count, *Escherichia coli*, and total colifo bacteria count reflected higher levels in the obese rats compared with the lean rats, opposed to *Clostridium* spp., *Bifidobacterium* spp., and *Lactobacillus* spp. that showed low levels in the obese rats compared with the lean rats. The effect of a high-fat diet on g microbiota variations has been extensively investigated, indicating a possible role in t development of obesity [51]. In addition, some studies have investigated the benefit onion and apple feeding in order to assess the effect of polyphenols on the modulati of intestinal microbiota [34,35,38,52]. In the current study, although in all obese grou at the beginning of the study, the total aerobic bacteria and *Escherichia coli* levels we higher in comparison with the LC group, the onion/apple feeding for 8 weeks cause significant reduction in both. In this sense, it is important to take into consideration th it has been demonstrated that increased levels of Gram-negative bacteria, which cou include *E. coli*, could be related to high plasma endotoxin concentrations and inflammat markers associated with obesity, as evidenced in animal models [53,54] and humans [55,5 In addition, although at the start of the feeding, *Enterobacter* spp. did not differ betwe the lean and obese groups, after 8 weeks of experimental feeding, OC rats showed high levels in comparison with LC rats, and the OO and OA groups showed reduced lev compared with the OC group. Our results are consistent with previous studies that ha reported an increase in Enterobacteriales in diet-induced obese rats [57]. Another stu focused on specific species, such as *Enterobacter cloacae*, in germfree mice induce obesity and insulin resistance [58]. Moreover, after administration of polyphenol resveratrol in a dextran sulfate sodium-induced colitis rat model, increased *Bifidobacterium* and *Lactobacillus* counts associated with a decrease in enterobacteria proliferation [59] were shown. In addition, O'Connor et al. (2019) [60] reported that ingestion of components from cranberry modulates the microbiota in a manner that may be beneficial by enriching *Bacteroidaceae* and reducing *Enterobacteriaceae* levels in the gut microbiota. Diet is one of the factors that can positively modulate intestinal microbiota composition. Particularly, polyphenols may exert anti-inflammatory, antioxidant, anticancer, and antidiabetic activities by positively modulating the gut microbiota. Polyphenols have been shown to positively modulate the bacterial component, increasing *Lactiplantibacillus* spp. and *Bifidobacterium* spp. involved in the protection of the intestinal barrier [61]. Hsu et al. (2020) [62] reported that perinatal resveratrol therapy in female Sprague–Dawley rats increased the abundance of *Lactobacillus* and *Bifidobacterium* in male adult offspring. The intestinal microbiota plays a fundamental role in improving their absorption, as intestinal bacteria seem to maintain a reciprocal relationship with them, increasing their bioavailability.

Short-chain fatty acids (SCFAs), which mainly include acetate, propionate, and butyrate, are the major metabolic products of fermentation by colonic intestinal bacteria through dietary fibre in the intestine and some amino acids [63]. Therefore, the presence of dietary fibre in the lumen reflects the activity of the bacterial flora, and SCFA receptors possibly monitor the bacteria for host defence [64]. Some studies suggest that both acetic acid and butyric acid are capable of binding to the anti-inflammatory receptor GPR43, greatly inhibiting the inflammatory response [65,66]. The evidence suggests that increased dietary fibre consumption can positively influence metabolic health by altering gut microbiota. The increased abundance of fermenting species may increase the production and availability of SCFAs, which, in turn, can reduce hyperlipidaemia, hyperglycaemia, hyperinsulinemia, and hypercholesterolemia in a wide range of cohorts, both healthy and metabolically challenged [67]. In this line, improved lipid profile and glucose metabolism, reduced biomarkers of hepatic injury, improved oxidative stress, inflammatory and vascular injury biomarkers, metabolic hormones, endothelial function, and a decreased proinflammatory gene expression of NLRP3, NFKβ1, and COX2 have been previously reported [33]. The microbial activity in the large intestine is generally increased by diets containing a high soluble fibre content. In the present study, the positive effects on the increase in the concentration of some SCFAs, such as acetic, propionic, isobutyric, and valeric acids in the obese groups that consumed the diet enriched with onion (OO) and apple (OA) compared to the obese control (OC) group, indicate the stimulation of the microbiota was probably caused by the fibre present in the onion and apple. In the same line as our findings, this effect has also been observed in other studies, which show an increase in microbial activity resulting in a higher production of SCFAs with a diet rich in apple fibre. Sembries et al. (2006) [68] reported a significant increase in acetic acid in the caecal content of rats fed with apple extraction juices for 4 weeks compared to the control group. At the same time, Aprikian et al. (2001) [69] reported a significant increase in butyrate in the caecal content of rats compared to controls, after the ingestion of freeze-dried pectin. Ravn-Haren et al. (2018) [70] showed a tendency toward an increased concentration of caecal acetate and significantly increased caecal butyrate concentration in rats fed with apple pomace for 4 weeks. Furthermore, Roldán-Marín et al. (2009) [39] showed a significant increase in the formation of caecal propionate and butyrate in rats fed different onion by-products for 4 weeks. In addition, Martinez et al. (2023) [71] showed that the intake of sorghum flour increased the propionic acid in male Wistar rats fed with a high-fat high-fructose diet and modulated the gut microbiota composition and the abundance of SCFA-producing bacteria. On the other hand, in the present study, it should be mentioned that low concentrations of caproic acid were also observed, without any significant difference between the groups, as a result of bacterial degradation of certain amino acids.

Although mouse models have long been used to study essential aspects of bile a metabolism, compared with humans, there is a clear species difference in the composit of BAs [72]. Human primary BAs are cholic acid (CA) and chenodeoxycholic acid (CDC While CDCA is abundant in human bile, it is not in mice because CDCA is converted muricholic acid (MCA); therefore, the primary BAs in rodents are α- and β-murich acids (α-MCA, β-MCA). Previous studies suggest that despite UDCA being a very mi constituent in the total bile acid pool in mice, it could be also an apparent biosynth precursor to β-MCA by a mechanism that remains unclear [73]. Therefore, these differen are an essential consideration when trying to extrapolate the findings in mouse BAs humans. Interestingly, in the present study, CDCA and some of the secondary BAs (H and UDCA) resulted in decreased excretion in the LC group in comparison with the group. The diet enriched with onion and apple also resulted in a decreased excretion some BAs, mainly of the primary (CA, α-MCA, and β-MCA) and secondary type (ω-M HDCA, NCA, DCA, and LCA). Some studies in animal models suggest that the hig fat content of an animal-based diet increases bile secretion, augmenting the concentrat of hydrophobic BAs in the intestine. Mainly CDCA leads to a decrease in *Lactobaci* and an increase in *Clostridium* subcluster XIVa, which generates a persistent degree chronic inflammation and tumour progression [74]. On the other hand, many studies ha generally shown the negative impact of BAs on the integrity of the membrane of intesti bacteria due to a detergent effect, which causes a decrease in the production of membra proteins and oxidative/nitrosative stress [75]. This triggers an increase in permeabi and cell death [76], as Gram-negative bacteria is more resistant to BAs than Gram-posit bacteria [77]. Consistent with these studies, Balderas et al. (2022) [33] found metab benefits accompanied by the systemic improvement of oxidative stress, inflammation, a vascular injury biomarkers in rats fed onion- and apple-enriched diets compared w obese control rats. It is important to highlight that, in the present study, obese rats ha shown higher levels of primary and secondary BAs and lower levels of *Lactibacillus* s It was reported that the proportion of CDCA and CA in faeces was associated with colon transit [78]. When the hepatic BA synthesis increases and the colon is exposed more BAs, they will promote fluid and electrolyte secretion. These alterations may expl the mechanisms underlying some cases of BA diarrhoea [79]. The alteration to the microbiota ecosystem is characterised by a reduced amount of probiotic bacteria and primarily determined by intrinsic and extrinsic factors, which can give rise to diseases only in the gastrointestinal tract but also at a systemic level.

In general, BAs (including DCA, LCA, NCA, HDCA, and ω-MCA) were increas in the OC group; all of them are secondary BAs. It is well known that the secondary B are derived from intestinal microbial biotransforming; in this context, different *Clostridi* species have been identified for their ability to transform primary BAs into secondary B in obesity models due to dysbiosis of the gut microbiota [80]. In this respect, our resu in terms of the *Clostridium* spp. levels are closely aligned with this existing knowled Generally, the hydrophobic bile acids LCA and DCA are cytotoxic, and the hydrophilic b acid UDCA and its derivative tauroursodeoxycholic acid (TUDCA) are cytoprotective [The more hydrophobic the BAs are, only a small amount will be reabsorbed back into enterohepatic circulation. Consequently, the amount of that BA in the faeces is higher. previous studies, some secondary BAs, especially DCA, when they are present in high lev in the enterohepatic circulation, are associated with the contributions of colorectal can gallstones, cirrhosis, and other gastrointestinal diseases [82–84]. Such is the case in a stu by Yoshimoto et al. (2013) [84] that showed that the gut bacterial metabolite DCA promo the development of obesity-associated hepatocellular carcinoma, which was induced w 7,12-dimethylbenzanthracene (DMBA) in a mouse model. In humans, high faecal excret of DCA and LCA has been previously linked with the incidence of colorectal cancer afte high-fat diet intake [85]. Animal experiments also revealed that the secondary BAs prom tumorigenesis compared to the primary BAs. Therefore, the relation between second BAs/primary BAs, for example, the DCA/CA ratio, helps determine cancer risk [86]. It h been demonstrated that the production of DCA is via the catalysis of the 7α-dehydroxylase that is present in bacteria, including species of *Eubacterium* and *Clostridium* genus. Therefore, the observed increase in faecal DCA in OC rats could be related to the enhanced population of 7α-dehydroxylase containing bacteria [6].

In the liver, secondary BAs are also conjugated with glycine and taurine, producing GCA, GDCA, GCDCA, GLCA, TCA, TUDCA, TCDA, TDCA, TLCA, T-α-MCA, and T-β-MCA. In the present study, the levels of conjugated BAs in faeces were decreased in the OC group, specifically, the levels of TUDCA, T-α-MCA, and GLCA. The conjugated BAs act as an emulsifier, aiding the absorption of ingested fat, which explains the reduced levels of conjugated BAs in faeces. In addition, the decreasing levels of conjugated BAs are contributed by gut microbiota that is capable of the hydrolysis of BAs, producing a range of secondary BAs [87,88]. Zheng et al. (2017) [89] found a correlative relationship between BAs and the high-fat diet-induced alteration of gut microbiota in obese mice with increased concentrations of conjugated BAs TCA, TCDCA, T-α-MCA, T-β-MCA, GCA, and GCDCA. In the present study, onion and apple diets had considerable effects on the concentration of both conjugated and unconjugated bile acids by increasing, in general, the amount of taurine- and glycine-conjugated BAs and decreasing the concentration mainly of the primary and secondary BAs. Consistent with the findings in the present study, González-Peña et al. (2017) [34] previously reported an increase in CA, MCA, CDCA, and UDCA concentrations in rats fed a high-cholesterol diet significantly prevented in rats fed the high-cholesterol enriched with onion diet. In addition, rats fed the high-cholesterol enriched with onion diet showed a tendency to augment the concentration of conjugated forms compared with the rats fed a high-cholesterol diet. Kumari and Augusti (2007) [90] studied the lipid-lowering action of *S*-methyl cysteine sulfoxide isolated from *Allium cepa* Linn in Sprague–Dawley rats fed a 1% cholesterol diet, reporting an increase in the excretion of BAs and sterols in rats treated with *S*-methyl cysteine sulfoxide attributed to a reduction in endogenous lipogenesis, an increased catabolism of lipids, and the subsequent excretion of metabolic by-products through the intestinal tract. In the same line, Hall et al. (2020) [91] reported an increased tauroursodeoxycholate in cats after consuming apple pomace compared with short-chain fructooligosaccharides, and Hosoyamada and Yamada (2017) [92] showed and increased excretion of BAs into the faeces of rats fed apple polyphenol, fish oil, or fish oil and apple polyphenol for 4 weeks, suggesting the beneficial effect of polyphenols.

5. Conclusions

This study provides new insights that onion and apple intervention has positive effects in obese Zucker (fa/fa) rats through a prebiotic effect, an increase in the concentration of some SCFAs, such as acetic, propionic, isobutyric, and valeric acids, a decrease in the excretion of some BAs, mainly of the primary (CA, α-MCA, and β-MCA) and secondary type (ω-MCA, HDCA, NCA, DCA, and LCA), and an increase in conjugated BAs by enhancing, in general, the amount of taurine- and glycine-conjugated BAs. Future research is needed into the contribution of specific bioactive compounds and their metabolites in the regulation of microbiome and the pathways linked to SCFA and BA formation and their relationship with some diseases. Solid insights into these underlying mechanisms to pave the way for preventive strategies targeting metabolites and regulation pathways are needed.

Supplementary Materials: The following supporting information can be downloaded at: https://www.mdpi.com/article/10.3390/nu15133035/s1, Supplementary Materials and Methods S2.1. Onion and Apple Powder Preparation, Table S1: Nutritional composition, phytochemical compounds, and antioxidant activity of onion and apple powder; S2.2. Animal Model and Experimental Design, Table S2: Composition of the experimental diets.

Author Contributions: Conceptualization, C.S.-M. and B.d.A.; methodology, C.B., B.d.A. and C.S.-M.; validation, C.B. and C.S.-M.; formal analysis, C.B., B.d.A. and C.S.-M.; investigation, C.B., B.d.A. and C.S.-M.; resources, C.S.-M. and B.d.A.; data curation, B.d.A. and C.S.-M.; writing—original draft preparation, C.B. and C.S.-M.; writing—review and editing, C.B., B.d.A. and C.S.-M.; visualization, C.S.-M. and B.d.A.; supervision, C.S.-M. and B.d.A.; project administration, C.S.-M. and B.d.A.; funding acquisition, C.S.-M. and B.d.A. All authors have read and agreed to the published version of the manuscript.

Funding: This work was supported by the grants AGL2016-76817-R funded by MCIN/AEI/10.13039/501100011033, by "ERDF A way of making Europe", and PID2019-107980RB-I00 funded by MCIN/AEI/10.13039/501100011033.

Institutional Review Board Statement: All the experiments were conducted according to the European Union (2010/63/EU) and Spanish (RD 53/2013) legislation and with the approval of the Animal Care and Use Committee of the Community of Madrid (PROEX 133/16).

Informed Consent Statement: Not applicable.

Data Availability Statement: Data are contained within the article.

Acknowledgments: The authors would like to thank the Analytical, Instrumental, and Microbiological Techniques Unit of ICTAN. We would also like to thank Lucía Giménez for her excellent laboratory technical assistance.

Conflicts of Interest: The authors declare no conflict of interest.

References

1. Chiang, J.Y. Regulation of bile acid synthesis. *Front. Biosci.* **1998**, *3*, 176–193. [CrossRef]
2. Meier, P.J.; Stieger, B. Bile salt transporters. *Annu. Rev. Physiol.* **2002**, *64*, 635–661. [CrossRef]
3. Russell, D.W. The enzymes, regulation, and genetics of bile acid synthesis. *Annu. Rev. Biochem.* **2003**, *72*, 137–174. [CrossRef]
4. Li, J.; Dawson, P.A. Animal models to study bile acid metabolism. *Biochim. Biophys. Acta Mol. Basis Dis.* **2019**, *1865*, 895–9 [CrossRef]
5. Hofmann, A.F. The continuing importance of bile acids in liver and intestinal disease. *Arch. Intern. Med.* **1999**, *159*, 2647–26 [CrossRef]
6. Lin, H.; An, Y.; Tang, H.; Wang, Y. Alterations of bile acids and gut microbiota in obesity induced by high fat diet in rat mode *Agric. Food Chem.* **2019**, *67*, 3624–3632. [CrossRef]
7. Winston, J.A.; Theriot, C.M. Diversification of host bile acids by members of te gut microbiota. *Gut Microbes* **2020**, *11*, 158–1 [CrossRef]
8. Li, R.; Andreu-Sánchez, S.; Kuipers, F.; Fu, J. Gut microbiome and bile acids in obesity-related diseases. *Best Pract. Res. C Endocrinol. Metab.* **2021**, *35*, 101493. [CrossRef]
9. Murphy, K.; O'Donovan, A.N.; Caplice, N.M.; Ross, R.P.; Stanton, C. Exploring the gut microbiota and cardiovascular dise *Metabolites* **2021**, *11*, 493. [CrossRef]
10. Islam, K.B.; Fukiya, S.; Hagio, M.; Fujii, N.; Ishizuka, S.; Ooka, T.; Yokota, A. Bile acid is a host factor that regulates the composit of the cecal microbiota in rats. *Gastroenterology* **2011**, *141*, 1773–1781. [CrossRef]
11. Yang, J.F.; Palmiotti, A.; Kuipers, F. Emerging roles of bile acids in control of intestinal functions. *Curr. Opin. Clin. Nutr. Met Care* **2021**, *24*, 127–133. [CrossRef]
12. Rinninella, E.; Raoul, P.; Cintoni, M.; Franceschi, F.; Miggiano, G.A.D.; Gasbarrini, A.; Mele, M.C. What is the healthy microbiota composition? A changing ecosystem across age, environment, diet, and diseases. *Microorganisms* **2019**, *7*, [CrossRef]
13. Wagenaar, C.A.; van de Put, M.; Bisschops, M.; Walrabenstein, W.; de Jonge, C.S.; Herrema, H.; van Schaardenburg, D. The Eff of dietary interventions on chronic inflammatory diseases in relation to the microbiome: A systematic review. *Nutrients* **20** *13*, 3208. [CrossRef]
14. Illiano, P.; Brambilla, R.; Parolini, C. The mutual interplay of gut microbiota, diet and human disease. *FEBS J.* **2020**, *287*, 833–8 [CrossRef]
15. Ghaffarzadegan, T.; Essen, S.; Verbrugghe, P.; Marungruang, N.; Hallenius, F.F.; Nyman, M.; Sandahl, M. Determination of f and conjugated bile acids in serum of Apoe(−/−) mice fed different lingonberry fractions by UHPLC-MS. *Sci. Rep.* **2019**, *9*, 38 [CrossRef]
16. Singh, R.K.; Chang, H.W.; Yan, D.; Lee, K.M.; Ucmak, D.; Wong, K.; Liao, W. Influence of diet on the gut microbiome a implications for human health. *J. Transl. Med.* **2017**, *15*, 73. [CrossRef]
17. Cummings, J.H.; Pomare, E.W.; Branch, W.J.; Naylor, C.P.; Macfarlane, G.T. Short chain fatty acids in human large intestine, por hepatic and venous blood. *Gut* **1987**, *28*, 1221–1227. [CrossRef]

Ramos Meyers, G.; Samouda, H.; Bohn, T. Short chain fatty acid metabolism in relation to gut microbiota and genetic variability. *Nutrients* **2022**, *14*, 5361. [CrossRef]
Kim, K.N.; Yao, Y.; Ju, S.Y. Short chain fatty acids and fecal microbiota abundance in humans with obesity: A systematic review and meta-analysis. *Nutrients* **2019**, *11*, 2512. [CrossRef]
O'Connor, K.M.; Lucking, E.F.; Bastiaanssen, T.F.S.; Peterson, V.L.; Crispie, F.; Cotter, P.D.; Clarke, G.; Cryan, J.F.; O'Halloran, K.D. Prebiotic administration modulates gut microbiota and faecal short-chain fatty acid concentrations but does not prevent chronic intermittent hypoxia-induced apnoea and hypertension in adult rats. *EBioMedicine* **2020**, *59*, 102968. [CrossRef]
Hsu, C.-N.; Chan, J.Y.H.; Wu, K.L.H.; Yu, H.-R.; Lee, W.-C.; Hou, C.-Y.; Tain, Y.-L. Altered gut microbiota and its metabolites in hypertension of developmental origins: Exploring differences between fructose and antibiotics exposure. *Int. J. Mol. Sci.* **2021**, *22*, 2674. [CrossRef] [PubMed]
Lai, W.-D.; Tung, T.-H.; Teng, C.-Y.; Chang, C.-H.; Chen, Y.-C.; Huang, H.-Y.; Lee, H.-C.; Huang, S.-Y. Fish oil ameliorates neuropsychiatric behaviors and gut dysbiosis by elevating selected microbiota–derived metabolites and tissue tight junctions in rats under chronic sleep deprivation. *Food Funct.* **2022**, *13*, 2662. [CrossRef] [PubMed]
Yi, Z.Y.; Chen, L.; Wang, Y.; He, D.; Zhao, D.; Zhang, S.H.; Yu, R.; Huang, J.H. The potential mechanism of Liu–Wei–Di–Huang Pills in treatment of type 2 diabetic mellitus: From gut microbiota to short-chain fatty acids metabolism. *Acta Diabetol.* **2022**, *59*, 1295–1308. [CrossRef]
Zhang, C.; Ma, S.; Wu, J.; Luo, L.; Qiao, S.; Li, R.; Xu, W.; Wang, N.; Zhao, B.; Wang, X.; et al. A specific gut microbiota and metabolomic profiles shifts related to antidiabetic action: The similar and complementary antidiabetic properties of type 3 resistant starch from *Canna edulis* and metformin. *Pharmacol. Res.* **2020**, *159*, 104985. [CrossRef]
Thipart, K.; Gruneck, L.; Phunikhom, K.; Sharpton, T.J.; Sattayasai, J.; Popluechai, S. Dark-purple rice extract modulates gut microbiota composition in acetic acid- and indomethacin-induced inflammatory bowel disease in rats. *Int. Microbiol.* **2023**, *26*, 423–434. [CrossRef]
Lee, J.; Mitchell, A.E. Quercetin and isorhamnetin glycosides in onion (*Allium cepa* L.): Varietal comparison, physical distribution, coproduct evaluation, and long-term storage stability. *J. Agric. Food Chem.* **2011**, *59*, 857–863. [CrossRef]
Bouayed, J.; Deußer, H.; Hoffmann, L.; Bohn, T. Bioaccessible and dialysable polyphenols in selected apple varieties following in vitro digestion vs. their native patterns. *Food Chem.* **2012**, *131*, 1466–1472. [CrossRef]
Diotallevi, C.; Fava, F.; Gobbetti, M.; Tuohy, K. Healthy dietary patterns to reduce obesity-related metabolic disease: Polyphenol-microbiome interactions unifying health effects across geography. *Curr. Opin. Clin. Nutr. Metab. Care* **2020**, *23*, 437–444. [CrossRef]
Aisara, J.; Wongputtisin, P.; Deejing, S.; Maneewong, C.; Unban, K.; Khanongnuch, C.; Kanpiengjai, A. Potential of inulin-fructooligosaccharides extract produced from red onion. *Plants* **2021**, *10*, 2401. [CrossRef]
Dhingra, D.; Michael, M.; Rajput, H.; Patil, R.T. Dietary fibre in foods: A review. *J. Food Sci. Technol.* **2012**, *49*, 255–266. [CrossRef]
Pascale, N.; Gu, F.; Larsen, N.; Jespersen, L.; Respondek, F. The potential of pectins to modulate the human gut microbiota evaluated by in vitro fermentation: A systematic review. *Nutrients* **2022**, *14*, 3629. [CrossRef]
Fang, W.; Zhang, L.; Meng, Q.; Wu, W.; Lee, Y.K.; Xie, J.; Zhang, H. Effects of dietary pectin on the profile and transport of intestinal bile acids in young pigs. *J. Anim. Sci.* **2018**, *96*, 4743–4754. [CrossRef]
Balderas, C.; Angulo, J.; Sevilleja-Ortiz, A.; Peiró, C.; Vallejo, S.; Dongil, P.; de Ancos, B.; Sánchez-Moreno, C. Onion and apple functional ingredients intake improves antioxidant and inflammatory status and vascular injury in obese Zucker rats. *Antioxidants* **2022**, *11*, 1953. [CrossRef]
González-Peña, D.; Giménez, L.; de Ancos, B.; Sánchez-Moreno, C. Role of dietary onion in modifying the faecal bile acid content in rats fed a high-cholesterol diet. *Food Funct.* **2017**, *8*, 2184–2192. [CrossRef] [PubMed]
Fernández-Jalao, I.; Balderas, C.; Calvo, M.V.; Fontecha, J.; Sánchez-Moreno, C.; de Ancos, B. Impact of high-pressure processed onion on colonic metabolism using a dynamic gastrointestinal digestion simulator. *Metabolites* **2021**, *11*, 262. [CrossRef] [PubMed]
Reeves, P.G. Components of the AIN-93 diets as improvements in the AIN-76A diet. *J. Nutr.* **1997**, *127*, 838S–841S. [CrossRef]
Reagan-Shaw, S.; Nihal, M.; Ahmad, N. Dose translation from animal to human studies revisited. *FASEB J.* **2008**, *22*, 659–661. [CrossRef]
Colina-Coca, C.; González-Peña, D.; de Ancos, B.; Sánchez-Moreno, C. Dietary onion ameliorates antioxidant defence, inflammatory response, and cardiovascular risk biomarkers in hypercholesterolemic Wistar rats. *J. Funct. Food* **2017**, *36*, 300–309. [CrossRef]
Roldán-Marín, E.; Krath, B.N.; Poulsen, M.; Binderup, M.L.; Nielsen, T.H.; Hansen, M.; Langkilde, S.; Cano, M.P.; Sánchez-Moreno, C.; Dragsted, L.O. Effects of an onion by-product on bioactivity and safety markers in healthy rats. *Br. J. Nutr.* **2009**, *102*, 1574–1582. [CrossRef]
Roldán-Marín, E.; Jensen, R.I.; Krath, B.N.; Kristensen, M.; Poulsen, M.; Cano, M.P.; Sánchez-Moreno, C.; Dragsted, L.O. An onion byproduct affects plasma lipids in healthy rats. *J. Agric. Food Chem.* **2010**, *58*, 5308–5314. [CrossRef]
Chong, J.; Wishart, D.S.; Xia, J. Using MetaboAnalyst 4.0 for comprehensive and integrative metabolomics data analysis. *Curr. Protoc. Bioinform.* **2019**, *68*, e86. [CrossRef]
Cho, K.D.; Han, C.K.; Lee, B.H. Loss of body weight and fat and improved lipid profiles in obese rats fed apple pomace or apple juice concentrate. *J. Med. Food* **2013**, *16*, 823–830. [CrossRef]

43. Yoshinari, O.; Shiojima, Y.; Igarashi, K. Anti-obesity effects of onion extract in Zucker diabetic fatty rats. *Nutrients* **2012** 1518–1526. [CrossRef]
44. Boyer, J.; Liu, R.H. Apple phytochemicals and their health benefits. *Nutr. J.* **2004**, *3*, 5. [CrossRef]
45. Denis, M.C.; Roy, D.; Yeganeh, P.R.; Desjardins, Y.; Varin, T.; Haddad, N.; Amre, D.; Sane, A.T.; Garofalo, C.; Furtos, A.; e Apple peel polyphenols: A key player in the prevention and treatment of experimental inflammatory bowel disease. *Clin.* **2016**, *130*, 2217–2237. [CrossRef]
46. Tilg, H. Obesity, metabolic syndrome, and microbiota: Multiple interactions. *J. Clin. Gastroenterol.* **2010**, *44* (Suppl. S1), S16– [CrossRef]
47. Abenavoli, L.; Scarpellini, E.; Colica, C.; Boccuto, L.; Salehi, B.; Sharifi-Rad, J.; Aiello, V.; Romano, B.; De Lorenzo, A.; Izzo, A et al. Gut microbiota and obesity: A role for probiotics. *Nutrients* **2019**, *11*, 2690. [CrossRef]
48. Cao, W.; Chin, Y.; Chen, X.; Mi, Y.; Xue, C.; Wang, Y.; Tang, Q. The role of gut microbiota in the resistance to obesity in mice fe high fat diet. *Int. J. Food Sci. Nutr.* **2020**, *71*, 453–463. [CrossRef]
49. Backhed, F.; Manchester, J.K.; Semenkovich, C.F.; Gordon, J.I. Mechanisms underlying the resistance to diet-induced obesity germ-free mice. *Proc. Natl. Acad. Sci. USA* **2007**, *104*, 979–984. [CrossRef]
50. Waldram, A.; Holmes, E.; Wang, Y.; Rantalainen, M.; Wilson, I.D.; Tuohy, K.M.; McCartney, A.L.; Gibson, G.R.; Nicholson. Top-down systems biology modeling of host metabotype-microbiome associations in obese rodents. *J. Proteome Res.* **2009** 2361–2375. [CrossRef]
51. Turnbaugh, P.J.; Ley, R.E.; Mahowald, M.A.; Magrini, V.; Mardis, E.R.; Gordon, J.I. An obesity-associated gut microbiome w increased capacity for energy harvest. *Nature* **2006**, *444*, 1027–1031. [CrossRef]
52. Molinari, R.; Merendino, N.; Costantini, L. Polyphenols as modulators of pre-established gut microbiota dysbiosis: State-of-the- *Biofactors* **2022**, *48*, 255–273. [CrossRef]
53. Cani, P.D.; Neyrinck, A.M.; Fava, F.; Knauf, C.; Burcelin, R.G.; Tuohy, K.M.; Gibson, G.R.; Delzenne, N.M. Selective increa of bifidobacteria in gut microflora improve high-fat-diet-induced diabetes in mice through a mechanism associated w endotoxaemia. *Diabetologia* **2007**, *50*, 2374–2383. [CrossRef]
54. Ramos-Romero, S.; Hereu, M.; Atienza, L.; Casas, J.; Jauregui, O.; Amezqueta, S.; Dasilva, G.; Medina, I.; Nogues, M.R.; Rom M.; et al. Mechanistically different effects of fat and sugar on insulin resistance, hypertension, and gut microbiota in rats. *Am Physiol.-Endocrinol. Metab.* **2018**, *314*, E552–E563. [CrossRef]
55. Santacruz, A.; Collado, M.C.; García-Valdés, L.; Segura, M.T.; Martín-Lagos, J.A.; Anjos, T.; Martí-Romero, M.; Lopez, R. Florido, J.; Campoy, C.; et al. Gut microbiota composition is associated with body weight, weight gain and biochemical paramet in pregnant women. *Br. J. Nutr.* **2010**, *104*, 83–92. [CrossRef]
56. Gao, X.; Jia, R.; Xie, L.; Kuang, L.; Feng, L.; Wan, C. Obesity in school-aged children and its correlation with gut *E. coli* a *Bifidobacteria*: A case-control study. *BMC Pediatr.* **2015**, *15*, 64. [CrossRef] [PubMed]
57. de La Serre, C.B.; Ellis, C.L.; Lee, J.; Hartman, A.L.; Rutledge, J.C.; Raybould, H.E. Propensity to high-fat diet-induced obesity rats is associated with changes in the gut microbiota and gut inflammation. *Am. J. Physiol. Gastroint. Liver Physiol.* **2010**, 2 G440–G448. [CrossRef]
58. Fei, N.; Zhao, L. An opportunistic pathogen isolated from the gut of an obese human causes obesity in germfree mice. *ISM* **2013**, *7*, 880–884. [CrossRef]
59. Larrosa, M.; Yanéz-Gascón, M.J.; Selma, M.V.; González-Sarrías, A.; Toti, S.; Cerón, J.J.; Tomás-Barberán, F.; Dolara, P.; Esp J.C. Effect of a low dose of dietary resveratrol on colon microbiota, inflammation and tissue damage in a DSS-induced colitis model. *J. Agric. Food Chem.* **2009**, *57*, 2211–2220. [CrossRef]
60. O'Connor, K.; Morrissette, M.; Strandwitz, P.; Ghiglieri, M.; Caboni, M.; Liu, H.; Khoo, C.; D'Onofrio, A.; Lewis, K. Cranbe extracts promote growth of *Bacteroidaceae* and decrease abundance of *Enterobacteriaceae* in a human gut simulator model. *Pl ONE* **2019**, *14*, e0224836. [CrossRef]
61. Lippolis, T.; Cofano, M.; Caponio, G.R.; De Nunzio, V.; Notarnicola, M. Bioaccessibility and bioavailability of diet polyphen and their modulation of gut microbiota. *Int. J. Mol. Sci.* **2023**, *24*, 3813. [CrossRef]
62. Hsu, C.-N.; Hou, C.-Y.; Chang-Chien, G.-P.; Lin, S.; Yang, H.-W.; Tain, Y.-L. Perinatal resveratrol therapy prevents hypertensi programmed by maternal chronic kidney disease in adult male offspring: Implications of the gut microbiome and their metaboli *Biomedicines* **2020**, *8*, 567. [CrossRef] [PubMed]
63. Cook, S.I.; Sellin, J.H. Review article: Short chain fatty acids in health and disease. *Aliment. Pharmacol. Ther.* **1998**, *12*, 499– [CrossRef]
64. Karaki, S.; Mitsui, R.; Hayashi, H.; Kato, I.; Sugiya, H.; Iwanaga, T.; Furness, J.B.; Kuwahara, A. Short-chain fatty acid recep GPR43, is expressed by enteroendocrine cells and mucosal mast cells in rat intestine. *Cell Tissue Res.* **2006**, *324*, 353–360. [CrossR
65. Maslowski, K.M.; Vieira, A.T.; Ng, A.; Kranich, J.; Sierro, F.; Yu, D.; Schilter, H.C.; Rolph, M.S.; Mackay, F.; Artis, D.; et Regulation of inflammatory responses by gut microbiota and chemoattractant receptor GPR43. *Nature* **2009**, *461*, 1282–12 [CrossRef]
66. Vinolo, M.A.; Rodrigues, H.G.; Hatanaka, E.; Hebeda, C.B.; Farsky, S.H.; Curi, R. Short-chain fatty acids stimulate the migrat of neutrophils to inflammatory sites. *Clin. Sci.* **2009**, *117*, 331–338. [CrossRef]
67. Cronin, P.; Joyce, S.A.; O'Toole, P.W.; O'Connor, E.M. Dietary fibre modulates the gut microbiota. *Nutrients* **2021**, *13*, 16 [CrossRef]

Sembries, S.; Dongowski, G.; Mehrländer, K.; Will, F.; Dietrich, H. Physiological effects of extraction juices from apple, grape, and red beet pomaces in rats. *J. Agric. Food Chem.* **2006**, *54*, 10269–10280. [CrossRef]
Aprikian, O.; Levrat-Verny, M.A.; Besson, C.; Busserolles, J.; Remesy, C.; Demigne, C. Apple favourably affects parameters of cholesterol metabolism and of anti-oxidative protection in cholesterol-fed rats. *Food Chem.* **2001**, *75*, 445–452. [CrossRef]
Ravn-Haren, G.; Krath, B.N.; Markowski, J.; Poulsen, M.; Hansen, M.; Kołodziejczyk, K.; Kosmala, M.; Dragsted, L.O. Apple pomace improves gut health in Fisher rats independent of seed content. *Food Funct.* **2018**, *9*, 2931. [CrossRef]
Martinez, O.D.M.; Gomes, M.J.C.; Grancieri, M.; de São José, V.P.B.; Toledo, R.C.L.; Queiroz, V.A.V.; da Silva, B.P.; Martino, H.S.D. Sorghum flour BRS 305 hybrid has the potential to modulate the intestinal microbiota of rats fed with a high-fat high-fructose diet. *Eur. J. Nutr.* **2023**, *62*, 647–657. [CrossRef] [PubMed]
Dawson, P.A.; Karpen, S.J. Intestinal transport and metabolism of bile acids. *J. Lipid Res.* **2015**, *56*, 1085–1099. [CrossRef] [PubMed]
Takahashi, S.; Fukami, T.; Masuo, Y.; Brocker, C.N.; Xie, C.; Krausz, K.W.; Wolf, C.R.; Henderson, C.J.; Gonzalez, F.J. Cyp2c70 is responsible for the species difference in bile acid metabolism between mice and humans. *J. Lipid Res.* **2016**, *57*, 2130–2137. [CrossRef] [PubMed]
Di Ciaula, A.; Garruti, G.; Lunardi Baccetto, R.; Molina-Molina, E.; Bonfrate, L.; Wang, D.Q.; Portincasa, P. Bile acid physiology. *Ann. Hepatol.* **2017**, *16*, s4–s14. [CrossRef]
Jones, B.V.; Begley, M.; Hill, C.; Gahan, C.G.; Marchesi. J.R. Functional and comparative metagenomic analysis of bile salt hydrolase activity in the human gut microbiome. *Proc. Natl. Acad. Sci. USA* **2008**, *105*, 13580–13585. [CrossRef]
De Boever, P.; Wouters, R.; Verschaeve, L.; Berckmans, P.; Schoeters, G.; Verstraete, W. Protective effect of the bile salt hydrolase-active *Lactobacillus reuteri* against bile salt cytotoxicity. *Appl. Microbiol. Biotechnol.* **2000**, *53*, 709–714. [CrossRef]
Begley, M.; Gahan, C.G.; Hill, C. The interaction between bacteria and bile. *FEMS Microbiol. Rev.* **2005**, *29*, 625–651. [CrossRef]
Johnston, I.; Nolan, J.; Pattni, S.S.; Walters, J.R. New insights into bile acid malabsorption. *Curr. Gastroenterol. Rep.* **2011**, *13*, 418–425. [CrossRef]
Walters, J.R. Defining primary bile acid diarrhea: Making the diagnosis and recognizing the disorder. *Expert Rev. Gastroenterol. Hepatol.* **2010**, *4*, 561–567. [CrossRef]
Mullish, B.H.; Allegretti, J.R. The contribution of bile acid metabolism to the pathogenesis of *Clostridioides difficile* infection. *Ther. Adv. Gastroenterol.* **2021**, *14*, 17562848211017725. [CrossRef]
Amaral, J.D.; Viana, R.J.; Ramalho, R.M.; Steer, C.J.; Rodrigues, C.M. Bile acids: Regulation of apoptosis by ursodeoxycholic acid. *J. Lipid Res.* **2009**, *50*, 1721–1734. [CrossRef]
Wells, J.E.; Berr, F.; Thomas, L.A.; Dowling, R.H.; Hylemon, P.B. Isolation and characterization of cholic acid 7α-dehydroxylating fecal bacteria from cholesterol gallstone patients. *J. Hepatol.* **2000**, *32*, 4–10. [CrossRef] [PubMed]
Kakiyama, G.; Pandak, W.M.; Gillevet, P.M.; Hylemon, P.B.; Heuman, D.M.; Daita, K.; Takei, H.; Muto, A.; Nittono, H.; Ridlon, J.M.; et al. Modulation of the fecal bile acid profile by gut microbiota in cirrhosis. *J. Hepatol.* **2013**, *58*, 949–955. [CrossRef] [PubMed]
Yoshimoto, S.; Loo, T.M.; Atarashi, K.; Kanda, H.; Sato, S.; Oyadomari, S.; Iwakura, Y.; Oshima, K.; Morita, H.; Hattori, M.; et al. Obesity-induced gut microbial metabolite promotes liver cancer through senescence secretome. *Nature* **2013**, *499*, 97–101. [CrossRef]
Liu, Y.; Zhang, S.; Zhou, W.; Hu, D.; Xu, H.; Ji, G. Secondary bile acids and tumorigenesis in colorectal cancer. *Front. Oncol.* **2022**, *12*, 813745. [CrossRef]
Thomas, C.E.; Luu, H.N.; Wang, R.; Xie, G.; Adams-Haduch, J.; Jin, A.; Koh, W.P.; Jia, W.; Behari, J.; Yuan, J.M. Association between pre-diagnostic serum bile acids and hepatocellular carcinoma: The Singapore Chinese health study. *Cancers* **2021**, *13*, 2648. [CrossRef] [PubMed]
Ley, R.E.; Bäckhed, F.; Turnbaugh, P.; Lozupone, C.A.; Knight, R.D.; Gordon, J.I. Obesity alters gut microbial ecology. *Proc. Natl. Acad. Sci. USA* **2005**, *102*, 11070–11075. [CrossRef]
Ridlon, J.M.; Kang, D.J.; Hylemon, P.B. Bile salt biotransformations by human intestinal bacteria. *J. Lipid Res.* **2006**, *47*, 241–259. [CrossRef]
Zheng, X.; Huang, F.; Zhao, A.; Lei, S.; Zhang, Y.; Xie, G.; Chen, T.; Qu, C.; Rajani, C.; Dong, B.; et al. Bile acid is a significant host factor shaping the gut microbiome of diet-induced obese mice. *BMC Biol.* **2017**, *15*, 120. [CrossRef]
Kumari, K.; Augusti, K.T. Lipid lowering effect of *S*-methyl cysteine sulfoxide from *Allium cepa* Linn in high cholesterol diet fed rats. *J. Ethnopharmacol.* **2007**, *109*, 367–371. [CrossRef]
Hall, J.A.; Jewell, D.E.; Ephraim, E. Changes in the fecal metabolome are associated with feeding fiber not health status in cats with chronic kidney disease. *Metabolites* **2020**, *10*, 281. [CrossRef] [PubMed]
Hosoyamada, Y.; Yamada, M. Effects of dietary fish oil and apple polyphenol on the concentration serum lipids and excretion of fecal bile acids in rats. *J. Nutr. Sci. Vitaminol.* **2017**, *63*, 21–27. [CrossRef] [PubMed]

nutrients

MDP

Article

Novel Multi-Ingredient Supplement Facilitates Weight Loss and Improves Body Composition in Overweight and Obese Individuals: A Randomized, Double-Blind, Placebo-Controllec Clinical Trial

Joshua P. Nederveen [1], Alexander J. Mastrolonardo [1], Donald Xhuti [1], Alessia Di Carlo [1], Katherine Manta [1], Matthew R. Fuda [1] and Mark A. Tarnopolsky [1,2,*]

[1] Department of Pediatrics, Faculty of Health Sciences, McMaster University Medical Center (MUMC), Hamilton, ON L8N 3Z5, Canada; nedervj@mcmaster.ca (J.P.N.)
[2] Exerkine Corporation, McMaster University Medical Center (MUMC), Hamilton, ON L8N 3Z5, Canada
* Correspondence: tarnopol@mcmaster.ca

Abstract: Background: Despite the growing recognition of the obesity crisis, its rates continue to r The current first-line therapies, such as dietary changes, energy restriction, and physical activity, typically met with poor adherence. Novel nutritional interventions can address the root causes obesity, including mitochondrial dysfunction, and facilitate weight loss. Objective: The object of this study was to investigate the effects of a multi-ingredient nutritional supplement design to facilitate mitochondrial function and metabolic health outcomes over a 12 wk period. Metho Fifty-five overweight and/or obese participants (age (mean ± SEM): 26 ± 1; body mass index (B (kg/m^2): 30.5 ± 0.6) completed this double-blind, placebo-controlled clinical trial. Participa were randomized to 12 wks of daily consumption of multi-ingredient supplement (MIS; n = containing 50 mg forskolin, 500 mg green coffee bean extract, 500 mg green tea extract, 500 mg b root extract, 400 mg α-lipoic acid, 200 IU vitamin E, and 200 mg CoQ10) or control placebo (PI $n = 27$; containing microcrystalline cellulose) matched in appearance. The co-primary outcomes w bodyweight and fat mass (kg) changes. The secondary outcomes included other body composit measures, plasma markers of obesity, fatty liver disease biomarkers, resting energy metabolis blood pressure, physical performance, and quality of life. The post-intervention differences betwe MIS and PLA were examined via ANCOVA which was adjusted for the respective pre-interventi variables. Results: After adjustment for pre-intervention data, there was a significant difference weight ($p < 0.001$) and fat mass ($p < 0.001$) post-intervention between the PLA and MIS treatm arms. Post-intervention weight and fat mass were significantly lower in MIS. Significant p intervention differences corrected for baseline were found in markers of clinical biochemistry (A $p = 0.017$; ALT, $p = 0.008$), molecular metabolism (GDF15, $p = 0.028$), and extracellular vesi associated miRNA species miR-122 and miR-34a in MIS ($p < 0.05$). Conclusions: Following t 12 wks of MIS supplementation, weight and body composition significantly improved, concomit with improvements in molecular markers of liver health and metabolism.

Keywords: nutraceutical; catechin; polyphenol; chlorogenic acids; biomarker; multi-ingredient supplem

Citation: Nederveen, J.P.; Mastrolonardo, A.J.; Xhuti, D.; Di Carlo, A.; Manta, K.; Fuda, M.R.; Tarnopolsky, M.A. Novel Multi-Ingredient Supplement Facilitates Weight Loss and Improves Body Composition in Overweight and Obese Individuals: A Randomized, Double-Blind, Placebo-Controlled Clinical Trial. *Nutrients* **2023**, *15*, 3693. https://doi.org/10.3390/nu15173693

Academic Editors: Sonia Ramos Rivero and María Angeles Martín

Received: 1 June 2023
Revised: 2 August 2023
Accepted: 3 August 2023
Published: 23 August 2023

1. Introduction

The obesity epidemic continues to grow unabated on a global scale. Over the l 30 years, the incidence of obesity has nearly tripled [1,2], with body mass index (B continuing to trend upward [3,4]. In North America, ~70% of the adult population overweight, with ~30% classified as obese (defined as BMI ≥ 30 kg/m^2) [5–7]. Emerg countries appear to be growing in risk as well, with evidence suggesting that an obesoge environment exists in both poor and affluent populations alike [8]. Obesity exacerbates t pathology of many other conditions, including type 2 diabetes, non-alcoholic fatty liver disease (NAFLD), a variety of cancers, cardiovascular disease, poor mobility, and all-cause mortality [9–14].

Ultimately, the development of obesity can stem from an increased energy intake concomitant with reduced energy expenditure [15]. An overwhelming percentage of the North American population does not meet the minimum guidelines for physical activity [16], and evidence suggests that individuals find exercise to be inconvenient, costly, challenging, or otherwise unenjoyable [17]. Consuming a healthy, well-balanced diet without excessive proportion size is a method of reducing energy intake to promote fat loss and maintain a state of good health [18]; however, there are many socioeconomic barriers to eating a healthy diet, such as a lack of access to, greater cost of, and greater time commitment required to prepare it, as well as the unpalatability of many healthy foods or meals in comparison to less healthy alternatives [19].

Another factor in the etiology of obesity is excessive energy intake, which can cause oxidative stress [20,21] and, subsequently, mitochondrial dysfunction. While mitochondria are a crucial aspect of metabolism and the production of ATP from food substrates, they are also a major source of reactive oxygen species (ROS) in the cell [22,23]. Obesity-induced alterations in mitochondrial function can subsequently lead to increases in ROS production [20,24,25]. The increased oxidative stress generated by mitochondrial dysregulation can in turn promote aberrant cytokine production [26], which can contribute to the phenotype of chronic, low-grade inflammatory profile prevalent in obesity [27,28]. Further, some of these cytokines, peptides, and other factors such as microRNAs [29–31], known to play a role in the development of obesity, have been shown to be enclosed within extracellular vesicles [32]. Due to this important bioactive content, extracellular vesicles (EVs) have been implicated in the development and propagation of obesity [33,34]. Altogether, obesity is a multifactorial disease, with roots in mitochondrial function, upregulated ROS production, and the dysregulation of circulating inflammatory factors and bioactive EVs.

To address the constellation of factors underlying obesity, a multi-faceted approach must be taken. To counteract the cycle of ROS-associated weight gain, mitochondrial dysfunction, and subsequent inflammation, several natural ingredients have exhibited antioxidant properties. Supplementation with coenzyme Q10, an intermediate in the electron transport chain (ETC), has been shown to rescue pro-oxidative markers in serum and lower lipid peroxidation [35,36]. Similarly, fat-soluble vitamin E [37–39] and α-lipoic acid (α-LA) [40,41] have been shown to facilitate improved antioxidant status. Nitrate-rich beetroot extract (derived from the root of *Beta vulgaris*) can improve the efficiency of mitochondria [42] and has been shown to have considerable antioxidant effects [43–45]. These ingredients, in combination with other nutraceuticals known to stimulate weight loss via a variety of biological pathways, such as forskolin (derived from *Coleus forskohlii*) [46,47], green coffee bean [48,49], and green tea (derived from the leaves of *Camellia sinensis*) extract [50], have been associated with improving body composition [51] and may simultaneously influence the main pathways underlying obesity.

While these ingredients have often been examined individually, there is limited utilization regarding the potential for a multi-ingredient supplementation strategy targeting the multiple pathways known to be involved in the pathogenesis of obesity. Our pre-clinical data suggested that such a strategy lowered body fat, preserved muscle mass, and improved mitochondrial function in white adipose tissue [52]. Therefore, the primary aim of this study was to examine the effects of a multi-ingredient, metabolism-enhancing supplement (comprising seven ingredients, termed 'MIS') on weight management in overweight and/or obese individuals compared to the placebo control. We also examined exploratory outcomes: circulating biomarkers such as EV-enclosed miRNA as well as markers of fatty liver disease, glucose regulation, and inflammation. We hypothesized that the multi-ingredient nutritional intervention would decrease fat mass, preserve muscle mass and function, and reduce clinically relevant circulating biomarkers.

2. Materials and Methods

2.1. Participant Screening & Recruitment

Overweight and obese men and women were recruited for the current study. tential participants were first screened by email or telephone to confirm that they w a male or female between the ages 18 and 50 y, were overweight (BMI 25–29.9 kg/ or obese (BMI > 30 kg/m^2), and had no serious exercise contraindications for physi testing. Exclusion criteria included smoking, use of assisted walking devices, chronic use analgesic or anti-inflammatory drugs, diagnosed diabetes mellitus, cardiovascular dise (recent myocardial infarction and/or medications hypertension requiring more than 2 m ications), congestive heart failure, renal disease, previous stroke, active musculoskele injury and/or severe osteoarthritis, significant weight change in the 12 wk period prior the study (no weight gain or loss >5 kg in the prior 12 wks), muscular dystrophy, sev peripheral neuropathy, severe osteoporosis, initiating medications known to affect prot or energy metabolism, a chronic obstructive pulmonary disorder, uncontrolled asth and/or asthma requiring more than 2 medications, and participation in another biomedi study and/or structured exercise training regime within 4 wks prior to the initiation intervention. A flowchart of the enrollment process is shown in Supplemental Figure Those who qualified upon initial screening were invited to the Neurometabolic Clinic in Department of Pediatrics (Faculty of Health Science) located in McMaster Children's Hos tal to confirm eligibility by completing the following: (i) a medical screening questionna (ii) a complete physical activity readiness questionnaire; (iii) anthropometric measureme (e.g., height, weight, waist circumference), which was then used to calculate body m index (BMI) for the purpose of classifying individuals as overweight or obese [53] and calculate the waist-to-height ratio (WHR), as an indicator or measure of health status [54, Participants completed the SF-36 Health Questionnaire [56]. Sixty-five overweight/ob men and women participated in this randomized, double-blind, placebo-controlled para group trial, which took place between March 2019 and April 2020.

2.2. Experimental Design

Participants were randomly allocated to receive either the multi-ingredient nutritio supplement (MIS) consisting of 7 ingredients (Table 1) or a control placebo (PLA) for 12 wk intervention period using minimization [57,58]. Participants, as well as investigat who were responsible for recruiting and/or testing participants, were blinded to individual group assignments. The MIS and PLA were identical in appearance, produc in a Good Manufacturing Practice (GMP), and coded and distributed by an independ third party licensed by Health Canada (Infinite Nutrition, Windsor, ON, Canada). T group randomization code was provided to the investigators by Infinite Nutrition followi the trial. Study products were packaged and labeled in a blinded fashion by Infinit Nutrit Canada (Windsor, ON, Canada), in which both products were encapsulated in an opaq casing. To maintain the double-blind nature of the experiment, the placebo (PLA), wh contained microcrystalline cellulose, was matched in size and appearance. Anthropome data lock and the final locked data with randomization codes were simultaneously provid to an external third party. Infinit Nutrition did not have any access to the data.

The following assessments were performed on eligible participants at the initiation the study and after completion of the 12 wk supplement period: DXA scan, blood press antecubital blood draw, grip strength, aerobic capacity (VO_{2max}), resting metabolic r (RMR), 3-day dietary recall, and 1 wk (i.e., 7-d) pedometer step count. Dietary inta tracking instructions (to perform a 3D dietary recall or 'food record') were provided to participants. Participants were given a standard pedometer (Omron HJ-321, Omron, Kyo Japan) and recorded their daily totals for 1 wk as an indicator of baseline activity lev concomitant with a self-reported 'minutes of physical activity per day' evaluation.

Table 1. Nutritional composition of the multi-ingredient supplement (MIS) and placebo (PLA) control supplement.

Ingredients	PLA	MIS
Forskolin (mg)	0	50
Green Coffee Bean Extract (mg)	0	500
Green Tea Extract (mg)	0	500
Beet Root Extract (mg)	0	500
α-Lipoic Acid (mg)	0	400
Vitamin E (IU)	0	200
Coenzyme Q10 (mg)	0	200
Microcrystalline Cellulose (g)	2.15	0

The quantities listed represent the complete daily serving of the MIS or PLA. Participants consumed servings of the supplement during the 12 wk intervention period (7 d/wk^{-1}). MIS; multi-ingredient supplement, PLA; placebo control. PLA and MIS were both encapsulated in an opaque casing matched for appearance.

To assess the progress between study initiation and completion (i.e., 12 wk), participants returned for testing at the 2 wk and 5 wk mark from the study initiation to perform a DXA scan and anthropometric measurements. Weekly phone calls and/or emails were conducted throughout the study period to ensure participant compliance. At the onset of the study, all participants received a copy of the Canada Food Guide and the evidence-based Canadian Society for Exercise Physiology (CSEP) Movement Guidelines [59]. During the entire intervention period, participants were neither encouraged nor discouraged to maintain their initial body weight and physical activity level. The general experimental design is presented schematically in Supplemental Figure S2.

2.3. Nutritional Supplementation

The exact composition of and nutrition information for the MIS and PLA capsules is provided in Table 1. The supplement provided an energetic value of ~10 kcal per capsule, in a capsule weight of ~0.6 g. Participants were asked to consume 2 capsules ~30 min prior to the consumption of a meal, three times per day (i.e., breakfast, lunch, and dinner). Therefore, participants were instructed to consume a total of 6 capsules of the study supplement per day (7 d/wk^{-1}) throughout the 12 wk intervention period. The MIS provided ~200 mg of naturally sourced caffeine (primarily from green coffee bean extract) per day. Participants were given a 4 wk supply during monthly visits and maintained the used package and turned it in at the check-ins as a marker of study compliance. The DER (Drug Extract Ratio) and extraction solvents for plant-based extracts are found in Supplemental Materials Table S1.

2.4. Anthropometric Measurements and Body Composition

Participants were weighed at baseline and at each follow-up study visit. Weight was measured using a Health-O-Meter 2600KL Digital Wheelchair Scale (Pelstar, McCook, IL, USA), and height was measured using a stadiometer (Perspective Enterprises, Portage, MI, USA). Waist and measurements for the calculation of WHR [54,55] were performed with a Gulick tape measure. A horizontal measure was obtained directly above the iliac crest and for hip circumference measurement, the participants stood erect with feet together, and a horizontal measure was taken at the maximal circumference of the buttocks [60]. Fat-free mass (FFM) and fat mass (FM) were assessed using a DXA scan (GE Lunar Prodigy, Madison, WI, USA) and a software program (encore Version 9.15.010). Whole body composition assessed via DXA was validated in vivo [61–63]. A single investigator performed regional body compartment analysis prior to unblinding. Participants underwent all anthropometric measurements and DXA scans in a clean hospital gown with the same undergarments for every visit.

2.5. *Measurement of Blood Pressure*

Participants were fitted with a FlexiPort™ reusable blood pressure cuff (WelchAll Inc., Okumoto, NY, USA) to measure blood pressure while the participant was relax and seated for 15 min. Measurements were obtained with an automated blood pressu machine (Spot Vital Signs Device, WelchAllyn).

2.6. *Venous Blood Sampling and Analysis*

Blood was collected in the morning following an overnight 10 h fast (no food caffeine) and the participants were instructed to consume 250 mL of water prior to arriv Blood was taken from the antecubital vein and drawn into evacuated tubes with heparin plasma collection, ethylenediaminetetraacetic acid (EDTA) for plasma collection, and n treated tubes were used to collect serum. Serum and plasma samples were immediate taken to the Core Laboratory at McMaster University Medical Centre for the analy of the following panel of tests for general blood biochemistry: complete blood cou (CBC), gamma-glutamyl transferase (GGT), bilirubin, alanine aminotransferase (AL creatine kinase (CK), creatinine, and C-reactive protein (CRP). Markers of dyslipiden (i.e., triglycerides, total cholesterol, LDL cholesterol, and HDL cholesterol) and blood sug control (i.e., glucose, insulin) were measured. Homeostatic Model Assessment of Insu Resistance (HOMA-IR) was calculated as the equation [fasting insulin (U/L) × fasti glucose (nmol/L)/22.5] [64]. If available, the remaining samples were stored at −80 °C subsequent analysis. To verify the ALT findings, aspartate transferase (AST) levels we also assessed via Core Laboratory during secondary analysis. For exploratory analys plasma samples were batch-analyzed via enzyme-linked immunosorbent assay (ELIS for leptin (Quantikine DLP00, R&D Systems, Minneapolis, MN, USA), IL-6 (Quantiki D6050, R&D Systems), CoQ10 (MBS165643, BioSourceLLC, San Diego, CA, USA), FGF (ab222506, Abcam, Cambridge, UK), and GDF15 (Quantikine DGD150, R&D Systems) p kit instructions, respectively. All investigators and laboratory technicians were unaware the participants' group assignment during the initial downstream analysis.

2.7. *Isolation of Extracellular Vesicles*

Exactly 400 μL of serum was initially subjected to two consecutive centrifugati spins at 2000× *g* and 20,000× *g* (Beckman Coulter Optima MAX-TL, Pasadena, CA, US for 30 min each at 4 °C to remove any large EVs. Following differential centrifugatio exactly 300 μL of supernatant containing small EVs was carefully retained and subject to 16 h of ultracentrifugation (Beckman Coulter Optima MAX-TL, CA, USA) at a speed 120,000× *g* at 4 °C. A total of 250 μL of supernatant was carefully removed, whereas t fluid pellet was washed with 250 μL of 1X Dulbecco's phosphate-buffered saline (D85 Sigma-Aldrich, Burlington, MA, USA) and briefly sonicated to ensure proper resuspensi of the fluid pellet. The solution was further spun at a similar speed for 60 min where a fi volume of 100 μL of fluid pellet was retained and used for miRNA extraction.

2.8. *miRNA Extraction and RT-PCR*

The MagMAX™ *mirVana*™ *Total RNA* Isolation Kit (A27828; Thermo Scientific, Waltha MA, USA) was utilized as per the manufacturer's instructions. Briefly, 100 μL of EV isola were plated onto extraction plates where magnetic beads aided in the isolation of 50 μL of to RNA. Prior to the extraction, *cel-miR-54*-3p (custom oligo; Thermo Fischer Scientific) spike was added to each sample at a concentration of 1×10^{-9} M. Two μL of eluted RNA w then used as input to synthesize cDNA using TaqMan™ Advanced miRNA cDNA Synthe Kit (A28007; Thermo Scientific). The expression of the miRNAs of interest (Supplemen Materials Table S2) was then measured using TaqMan™ Fast Advanced Master Mix (44445 Thermo Fisher Scientific) and the CFX Connect Real-Time PCR Detection System (BioRa Hercules, CA, USA). The *cel-54-3p* was used as a reference gene for normalization and fo change between baseline (Day 0) and 12 wk period (end of intervention) was calculated usi the $2^{-\Delta\Delta C_T}$ method [65].

2.9. Metabolic Measures and Resting Metabolic Rate (RMR)

Participants were placed in a supine position while connected to a metabolic cart with an online gas collection system (Moxus Modular Oxygen Uptake System, AEI Technologies, Pittsburgh, PA, USA), and the system acquired oxygen consumption (VO_2) and carbon dioxide (CO_2) production data, which was recorded after 30 min of rest. For the duration of the test, the participants were instructed to remain relaxed without falling asleep. From this data, the abbreviated Weir equation was used to calculate RMR [RMR = $1440 \times (VO_2 \times 3.941) + (VCO_2 \times 1.11)$] [66,67]. The RMR was also estimated (eRMR) using the standard Mifflin-St. Jeor equation [68], which was validated in overweight/obese populations [69,70]. For females, RMR was calculated as the equation [RMR = $9.99 \times$ weight + $6.25 \times$ height $- 4.92 \times$ age $- 161$]. For males, RMR was calculated as the equation [RMR = $9.99 \times$ weight + $6.25 \times$ height $- 4.92 \times$ age + 5]).

2.10. Maximal Voluntary VO_{2peak} Fitness Test

Participants completed a double-leg incremental peak oxygen uptake (VO_{2peak}) test on a cycle ergometer (Lode, Groningen, The Netherlands). A metabolic cart with an online gas collection system (Moxus Modular Oxygen Uptake System, AEI Technologies, Pittsburgh, PA, USA) acquired oxygen consumption (VO_2) and carbon dioxide (CO_2) production data. The test began with a 2 min warm-up at 30 watts (W), after which the power was increased by 30 W every minute until volitional exhaustion or when the pedal cadence fell below 55 rpm. Heart rate was monitored continuously throughout the test via telemetry with a heart rate monitor (Polar A3, Lake Success, NY, USA). VO_{2peak} was defined as the highest oxygen consumption achieved over a 30 s period. Maximal workload (W_{max}) was the highest power output achieved during the test. After the completion of the test, the participant performed a 3 min active recovery pedaling against 30 W.

2.11. Maximal Voluntary Handgrip Assessment

Handgrip strength was measured using an isometric dynamometer (JAMAR®, Sammons, Bolingbrook, IL, USA). The grip width was adjusted to hand size and with the arm flexed at 90° while seated. The participant performed three separate maximal 3 s trials with a 1 min rest period between trials, and the mean value was taken.

2.12. Questionnaires

The RAND 36-Item Health Survey Version 1.0 is used extensively as a survey instrument for assessing participant health-related quality of life (HRQOL). We examined scales pertaining to physical functioning and role limitations due to health problems instead of the total score [71]. These domains were scored on a 0 to 100 range such that a higher score was representative of a higher quality of life [72]. These scales are validated [73] and have been utilized in the context of weight loss [74].

2.13. Dietary Intake

Three-day food records (recording 2 weekdays and 1 weekend day) were collected and analyzed using the MyFitnessPal smartphone app (MyFitnessPal, Under Armour, Baltimore, MD, USA.) and website that tracks diet and exercise. If participants were not comfortable with utilizing the smartphone application, they were provided with a paper diet log. Participants were given instructions on how to record their intake of any food and beverages. Diet logs were completed at periods before the participant began the study protocol, at the study midpoint (5 wk), and final timepoint (12 wk).

2.14. Activity Tracking

Study participants were provided with a pedometer (Omron HJ-321, Omron, Kyoto, Japan) to record their daily step counts for three separate 7 d periods within the study to determine average daily step count. Daily step count for a 7 d period was recorded at periods before the participant began the study protocol, at the study midpoint (5 wk), a final timepoint (12 wk).

2.15. *Sample Size Calculation*

The MIS was comprised of 7 separate ingredients, with aspects featuring a pot tial varied or synergistic [52] impact on body composition. Sample sizes were calcula based on a detection difference in fat mass (kg) loss of 4.01 kg (sample standard deviati 4.5 kg) modified from observations of the effect of oral ingestion of forskolin (derived fr *Coleus forskohlii*) on body composition [46]. The sample size was calculated with a power 80% and a significance level of 0.05 (2-tailed), which yielded an estimate of approximat 20 participants per group. To account for an ~25% drop-out rate [75] observed in similar als using nutritional supplementation and inherent variability in weight management wh participants performed some levels of baseline physical activity [76], at least 25 participa were recruited.

2.16. *Statistical Analysis*

Post-intervention differences between treatment arms in outcome variables were co pared using a one-way analysis of covariance (ANCOVA), implementing the correspondi pre-intervention variables as covariates. If data were not normally distributed, logari mic transformations were performed. Endpoints that were intractably non-normal we assessed using the Mann–Whitney U test. Within-group analysis was performed utilizi a paired *t*-test or, in the case of intractable non-normality, the Wilcoxon sign rank te Exclusively for co-primary outcomes (i.e., weight and fat mass), statistical analysis w performed using a modified intention-to-treat (ITT) population, wherein all participa who were randomized to receive PLA or MIS, and on whom any post-randomizati information prior to dropout or completion was available. This is in accordance w recently suggested practices for randomized weight loss trials [77]. In this case, all missi values were imputed using the last observation carried forward (LCOF) method. Followi the co-primary outcome analysis, statistical analysis was performed on all anthropor try, clinical biochemistry, and lifestyle data (e.g., step count and diet logs) exclusively participants who completed the intervention. Statistical significance was set at $\alpha < 0.$ SPSS (version 28; SPSS, Inc., Chicago, IL, USA) was used for all statistical analyses. Duri the intervention period, the SARS-CoV-2 virus pandemic impacted Canada. Coronavi disease (COVID-19) is an infectious disease caused by the SARS-CoV-2 virus. At the ti (April 2020), the infection vector had not been well described, and to reduce close cont and/or infectious aerosol particulates transported between research staff and participa in the trial, the final participant cohort (n = 5 PLA, n = 7 MIS) did not have a final blo draw/analysis, RMR, resting HR, grip strength, resting heart rate, VO_{2peak}, SBP, or D measurements taken at the endpoint (i.e., 12 wk mark). The LCOF method was not used replace data for these measurements in this cohort.

3. Results

3.1. *Study Information and Compliance*

A total of 65 participants were randomized: 55 completed the study, five dropped (n = 3, and n = 2 dropouts in the PLA and MIS groups, respectively), and three were l to follow-up (n = 3 MIS). One participant was removed from the study due to initiatin medication known to alter protein metabolism during the intervention period (n = 1 M and one participant was removed due to non-compliance with study visits (n = 1 M Of the participants who withdrew from the study or were lost to follow-up followi randomization, n = 2 withdrew prior to Visit 2, n = 3 withdrew prior to Visit 3, and n = withdrew prior to Visit 4 (endpoint). The reasons for participant withdrawal or removal listed in Supplemental Figure S1. Treatment arm compliance was assessed by pill counti from the returned supplement package per visit (monthly). Supplement consumpti compliance over the duration of the intervention was 82.7 ± 4.6% and 85.4 ± 3.1% in t

PLA and MIS groups, respectively. No significant adverse events directly related to the treatment were recorded or shared by the participants. Three participants reported mild gastric reflux immediately after the intervention ($n = 1$ PLA; $n = 2$ MIS). These symptoms were not reported following the first week (subsiding by Visit 2; ~2 weeks following initiation of intervention), and no participants withdrew from the study due to events or concerns related to consuming the PLA or MIS.

3.2. Baseline Characteristics

Fifty-five participants (23 men, 32 women) with a mean age of 25.9 ± 1.1 y (mean ± SEM) completed the study. The participants were overweight to obese, with a mean BMI of 30.5 ± 0.6 kg/m^2. Pre-intervention, there were no significant differences between treatment arms in weight, BMI, total fat mass, total fat-free mass, or waist-to-height ratio ($p \geq 0.05$, Table 2).

Table 2. Baseline participant characteristics in both treatment arms.

Variable (Unit)	Placebo (PLA; $n = 27$)	Active (MIS; $n = 28$)
Sex, n (M/F)	12/15	11/17
Age (years)	25.5 ± 1.5	26.4 ± 1.6
Body weight (kg)	87.8 ± 3.5	89.4 ± 2.8
BMI (kg/m^2)	29.8 ± 1.0	31.1 ± 0.8
Waist-to-Height Ratio (a.u)	0.57 ± 0.01	0.59 ± 0.01
Body Fat percentage (%)	40.2 ± 1.7	40.3 ± 1.5
Fat mass (kg)	34.2 ± 2.3	34.6 ± 1.7
Fat-free mass (kg)	50.2 ± 0.1	51.4 ± 0.1

Data are presented as means ± SE. BMI: body mass index.

3.3. Co-Primary Outcomes Utilizing Modified ITT

For the examination of weight and fat mass (co-primary outcomes) exclusively, we utilized a modified ITT analysis. After adjustment for pre-intervention body weight, there was a statistically significant difference in post-intervention body weight between the PLA and MIS interventions, $F(1, 52) = 79.090$, $p = 0.001$, partial $\eta^2 = 0.183$. In the PLA group, there was a significant increase in body weight at the endpoint compared to the baseline ($p = 0.033$). In the MIS group, there was a significant decrease at the endpoint compared to the baseline ($p < 0.001$). After adjustment for pre-intervention fat mass, there was a statistically significant difference in post-intervention fat mass between the PLA and MIS interventions: $F(1, 52) = 31.871$, $p = 0.031$, and partial $\eta^2 = 0.087$. During the intervention, there was a significant within-group difference from baseline (Day 0) to endpoint in both PLA (increased, $p = 0.049$) and MIS (decreased, $p = 0.034$).

3.4. Anthropometry and Body Composition

After adjustment for pre-intervention body weight, there was a statistically significant difference in post-intervention weight between the treatment arms ($p \leq 0.001$, Figure 1A). Participants in the MIS treatment arm significantly decreased in weight from baseline ($p \leq 0.01$, $\Delta -2.2 \pm 2.8$ kg, mean ± SD), and those in the PLA treatment arm tended to increase in weight from baseline ($p = 0.052$, Δ 0.9 ± 2.4 kg). All individual participant changes in weight over the period of the intervention are shown for the PLA and MIS treatment arms, respectively (Figure 1B).

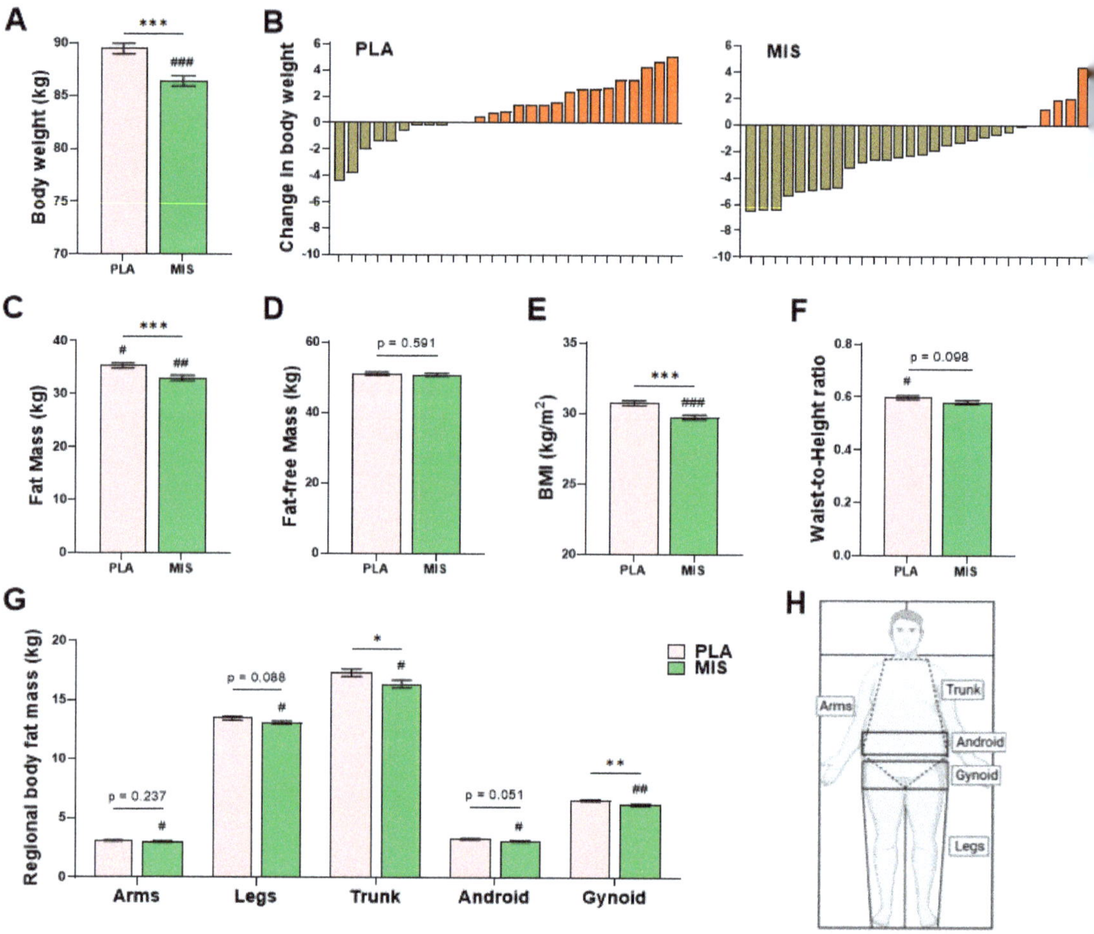

Figure 1. Characterization of anthropometric changes following 12 wks of a multi-ingredient s plement (MIS, $n = 28$, green bars) compared to placebo (PLA, $n = 27$, pink bars). Post-interventi differences between treatment arms were compared using one-way ANCOVA, implementing corresponding pre-intervention variables as covariates. Post-intervention data are presented adjusted means ± SEs. Asterisks (*) denote significant ($p < 0.05$) between-group differences vers PLA, $p \geq 0.05$, *; $p \leq 0.05$, **; $p \leq 0.01$, ***; $p \leq 0.001$. Number signs (#) denote significant with group differences relative to Day 0 (baseline), assessed via paired *t*-test, #; $p \leq 0.05$, ##; $p \leq 0.$ ###; $p \leq 0.001$. (**A**) Post-intervention weight (kg). (**B**) Individual change (Δ) in body weight (following 12 wk intervention in the PLA and MIS group. (**C**) Post-intervention DXA-derived mass (kg). (**D**) Post-intervention DXA-derived fat-free mass (kg). (**E**) Post-intervention BMI (kg/m (**F**) Post-intervention waist-to-height ratio (a.u.). (**G**) Post-intervention DXA-derived regional bo fat mass (kg). (**H**) Representative image of regional body fat mass demarcations.

After adjustment for pre-intervention fat mass, there was a significant difference post-intervention fat mass between the treatment arms ($p \leq 0.001$, Figure 1C). Followi the intervention period, participants in the MIS arm had significantly decreased fat ma ($p \leq 0.01$, $\Delta -1.4 \pm 2.5$ kg), whereas those in the PLA group tended to have increased mass ($p = 0.052$, $\Delta\ 0.9 \pm 2.7$ kg).

There was no significant difference in the post-intervention total fat-free mass between the treatment arms ($p = 0.591$, Figure 1E), adjusted for pre-intervention data. No changes were observed in either treatment arms ($p \geq 0.05$, Table 3).

Table 3. Anthropometric measurements following 12 wks of treatment, adjusted for baseline variables.

Variable (Unit)	Placebo (*n*)	Active (*n*)	*p* Value
Weight (kg)	89.5 ± 0.5 (27)	86.4 ± 0.5 (28) ***	**<0.001**
BMI (kg/m^2)	30.7 ± 0.2 (27)	29.7 ± 0.2 (28) ***	**<0.001**
Waist to Height ratio (a.u.)	0.60 ± 0.01 (27) *	0.58 ± 0.01 (28)	0.098
Total Fat-Free Mass (kg)	51.0 ± 0.4 (27)	50.7 ± 0.4 (28)	0.591
Total Fat Mass (kg) §,†	35.4 ± 0.5 (27)	33.0 ± 0.5 (28) **	**<0.001**
BCI (a.u.)	1.57 ± 0.04 (27)	1.69 ± 0.04 (28) *	**0.034**
Arm Fat Mass (kg)	3.10 ± 0.06 (27)	3.01 ± 0.06 (28) *	0.237
Leg Fat Mass (kg)	13.5 ± 0.2 (27)	13.1 ± 0.2 (28) *	0.088
Trunk Fat Mass (kg)	17.4 ± 0.3 (27)	16.4 ± 0.3 (28) *	**0.039**
Android Fat (kg)	3.31 ± 0.06 (27)	3.14 ± 0.06 (28) *	0.051
Gynoid Fat (kg)	6.60 ± 0.09 (27)	6.22 ± 0.08 (28) **	**0.002**

Post-intervention data are presented as adjusted means ± SEs. Post-intervention variables were compared using one-way ANCOVA, with corresponding baseline variables as covariates. Within-group comparisons were made using a paired *t*-test. *; $p < 0.05$, **; $p \leq 0.01$, ***; $p \leq 0.001$, within-group comparison. § Between-group comparisons were made using the Mann–Whitney U test. † Within-group comparisons were made using the non-parametric sign rank test. BCI: body composition index; BMI: body mass index.

The ratio of fat-free mass (kg) to fat mass (kg), termed herein the Body Composition Index (BCI), was significantly different between the intervention arms ($p = 0.034$, Table 3) following the intervention. The BCI of the MIS treatment arm was significantly increased compared to baseline ($p < 0.05$, Δ 0.1 ± 0.2 a.u.), corresponding with the relative loss of fat mass and the relative preservation of fat-free mass.

The BMI adjusted for pre-intervention values was significantly different between the treatment arms ($p \leq 0.001$, Figure 1F), with the MIS group having significantly decreased BMI from baseline ($p \leq 0.001$, Δ −0.7 ± 0.9 kg/m^2), but there was no difference in the PLA group ($p = 0.086$, Δ 0.3 ± 0.8 kg/m^2).

Following adjustment for pre-intervention data, there was no significant difference in the post-intervention waist-to-height ratio between the treatment arms ($p = 0.098$, Figure 1G), with participants in the PLA treatment arm significantly increasing from baseline ($p < 0.05$, Δ 0.02 ± 0.04 a.u).

There were significant differences detected between the treatment arms in the post-intervention regional fat mass in the trunk and gynoid regions ($p = 0.039$ and $p = 0.002$, respectively), adjusted for pre-intervention values (Figure 1G). In the MIS group, the trunk ($p = 0.014$, Δ −0.6 ± 1.3 kg) and the gynoid ($p \leq 0.001$, Δ −0.3 ± 0.4 kg) regional fat masses were significantly lower following the intervention period. Following adjustment for the respective baseline variables, there were no significant differences in post-intervention arm fat mass ($p = 0.237$), leg fat mass ($p = 0.088$), or android fat mass ($p = 0.051$) between the treatment arms (Table 3). Following the intervention, participants in the MIS treatment arm had significantly reduced arm fat mass ($p < 0.05$, Δ −0.3 ± 1.3 kg), leg fat mass ($p < 0.05$, Δ −0.4 ± 0.8 kg), and android fat mass ($p < 0.05$, Δ −0.1 ± 0.2 kg) respectively. At baseline, there were no significant differences in regional fat mass (kg) between the treatment arms ($p \geq 0.05$).

3.5. Clinical Biochemistry

Following adjustment for pre-intervention ALT, there was a significant difference in post-intervention ALT between the treatment arms ($p = 0.007$, Table 4). Following the intervention, participants in the MIS group showed a significant decrease from baseline ($p < 0.05$, Δ −7.0 ± 13.2 U/L). Correspondingly, significant post-intervention differences were detected in AST between treatment arms ($p = 0.017$, Table 4), adjusted for pre-intervention data, with the MIS group significantly decreasing from baseline ($p < 0.05$, Δ −4.5 ± 8.7 U/L).

Table 4. Post-intervention clinical biochemistry, fasting glucose, and blood lipids following 12 wk treatment, adjusted for baseline variables.

Variable (Unit)	Placebo (*n*)	Active (*n*)	*p* Value
Clinical biochemistry			
ALT (U/L)	30.3 ± 2.4 (21)	21.1 ± 2.4 (21) *	**0.008**
AST (U/L)	32.9 ± 1.6 (18)	27.3 ± 1.6 (18) *	**0.017**
GGT (U/L) ‡	29.3 ± 1.4 (21)	29.5 ± 1.4 (21)	0.913
CRP (mg/L) ‡	3.4 ± 1.0 (21)	4.2 ± 1.0 (21)	0.804
Creatinine (μmol/L) §,†	80.0 ± 1.4 (21)	83.4 ± 1.4 (21) *	0.089
Bilirubin (μmol/L) ‡	12.3 ± 1.0 (21)	11.3 ± 1.0 (21)	0.362
Fasting blood lipids & glucose homeostasis			
Total Cholesterol (mmol/L)	4.8 ± 0.1 (21)	4.9 ± 0.1 (21)	0.526
LDL (mmol/L)	2.9 ± 0.1 (21)	3.0 ± 0.1 (21)	0.342
HDL (mmol/L)	1.3 ± 0.03 (21)	1.2 ± 0.03 (21)	0.311
Triglycerides (mmol/L)	1.4 ± 0.1 (21)	1.4 ± 0.1 (21)	0.924
Glucose (mmol/L)	5.0 ± 0.1 (21)	5.0 ± 0.1 (21)	0.942
Insulin (pmol/L)	80.1 ± 5.5 (21)	81.0 ± 5.5 (21)	0.908
HOMA IR (a.u.)	2.6 ± 0.2 (21)	2.6 ± 0.2 (21)	0.991

Post-intervention data are presented as adjusted means ± SEs. Post-intervention variables were compa using one-way ANCOVA, with corresponding baseline variables as covariates. Within-group comparisons w made using a paired *t*-test. ‡ Logarithmic transformation was required to achieve normality. § Between-gr comparisons were made using the Mann–Whitney U test. † Within-group comparisons were made using non-parametric sign rank test. * Denotes statistically significant ($p < 0.05$) within-group difference. HOMA Homeostatic Model Assessment for Insulin Resistance.

Following adjustment for the respective baseline variables, there were no sign cant differences in post-intervention serum creatinine ($p = 0.089$, Table 4) between treatment arms. Following the intervention, participants in the MIS treatment arm h significantly increased endpoint creatinine levels ($p < 0.05$, Δ 4.0 ± 5.6 μmol/L) compa to baseline, respectively.

There were no significant differences between the treatment arms in GGT, CRP, bilirubin ($p \geq 0.05$, Table 4) between the two treatment arms, adjusted for their respecti pre-intervention values, nor were there any significant differences within the treatm arm at the endpoint compared to baseline ($p \geq 0.05$, Table 4).

3.6. *Markers of Dyslipidemia and Glucose Metabolism*

There were no significant differences between the treatment arms with respect to to cholesterol, low-density lipoprotein cholesterol (LDL), high-density lipoprotein choleste (HDL), and triglycerides ($p \geq 0.05$, Table 4) between the two treatment arms, adjusted their respective pre-intervention values. No significant post-intervention differences w detected in fasting glucose, insulin, or derived HOMA-IR, adjusted for pre-interventi values ($p \geq 0.05$, Table 4). At baseline, there were no significant differences in circulati lipid markers or glucose metabolism outcomes between ($p \geq 0.05$) or within ($p \geq 0.05$) treatment arms.

3.7. *Hematology*

After adjustment for pre-intervention circulating eosinophil cell population, there w a statistically significant difference in post-intervention data between the treatment ar ($p \leq 0.01$, Table 5), with the PLA group significantly increasing from baseline ($p <$ 0. Δ $0.03 \pm 0.1 \times 10^9$/L). There were no significant differences between the treatment ar in hemoglobin (Hb), hematocrit (HCT), mean corpuscular volume (MCV), mean corp cular hemoglobin (MCH), mean corpuscular hemoglobin concentration (MCHC), me platelet volume (MPV), or red cell distribution width (RDW) after adjustment for c responding pre-intervention data ($p \geq 0.05$, Table 5). Cell quantification of leukocyt erythrocytes, platelets, neutrophils, lymphocytes, monocytes, and basophils did not c fer post-intervention between the two treatment arms, adjusted for their correspondi pre-intervention values ($p \geq 0.05$, Table 5). These hematology outcomes were unaltered following the intervention in both treatment arms ($p \geq 0.05$).

Table 5. Post-intervention hematology following 12 wks of treatment, adjusted for baseline variables.

Variable (Unit)	Placebo (*n*)	Active (*n*)	*p* Value
Hemoglobin Hb (g/L)	142.9 ± 1.19 (21)	143.7 ± 1.19 (21)	0.650
Hematocrit HCT (%)	0.4 ± 0.0 (21)	0.4 ± 0.0 (21)	0.977
Mean Corpuscular Volume MCV (fL)	87.3 ± 0.3 (21)	86.9 ± 0.3 (21)	0.309
Mean Corpuscular Hb MCH (pg)	28.8 ± 0.1 (21)	28.9 ± 0.1 (21)	0.694
Mean Corpuscular Hb Conc. MCHC (g/L)	330.9 ± 1.5 (21)	331.3 ± 1.5 (21)	0.835
Mean Platelet Volume §,† MPV (fL)	10.5 ± 0.1 (21)	10.6 ± 0.1 (21)	0.472
Red Cell Distribution Width RDW (%)	12.9 ± 0.09 (21)	13.1 ± 0.09 (21)	0.172
Leukocyte Count LKCS ($\times 10^9$/L)	6.4 ± 0.3 (21)	6.5 ± 0.3 (21)	0.783
Erythrocyte Count ERCS ($\times 10^{12}$/L)	5.0 ± 0.04 (21)	5.0 ± 0.04 (21)	0.864
Platelet Count ‡ ($\times 10^9$/L)	253.6 ± 6.5 (21)	254.9 ± 6.5 (21)	0.880
Neutrophil Count ($\times 10^9$/L)	3.5 ± 0.2 (21)	3.8 ± 0.2 (21)	0.370
Lymphocyte Count ($\times 10^9$/L)	2.1 ± 0.1 (21)	2.0 ± 0.1 (21)	0.273
Monocyte Count ‡ ($\times 10^9$/L)	0.5 ± 0.0 (21)	0.5 ± 0.0 (21)	0.555
Eosinophil Count §,† ($\times 10^9$/L)	0.19 ± 0.02 (21) *	0.13 ± 0.02 (21)	**0.007**
Basophil Count §,† ($\times 10^9$/L)	0.03 ± 0.01 (21)	0.04 ± 0.01 (21)	0.521

Post-intervention data are presented as adjusted means ± SEs. Post-intervention variables were compared using one-way ANCOVA, with corresponding baseline variables as covariates. Within-group comparisons were made using a paired *t*-test. ‡ Logarithmic transformation was required to achieve normality. § Between-group comparisons were made using the Mann–Whitney U test. † Within-group comparisons were made using the non-parametric sign rank test. * Denotes statistically significant ($p < 0.05$) within-group difference.

3.8. Indices of Physical Health, Metabolism and Function

No statistically significant differences were found between the two treatment arms in post-intervention systolic blood pressure, diastolic blood pressure, resting heart rate, or bone mineral density, adjusted for the corresponding pre-intervention variable data (Table 6). Following adjustment for pre-intervention relative grip strength, there was a significant post-intervention difference between the treatment arms ($p = 0.002$, Table 6). Following the intervention, participants in the MIS group significantly increased from baseline ($p < 0.05$, Δ 0.07 ± 0.06 kg/kg bw^{-1}), whereas participants in the PLA group significantly decreased from baseline ($p < 0.05$, Δ 0.02 ± 0.04 kg/kg bw^{-1}), in line with changes in total body weight observed following the intervention period. Relative VO_{2peak} was not significantly different post-intervention between the treatment arms after correcting for baseline data ($p \geq 0.05$, Table 6). The resting metabolic rate (RMR), as measured by gas exchange data *in vivo*, was not significantly different post-intervention between the treatment arms after correcting for baseline data ($p \geq 0.05$, Table 6).

Table 6. Post-intervention indices of physical health, function, and metabolism following 12 wks treatment, adjusted for baseline variables.

Variable (Unit)	Placebo (*n*)	Active (*n*)	*p* Value
Indices of health			
Systolic Blood Pressure (mmHg)	121 ± 2 (21)	127 ± 2 (21)	0.08
Diastolic Blood Pressure (mmHg)	78 ± 2 (17)	81 ± 2 (17)	0.101
Resting Heart Rate (bpm)	77 ± 3 (21)	78 ± 3 (21)	0.655
Bone mineral density (g/cm^2)	1.24 ± 0.01 (27)	1.24 ± 0.01 (28)	0.734
Indices of physical fitness			
Relative VO_{2peak} §,† ($mL/kg/min^{-1}$)	29.6 ± 1.2 (19)	30.1 ± 1.1 (21)	0.526
Relative Grip Strength ($kg/kg\ bw^{-1}$)	0.42 ± 0.01 (21) *	0.47 ± 0.01 (21) *	**0.002**
Indices of metabolic rate			
Resting metabolic rate (Kcal/day)	1808.3 ± 44.5 (21)	1827.6 ± 44.5 (21)	0.761
Estimated metabolic rate (Kcal/day)	1736.3 ± 5.0 (27)	1705.0 ± 4.9 (28) *	**<0.001**

Post-intervention data are presented as adjusted means ± SEs. Post-intervention variables were compared us one-way ANCOVA, with corresponding baseline variables as covariates. Within-group comparisons were m using a paired *t*-test. § Between-group comparisons were made using the Mann–Whitney U test. † Within-grc comparisons were made using the non-parametric sign rank test. * Denotes statistically significant (p < 0. within-group difference.

3.9. Diet, Activity Level, and Self-Reported Quality of Life

At baseline, there was no significant difference between the treatment arms in ter of energy intake, caffeine consumption, and 7-day pedometer-derived average daily st count ($p \geq 0.05$). Energy intake derived from the 3-d dietary recall log (kcal/d) was r different between the treatment arms (p = 0.856) after adjustment for pre-interventi energy intake. The adjusted mean of energy intake was 1802.1 ± 98.9 $kcal/d^{-1}$ in the Pl arm compared to 1761.8 ± 97.1 $kcal/d^{-1}$ in the MIS arm. There were no significant diff ences in energy intake per day at the endpoint compared to baseline within either gro ($p \geq 0.05$). Caffeine consumption was not different between the treatment arms (p = 0.5 after adjustment for the corresponding pre-intervention data. The pre-intervention adjust mean dietary caffeine consumption was 66.4 ± 13.9 mg/d^{-1} in the PLA arm compar to 67.1 ± 13.4 mg/d^{-1} in the MIS arm. The MIS group also consumed an additio ~200 mg/d^{-1} of caffeine contained as part of the MIS formulation (primarily from G(extract). There were no significant differences in caffeine intake at the endpoint compar to baseline within either group ($p \geq 0.05$). Habitual alcohol consumption was self-report at the onset of the intervention. In the PLA treatment arm, 51.9% (14/27) never consum alcohol, 22.2% (6/27) occasionally consumed alcohol, 18.5% (5/27) consumed alcohol a weekly basis (less than 5d per week), and 7.4% (2/27) consumed one unit of alcol daily. In the MIS treatment arm, similar proportions were observed with 39.3% (11/ never having consumed alcohol, and 28.6% (8/28), 21.4% (6/28), and 10.7% (3/28) cc suming occasional, weekly, or daily alcohol, respectively. The step count was not differe between the treatment arms (p = 0.320) following adjustment for pre-intervention da The adjusted mean for step count was 6152.7 ± 822.7 $steps/d^{-1}$ in the PLA arm compar to 6156.5 ± 807.4 $steps/d^{-1}$ in the MIS arm. There were no significant differences in t daily step count at the endpoint compared to the baseline within either MIS or PLA gro ($p \geq 0.05$). The SF-36 scales associated with physical functioning were not significan different between treatment arms at baseline (PLA; 90.6 ± 3.7, MIS; 93.0 ± 2.3) and were r different between groups (p = 0.262) or within groups ($p \geq 0.05$) following the interventi period. The scores associated with role limitations due to physical health problems were not significantly different at baseline between the PLA (90.4 ± 4.1) and MIS (95.5 ± 2.8) and were not significantly different between the treatment arms ($p = 0.429$) or within groups ($p \geq 0.05$).

3.10. Molecular Signalling Factors and Antioxidant Capacity

Post-intervention growth differentiation factor 15 (GDF15) was significantly different between the two treatment arms, following adjustment for pre-intervention data ($p = 0.028$, Figure 2A). Circulating GDF15 protein content was significantly increased in the MIS group compared to baseline ($p \leq 0.01$, Δ 42.6 ± 53.8 pg/mL). There were no significance differences between the treatment arms in circulating fibroblast growth factor 21 (FGF21, Figure 2B), interleukin 6 (IL-6, Figure 2C), or leptin ($p \geq 0.05$, Figure 2D) between the two treatment arms, adjusted for their respective pre-intervention values (Table 7).

The circulating EV-associated miRNA species miR-34a ($p = 0.045$) and miR-122 ($p = 0.016$) were significantly different between the treatment arms post-intervention (Figure 2E). EV miRNA miR-34a (~0.5-fold) and miR-122 expression (~0.6-fold) were significantly downregulated in the MIS treatment arm following the intervention ($p < 0.05$, Figure 2E). In the PLA group, miR-34a ($p = 0.570$) and miR-122 expression ($p = 0.061$) were not statistically different from the baseline (Figure 2E). While there were no significant differences between the treatment arms post-intervention ($p \geq 0.05$), EV miRNA-143 was significantly upregulated in the PLA group ($p < 0.05$, Figure 2E, ~1.9-fold). miR-29a and miR-146a were not statistically different between treatment arms following the intervention ($p \geq 0.05$, Figure 2E) and did not change over the intervention period within-the group.

Oxygen Radical Absorbance Capacity (ORAC) within the blood plasma was not significantly different the between treatment arms ($p = 0.072$, Figure 3A) adjusted for pre-intervention values. The MIS group significantly increased plasma ORAC from baseline compared to the endpoint ($p < 0.05$, Δ 6.2 ± 12.8 μM trolox equivalents/mL). Circulating CoQ10 was not significantly different between the intervention arms ($p = 0.137$, Figure 3B) after adjustment for pre-intervention data. Plasma CoQ10 levels were significantly increased in the MIS arm after treatment ($p < 0.05$, Δ 0.09 ± 0.17 μg/mL).

Table 7. Post-intervention indices of molecular markers of inflammation, metabolism, and antioxidant capacity following 12 wks of treatment, adjusted for baseline variables.

Variable (Unit)	Placebo (*n*)	Active (*n*)	*p* Value
GDF15 (pg/mL)	411.1 ± 17.0 (20)	466.0 ± 17.0 (20) *	**0.028**
FGF21 (μg/mL) §,†	253.5 ± 48.0 (20)	313.4 ± 48.0 (20)	0.192
IL-6 (pg/mL) §,†	110.1 ± 11.3 (18)	103.8 ± 10.7 (20)	0.633
Leptin (ng/mL)	28.5 ± 2.0 (20)	25.1 ± 1.9 (19)	0.225
ORAC (μM trolox equivalents/mL)	129.0 ± 2.9 (20)	136.6 ± 2.9 (20) *	0.072
CoQ10 (μg/mL) §,†	1405.3 ± 58.9 (16)	1544.5 ± 51.2 (21) *	0.137

Post-intervention data are presented as adjusted means ± SEs. Post-intervention variables were compared using one-way ANCOVA, with corresponding baseline variables as covariates. Within-group comparisons were made using a paired *t*-test. § Between-group comparisons were made using the Mann–Whitney U test. † Within-group comparisons were made using the non-parametric sign rank test. GDF15, growth differentiation factor 15; FGF21, fibroblast growth factor 21; IL-6, interleukin 6; ORAC, Oxygen Radical Absorbance Capacity. * Denotes statistically significant ($p < 0.05$) within-group difference relative to Day 0 (baseline).

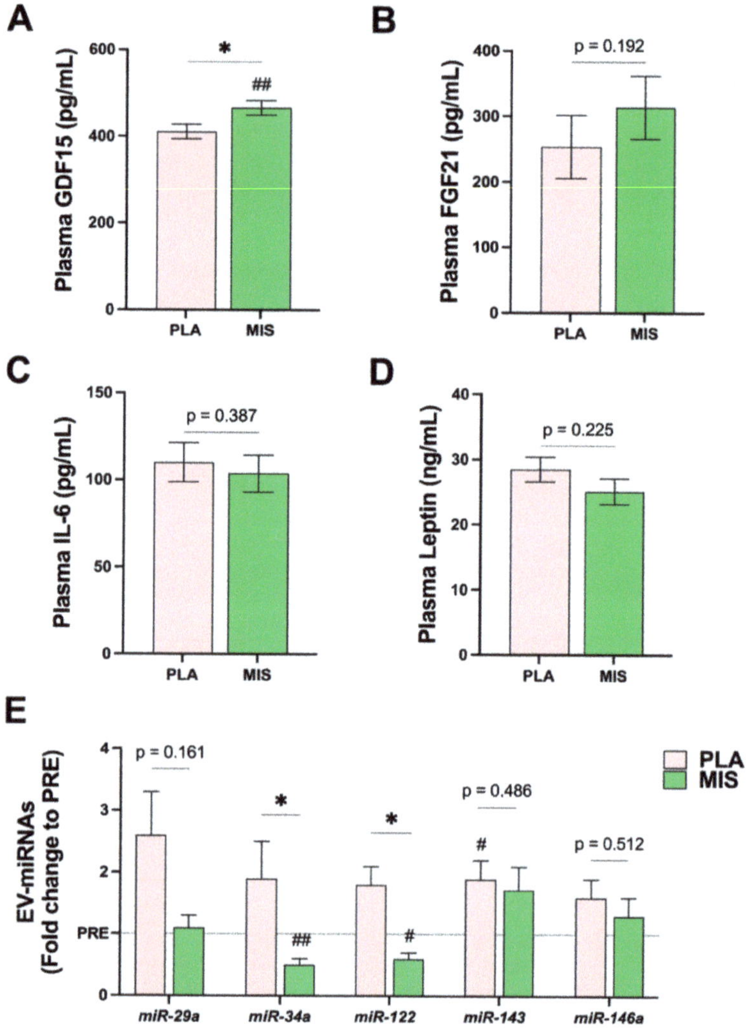

Figure 2. Characterization of plasma and circulating molecular signaling factors following 12 wks a multi-ingredient supplement (MIS, green bars) compared to placebo (PLA, pink bars). (**A**) P intervention plasma GDF15 (pg/mL). Post-intervention differences between treatment arms w compared using one-way ANCOVA, implementing the corresponding pre-intervention variables covariates. Within-group comparisons were made using a paired *t*-test. (**B**) Post-intervention plas FGF21 (pg/mL). (**C**) Post-intervention plasma IL-6 (pg/mL). (**D**) Post-intervention plasma lep (ng/mL). (**E**) Extracellular vesicle (EV)-associated miRNA species shown as fold change from D 0 (baseline), $n = 15$ per group. miRNAs are normalized to exogenous miRNA, *cel-miR-54* add prior to RNA isolation. Between-group comparisons were made using the Mann–Whitney U t Within-group comparisons were made using the non-parametric sign rank test. Asterisks (*) den significant ($p < 0.05$) between-group differences versus PLA, $p \geq 0.05$, *; $p < 0.05$. Number si (#) denote significant within-group differences relative to Day 0 (baseline), #; $p < 0.05$, ##; $p \leq 0$ Post-intervention data are presented as adjusted means ± SEs. GDF15, Growth differentiation fac 15; FGF21, fibroblast growth factor 21; IL-6, interleukin 6; miRNA, microRNA.

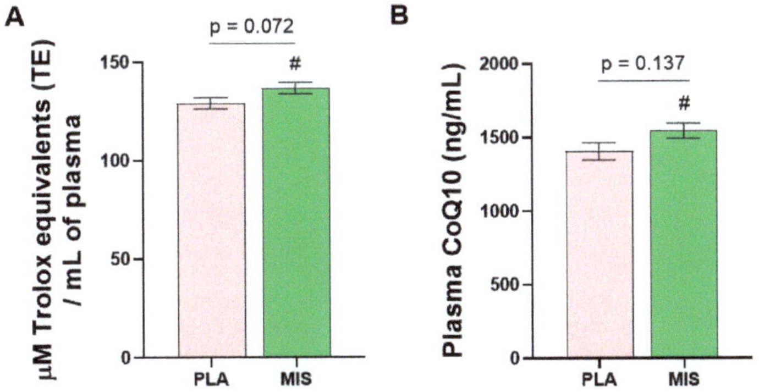

Figure 3. Characterization of plasma antioxidant content and capacity following 12 wks of a multi-ingredient supplement (MIS, green bars) compared to placebo (PLA, pink bars). (**A**) Post-intervention plasma Trolox equivalents. Post-intervention variables were compared using one-way ANCOVA, with corresponding baseline variables as covariates. Within-group differences relative to Day 0 (baseline) were assessed via paired *t*-test (**B**) Post-intervention plasma CoQ10 content. Between-group comparisons were made using the Mann–Whitney U test. Within-group comparisons were made using the non-parametric sign rank test. Number signs (#) denote significant within-group differences relative to Day 0 (baseline), #; $p < 0.05$. Post-intervention data are presented as adjusted means ± SEs. CoQ10; coenzyme Q10.

4. Discussion

We report that the daily consumption of a multi-ingredient supplement (MIS) designed to facilitate weight loss and mitochondrial function [52] resulted in significant loss of body weight, being driven predominately by a reduction in fat mass with the preservation of fat-free mass. In addition, other clinically significant outcomes, such as BMI and WHR, were significantly reduced following a 12 week intervention period in the MIS treatment arm. These changes occurred in the absence of major dietary changes and/or changes in step count (as a marker of physical activity) by participants during the experimental period, suggesting that the observed changes were driven by MIS. We observed a significant improvement in blood biochemical outcomes, including AST and ALT, and a reduction in extracellular vesicle-derived miRNAs associated with obesity, liver dysfunction, and inflammation. Taken together, our findings suggest that a seven-component, multi-ingredient nutritional supplement (MIS) results in significant weight and total fat mass loss in comparison with PLA in overweight and/or obese men and women.

Amongst an overweight population, there is a significantly lower mortality risk for individuals who are classified as overweight (BMI of 25.0–29.9 kg/m^2), as compared to a greater mortality risk for individuals who have a BMI range of 35.0–35.9 [7,78,79]. Similarly, higher body fat content has been shown to be associated with a higher risk of mortality [80]. To counteract the insidious risk of developing co-morbidities associated with being overweight and/or a progression to obesity, a significant weight loss of 5–10% is suggested by health care practitioners [81]. Concomitant with a clinically relevant reduction in weight of ~2.5% (−2.2 kg), we report a fat mass reduction of ~4% in the MIS treatment arm. Despite this significant reduction in weight and fat mass, we observe no overall reduction in FFM or bone mineral density. These findings highlight the potential therapeutic value of nutritional MIS, given the loss of FFM as a part of the total weight loss observed with other weight management strategies. Indeed, bariatric surgery leads to significant, if not massive, weight and fat loss, as well as a significant loss (~30%) of muscle and bone mass [82]. Recent pharmacological interventions, such as glucagon-like peptide-1 receptor agonists and sodium-glucose co-transporter-2 inhibitors (SGLT2i), can facilitate clinically relevant weight loss in individuals with type 2 diabetes [83–85].

There is evidence to suggest that both semaglutide (i.e., an approved GLP-1RA) a canagliflozin (i.e., an approved SGLT-2i) can reduce total FFM [86,87], though whether t has a significant impact on the proportion of lean mass to total mass is still being debat in the literature [87–89]. Even less robust means of weight loss such as dietary ener restriction can present challenges for the preservation of lean body mass. In individu who are overweight or obese, the loss of FFM can be ~20–30% of the total observed weig loss [90–92]. From a mechanistic perspective, Hector and colleagues observed reductic in measured integrated muscle protein synthesis (MPS) assessed via muscle biopsy response to an energy deficit [93] in overweight adult men. Previous work suggested t post-absorptive muscle protein breakdown (MPB) increased by ~60% following 10 da of 20% energy restriction [94], underscoring the findings that extensive dietary ener restriction can lead to a reduction in body mass, with ~25% of FFM/muscle [91]. Giv that skeletal muscle features high metabolic activity and is the largest component of FF a loss of muscle can lead to a reduction in mobility and/or metabolic health. Thus, ca must be taken to preserve the muscle during weight loss [95]. Collectively, our da suggest that MIS can facilitate weight loss while promoting the maintenance of FFM clinically relevant feature of therapeutic strategies to combat obesity and obesity-relat comorbidities. Given that the MIS used in the current study predominantly acts vi different mechanism compared to GLP-1RA medications (browning of white adipose tiss vs. appetite suppression, respectively), future studies should evaluate the potential weight loss synergy and the potential for FFM preservation with concurrent use.

By design, it is not possible to isolate which individual aspects of the seven-compon MIS contributed to our observations; each has been shown to independently facilit weight loss and/or improve mitochondrial function. We have previously reported t 5-, 7-, and 10-component MIS significantly reduced body weight and fat mass, and i proved markers of in vivo metabolism in a murine model [52]. Notably, green tea catech (particularly epigallocatechin-3-gallate) [96] have been well-documented to facilitat reduction in body weight and body fat tissue in humans over a similar experimental riod [97–101]. Mechanistically, green tea (and/or its extracts) may alter SNS activati and lipolysis [102,103], downregulate fatty acid synthase [104], and induce thermoge sis [105]. Similarly, both forskolin [46,106] and green coffee beans [48,107] have been sho to have beneficial effects on weight loss and body re-composition in favor of lower fat ma Forskolin has been shown in vitro and in animal models to activate cyclic AMP, stimul fat oxidation, directly activate hormone-sensitive lipase, and reduce body weight [10 and has been recently shown to improve insulin resistance in obese adults [109]. Chlo genic acid (CA), a bioactive component of coffee beans that is greater in green bean cof compared to the roasted counterpart [110], has been shown to inhibit hepatic glucose phosphatase, the rate-limiting enzyme involved in gluconeogenesis [111], and promote t reduction in hepatic triglycerides in conjunction with caffeine in obese rats [112].

The composition of MIS was designed to also include bioactive compounds that cou facilitate mitochondrial function, given the established relationship between mitochondr dysfunction and the pathogenesis of obesity. Increased consumption of food can le to increased circulating lipids and carbohydrates, providing an 'over-supply' of ener substrates in metabolic tissues, which in turn stimulates the production of ROS [22]. have previously utilized these aspects in a clinical trial in patients with mitochondr disease, ultimately improving mitochondrial function (i.e., lower lactate) and reduci oxidative stress [113]. Vitamin E is an important lipid-soluble antioxidant [114] and may crease fatty liver development [38,115]. Some clinical trials have observed that α-facilita weight loss in humans [116]; whereas, others have observed no significant effect of α-alo on weight loss. However, α-is unequivocally important for mitochondrial function [11 and is a significant scavenger of free radicals [118]. Plasma antioxidant capacity inta has been shown to be correlated with the dietary intake of antioxidants [119]; therefo we confirmed our hypothesis that antioxidant capacity in plasma (as measured by ox gen radical absorbance capacity) would significantly increase in the MIS treatment ar over time. In line with the provision of CoQ10 as an aspect of the 7-component MIS, we observed that participants in the MIS treatment arm had a trend for an elevated CoQ10 following the supplementation period. Absorbed CoQ10, when provided via nutritional supplementation, has a half-life of ~34 h, with T_{max} occurring at ~6 hrs [120]. Given that blood collection was performed ~10–12 h following supplementation (i.e., the last meal), it stands to reason that CoQ10 and antioxidant capacity in the blood would still be elevated if measured earlier following ingestion. Using a pre-clinical model of obesity, we have previously reported significant improvement in body composition (i.e., lower relative fat and higher relative muscle mass) in mice fed a high-fat diet using components including weight loss-inducing agents with concomitant inclusion of ingredients known to facilitate mitochondrial function [52]. Irrespective of the individual contributions of the MIS components, pre-clinical evaluation of the MIS ingredients did lead to an increase in mitochondria in murine white adipose tissue. Taken together, the combined approach utilizing mitochondrial- and fat-loss-targeting ingredients may be uniquely beneficial.

Fatty liver disease, when the triglyceride content of the liver organ weight surpasses 5%, is the most common liver disease associated with individuals who are obese [121]. When independent of alcohol consumption, this disease can be classified as non-alcoholic fatty liver disease (NAFLD). Our examination of alcohol consumption in both the PLA and the MIS treatment arms revealed no significant differences between the groups and a relatively low consumption across participants in the study. The prevalence of NAFLD and the subsequent progression into non-alcoholic steatohepatitis (NASH) is linked to a large degree of obesity, with 65% of individuals presenting with grade I-II obesity (BMI = 30–39.9 kg/m^2) also having hepatic steatosis, increasing to 85% in individuals with grade III obesity (BMI = 40–59 kg/m^2) [122]. ALT is a serum liver enzyme that is mainly concentrated in the cytosol of hepatocytes and is associated with metabolism, and indicates hepatocellular damage [123]. Being overweight has been established as a major risk factor for increased ALT levels and correlates with the high ALT levels observed in those with general obesity [124] and liver diseases [125]. Herein, we found that AST and ALT levels were significantly lower after the intervention period in the MIS treatment arm. Arora and colleagues (2022) demonstrated that ALT levels decrease in correlation with weight loss in patients with obesity and NAFLD, who followed a lifestyle intervention for six months [126]. Similarly, weight loss stemming from low energy intake has been shown to facilitate an improvement in AST [127]. Taken together, while we did not directly examine NAFLD and/or liver health via abdominal ultrasound, we observed a significant improvement in NAFLD-associated clinical biomarkers.

To examine the potential mechanisms of improved hepatic biomarkers, we performed additional molecular analysis. Growth differentiation factor-15 (GDF-15) is a novel cytokine member of the transforming growth factor β superfamily, whose expression increases in response to a variety of stimuli, including metabolic stress [128]. In the current study, 12 wks of MIS significantly increased circulating levels of GDF15 in the active group compared to those in the PLA treatment arm. Interestingly, higher levels of GDF15 have been associated with a higher prevalence of NAFLD in youth with obesity, as well as in overweight adults, with higher levels of circulating GDF15 linked to worsening of liver fibrosis [129,130]. However, this relationship is complex. Upon activation, the GDNF family receptor α–like (GFRAL), located in the hindbrain, can drive a desire to reduce food intake-making interpretation of circulating levels challenging [128]. To compound this challenge, evidence suggests that GFRAL and GDF15 are necessary for the weight loss benefits of metformin [131,132]. Recent studies in murine models have shown that GDF15 promotes weight loss by increasing energy expenditure via a GFRAL–β-adrenergic signaling and counteracts compensatory reductions in energy expenditure during periods of caloric restriction [133]. In humans, increases in GDF15 are seen with metformin-induced weight loss [132] and exercise training [134], suggesting that higher circulating GDF15 levels linked to metabolic disease are not causative, but rather because of the metabolic stress from obesity and NAFLD. In addition, an increase in GDF15 following weight loss in obese adults [135] further confirms the findings that an increase in GDF15 is associat with metabolic improvements following weight loss.

Another potential mechanism via which the MIS may elicit metabolic and hepa benefits is via circulating miRNAs contained within extracellular vesicles. EVs are emergi mechanisms of intercellular communication through the release or shedding of vesic by various cells [136–138], and can be critical for maintaining the health of adipocyte a hepatic cell homeostasis. Obesity [31] and NAFLD can alter miRNAs in adipose tiss and circulation, which may further exacerbate the disease pathology [139]. We founc significant reduction in EV-associated miR-34a and miR-122 following 12 wk interventi in the MIS treatment arm. In mice with fat-induced liver injury, both miR-34a and m 122 have been correlated with hepatic histopathology [140]. In clinical studies, miR3 has been associated with NAFLD, with higher circulating miR34a levels prevalent higher-severity disease phenotypes [141]. Similarly, miR-122 has also been associated w NAFLD, obesity, and insulin resistance [142], and the EVs of mice with fatty liver dise have a significant enrichment of mir-122 [143]. Interestingly, we observed an increas expression of miRNA-143 in the PLA treatment arm over the duration of the interventi period. miRNA-143 has been previously associated with fatty liver disease in obese adu undergoing bariatric surgery [144], as well as patients with NAFLD and coronary art disease [145]. Taken together, a reduction in specific EV-associated miRNAs in the M treatment arm, in conjunction with observed improvements in other indices of hepa function, suggests that altered intercellular communication following treatment may con some hepatic and metabolic benefits.

Our examination of clinical hematology revealed a significant difference betwe the MIS and PLA treatment arms, with a significantly lower eosinophil count followi 12 wks of MIS. Interestingly, eosinophil count has been previously shown to be positiv correlated with BMI up to a plateau of 40 kg/m^2, suggesting that obese patients have elevated eosinophil count [146], with a similar relationship between patients with metabc syndrome and elevated eosinophil count [147]. Future studies should examine whether t provision of MIS can alter allergic diseases such as asthma in obese patients.

An impairment of kidney function, specifically obesity-related glomerulopathy [148–1 suggests that obesity is associated with kidney disease [151,152]. Indeed, large-scale stud suggest that a BMI >30 kg/m^2 with the concurrent risk factors of metabolic syndrome and insu resistance is associated with a higher risk of developing kidney disease [153]. Consequently, small but significant increase in creatinine in the MIS group, despite weight and fat loss, wo seem paradoxical. First, it is important to note that the change in serum creatinine is not clinica significant, with the MIS treatment arm having a post-intervention serum creatine concentrati of ~83 µmol/L. Further, central obesity has been associated with renal hyperfiltration [1 or an abnormally high glomerular filtration rate due to systemic hypertension, which oft coexists in obese individuals [149,155]. Significant weight loss can improve glomerular fl hyperfiltration in obese individuals without overt renal disease [156]. Furthermore, it is w known that creatinine is also proportional to FFM, independent of GFR [150,157], and with preservation of FFM concomitant with a lower total body mass, the ratio of fat-free mass to to mass may increase creatinine.

We acknowledge several limitations of the current study. A possible limitation of t study is that we did not have full control of physical activity and/or dietary intake duri the experimental period. At the onset of the study, we provided simple diet informati (the Canadian Food Guide) and only provided diet and activity guidelines at the incepti to determine the potential effect of an MIS in a free-living scenario without intensi behavioral intervention. The lack of strict dietary and/or exercise controls and the imp of COVID-19 in the third inception cohort were also factors that may have contributed the increase in body weight in the control group and may represent a potential attenuati of weight loss in the MIS group. Finally, the current study was conducted for 12 wks a thus we could not observe the sustainability of the effects over a longer period. Futu studies should address interventions with longer durations.

5. Conclusions

In conclusion, the provision of 12 wks of MIS comprising ingredients known to facilitate mitochondrial function, increase lipid metabolism, and enhance body re-composition [52] was able to significantly reduce total body weight and fat mass, while preserving fat-free mass. Alterations in body composition were independent of alterations in energy intake or physical activity patterns. Further, we found that the MIS treatment arm had improved biochemical and EV-associated miRNA profiles associated with improved hepatic function. Future studies should examine the impact of physical activity in conjunction with MIS to achieve synergistic metabolic benefits.

Supplementary Materials: The following supporting information can be downloaded at: https://www.mdpi.com/article/10.3390/nu15173693/s1, Figure S1: Enrollment process. Figure S2: Study Schematic. Table S1: Drug Extract Ratio (DER) and solvents. Table S2: miRNA assay IDs.

Author Contributions: J.P.N.: Conceptualization, Methodology, Validation, Formal Analysis, Investigation, Writing—Original Draft, Visualization, and Project Administration. A.J.M.: Formal Analysis, Investigation, Reviewing, and Project Administration. D.X.: Formal Analysis, Investigation, Writing, and Reviewing. A.D.C.: Formal Analysis, Investigation, and Reviewing. K.M.: Formal Analysis, Investigation, and Reviewing. M.R.F.: Formal Analysis, Investigation, and Reviewing M.A.T.: Supervision of all aspects, Writing, Reviewing, and Editing. All authors have read and agreed to the published version of the manuscript.

Funding: Funding for the study was provided by the Canadian Institutes of Health Research (CIHR) Foundation Grant (# 143325) awarded to M.A.T., J.P.N. was supported by a CIHR Postdoctoral Fellowship, and K.M. was supported by an Ontario Graduate Scholarship (OGS).

Institutional Review Board Statement: The study was approved by the Hamilton Health Sciences Integrated Research Ethics Board (HIREB-5754), conformed to the guidelines outlined in the Declaration of Helsinki, and complied with the guidelines set out in the Canadian Tri-Council policy statement on ethical conduct for research involving humans. The trial was registered at clinicaltrials.gov (NCT03812211). The study was conducted under Good Clinical Practice (GCP) guidelines.

Informed Consent Statement: All participants were informed of the nature of the study and its potential risks before providing their written informed consent.

Data Availability Statement: Not applicable.

Acknowledgments: The authors would like to thank the Hamilton Regional Laboratory Medicine Program–McMaster Site Core Laboratory for their assistance in clinical blood analysis. We would like to thank Mahek Minhas for her technical assistance. Data are available upon reasonable request to the corresponding author. Figures were created with Biorender.

Conflicts of Interest: Exerkine Corporation is a biotechnology company that develops and commercializes therapies based on supplements, exercise-derived factors ('exerkines'), and extracellular vesicles to treat genetic disorders, chronic diseases, and aging. MAT is the founder, CEO, and CSO of Exerkine Corporation. At the time of publication submission, a Canadian patent (CA 3050823) was granted for the supplement described herein. The author warrants that the COI did not impact the execution of the research, including data collection, analyses, and interpretation, or in the decision to publish the manuscript.

erences

Chooi, Y.C.; Ding, C.; Magkos, F. The epidemiology of obesity. *Metabolism* **2019**, *92*, 6–10. [CrossRef]
Hales, C.M.; Carroll, M.D.; Fryar, C.D.; Ogden, C.L. *Prevalence of Obesity and Severe Obesity among Adults: United States, 2017–2018 Key Findings Data from the National Health and Nutrition Examination Survey*; Centers for Disease Control and Prevention: Antlanta, GA, USA, 2017.
Roberto, C.A.; Swinburn, B.; Hawkes, C.; Huang, T.T.K.; Costa, S.A.; Ashe, M.; Zwicker, L.; Cawley, J.H.; Brownell, K.D. Patchy progress on obesity prevention: Emerging examples, entrenched barriers, and new thinking. *Lancet* **2015**, *385*, 2400–2409. [CrossRef] [PubMed]
Di Cesare, M.; Bentham, J.; Stevens, G.A.; Zhou, B.; Danaei, G.; Lu, Y.; Bixby, H.; Cowan, M.J.; Riley, L.M.; Hajifathalian, K.; et al. Trends in adult body-mass index in 200 countries from 1975 to 2014: A pooled analysis of 1698 population-based measurement studies with 19.2 million participants. *Lancet* **2016**, *387*, 1377–1396. [CrossRef]

5. Afshin, A.; Forouzanfar, M.H.; Reitsma, M.B.; Sur, P.; Estep, K.; Lee, A.; Marczak, L.; Mokdad, A.H.; Moradi-Lakeh, M.; Nagh M.; et al. Health effects of overweight and obesity in 195 countries over 25 years. *N. Engl. J. Med.* **2017**, *377*, 13–27. [CrossR [PubMed]
6. Bancej, C.; Jayabalasingham, B.; Wall, R.W.; Rao, D.P.; Do, M.T.; de Groh, M.; Jayaraman, G.C. Trends and projections of obe among Canadians. *Health Promot. Chronic Dis. Prev. Can.* **2015**, *35*, 109–112. [CrossRef] [PubMed]
7. Flegal, K.M.; Kruszon-Moran, D.; Carroll, M.D.; Fryar, C.D.; Ogden, C.L. Trends in obesity among adults in the United Sta 2005 to 2014. *JAMA J. Am. Med. Assoc.* **2016**, *315*, 2284–2291. [CrossRef]
8. Ng, M.; Fleming, T.; Robinson, M.; Thomson, B.; Graetz, N.; Margono, C.; Mullany, E.C.; Biryukov, S.; Abbafati, C.; Abera, S et al. Global, regional, and national prevalence of overweight and obesity in children and adults during 1980–2013: A system analysis for the Global Burden of Disease Study 2013. *Lancet* **2014**, *384*, 766–781. [CrossRef]
9. Willett, W.C.; Manson, J.E.; Stampfer, M.J.; Colditz, G.A.; Rosner, B.; Speizer, F.E.; Hennekens, C.H. Weight, weight change, a coronary heart disease in women. Risk within the "normal" weight range. *JAMA* **1995**, *273*, 461–465. [CrossRef]
10. Manson, J.E.; Willett, W.C.; Stampfer, M.J.; Colditz, G.A.; Hunter, D.J.; Hankinson, S.E.; Hennekens, C.H.; Speizer, F.E. Bo Weight and Mortality among Women. *N. Engl. J. Med.* **1995**, *333*, 677–685. [CrossRef]
11. Pajares, F. Self-Efficacy Beliefs in Academic Settings. *Rev. Educ. Res.* **1996**, *66*, 543–578. [CrossRef]
12. Calle, E.E.; Rodriguez, C.; Walker-Thurmond, K.; Thun, M.J. Overweight, Obesity, and Mortality from Cancer in a Prospectiv Studied Cohort of U.S. Adults. *N. Engl. J. Med.* **2003**, *348*, 1625–1638. [CrossRef]
13. Yusuf, S.; Hawken, S.; Ôunpuu, S.; Bautista, L.; Franzosi, M.G.; Commerford, P.; Lang, C.C.; Rumboldt, Z.; Onen, C.L.; Lisheng, et al. Obesity and the risk of myocardial infarction in 27,000 participants from 52 countries: A case-control study. *Lancet* **2005**, 3 1640–1649. [CrossRef] [PubMed]
14. Brawer, R.; Brisbon, N.; Plumb, J. Obesity and Cancer. *Prim. Care Clin. Off. Pract.* **2009**, *36*, 509–531. [CrossRef] [PubMed]
15. Bray, G.A. From farm to fat cell: Why aren't we all fat? *Metabolism* **2015**, *64*, 349–353. [CrossRef]
16. Colley, R.C.; Garriguet, D.; Janssen, I.; Craig, C.L.; Clarke, J.; Tremblay, M.S. Physical Activity of Canadian Adults:Accelerome Results from the 2007 to 2009 Canadian Health Measures Survey. *Health Rep.* **2011**, *22*, 7.
17. Tappe, M.K.; Duda, J.L.; Ehrnwald, P.M. Perceived barriers to exercise among adolescents. *J. Sch. Health* **1989**, *59*, 153–1 [CrossRef]
18. Muktabhant, B.; Lawrie, T.A.; Lumbiganon, P.; Laopaiboon, M. Diet or exercise, or both, for preventing excessive weight gain pregnancy. *Cochrane Database Syst. Rev.* **2015**, *2015*, CD007145. [CrossRef]
19. Mc Morrow, L.; Ludbrook, A.; Macdiarmid, J.I.; Olajide, D. Perceived barriers towards healthy eating and their association w fruit and vegetable consumption. *J. Public Health* **2016**, *39*, fdw038. [CrossRef]
20. Matsuda, M.; Shimomura, I. Increased oxidative stress in obesity: Implications for metabolic syndrome, diabetes, hypertensi dyslipidemia, atherosclerosis, and cancer. *Obes. Res. Clin. Pract.* **2013**, *7*, e330–e341. [CrossRef]
21. Bondia-Pons, I.; Ryan, L.; Martinez, J.A. Oxidative stress and inflammation interactions in human obesity. *J. Physiol. Bioch* **2012**, *68*, 701–711. [CrossRef] [PubMed]
22. McMurray, F.; Patten, D.A.; Harper, M.E. Reactive Oxygen Species and Oxidative Stress in Obesity—Recent Findings a Empirical Approaches. *Obesity* **2016**, *24*, 2301–2310. [CrossRef]
23. Bonomini, F.; Rodella, L.F.; Rezzani, R. Metabolic syndrome, aging and involvement of oxidative stress. *Aging Dis.* **2015** 109–120. [CrossRef] [PubMed]
24. Lin, Y.; Berg, A.H.; Iyengar, P.; Lam, T.K.T.; Giacca, A.; Combs, T.P.; Rajala, M.W.; Du, X.; Rollman, B.; Li, W.; et al. T hyperglycemia-induced inflammatory response in adipocytes: The role of reactive oxygen species. *J. Biol. Chem.* **2005**, 2 4617–4626. [CrossRef]
25. Talior, I.; Yarkoni, M.; Bashan, N.; Eldar-Finkelman, H. Increased glucose uptake promotes oxidative stress and PKC-δ activati in adipocytes of obese, insulin-resistant mice. *Am. J. Physiol. Endocrinol. Metab.* **2003**, *285*, E295–E302. [CrossRef] [PubMed]
26. Harijith, A.; Ebenezer, D.L.; Natarajan, V. Reactive oxygen species at the crossroads of inflammasome and inflammation. *Fr Physiol.* **2014**, *5*, 352. [CrossRef]
27. Herder, C.; Schneitler, S.; Rathmann, W.; Haastert, B.; Schneitler, H.; Winkler, H.; Bredahl, R.; Hahnloser, E.; Martin, S. Low-gra inflammation, obesity, and insulin resistance in adolescents. *J. Clin. Endocrinol. Metab.* **2007**, *92*, 4569–4574. [CrossRef]
28. Johnson, A.R.; Justin Milner, J.; Makowski, L. The inflammation highway: Metabolism accelerates inflammatory traffic in obes *Immunol. Rev.* **2012**, *249*, 218–238. [CrossRef] [PubMed]
29. Iacomino, G.; Siani, A. Role of microRNAs in obesity and obesity-related diseases. *Genes Nutr.* **2017**, *12*, 23. [CrossRef]
30. Ji, C.; Guo, X. The clinical potential of circulating microRNAs in obesity. *Nat. Rev. Endocrinol.* **2019**, *15*, 731–743. [CrossR [PubMed]
31. Landrier, J.F.; Derghal, A.; Mounien, L. MicroRNAs in Obesity and Related Metabolic Disorders. *Cells* **2019**, *8*, 859. [CrossRef
32. Yáñez-Mó, M.; Siljander, P.R.M.R.-M.; Andreu, Z.; Bedina Zavec, A.; Borràs, F.E.; Buzas, E.I.; Buzas, K.; Casal, E.; Cappel F.; Carvalho, J.; et al. Biological properties of extracellular vesicles and their physiological functions. *J. Extracell. Vesicles* **20** *4*, 27066. [CrossRef]
33. Lei, L.M.; Lin, X.; Xu, F.; Shan, S.K.; Guo, B.; Li, F.X.Z.; Zheng, M.H.; Wang, Y.; Xu, Q.S.; Yuan, L.Q. Exosomes and Obesity-Rela Insulin Resistance. *Front. Cell Dev. Biol.* **2021**, *9*, 588. [CrossRef] [PubMed]

Kim, A.; Shah, A.; Nakamura, T. Extracellular Vesicles: A Potential Novel Regulator of Obesity and Its Associated Complications. *Children* **2018**, *5*, 152. [CrossRef]
Cordero, M.D.; Alcocer-Gómez, E.; De Miguel, M.; Cano-García, F.J.; Luque, C.M.; Fernández-Riejo, P.; Fernández, A.M.M.; Sánchez-Alcazar, J.A. Coenzyme Q10: A novel therapeutic approach for Fibromyalgia? Case series with 5 patients. *Mitochondrion* **2011**, *11*, 623–625. [CrossRef] [PubMed]
Sangsefidi, Z.S.; Yaghoubi, F.; Hajiahmadi, S.; Hosseinzadeh, M. The effect of coenzyme Q10 supplementation on oxidative stress: A systematic review and meta-analysis of randomized controlled clinical trials. *Food Sci. Nutr.* **2020**, *8*, 1766–1776. [CrossRef] [PubMed]
Chow, C.K. Vitamin E Regulation of Mitochondrial Superoxide Generation. *Neurosignals* **2001**, *10*, 112–124. [CrossRef]
Alcalá, M.; Sánchez-Vera, I.; Sevillano, J.; Herrero, L.; Serra, D.; Ramos, M.P.; Viana, M. Vitamin E reduces adipose tissue fibrosis, inflammation, and oxidative stress and improves metabolic profile in obesity. *Obesity* **2015**, *23*, 1598–1606. [CrossRef]
Miyazawa, T.; Burdeos, G.C.; Itaya, M.; Nakagawa, K.; Miyazawa, T. Vitamin E: Regulatory Redox Interactions. *IUBMB Life* **2019**, *71*, 430–441. [CrossRef]
Zembron-Lacny, A.; Slowinska-Lisowska, M.; Szygula, Z.; Witkowski, K.; Stefaniak, T.; Dziubek, W. Assessment of the antioxidant effectiveness of alpha-lipoic acid in healthy men exposed to muscle-damaging exercise. *J. Physiol. Pharmacol.* **2009**, *60*, 139–143.
Martins, V.D.; Manfredini, V.; Peralba, M.C.R.; Benfato, M.S. Alpha-lipoic acid modifies oxidative stress parameters in sickle cell trait subjects and sickle cell patients. *Clin. Nutr.* **2009**, *28*, 192–197. [CrossRef]
Larsen, F.J.; Schiffer, T.A.; Borniquel, S.; Sahlin, K.; Ekblom, B.; Lundberg, J.O.; Weitzberg, E. Dietary inorganic nitrate improves mitochondrial efficiency in humans. *Cell Metab.* **2011**, *13*, 149–159. [CrossRef] [PubMed]
Clifford, T.; Howatson, G.; West, D.J.; Stevenson, E.J. The potential benefits of red beetroot supplementation in health and disease. *Nutrients* **2015**, *7*, 2801–2822. [CrossRef] [PubMed]
Wootton-Beard, P.C.; Brandt, K.; Fell, D.; Warner, S.; Ryan, L. Effects of a beetroot juice with high neobetanin content on the early-phase insulin response in healthy volunteers. *J. Nutr. Sci.* **2014**, *3*, e9. [CrossRef]
Tesoriere, L.; Fazzari, M.; Angileri, F.; Gentile, C.; Livrea, M.A. In vitro digestion of betalainic foods. Stability and bioaccessibility of betaxanthins and betacyanins and antioxidative potential of food digesta. *J. Agric. Food Chem.* **2008**, *56*, 10487–10492. [CrossRef]
Godard, M.P.; Johnson, B.A.; Richmond, S.R. Body Composition and Hormonal Adaptations Associated with Forskolin Consumption in Overweight and Obese Men. *Obes. Res.* **2005**, *13*, 1335–1343. [CrossRef] [PubMed]
Litosch, I.; Hudson, T.H.; Mills, I.; Li, S.Y.; Fain, J.N. Forskolin as an activator of cyclic AMP accumulation and lipolysis in rat adipocytes. *Mol. Pharmacol.* **1982**, *22*, 109–115.
Thom, E. The Effect of Chlorogenic Acid Enriched Coffee on Glucose Absorption in Healthy Volunteers and Its Effect on Body Mass When Used Long-term in Overweight and Obese People. *J. Int. Med. Res.* **2007**, *35*, 900–908. [CrossRef]
Revuelta-Iniesta, R.; Al-Dujaili, E.A.S. Consumption of Green Coffee Reduces Blood Pressure and Body Composition by Influencing 11 β-HSD1 Enzyme Activity in Healthy Individuals: A Pilot Crossover Study Using Green and Black Coffee. *BioMed Res. Int.* **2014**, *2014*, 482704. [CrossRef]
Huang, J.; Wang, Y.; Xie, Z.; Zhou, Y.; Zhang, Y.; Wan, X. The anti-obesity effects of green tea in human intervention and basic molecular studies. *Eur. J. Clin. Nutr.* **2014**, *68*, 1075–1087. [CrossRef]
Barrea, L.; Altieri, B.; Polese, B.; De Conno, B.; Muscogiuri, G.; Colao, A.; Savastano, S. Nutritionist and obesity: Brief overview on efficacy, safety, and drug interactions of the main weight-loss dietary supplements. *Int. J. Obes. Suppl.* **2019**, *9*, 32–49. [CrossRef]
Nederveen, J.P.; Manta, K.; Bujak, A.L.; Simone, A.C.; Fuda, M.R.; Nilsson, M.I.; Hettinga, B.P.; Hughes, M.C.; Perry, C.G.R.; Tarnopolsky, M.A. A Novel Multi-Ingredient Supplement Activates a Browning Program in White Adipose Tissue and Mitigates Weight Gain in High-Fat Diet-Fed Mice. *Nutrients* **2021**, *13*, 3726. [CrossRef]
Pi-Sunyer, F.X.; Becker, D.M.; Bouchard, C.; Carleton, R.A.; Colditz, G.A.; Dietz, W.H.; Foreyt, J.P.; Garrison, R.J.; Grundy, S.M.; Hansen, B.C.; et al. Clinical guidelines on the identification, evaluation, and treatment of overweight and obesity in adults: Executive summary. *Am. J. Clin. Nutr.* **1998**, *68*, 899–917.
Ashwell, M.; Gunn, P.; Gibson, S. Waist-to-height ratio is a better screening tool than waist circumference and BMI for adult cardiometabolic risk factors: Systematic review and meta-analysis. *Obes. Rev.* **2012**, *13*, 275–286. [CrossRef] [PubMed]
De Koning, L.; Merchant, A.T.; Pogue, J.; Anand, S.S. Waist circumference and waist-to-hip ratio as predictors of cardiovascular events: Meta-regression analysis of prospective studies. *Eur. Heart J.* **2007**, *28*, 850–856. [CrossRef] [PubMed]
Brazier, J.E.; Harper, R.; Jones, N.M.B.; O'Cathain, A.; Thomas, K.J.; Usherwood, T.; Westlake, L. Validating the SF-36 health survey questionnaire: New outcome measure for primary care. *Br. Med. J.* **1992**, *305*, 160–164. [CrossRef] [PubMed]
Scott, N.W.; McPherson, G.C.; Ramsay, C.R.; Campbell, M.K. The method of minimization for allocation to clinical trials: A review. *Control. Clin. Trials* **2002**, *23*, 662–674. [CrossRef]
Altman, D.G.; Bland, J.M. Treatment allocation by minimisation. *BMJ* **2005**, *330*, 843. [CrossRef]
Tremblay, M.S.; Warburton, D.E.R.; Janssen, I.; Paterson, D.H.; Latimer, A.E.; Rhodes, R.E.; Kho, M.E.; Hicks, A.; LeBlanc, A.G.; Zehr, L.; et al. New Canadian physical activity guidelines. *Appl. Physiol. Nutr. Metab.* **2011**, *36*, 36–46. [CrossRef]
Nishida, C.; Ko, G.T.; Kumanyika, S. Body fat distribution and noncommunicable diseases in populations: Overview of the 2008 WHO Expert Consultation on Waist Circumference and Waist-Hip Ratio. *Eur. J. Clin. Nutr.* **2010**, *64*, 2–5. [CrossRef]
Albanese, C.V.; Diessel, E.; Genant, H.K. Clinical applications of body composition measurements using DXA. *J. Clin. Densitom.* **2003**, *6*, 75–85. [CrossRef]

62. Prior, B.M.; Cureton, K.J.; Modlesky, C.M.; Evans, E.M.; Sloniger, M.A.; Saunders, M.; Lewis, R.D. In vivo validation of wh body composition estimates from dual-energy X-ray absorptiometry. *J. Appl. Physiol.* **1997**, *83*, 623–630. [CrossRef] [PubMed
63. Garg, M.; Kharb, S. Dual energy X-ray absorptiometry: Pitfalls in measurement and interpretation of bone mineral density. *Ind J. Endocrinol. Metab.* **2013**, *17*, 203. [CrossRef]
64. Salgado, A.L.F.D.A.; De Carvalho, L.; Oliveira, A.C.; Dos Santos, V.N.; Vieira, J.G.; Parise, E.R. Insulin resistance index (HOMA- in the differentiation of patients with non-alcoholic fatty liver disease and healthy individuals. *Arq. Gastroenterol.* **2010**, 165–169. [CrossRef] [PubMed]
65. Schmittgen, T.D.; Livak, K.J. Analyzing real-time PCR data by the comparative C(T) method. *Nat. Protoc.* **2008**, *3*, 1101–1 [CrossRef]
66. Weir, J.B.D.V. New methods for calculating metabolic rate with special reference to protein metabolism. *J. Physiol.* **1949**, *109*, [CrossRef] [PubMed]
67. Ferrannini, E. The theoretical bases of indirect calorimetry: A review. *Metabolism* **1988**, *37*, 287–301. [CrossRef]
68. Mifflin, M.D.; St Jeor, S.T.; Hill, L.A.; Scott, B.J.; Daugherty, S.A.; Koh, Y.O. A new predictive equation for resting ene expenditure in healthy individuals. *Am. J. Clin. Nutr.* **1990**, *51*, 241–247. [CrossRef]
69. Weijs, P.J.M. Validity of predictive equations for resting energy expenditure in US and Dutch overweight and obese class I an adults aged 18–65 y. *Am. J. Clin. Nutr.* **2008**, *88*, 959–970. [CrossRef]
70. Frankenfield, D.C.; Rowe, W.A.; Smith, J.S.; Cooney, R.N. Validation of several established equations for resting metabolic rat obese and nonobese people. *J. Am. Diet. Assoc.* **2003**, *103*, 1152–1159. [CrossRef]
71. Lins, L.; Carvalho, F.M. SF-36 total score as a single measure of health-related quality of life: Scoping review. *SAGE Open M* **2016**, *4*, 205031211667172. [CrossRef]
72. Hays, R.D.; Morales, L.S. The RAND-36 measure of health-related quality of life. *Ann. Med.* **2001**, *33*, 350–357. [CrossR [PubMed]
73. Ware, J.E.J.; Sherbourne, C.D. The MOS 36-item short-form health survey (SF-36). I. Conceptual framework and item selecti *Med. Care* **1992**, *30*, 473–483. [CrossRef] [PubMed]
74. Blissmer, B.; Riebe, D.; Dye, G.; Ruggiero, L.; Greene, G.; Caldwell, M. Health-related quality of life following a clinical wei loss intervention among overweight and obese adults: Intervention and 24 month follow-up effects. *Health Qual. Life Outcon* **2006**, *4*, 43. [CrossRef]
75. Toolsee, N.A.; Aruoma, O.I.; Gunness, T.K.; Kowlessur, S.; Dambala, V.; Murad, F.; Googoolye, K.; Daus, D.; Indelicato Rondeau, P.; et al. Effectiveness of green tea in a randomized human cohort: Relevance to diabetes and its complications. *BioM Res. Int.* **2013**, *2013*, 412379. [CrossRef]
76. Verheggen, R.J.H.M.; Maessen, M.F.H.; Green, D.J.; Hermus, A.R.M.M.; Hopman, M.T.E.; Thijssen, D.H.T. A systematic revi and meta-analysis on the effects of exercise training versus hypocaloric diet: Distinct effects on body weight and visceral adip tissue. *Obes. Rev.* **2016**, *17*, 664–690. [CrossRef]
77. Elobeid, M.A.; Padilla, M.A.; McVie, T.; Thomas, O.; Brock, D.W.; Musser, B.; Lu, K.; Coffey, C.S.; Desmond, R.A.; St-Onge, M et al. Missing data in randomized clinical trials for weight loss: Scope of the problem, state of the field, and performance statistical methods. *PLoS ONE* **2009**, *4*, e6624. [CrossRef]
78. The Emerging Risk Factors Collaboration; Wormser, D.; Kaptoge, S.; Di Angelantonio, E.; Wood, A.M.; Pennells, L.; Thomps A.; Sarwar, N.; Kizer, J.R.; Lawlor, D.A.; et al. Separate and combined associations of body-mass index and abdominal adipos with cardiovascular disease: Collaborative analysis of 58 prospective studies. *Lancet* **2011**, *377*, 1085–1095. [CrossRef]
79. MacMahon, S.; Baigent, C.; Duffy, S.; Rodgers, A.; Tominaga, S.; Chambless, L.; De Backer, G.; De Bacquer, D.; Kornitzer, Whincup, P.; et al. Body-mass index and cause-specific mortality in 900,000 adults: Collaborative analyses of 57 prospect studies. *Lancet* **2009**, *373*, 1083–1096. [CrossRef]
80. Jayedi, A.; Khan, T.A.; Aune, D.; Emadi, A.; Shab-Bidar, S. Body fat and risk of all-cause mortality: A systematic review a dose-response meta-analysis of prospective cohort studies. *Int. J. Obes.* **2022**, *46*, 1573–1581. [CrossRef] [PubMed]
81. Jensen, M.D.; Ryan, D.H.; Apovian, C.M.; Ard, J.D.; Comuzzie, A.G.; Donato, K.A.; Hu, F.B.; Hubbard, V.S.; Jakicic, J.M.; Kush R.F.; et al. 2013 AHA/ACC/TOS guideline for the management of overweight and obesity in adults: A report of the Americ college of cardiology/American heart association task force on practice guidelines and the obesity society. *J. Am. Coll. Card* **2014**, *63*, 2985–3023. [CrossRef]
82. Chaston, T.B.; Dixon, J.B.; O'Brien, P.E. Changes in fat-free mass during significant weight loss: A systematic review. *Int. J. O* **2007**, *31*, 743–750. [CrossRef] [PubMed]
83. Van Gaal, L.; Scheen, A. Weight management in type 2 diabetes: Current and emerging approaches to treatment. *Diabetes C* **2015**, *38*, 1161–1172. [CrossRef]
84. Sorli, C.; Harashima, S.; Tsoukas, G.M.; Unger, J.; Karsbøl, J.D.; Hansen, T.; Bain, S.C. Efficacy and safety of once-wee semaglutide monotherapy versus placebo in patients with type 2 diabetes (SUSTAIN 1): A double-blind, randomised, place controlled, parallel-group, multinational, multicentre phase 3a trial. *Lancet Diabetes Endocrinol.* **2017**, *5*, 251–260. [CrossR [PubMed]
85. Jakher, H.; Chang, T.I.; Tan, M.; Mahaffey, K.W. Canagliflozin review—Safety and efficacy profile in patients with T2DM. *Diabe Metab. Syndr. Obes. Targets Ther.* **2019**, *12*, 209–215. [CrossRef] [PubMed]

McCrimmon, R.J.; Catarig, A.M.; Frias, J.P.; Lausvig, N.L.; le Roux, C.W.; Thielke, D.; Lingvay, I. Effects of once-weekly semaglutide vs once-daily canagliflozin on body composition in type 2 diabetes: A substudy of the SUSTAIN 8 randomised controlled clinical trial. *Diabetologia* **2020**, *63*, 473–485. [CrossRef]
Wilding, J.P.H.; Batterham, R.L.; Calanna, S.; Davies, M.; Van Gaal, L.F.; Lingvay, I.; McGowan, B.M.; Rosenstock, J.; Tran, M.T.D.; Wadden, T.A.; et al. Once-Weekly Semaglutide in Adults with Overweight or Obesity. *N. Engl. J. Med.* **2021**, *384*, 989–1002. [CrossRef]
Ida, S.; Kaneko, R.; Imataka, K.; Okubo, K.; Shirakura, Y.; Azuma, K.; Fujiwara, R.; Murata, K. Effects of Antidiabetic Drugs on Muscle Mass in Type 2 Diabetes Mellitus. *Curr. Diabetes Rev.* **2020**, *17*, 293–303. [CrossRef]
Hong, Y.; Lee, J.H.; Jeong, K.W.; Choi, C.S.; Jun, H.S. Amelioration of muscle wasting by glucagon-like peptide-1 receptor agonist in muscle atrophy. *J. Cachexia Sarcopenia Muscle* **2019**, *10*, 903–918. [CrossRef]
Ross, R. Reduction in Obesity and Related Comorbid Conditions after Diet-Induced Weight Loss or Exercise-Induced Weight Loss in Men. *Ann. Intern. Med.* **2000**, *133*, 92. [CrossRef]
Wolfe, R.R. The underappreciated role of muscle in health and disease. *Am. J. Clin. Nutr.* **2006**, *84*, 475–482. [CrossRef]
Mittendorfer, B.; Magkos, F.; Fabbrini, E.; Mohammed, B.S.; Klein, S. Relationship between body fat mass and free fatty acid kinetics in men and women. *Obesity* **2009**, *17*, 1872–1877. [CrossRef] [PubMed]
Hector, A.J.; McGlory, C.; Damas, F.; Mazara, N.; Baker, S.K.; Phillips, S.M. Pronounced energy restriction with elevated protein intake results in no change in proteolysis and reductions in skeletal muscle protein synthesis that are mitigated by resistance exercise. *FASEB J.* **2018**, *32*, 265–275. [CrossRef]
Carbone, J.W.; Pasiakos, S.M.; Vislocky, L.M.; Anderson, J.M.; Rodriguez, N.R. Effects of short-term energy deficit on muscle protein breakdown and intramuscular proteolysis in normal-weight young adults. *Appl. Physiol. Nutr. Metab.* **2014**, *39*, 960–968. [CrossRef] [PubMed]
Cava, E.; Yeat, N.C.; Mittendorfer, B. Preserving healthy muscle during weight loss. *Adv. Nutr.* **2017**, *8*, 511–519. [CrossRef]
Yang, C.S.; Landau, J.M. Effects of tea consumption nutrition health. *J. Nutr.* **2000**, *130*, 2409–2412. [CrossRef]
Chantre, P.; Lairon, D. Recent findings of green tea extract AR25 (exolise) and its activity for the treatment of obesity. *Phytomedicine* **2002**, *9*, 3–8. [CrossRef] [PubMed]
Nagao, T.; Komine, Y.; Soga, S.; Meguro, S.; Hase, T.; Tanaka, Y.; Tokimitsu, I. Ingestion of a tea rich in catechins leads to a reduction in body fat and malondialdehyde-modified LDL in men. *Am. J. Clin. Nutr.* **2005**, *81*, 122–129. [CrossRef]
Nagao, T.; Hase, T.; Tokimitsu, I. A green tea extract high in catechins reduces body fat and cardiovascular risks in humans. *Obesity* **2007**, *15*, 1473–1483. [CrossRef]
Hase, T.; Komine, Y.; Meguro, S.; Takeda, Y.; Takahashi, H.; Matsui, Y.; Inaoka, S.; Katsuragi, Y.; Tokimitsu, I.; Shimasaki, H.; et al. Anti-obesity Effects of Tea Catechins in Humans. *J. Oleo Sci.* **2001**, *50*, 599–605. [CrossRef]
Wang, H.; Wen, Y.; Du, Y.; Yan, X.; Guo, H.; Rycroft, J.A.; Boon, N.; Kovacs, E.M.R.; Mela, D.J. Effects of catechin enriched green tea on body composition. *Obesity* **2010**, *18*, 773–779. [CrossRef]
Dulloo, A.G.; Seydoux, J.; Girardier, L.; Chantre, P.; Vandermander, J. Green tea and thermogenesis: Interactions between catechin-polyphenols, caffeine and sympathetic activity. *Int. J. Obes.* **2000**, *24*, 252–258. [CrossRef] [PubMed]
Dulloo, A.G.; Duret, C.; Rohrer, D.; Girardier, L.; Mensi, N.; Fathi, M.; Chantre, P.; Vandermander, J. Efficacy of a green tea extract rich in catechin polyphenols and caffeine in increasing 24-h energy expenditure and fat oxidation in humans. *Am. J. Clin. Nutr.* **1999**, *70*, 1040–1045. [CrossRef]
Zhang, R.; Xiao, W.; Wang, X.; Wu, X.; Tian, W. Novel inhibitors of fatty-acid synthase from green tea (*Camellia sinensis* Xihu Longjing) with high activity and a new reacting site. *Biotechnol. Appl. Biochem.* **2006**, *43*, 1–7. [CrossRef] [PubMed]
Diepvens, K.; Kovacs, E.M.R.; Nijs, I.M.T.; Vogels, N.; Westerterp-Plantenga, M.S. Effect of green tea on resting energy expenditure and substrate oxidation during weight loss in overweight females. *Br. J. Nutr.* **2005**, *94*, 1026–1034. [CrossRef]
Henderson, S.; Magu, B.; Rasmussen, C.; Lancaster, S.; Kerksick, C.; Smith, P.; Melton, C.; Cowan, P.; Greenwood, M.; Earnest, C.; et al. Effects of Coleus Forskohlii Supplementation on Body Composition and Hematological Profiles in Mildly Overweight Women. *J. Int. Soc. Sports Nutr.* **2005**, *2*, 54. [CrossRef]
Stohs, S.J.; Badmaev, V. A Review of Natural Stimulant and Non-stimulant Thermogenic Agents. *Phytother. Res.* **2016**, *30*, 732–740. [CrossRef]
Shivaprasad, H.N.; Gopalakrishna, S.; Mariyanna, B.; Thekkoot, M.; Reddy, R.; Tippeswamy, B.S. Effect of Coleus forskohliiextract on cafeteria diet-induced obesity in rats. *Pharmacogn. Res.* **2014**, *6*, 42–45. [CrossRef]
Loftus, H.L.; Astell, K.J.; Mathai, M.L.; Su, X.Q. Coleus forskohlii extract supplementation in conjunction with a hypocaloric diet reduces the risk factors of metabolic syndrome in overweight and obese subjects: A randomized controlled trial. *Nutrients* **2015**, *7*, 9508–9522. [CrossRef]
Del Castillo, M.D.; Ames, J.M.; Gordon, M.H. Effect of roasting on the antioxidant activity of coffee brews. *J. Agric. Food Chem.* **2002**, *50*, 3698–3703. [CrossRef]
Arion, W.J.; Canfield, W.K.; Ramos, F.C.; Schindler, P.W.; Burger, H.J.; Hemmerle, H.; Schubert, G.; Below, P.; Herling, A.W. Chlorogenic acid and hydroxynitrobenzaldehyde: New inhibitors of hepatic glucose 6-phosphatase. *Arch. Biochem. Biophys.* **1997**, *339*, 315–322. [CrossRef] [PubMed]
Shimoda, H.; Seki, E.; Aitani, M. Inhibitory effect of green coffee bean extract on fat accumulation and body weight gain in mice. *BMC Complement. Altern. Med.* **2006**, *6*, 9. [CrossRef] [PubMed]

113. Rodriguez, M.C.; MacDonald, J.R.; Mahoney, D.J.; Parise, G.; Beal, M.F.; Tarnopolsky, M.A. Beneficial effects of creatine, CoQ and lipoic acid in mitochondrial disorders. *Muscle Nerve* **2007**, *35*, 235–242. [CrossRef]
114. Traber, M.G.; Atkinson, J. Vitamin E, antioxidant and nothing more. *Free Radic. Biol. Med.* **2007**, *43*, 4–15. [CrossRef]
115. Alcala, M.; Calderon-Dominguez, M.; Serra, D.; Herrero, L.; Ramos, M.P.; Viana, M. Short-term vitamin E treatment impa reactive oxygen species signaling required for adipose tissue expansion, resulting in fatty liver and insulin resistance in ob mice. *PLoS ONE* **2017**, *12*, e0186579. [CrossRef] [PubMed]
116. Carbonelli, M.; Renzo, L.; Bigioni, M.; Daniele, N.; De Lorenzo, A.; Fusco, M. α-Lipoic Acid Supplementation: A Tool for Obes Therapy? *Curr. Pharm. Des.* **2010**, *16*, 840–846. [CrossRef]
117. Salehi, B.; Berkay Yılmaz, Y.; Antika, G.; Boyunegmez Tumer, T.; Fawzi Mahomoodally, M.; Lobine, D.; Akram, M.; Riaz, Capanoglu, E.; Sharopov, F.; et al. Insights on the use of α-lipoic acid for therapeutic purposes. *Biomolecules* **2019**, *9*, 356. [CrossR [PubMed]
118. Shay, K.P.; Moreau, R.F.; Smith, E.J.; Smith, A.R.; Hagen, T.M. Alpha-lipoic acid as a dietary supplement: Molecular mechanis and therapeutic potential. *Biochim. Biophys. Acta Gen. Subj.* **2009**, *1790*, 1149–1160. [CrossRef]
119. Wang, Y.; Yang, M.; Lee, S.G.; Davis, C.G.; Kenny, A.; Koo, S.I.; Chun, O.K. Plasma total antioxidant capacity is associated w dietary intake and plasma level of antioxidants in postmenopausal women. *J. Nutr. Biochem.* **2012**, *23*, 1725–1731. [CrossRef]
120. Bhagavan, H.N.; Chopra, R.K. Coenzyme Q10: Absorption, tissue uptake, metabolism and pharmacokinetics. *Free Radic. R* **2006**, *40*, 445–453. [CrossRef]
121. Chalasani, N.; Younossi, Z.; Lavine, J.E.; Charlton, M.; Cusi, K.; Rinella, M.; Harrison, S.A.; Brunt, E.M.; Sanyal, A.J. The diagno and management of nonalcoholic fatty liver disease: Practice guidance from the American Association for the Study of Liv Diseases. *Hepatology* **2018**, *67*, 328–357. [CrossRef]
122. Younossi, Z.M.; Koenig, A.B.; Abdelatif, D.; Fazel, Y.; Henry, L.; Wymer, M. Global epidemiology of nonalcoholic fatty li disease—Meta-analytic assessment of prevalence, incidence, and outcomes. *Hepatology* **2016**, *64*, 73–84. [CrossRef]
123. McGill, M.R. The past and present of serum aminotransferases and the future of liver injury biomarkers. *EXCLI J.* **2016**, 817–828. [CrossRef]
124. Ali, N.; Sumon, A.H.; Fariha, K.A.; Asaduzzaman, M.; Kathak, R.R.; Molla, N.H.; Mou, A.D.; Barman, Z.; Hasan, M.; Miah, et al. Assessment of the relationship of serum liver enzymes activity with general and abdominal obesity in an urban Banglade population. *Sci. Rep.* **2021**, *11*, 6640. [CrossRef]
125. Ruhl, C.E.; Everharty, J.E. Determinants of the association of overweight with elevated serum alanine aminotransferase activity the United States. *Gastroenterology* **2003**, *124*, 71–79. [CrossRef] [PubMed]
126. Arora, C.; Malhotra, A.; Ranjan, P.; Singh, V.; Singh, N.; Shalimar; Dwivedi, S.N.; Vikram, N.K. Effect of intensive weight-l intervention on metabolic, ultrasound and anthropometric parameters among patients with obesity and non-alcoholic fatty li disease: An RCT. *Eur. J. Clin. Nutr.* **2022**, *76*, 1332–1338. [CrossRef]
127. Gasteyger, C.; Larsen, T.M.; Vercruysse, F.; Astrup, A. Effect of a dietary-induced weight loss on liver enzymes in obese subje *Am. J. Clin. Nutr.* **2008**, *87*, 1141–1147. [CrossRef]
128. Wang, D.; Day, E.A.; Townsend, L.K.; Djordjevic, D.; Jørgensen, S.B.; Steinberg, G.R. GDF15: Emerging biology and therapeu applications for obesity and cardiometabolic disease. *Nat. Rev. Endocrinol.* **2021**, *17*, 592–607. [CrossRef] [PubMed]
129. Koo, B.K.; Um, S.H.; Seo, D.S.; Joo, S.K.; Bae, J.M.; Park, J.H.; Chang, M.S.; Kim, J.H.; Lee, J.; Jeong, W., II; et al. Grow differentiation factor 15 predicts advanced fibrosis in biopsy-proven non-alcoholic fatty liver disease. *Liver Int.* **2018**, *38*, 695–7 [CrossRef] [PubMed]
130. Galuppo, B.; Agazzi, C.; Pierpont, B.; Chick, J.; Li, Z.; Caprio, S.; Santoro, N. Growth differentiation factor 15 (GDF15) is associa with non-alcoholic fatty liver disease (NAFLD) in youth with overweight or obesity. *Nutr. Diabetes* **2022**, *12*, 9. [CrossRef]
131. Coll, A.P.; Chen, M.; Taskar, P.; Rimmington, D.; Patel, S.; Tadross, J.A.; Cimino, I.; Yang, M.; Welsh, P.; Virtue, S.; et al. GDF mediates the effects of metformin on body weight and energy balance. *Nature* **2020**, *578*, 444–448. [CrossRef]
132. Day, E.A.; Ford, R.J.; Smith, B.K.; Mohammadi-Shemirani, P.; Morrow, M.R.; Gutgesell, R.M.; Lu, R.; Raphenya, A.R.; Kabiri, McArthur, A.G.; et al. Metformin-induced increases in GDF15 are important for suppressing appetite and promoting weight lo *Nat. Metab.* **2019**, *1*, 1202–1208. [CrossRef] [PubMed]
133. Wang, D.; Townsend, L.K.; DesOrmeaux, G.J.; Frangos, S.M.; Batchuluun, B.; Dumont, L.; Kuhre, R.E.; Ahmadi, E.; Hu, S.; Rebal I.A.; et al. GDF15 promotes weight loss by enhancing energy expenditure in muscle. *Nature* **2023**, *619*, 143–150. [CrossR [PubMed]
134. Kleinert, M.; Clemmensen, C.; Sjøberg, K.A.; Carl, C.S.; Jeppesen, J.F.; Wojtaszewski, J.F.P.; Kiens, B.; Richter, E.A. Exerc increases circulating GDF15 in humans. *Mol. Metab.* **2018**, *9*, 187–191. [CrossRef]
135. Cai, L.; Li, C.; Wang, Y.; Mo, Y.; Yin, J.; Ma, X. Increased serum GDF15 related to improvement in metabolism by lifest intervention among young overweight and obese adults. *Diabetes Metab. Syndr. Obes. Targets Ther.* **2021**, *14*, 1195–1202. [CrossR
136. Stahl, P.D.; Raposo, G. Extracellular Vesicles: Exosomes and Microvesicles, Integrators of Homeostasis. *Physiology* **2019**, 169–177. [CrossRef]
137. Nederveen, J.P.; Warnier, G.; Di Carlo, A.; Nilsson, M.I.; Tarnopolsky, M.A. Extracellular Vesicles and Exosomes: Insights Fro Exercise Science. *Front. Physiol.* **2021**, *11*, 1757. [CrossRef] [PubMed]

Verweij, F.J.; Bebelman, M.P.; Jimenez, C.R.; Garcia-Vallejo, J.J.; Janssen, H.; Neefjes, J.; Knol, J.C.; de Goeij-de Haas, R.; Piersma, S.R.; Baglio, S.R.; et al. Quantifying exosome secretion from single cells reveals a modulatory role for GPCR signaling. *J. Cell Biol.* **2018**, *217*, 1129–1142. [CrossRef]
Heyn, G.S.; Corrêa, L.H.; Magalhães, K.G. The Impact of Adipose Tissue–Derived miRNAs in Metabolic Syndrome, Obesity, and Cancer. *Front. Endocrinol.* **2020**, *11*, 563816. [CrossRef]
Tryndyak, V.P.; Latendresse, J.R.; Montgomery, B.; Ross, S.A.; Beland, F.A.; Rusyn, I.; Pogribny, I.P. Plasma microRNAs are sensitive indicators of inter-strain differences in the severity of liver injury induced in mice by a choline- and folate-deficient diet. *Toxicol. Appl. Pharmacol.* **2012**, *262*, 52–59. [CrossRef]
Yamada, H.; Suzuki, K.; Ichino, N.; Ando, Y.; Sawada, A.; Osakabe, K.; Sugimoto, K.; Ohashi, K.; Teradaira, R.; Inoue, T.; et al. Associations between circulating microRNAs (miR-21, miR-34a, miR-122 and miR-451) and non-alcoholic fatty liver. *Clin. Chim. Acta* **2013**, *424*, 99–103. [CrossRef]
Wang, R.; Hong, J.; Cao, Y.; Shi, J.; Gu, W.; Ning, G.; Zhang, Y.; Wang, W. Elevated circulating microRNA-122 is associated with obesity and insulin resistance in young adults. *Eur. J. Endocrinol.* **2015**, *172*, 291–300. [CrossRef]
Povero, D.; Eguchi, A.; Li, H.; Johnson, C.D.; Papouchado, B.G.; Wree, A.; Messer, K.; Feldstein, A.E. Circulating extracellular vesicles with specific proteome and liver microRNAs are potential biomarkers for liver injury in experimental fatty liver disease. *PLoS ONE* **2014**, *9*, e113651. [CrossRef] [PubMed]
Guo, Y.; Xiong, Y.; Sheng, Q.; Zhao, S.; Wattacheril, J.; Flynn, C.R. A micro-RNA expression signature for human NAFLD progression. *J. Gastroenterol.* **2016**, *51*, 1022–1030. [CrossRef] [PubMed]
Mehta, R.; Otgonsuren, M.; Younoszai, Z.; Allawi, H.; Raybuck, B.; Younossi, Z. Circulating miRNA in patients with non-alcoholic fatty liver disease and coronary artery disease. *BMJ Open Gastroenterol.* **2016**, *3*, e000096. [CrossRef]
Sunadome, H.; Matsumoto, H.; Izuhara, Y.; Nagasaki, T.; Kanemitsu, Y.; Ishiyama, Y.; Morimoto, C.; Oguma, T.; Ito, I.; Murase, K.; et al. Correlation between eosinophil count, its genetic background and body mass index: The Nagahama Study. *Allergol. Int.* **2020**, *69*, 46–52. [CrossRef] [PubMed]
Shim, W.S.; Kim, H.J.; Kang, E.S.; Ahn, C.W.; Lim, S.K.; Lee, H.C.; Cha, B.S. The association of total and differential white blood cell count with metabolic syndrome in type 2 diabetic patients. *Diabetes Res. Clin. Pract.* **2006**, *73*, 284–291. [CrossRef]
Tsuboi, N.; Koike, K.; Hirano, K.; Utsunomiya, Y.; Kawamura, T.; Hosoya, T. Clinical features and long-term renal outcomes of Japanese patients with obesity-related glomerulopathy. *Clin. Exp. Nephrol.* **2013**, *17*, 379–385. [CrossRef] [PubMed]
Kambham, N.; Markowitz, G.S.; Valeri, A.M.; Lin, J.; D'Agati, V.D. Obesity-related glomerulopathy: An emerging epidemic. *Kidney Int.* **2001**, *59*, 1498–1509. [CrossRef] [PubMed]
Kalantar-Zadeh, K.; Streja, E.; Molnar, M.Z.; Lukowsky, L.R.; Krishnan, M.; Kovesdy, C.P.; Greenland, S. Mortality prediction by surrogates of body composition: An examination of the obesity paradox in hemodialysis patients using composite ranking score analysis. *Am. J. Epidemiol.* **2012**, *175*, 793–803. [CrossRef]
Foster, M.C.; Hwang, S.J.; Larson, M.G.; Lichtman, J.H.; Parikh, N.I.; Vasan, R.S.; Levy, D.; Fox, C.S. Overweight, Obesity, and the Development of Stage 3 CKD: The Framingham Heart Study. *Am. J. Kidney Dis.* **2008**, *52*, 39–48. [CrossRef]
Hall, M.E.; Cohen, J.B.; Ard, J.D.; Egan, B.M.; Hall, J.E.; Lavie, C.J.; Ma, J.; Ndumele, C.E.; Schauer, P.R.; Shimbo, D. Weight-Loss Strategies for Prevention and Treatment of Hypertension: A Scientific Statement from the American Heart Association. *Hypertension* **2021**, *78*, E38–E50. [CrossRef] [PubMed]
Rhee, C.M.; Ahmadi, S.F.; Kalantar-Zadeh, K. The dual roles of obesity in chronic kidney disease: A review of the current literature. *Curr. Opin. Nephrol. Hypertens.* **2016**, *25*, 208–216. [CrossRef] [PubMed]
Stefansson, V.T.N.; Schei, J.; Jenssen, T.G.; Melsom, T.; Eriksen, B.O. Central obesity associates with renal hyperfiltration in the non-diabetic general population: A cross-sectional study. *BMC Nephrol.* **2016**, *17*, 172. [CrossRef] [PubMed]
Brenner, B.M.; Lawler, E.V.; Mackenzie, H.S. The hyperfiltration theory: A paradigm shift in nephrology. *Kidney Int.* **1996**, *49*, 1774–1777. [CrossRef]
Chagnac, A.; Weinstein, T.; Herman, M.; Hirsh, J.; Gafter, U.; Ori, Y. The effects of weight loss on renal function in patients with severe obesity. *J. Am. Soc. Nephrol.* **2003**, *14*, 1480–1486. [CrossRef]
Gerchman, F.; Tong, J.; Utzschneider, K.M.; Zraika, S.; Udayasankar, J.; McNeely, M.J.; Carr, D.B.; Leonetti, D.L.; Young, B.A.; De Boer, I.H.; et al. Body mass index is associated with increased creatinine clearance by a mechanism independent of body fat distribution. *J. Clin. Endocrinol. Metab.* **2009**, *94*, 3781–3788. [CrossRef]

nutrients

MDPI

Article

1,2-Dicinnamoyl-*sn*-glycero-3-phosphocholine Improves Insulin Sensitivity and Upregulates mtDNA-Encoded Genes in Insulin-Resistant 3T3-L1 Adipocytes: A Preliminary Study

Aneta Cierzniak [1], Anna Gliszczyńska [2,*] and Małgorzata Małodobra-Mazur [1,*]

1 Department of Forensic Medicine, Division of Molecular Techniques, Wroclaw Medical University, Sklodowskiej-Curie 52, 50-369 Wrocław, Poland; aneta.cierzniak@umw.edu.pl

2 Department of Food Chemistry and Biocatalysis, Wrocław University of Environmental and Life Science Norwida 25, 50-375 Wrocław, Poland

* Correspondence: anna.gliszczynska@upwr.edu.pl (A.G.); malgorzata.malodobra-mazur@umw.edu.pl (M.M.-M.) Tel.: +48-717-841-587 (M.M.-M.)

Abstract: Background: Insulin resistance is a condition characterized by a reduced biological respo to insulin. It is one of the most common metabolic diseases in modern civilization. Numerous natu substances have a positive effect on metabolism and energy homeostasis including restoring proper sensitivity to insulin. There may be several possible mechanisms of action. In the present stu we elucidated two natural compounds with an impact on insulin signaling in IR adipocytes involv mitochondria. Methods: Mature 3T3-L1 adipocytes with artificially induced insulin resista by palmitic acid (16:0) were used for the study. Cinnamic acid and 1,2-dicinnamoyl-*sn*-glycer phosphocholin (1,2-diCA-PC) were tested at three concentrations: 25 μM, 50 μM, and 125 μM. T number of mitochondria and the expression of genes encoded by mtDNA were elucidated in cont and experimental cells. Results: Experimental cells treated with 1,2-diCA-PC displayed increas insulin-stimulated glucose uptake in a dose-dependent manner, accompanied by an increase mtDNA copy number. Moreover, in experimental cells treated with 1,2-diCA-PC at a concentrati of 125 μM, a significant increase in the expression level of all analyzed genes encoded by mtD compared to control cells was observed. Our study showed a relationship between improved cellu sensitivity to insulin by 1,2-diCA-PC and an increase in the number of mitochondria and expressi levels of genes encoded by mtDNA. Conclusions: To summarize, the results suggest the therape potential of cinnamic acid derivative 1,2-diCA-PC to enhance the insulin sensitivity of adipocyte

Keywords: insulin resistance; cinnamic acid; phospholipid derivative; phenolic acid; 1,2-di-CA-I

Citation: Cierzniak, A.; Gliszczyńska, A.; Małodobra-Mazur, M. 1,2-Dicinnamoyl-*sn*-glycero-3-phosphocholine Improves Insulin Sensitivity and Upregulates mtDNA-Encoded Genes in Insulin-Resistant 3T3-L1 Adipocytes: A Preliminary Study. *Nutrients* **2024**, *16*, 3163. https://doi.org/10.3390/nu16183163

Academic Editors: Sonia Ramos Rivero, Laura Sturla and María Angeles Martín

Received: 14 August 2024
Revised: 13 September 2024
Accepted: 17 September 2024
Published: 19 September 2024

1. Introduction

Insulin is a key hormone regulating energy homeostasis and metabolic process within the human body, primarily affecting adipose tissue, skeletal muscle, and the liver Insulin resistance (IR) is a major public health problem worldwide. It is a conditi characterized by a reduced biological response of peripheral tissues to insulin and genera occurs in conjunction with numerous metabolic disorders as well as hormonal disord such as polycystic ovary syndrome (PCOS) and thyroid dysfunction [2]. A comm pathological feature in various forms of IR is the disruption of cellular insulin sig transmission at the molecular level, which is responsible for the development of IR. T main mechanism of insulin resistance induction is connected with impairments in insu signal transduction, primarily through Akt and phosphatidylinositol 3 (PI3) kinases, whi mediate the appropriate effects in the aforementioned tissues. Furthermore, IR induc hepatic gluconeogenesis, which further increases glycemia [3].

Additionally other mechanisms are implicated in the development of IR. Disrupti in lipid metabolism is the most common cause of IR development. Moreover, obes chronic low-grade inflammation in various tissues, hormonal dysregulation, and other factors also impair proper signal transduction, ultimately leading to disturbances in the mechanisms of insulin action [3]. Generally, lipid metabolism and the inflammatory pathway are intertwined processes which can lead to development of insulin resistance and other metabolic disorders. One of the most important links between lipids, inflammation, and insulin resistance is ceramide biosynthesis [4]. Obesity, besides the elevated lipid accumulation in adipose tissue, is characterized by lipid accumulation in other organs and elevated level of plasma free fatty acids (FFAs). On the other hand, the increase in FFAs in obesity is linked with the activation of Toll-like receptor 4 (TLR4) signaling in adipocytes, which induces the inflammatory process [5]. Furthermore, FFA excess affects mitochondria function by producing ROS, increasing mitochondrial proton conductance, and provoking the permeabilization of the outer mitochondrial membrane [6].

Mitochondria play a crucial role in cell metabolism and are responsible for controlling numerous cellular processes that require ATP. It has also been shown that in humans and other organisms, IR strongly correlates with reduced mitochondrial function, manifested by a reduced mitochondrial copy number, reduced mitochondrial biogenesis, and disruption in oxidative phosphorylation. Moreover, IR leads to impaired energy production in the form of ATP, caused by reduced mitochondrial respiratory capacity observed in insulin-resistant patients, type 2 diabetic patients, and obese individuals [7,8]. The main consequence of reduced mitochondrial capacity is an increase in reactive oxygen species (ROS) production, which is believed to be the root cause of IR development [9].

The current treatment of IR involves several pharmaceutics of which metformin is the first choice of pharmacotherapy. Unfortunately, it is not effective for some individuals; on the other hand, other patients suffer from numerous side effects. Numerous natural compounds have been shown to be effective agents in IR treatment, including various phenolic acids.

Cinnamic acid belongs to the class of aromatic acids and is a member of the phenylpropanoid family of natural compounds. It occurs naturally in various plants and essential oils and is an important intermediate in the biosynthesis of many secondary metabolites. Cinnamic acid exhibits various biological activities, including antioxidant and anti-inflammatory properties, and modulates glycogenesis and gluconeogenesis [10,11]. It has been observed that cinnamic acid exhibits antidiabetic effects by improving glucose tolerance and stimulating insulin secretion in rat islets and insulin action in mouse hepatocytes [12,13]. Mitochondrial functionality directly translates into the functionality of the entire cell, including insulin signal transduction and thus insulin sensitivity. With the excessive supply of free fatty acids seen in obesity, there is an increase in the rate of mitochondrial metabolism, and prolonged exposure of cells to excess nutrients leads to mitochondrial overload. This mitochondrial dysfunction can contribute to chronic oxidative stress, which plays a significant role in the pathogenesis of insulin resistance in adipocytes [14–16]. Restoring normal mitochondrial function can have a therapeutic effect.

In our previous paper, we reported the therapeutic effect of cinnamic acid and its derivatives on insulin-stimulated glucose uptake (ISGU) in insulin-resistant 3T3-L1 adipocytes [17]. However, trying to elucidate the possible mechanism of action through which these compounds increase insulin sensitivity, we did not observe any changes in the expression of insulin pathway genes in treated cells compared to untreated cells. For this reason, we further examined the possible mechanisms through which these compounds overcome insulin resistance. Considering the significant role of mitochondria in overall cell metabolism, we hypothesized that the mechanism might directly involve mitochondria. To verify this assumption, we investigated the number of mitochondria and the expression of genes encoded by mtDNA in insulin-resistant adipocytes treated with the examined substances to identify a potential mechanism underlying increased cellular sensitivity to insulin.

2. Materials and Methods

2.1. Study Reagents

Cinnamic acid (CA) was purchased from Sigma Aldrich (Saint Louis, MO, US The stock solution was prepared by dissolving it in 1 mL of ethanol, obtaining a 235 m concentration of working solution, and it was further diluted in cell culture media to t experimental concentrations (25 µM, 50 µM, and 125 µM).

1,2-Dicinnamoyl-*sn*-glycero-3-phosphocholine (1,2-diCA-PC) was synthesized in t Department of Food Chemistry and Biocatalysis at Wrocław University of Environmen and Life Sciences, following the previously reported procedure [18,19]. CA was esterifi with the cadmium complex of *sn*-glycero-3-phosphocholine in the presence of 4-(*N*, dimethylamino)pyridine (DMAP) as a catalyst and *N,N'*-dicyclohexylcarbodiimide (DC as a coupling agent. After purification, the pure product stock solution was prepared dissolving it in 1 mL of DMSO, obtaining a 44 mM concentration of working solution, a then it was further diluted in cell culture media to the experimental concentrations (25 µ 50 µM, and 125 µM). Figure 1 illustrates the chemical structures of the analyzed compoun

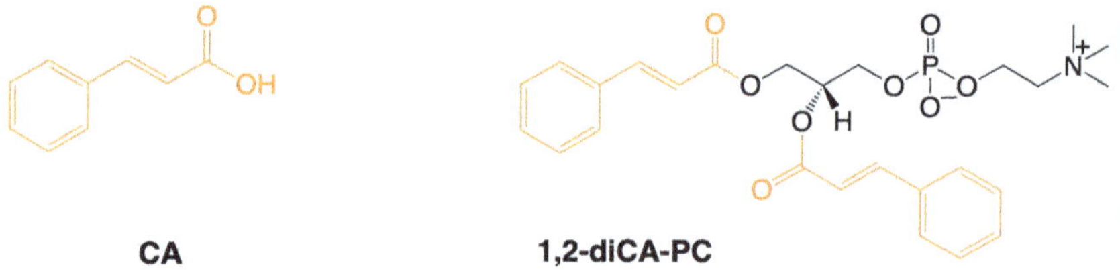

CA **1,2-diCA-PC**

Figure 1. Chemical structure of tested compounds. CA—cinnamic acid, 1,2-diCA-PC—1 dicinnamoyl-*sn*-glycero-3-phosphocholine.

2.2. The 3T3-L1 Cell Line Culturing and Differentiation

Mouse fibroblasts 3T3-L1 were purchased from ATCC (Manassas, VA, USA, CL-173 Cells were cultured in DMEM (Dulbecco's Modified Eagle's Medium, 4.5 g/L of gluco Corning Incorporated, New York, NY, USA) supplemented with 10% fetal calf serum (FC Sigma-Aldrich, Saint Louis, MI, USA) and antibiotics (penicillin, 50 U/mL; streptomyc 50 µg/mL, Corning Incorporated, New York, NY, USA) in a humidified incubator at 37 and with 5% CO_2.

After achieving 100% confluence, differentiation into mature adipocytes was induc by changing the medium to DMEM (4.5 g/L of glucose) supplemented with 10% fe bovine serum (FBS, Corning Incorporated, New York, NY, USA), antibiotics (penicill 50 U/mL; streptomycin, 50 µg/mL), 3-isobutyl-1-methylxanthine (115 µg/mL), dexa ethasone (390 ng/mL), and insulin (10 µg/mL). After three days, the medium was chang to DMEM (4.5 g/L of glucose) with antibiotics, 10% FBS, and insulin (10 µg/mL). Af an additional three days, the medium was changed to DMEM (4.5 g/L of glucose) wi antibiotics and 10% FBS, and the cells were further cultured for an additional two da The cells reached maturity 8 days after the initiation of differentiation.

2.3. Insulin Resistance Induction and the Effect of Phospholipid Derivatives

Insulin resistance was induced in mature adipocytes by culturing the cells in DME (4.5 g/L of glucose) supplemented with 10% FBS, antibiotics (penicillin, 50 U/mL; stre tomycin, 50 µg/mL), and palmitic acid 16:0 (Sigma-Aldrich, Saint Louis, MO, USA) a concentration of 0.5 mM for 48 h [9].

After inducing insulin resistance, the analyzed compounds, cinnamic acid (CA) a 1,2-dicinnamoyl-*sn*-glycero-3-phosphocholine (1,2-diCA-PC), were added at three vario concentrations (25 µM, 50 µM, and 125 µM). The concentrations were chosen based or viability test described previously [14]. The analyzed compounds were added to DME (4.5 g/L of glucose) supplemented with 10% FBS, antibiotics (penicillin, 50 U/mL; stre tomycin, 50 μg/mL), and palmitic acid (0.5 mM). For the controls, ethanol and DMSO were added to a final concentration of 0.25% in cultured medium. The adipocytes were then further cultured for an additional 48 h. Palmitic acid was included in the culture medium to maintain insulin resistance in the adipocytes. After incubation with the specific compounds, the cells were evaluated for insulin responses using a glucose uptake test, measuring both basal glucose uptake (BGU) and insulin-stimulated glucose uptake (ISGU).

2.4. Glucose Uptake Test

The Glucose Uptake-Glo Assay (Promega Corporation, Madison, WI, USA) was performed after 48 h of incubation with cinnamic acid (CA) and 1,2-dicinnamoyl-sn-glycero-3-phosphocholine (1,2-diCA-PC). The experiment was performed in four replicates. Cells were plated, differentiated, and the glucose uptake test was carried out on a 96-well plate. The day before the glucose uptake test, cells were starved in serum-free culture medium overnight. Just before starting the experiment, the culture medium was discarded, and the cells were washed twice with PBS to remove any residual glucose. Next, a portion of the experimental cells was stimulated with 1 μM insulin in PBS for 10 min (INS+), while the remaining adipocytes were left unstimulated (INS−). After insulin stimulation, 10 mM of 2-deoxyglucose (2DG6P) was added to all cells, and they were further incubated for an additional 10 min to assess both basal glucose uptake and insulin-stimulated glucose uptake. The cells were then processed according to the manufacturer's protocol. Luminescence was measured using the Victor3, 1420 Multilabel Plate Reader from PerkinElmer.

2.5. DNA and RNA Extraction from Experimental and Control Cells

DNA from the cells was isolated using the phenol/chloroform method (Sigma-Aldrich, St. Louis, MO, USA). Total RNA from the cells was isolated using the trizol method (Sigma-Aldrich, St. Louis, MO, USA). DNA and RNA concentration was measured using a NanoDrop 2000 spectrophotometer (Thermo Scientific™, Waltham, MA, USA).

RNA samples were treated with DNase before the reverse transcription reaction using RNase-Free DNase (Promega Corporation, Madison, WI, USA) according to the manufacturer's protocol. The experiment was performed in triplicate, and cells were cultured on 6-well plates.

2.6. Mitochondrial DNA Copy Number Quantification

Mitochondrial DNA copy numbers were quantified using the Absolute Mouse Mitochondrial DNA Copy Number Quantification qPCR Assay Kit (ScienCell™ Research Laboratories, Carlsbad, CA, USA). Procedures and calculations were performed according to the manufacturer's protocol. Real-time PCR was used to quantify mtDNA and gDNA of the analyzed samples and reference human genomic DNA using the known mtDNA copy number quantified per diploid cell. The mtDNA copy number was calculated based on Cq values using the ΔΔCq formula in reference to gDNA using SCR primers, provided by the producer. Next, the mtDNA was quantified by normalization to quantified standard concentration provided by the producer (724 copies ± 16 per diploid cell) using the following formulas:

$$\Delta\Delta Cq = \Delta Cq\ (mtDNA) - \Delta Cq\ (SCR), \quad (1)$$

where SCR represents primers specific for gDNA.

$$mtDNA\ copy\ number = (724 \pm 16) \times 2^{-\Delta\Delta Cq}, \quad (2)$$

where (724 ± 16) is the standard concentration.

2.7. cDNA Synthesis and Quantification of Gene Expression

Reverse transcription was performed using the High-Capacity cDNA Reverse Transcription Kit (Applied Biosystems, Waltham, MA, USA). A total of 200 ng of isolated RNA was used for cDNA synthesis. Gene expression analysis was conducted by real-time PCR using Fast SYBR Green Master Mix (Applied Biosystems, Waltham, MA, USA). Primers mtDNA-encoded genes—*Nd3* (NADH dehydrogenase subunit 3), *Nd4* (NADH dehydro nase subunit 4), *Nd5* (NADH dehydrogenase subunit 5), *Cytb* (cytochrome b), *Atp6* (A synthase 6), *Atp8* (ATP synthase 8), *Cox1* (cytochrome c oxidase 1), and *Cox2* (cytochro c oxidase 2)—as well as for the nuclear DNA (nDNA)-encoded reference gene *β-ac* (beta-actin) were manually designed (Table 1).

Table 1. The sequences of primers used for mitochondrial gene expression analysis.

Gene	Official Full Name of Gene	Primer's Sequence (5′→3′)	R^2 Valu
Nd3	NADH dehydrogenase subunit 3	F *: TAACCTGTACACTGTTATCTTC R *: GAGTACAGATTTATTTGGGGG	0.99
Nd4	NADH dehydrogenase subunit 4	F: TCCTCAGTTAGCCACATAGC R: ATAAGTGGGAAGACCATTTGAA	0.99
Nd5	NADH dehydrogenase subunit 5	F: CCCATGACTACCATCAGCAA R: ATAATGTGGTTAGGGCTCCG	0.99
Cytb	Cytochrome b	F: CATACGAAAAACACACCCATTA R: GTAGTGTATGGCTAAGAAAAGA	0.99
Atp6	ATP synthase 6	F: TTTACACCTACTACCCAACTAT R: GGAATTAGTGAAATTGGAGTTC	0.99
Atp8	ATP synthase 8	F: GCCACAACTAGATACATCAAC R: GAAGGTGCCAGTGGGAATG	0.98
Cox1	Cytochrome c oxidase 1	F: TATCTACTATTCGGAGCCTGA R: GCATGGGCAGTTACGATAAC	0.99
Cox2	Cytochrome c oxidase 2	F: GAAGACCTATGCTTTGATTCAT R: GGATTGGAAGTTCTATTGGCA	0.99
B-actin	Beta-actin	F: CGTTGACATCCGTAAAGACC R: CTAGGAGCCAGAGCAGTAAT	0.99

* F: forward primer; R: reverse primer.

The specificity of the primers was checked using Primer v.3-BLAST (NCBI), and s ondary structures were analyzed using OligoAnalyzer (Integrated DNA Technology). T acceptance criteria were DeltaG > −7 kcal/mol. Prior to real-time PCR, the efficiency of t primers was assessed using the standard curve method. Specificity was further confirm based on the melting curve. Only primers with efficiency values higher than $R^2 \geq 0.95$ a confirmed specificity through the melting curve were used for gene expression studi Relative gene expression levels, normalized to the housekeeping gene *β-actin*, were cal lated using the delta–delta Ct (ΔΔCt) method. The fold change was determined using t $2^{(-\Delta\Delta Ct)}$ algorithm.

2.8. Statistical Analysis

Statistical analyses were performed using Microsoft Office Excel 2007 and Statistica 1 (StatSoft, Tulsa, OK, USA). The normality of the distribution was assessed using t Shapiro–Wilk test. Based on the results, parametric tests (Student's *t*-test) were used statistical calculations and analysis of differences between the studied groups. The resu show the mean ± SD, and the number of repeats (*n*) is specified in the legends. Statisti significance was set at $p < 0.05$.

3. Results

3.1. Measurements of Glucose Uptake by Insulin-Resistant 3T3-L1 Adipocytes Cultured with Analyzed Compounds

First of all, we tried to determine if the analyzed compounds at the working co centrations were able to abolish insulin resistance in 3T3-L1 adipocytes. The experime was conducted in four repetitions. As a reference control, insulin-resistant cells incubated with the appropriate solvent (ethanol and DMSO for CA and 1,2-diCA-PC dissolving, respectively) were used. In our study, we did not detect any influence of ethanol or DMSO on insulin-stimulated glucose uptake by insulin-sensitive (IR−) and insulin-resistant (IR+) adipocytes. Cells with developed insulin resistance (IR+ ethanol, IR + DMSO) showed only a slight increase in ISGU in insulin-treated cells (INS+) compared to BGU (INS−). On the other hand, adipocytes with proper insulin sensitivity (IR—ethanol, IR—DMSO) displayed significantly increased ISGU compared to BGU (IR—ethanol, $p = 0.024$; IR—DMSO, $p = 0.043$; Figure 2).

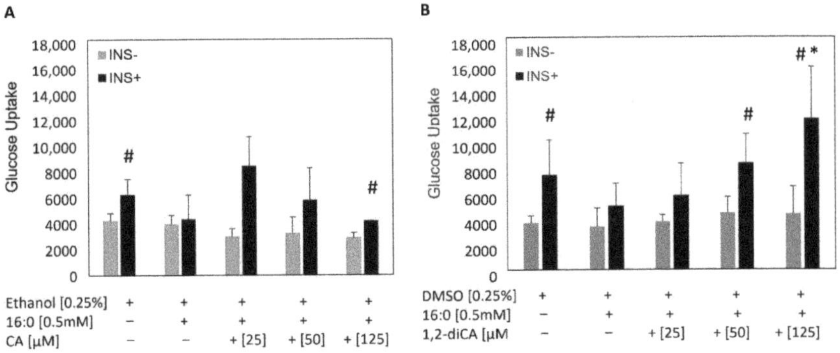

Figure 2. Glucose uptake in insulin-resistant 3T3-L1 adipocytes treated with the analyzed compounds (**A**) CA (cinnamic acid) and (**B**) 1,2-diCA-PC (1,2-dicinnamoyl-*sn*-glycero-3-phosphocholine) at concentrations of 25 µM, 50 µM, and 125 µM. INS− represents basal glucose uptake, and INS+ represents insulin-stimulated glucose uptake. Two controls were included: insulin-sensitive cells not treated with 16:0 and insulin-resistant cells treated with 16:0. * $p < 0.05$ indicates a statistical difference between experimental cells and IR adipocytes; # $p < 0.05$ indicates a statistical difference between ISGU and BGU. $n = 4$ repeats.

Cinnamic acid stimulated insulin-stimulated glucose uptake at all studied concentrations compared to basal glucose uptake. The increase was 2.3-fold for 25 µM, 2-fold for 50 µM, and 1.6-fold for 125 µM (25 µM: $p = 0.062$; 50 µM: $p = 0.093$; 125 µM: $p = 0.009$, Figure 2); however, the increase was only statistically significant for 125 µM. When compared to the insulin-stimulated insulin-resistant adipocytes (INS+, IR+), we did not observe any significant changes in glucose uptake in IR adipocytes treated or not treated with CA, and the difference was not noticeable at any of the analyzed concentrations.

The 1,2-diCA-PC significantly increased ISGU at the two highest concentrations in cells stimulated by insulin compared to nonstimulated cells. For the concentrations of 50 µM and 125 µM, the increase was 2.9-fold and 3.5-fold, respectively (50 µM: $p = 0.047$; 125 µM: $p = 0.013$, Figure 2), showing a dose-dependent effect. When comparing the ISGU in adipocytes treated with 1,2-diCA-PC to insulin-resistant adipocytes (IR+), we observed a significant increase in glucose uptake at the highest concentration, which was almost three times higher ($p = 0.034$). The medium concentration of 1,2-diCA-PC (50 µM) increased ISGU by about two times ($p = 0.108$). The lowest concentration of 1,2-diCA-PC (25 µM) did not increase ISGU in insulin-resistant adipocytes when compared to IR adipocytes not treated with this compound.

3.2. Analysis of mtDNA Copy Number in Experimental Cells Treated with Analyzed Compounds

After confirming the effect of analyzed compounds on insulin signaling and glucose uptake, next, we wanted to verify if the mechanism of action of analyzed compounds influenced mitochondria metabolism. Thus, first, we evaluated the number of mitochondria by mtDNA quantification. The mtDNA copy numbers obtained for experimental cells treated with CA and 1,2-diCA-PC at appropriate concentrations were compared to cont insulin-resistant adipocytes (IR+) as well as to adipocytes with proper insulin sensitiv (IR−) incubated solely with an appropriate solvent (ethanol and DMSO).

First, we focused on the evaluation of mtDNA copy number between insulin-resist and insulin-sensitive adipocytes. In cells where ethanol was used as the solvent, observed a slightly higher amount of mtDNA in control cells with induced insulin resista (IR+ ethanol) compared to cells with normal insulin sensitivity (IR− ethanol); howe the increase was not statistically significant. On the other hand, we noticed a statistica significant reduction in the amount of mtDNA in insulin-resistant adipocytes (IR+ DM compared to cells with normal insulin sensitivity (IR− DMSO) ($p = 0.013$) (Figure 3A,B

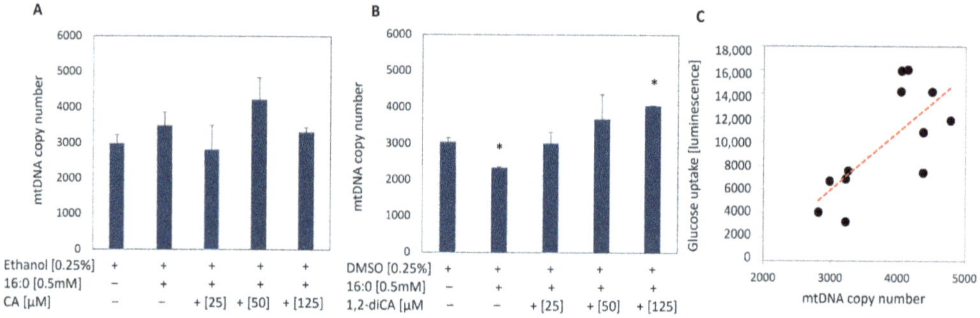

Figure 3. mtDNA copy number assessed by real-time PCR in insulin-resistant adipocytes with pro insulin sensitivity (not treated with 16:0) and with induced insulin resistance (treated with 16 as well as in adipocytes treated with (**A**) cinnamic acid (CA) and (**B**) 1,2-diCA-PC at appropri concentrations (25 μM, 50 μM, and 125 μM). (**C**) The correlation between mtDNA copy number a ISGU observed for 1,2-diCA-PC (125 μM), * $p < 0.05$. $n = 3$ repeats.

No substantial differences in the mtDNA copy number were observed in experimen cells treated with CA at any of the tested concentrations when compared to IR+ ce (Figure 3A).

In adipocytes treated with the 1,2-diCA-PC, we observed a significant increase the amount of mtDNA at all tested concentrations in a dose-dependent manner. T largest increase, up to 73% compared to the control adipocytes (IR+), was observed a concentration of 125 μM ($p = 0.000$), followed by a 57% increase at 50 μM ($p = 0.078$) an 29% increase at 25 μM ($p = 0.081$, Figure 3B).

We also searched for any correlation between glucose uptake rate after insulin stin lation (ISGU) and mtDNA copy number. We noticed a strong correlation between mtD copy number and ISGU in insulin-resistant adipocytes treated with 1,2-diCA-PC (R = 0. $p = 0.008$; Figure 3C). No statistically significant correlation has been observed for c namic acid.

3.3. *The Measurements of Expression of mtDNA-Encoded Genes in Experimental Cells Treated with Analyzed Compounds*

Besides the number of mitochondria, we also evaluated the expression of mtDN encoded genes as part of the mitochondrial function. Similar to mtDNA copy numb determination, first, the differences between control adipocytes with induced insu resistance (IR+) and insulin-sensitive (IR−) adipocytes cultured with the appropri solvent were evaluated to detect any differences in mtDNA-encoded gene expressi In other words, we aimed to determine if insulin resistance caused any changes in t expression of mitochondrial genes. The expression levels of all analyzed genes we markedly reduced in IR adipocytes compared to insulin-sensitive control cells. Surprising the downregulation of mitochondrial-encoded genes was observed only in cells cultur with DMSO (used as the solvent). In control cells treated with ethanol, we did not obser a reduction in expression rates (Figure 4A). It should also be noted that the reduction in expression rates in cells cultured with DMSO was not statistically significant, although the downregulation of several genes approached significance (Nd3 $p = 0.063$; Cytb $p = 0.118$; Atp6 $p = 0.063$; Atp8 $p = 0.197$; Cox1 $p = 0.110$; Cox2 $p = 0.158$, Figure 4B).

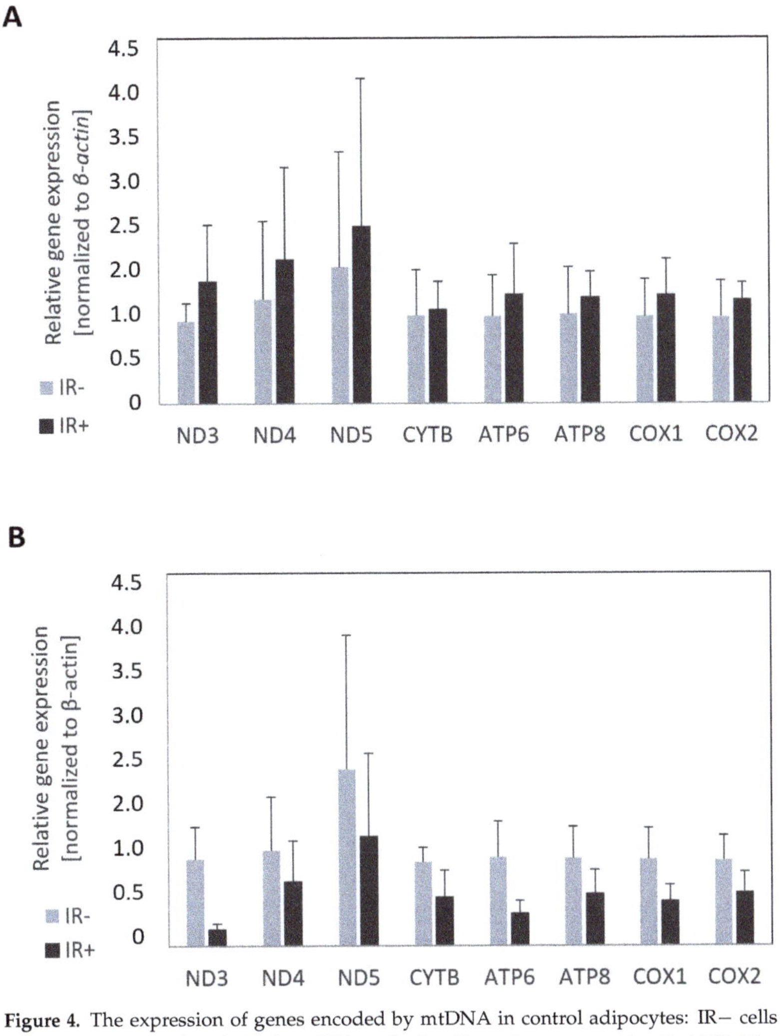

Figure 4. The expression of genes encoded by mtDNA in control adipocytes: IR− cells (insulin-sensitive, not treated with 16:0) and IR+ cells (insulin-resistant, treated with 16:0). The analysis was conducted for two types of control cells: those treated with ethanol (**A**) and those treated with DMSO (**B**). $n = 3$ repeats.

Next, we verified whether cinnamic acid or its derivative 1,2-diCA-PC influence mtDNA-encoded gene expression. Cinnamic acid did not influence the expression of genes encoded by mtDNA in experimental cells compared to insulin-resistant adipocytes (IR+). Cells incubated with CA at three tested concentrations (25 μM, 50 μM, and 125 μM) did not exhibit any significant changes in the expression levels of almost all analyzed genes (Figure 5A). Opposite to CA, insulin-resistant adipocytes treated with 1,2-diCA-PC exhibited a significant increase in the expression of all analyzed genes compared to insulin-resistant cells (IR+). However, the increase was notable only at the concentration of

125 μM of 1,2-diCA-PC. We observed the upregulation of the following genes: a 4.3-f
increase for Nd3 ($p = 0.017$), a 3.8-fold increase for Atp6 ($p = 0.022$), and an approximat
2-fold increase for Nd4 ($p = 0.106$), Nd5 (no statistical significance), Cytb ($p = 0.021$), A
($p = 0.134$), Cox1 ($p = 0.050$), and Cox2 ($p = 0.022$) (Figure 5B). Cells treated with lov
concentrations of 1,2-diCA-PC (25 and 50 μM) displayed a slight increase in expression
Nd3, Cytb, and Atp6 only, but these changes were not statistically significant (Figure 5

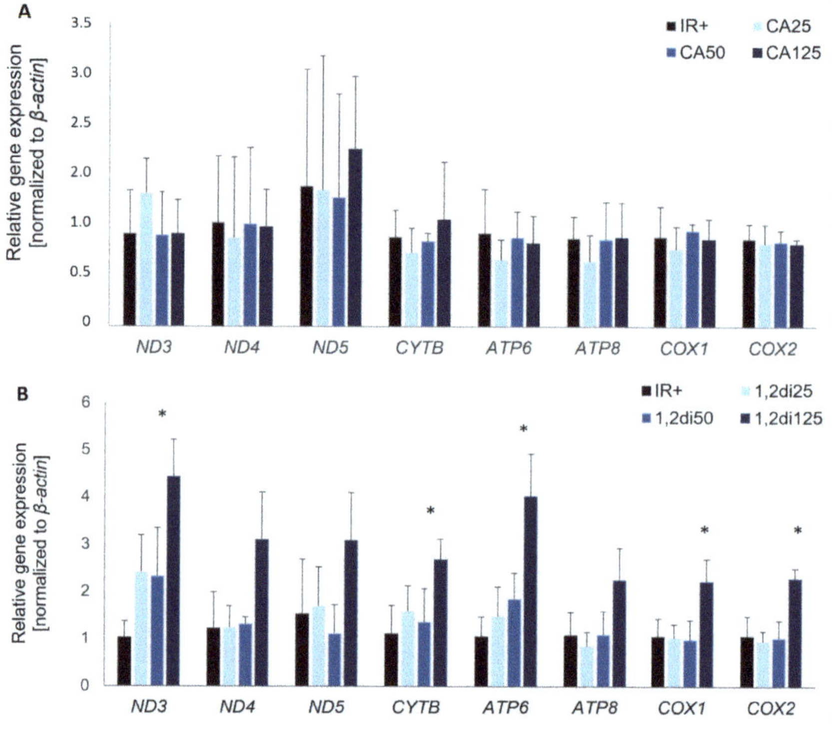

Figure 5. The expression of mitochondrial genes in insulin-resistant adipocytes treated with (**A**)
namic acid at three concentrations (25 μM, 50 μM, and 125 μM) compared to insulin-resistant con
adipocytes (IR+); (**B**) 1,2-diCA-PC at three concentrations (25 μM, 50 μM, and 125 μM) compared
insulin-resistant control adipocytes (IR+). * $p < 0.05$. $n = 3$ repeats.

4. Discussion

In the present study, we evaluated the influence of cinnamic acid and its phospholi
derivative, 1,2-dicinnamoyl-sn-glycero-3-phosphocholine (1,2-diCA-PC), on mitochondr
function in insulin-resistant 3T3-L1 adipocytes by investigating the number of mitochond
and the expression of genes encoded by mtDNA.

Firstly, we confirmed the successful development of insulin resistance in matu
adipocytes using palmitic acid (16:0), assessed by insulin-stimulated glucose uptake. W
observed a lower glucose uptake rate in cells treated with palmitic acid in both typ
of insulin-resistant controls—those incubated with ethanol or DMSO, the solvents us
for resuspending CA and 1,2-diCA-PC, respectively—compared to cells where insu
resistance was not induced.

Next, after incubation with the examined compounds, we performed insulin-stimulat
glucose uptake measurements in insulin-resistant adipocytes. For both analyzed co
pounds, CA and 1,2-diCA-PC, we confirmed previous observations that CA and 1,2-diC
PC increase insulin-stimulated glucose uptake in adipocytes with previously develop
insulin resistance [17]. It was previously reported that CA moderately increased g

cose uptake in insulin-resistant adipocytes, while 1,2-diCA-PC significantly increased the glucose uptake rate, effectively doubling the level of glucose uptake in insulin-sensitive adipocytes [17]. However, in the present study, we used a lower concentration of analyzed compounds in order to investigate dose dependency. Despite the variance, the observation obtained previously, but for higher concentrations, was replicated. As the possible mechanisms of action had not been previously revealed, we further investigated how cinnamic acid and its derivative could enhance glucose uptake and restore proper insulin signaling.

The number and size of mitochondria are crucial for the proper functioning of these organelles, and they are correlated with mitochondrial oxidative capacity and energy efficiency [20]. A reduced content of mtDNA has been observed in blood cells of obese subjects. Studies have demonstrated a strong negative correlation between mtDNA content and BMI, as well as between mtDNA content and the amount of visceral fat [21]. Furthermore, a reduced amount of mtDNA has also been observed in obese subjects with concomitant insulin resistance. The mtDNA/nDNA (nuclear DNA) ratio was inversely correlated with HOMA, glucose, and uric acid levels [22,23]. Additionally, in obese subjects with diagnosed T2D, a negative correlation was observed between mtDNA levels and BMI, fasting plasma glucose, fasting plasma insulin, LDL, and TG values in the blood [24]. A reduction in electron transport chain activity, which correlated with a reduced number of mitochondria, was observed in the skeletal muscle of obese individuals [25,26].

All these reports highlight the significant role of mitochondria not only in the pathogenesis of metabolic diseases but also in their interrelation. In our study, we observed a reduced number of mitochondria in insulin-resistant adipocytes compared to cells with proper insulin sensitivity. However, this decrease was observed only in cells cultured with DMSO and palmitic acid (used as the control for 1,2-diCA-PC). Slightly different results were obtained for adipocytes treated with palmitic acid and ethanol. In these cells, the number of mtDNA copies was even slightly higher in insulin-resistant adipocytes compared to adipocytes with proper insulin sensitivity.

Similarly, when comparing the expression of mitochondrial-encoded genes between insulin-resistant adipocytes and adipocytes with proper insulin sensitivity, we noticed downregulation in insulin-resistant adipocytes treated with palmitic acid and DMSO. However, no changes in expression or even slightly increased expression of mitochondrial-encoding genes were observed in adipocytes co-treated with palmitic acid and ethanol, suggesting some influence of ethanol on mitochondrial biogenesis. Further research is needed to elucidate the effects of ethanol on mitochondrial biogenesis, especially given that some of the literature provides evidence of ethanol's negative impact on mitochondrial metabolism [27]. This may take time as most studies refer to chronic alcohol intake, whereas our ethanol stimulation lasted only 48 h.

Despite these observations, we can strongly conclude that our results support previous findings reported by others, confirming that mitochondrial numbers are decreased in insulin resistance. The influence of ethanol on mitochondrial biogenesis and metabolism certainly requires more in-depth research.

Similar to our previous report, we observed a positive effect of the examined compounds on insulin-stimulated glucose uptake in insulin-resistant adipocytes. However, similar to previous reports, the glucose uptake stimulated by cinnamic acid (CA) was moderate across all studied concentrations [9]. In contrast, the results obtained for the derivative 1,2-dicinnamoyl-sn-glycero-3-phosphocholine (1,2-diCA-PC) demonstrated a dose-dependent increase in insulin-stimulated glucose uptake. At the highest tested concentration, 1,2-diCA-PC doubled the glucose uptake rate measured in adipocytes with proper insulin sensitivity. Similarly, we observed a dose-dependent increase in the number of mitochondria in insulin-resistant cells treated with 1,2-diCA-PC at all analyzed concentrations (25 μM, 50 μM, and 125 μM) compared to control cells without 1,2-diCA-PC. Moreover, we observed a positive correlation between the number of mtDNA copies and insulin-stimulated glucose uptake in adipocytes treated with 1,2-diCA-PC at the concentration of 125 μM, which supports the effect of increased mitochondrial copy number enhancing cellular sensitivity to insulin.

To further study the relationship between mitochondria and insulin sensitivity, measured the expression rate of genes encoded by mtDNA in experimental cells. T mtDNA genes encode proteins that are crucial for the enzymatic activity of the electr transport chain; thus, the regulation of these genes is vital for the proper functioning cellular metabolism. A decreased expression of *Nd1* and *Cox2* has been observed in ce from Zucker diabetic fatty (ZDF) rats, an animal model of type 2 diabetes, compared to le animals. Moreover, the expression levels of these genes are inversely correlated with plas glucose levels [28]. A 50% decrease in the expression of *Cox1*, a subunit of mitochondr complex IV, has been observed in insulin-resistant offspring of T2D parents [29]. In c study, we observed a significant increase in the expression levels of all analyzed mtDN encoded genes in cells treated with 1,2-diCA-PC at a concentration of 125 μM compared control cells.

The results described above suggest a potential mechanism by which 1,2-diCA-I at a concentration of 125 μM, reverses insulin resistance in adipocytes by enhancing mi chondrial biogenesis, as indicated by increased mtDNA copy numbers. A comprehensi understanding of 1,2-diCA-PC's effect on reversing insulin resistance can be gained by c sidering the results of gene expression, where the highest tested concentration significan increased the expression of numerous mtDNA-encoded genes.

Cinnamic acid and its numerous derivatives have a positive effect on carbohydr metabolism by regulating lipid metabolism. In general these compounds are consider as antioxidant and anti-inflammatory agents, which improve the antioxidant status reducing oxidative stress in patients, mostly influencing the LDL oxidation rate [30]. It h been shown that cinnamic acid and its derivatives lower plasma and liver triglyceride a cholesterol levels by reducing the HMG-CoA reductase and ACAT activity in experimen animals, thus influencing the FFA level [31,32]. FFA level has both positive and negati effects on mitochondria content and metabolism. FFAs are essential dietary nutrier however, when present in excess, they inhibit mitochondrial biogenesis and functi mainly via the PPARα pathway which activates the uncoupling protein (UCP) [33]. T might be another aspect that should be taken into account during further evaluation cinnamic acid's effect on mitochondria metabolism, especially in terms of 1,2-diCA-PC.

In 3T3-L1 adipocytes with developed insulin resistance, treatment with 1,2-diCA-PC 125 μM led to increased glucose uptake, indicating improved cellular sensitivity to insu This effect was accompanied by a significant increase in the number of mitochondria a the expression levels of all analyzed mtDNA-encoded genes compared to control ce Our studies demonstrate a relationship between improved cellular insulin sensitivity a an increase in both mitochondrial numbers and mtDNA-encoded gene expression. analyzed mtDNA genes encode proteins that are components of the electron transport cha and are essential for oxidative phosphorylation. Furthermore, these results highlight t significant therapeutic potential of the cinnamic acid derivative 1,2-diCA-PC in enhanci the insulin sensitivity of adipocytes.

The results presented also have some limitations. The most significant limitation this study is the lack of a detailed molecular mechanism explaining how 1,2-diCA-I enhances mitochondrial biogenesis, improves insulin response, and reverses insulin res tance. Further research is needed to fully elucidate this mechanism. Additionally, it wou be valuable to investigate whether similar effects are observed in other insulin-depende cells, such as skeletal muscle cells or hepatocytes, and to explore if higher concentrations 1,2-diCA-PC might produce stronger effects. Moreover, the expression of mtDNA-encod genes could be further validated by quantifying protein levels using Western blot analy Finally, the effects of expression rates in control cells treated with solvents alone yield various results. A significantly lower expression of mtDNA-encoded genes was observ in insulin-resistant cells treated with DMSO compared to insulin-sensitive controls, wi no changes in the expression of those genes in control cells where ethanol was used. Si larly, further exploration of this effect would provide additional insight into mitochondrial metabolism. Despite these limitations, this study is the first to report that 1,2-diCA-PC can reverse insulin resistance by increasing mtDNA copy number and upregulating key genes involved in the enzymatic complexes of oxidative phosphorylation.

5. Conclusions

In the present study, we demonstrated changes in both mtDNA copy number and the expression of mitochondrial genes in adipocytes as possible mechanisms through which 1,2-diCA-PC enhances the insulin response. We concluded that 1,2-diCA-PC reverses insulin resistance by promoting mitochondrial biogenesis and upregulating mitochondrial gene expression, but this effect is only observed at concentrations above 125 μM.

Author Contributions: Conceptualization, M.M.-M. and A.G.; methodology, M.M.-M. and A.G.; formal analysis, M.M.-M. and A.C.; investigation, M.M.-M., A.C. and A.G.; resources, M.M.-M. and A.G.; writing—original draft preparation, A.C.; writing—review and editing, M.M.-M. and A.G.; visualization, A.C.; supervision, M.M.-M.; project administration, M.M.-M.; funding acquisition, A.G. and A.C. All authors have read and agreed to the published version of the manuscript.

Funding: This work was supported by Wroclaw Medical University as part of the Subside Funds for Scientific Projects. No. SUBK.A120.22.072. This work was supported by the Wrocław University of Environmental and Life Sciences (Poland) as part of the Ph.D. research program "MISTRZ", No. N090/0017/22. The APC is co-financed by Wrocław University of Environmental and Life Sciences.

Institutional Review Board Statement: Not applicable.

Informed Consent Statement: Not applicable.

Data Availability Statement: The original contributions presented in the study are included in the article; further inquiries can be directed to the corresponding authors.

Conflicts of Interest: The authors declare no conflicts of interest.

erences

Lebovitz, H.E. Insulin Resistance: Definition and Consequences. *Exp. Clin. Endocrinol. Diabetes* **2001**, *109* (Suppl. S2), S135–S148. [CrossRef] [PubMed]
Lankarani, M.; Valizadeh, N.; Heshmat, R.; Peimani, M.; Sohrabvand, F. Evaluation of Insulin Resistance and Metabolic Syndrome in Patients with Polycystic Ovary Syndrome. *Gynecol. Endocrinol.* **2009**, *25*, 504–507. [CrossRef] [PubMed]
Petersen, M.C.; Shulman, G.I. Mechanisms of Insulin Action and Insulin Resistance. *Physiol. Rev.* **2018**, *98*, 2133–2223. [CrossRef]
Holland, W.L.; Bikman, B.T.; Wang, L.-P.; Yuguang, G.; Sargent, K.M.; Bulchand, S.; Knotts, T.A.; Shui, G.; Clegg, D.J.; Wenk, M.R.; et al. Lipid-Induced Insulin Resistance Mediated by the Proinflammatory Receptor TLR4 Requires Saturated Fatty Acid-Induced Ceramide Biosynthesis in Mice. *J. Clin. Investig.* **2011**, *121*, 1858–1870. [CrossRef] [PubMed]
Shi, H.; Kokoeva, M.V.; Inouye, K.; Tzameli, I.; Yin, H.; Flier, J.S. TLR4 Links Innate Immunity and Fatty Acid–Induced Insulin Resistance. *J. Clin. Investig.* **2006**, *116*, 3015–3025. [CrossRef] [PubMed]
Rial, E.; Rodríguez-Sánchez, L.; Gallardo-Vara, E.; Zaragoza, P.; Moyano, E.; González-Barroso, M.M. Lipotoxicity, Fatty Acid Uncoupling and Mitochondrial Carrier Function. *Biochim. Biophys. Acta (BBA)—Bioenerg.* **2010**, *1797*, 800–806. [CrossRef]
Asmann, Y.W.; Stump, C.S.; Short, K.R.; Coenen-Schimke, J.M.; Guo, Z.; Bigelow, M.L.; Nair, K.S. Skeletal Muscle Mitochondrial Functions, Mitochondrial DNA Copy Numbers, and Gene Transcript Profiles in Type 2 Diabetic and Nondiabetic Subjects at Equal Levels of Low or High Insulin and Euglycemia. *Diabetes* **2006**, *55*, 3309–3319. [CrossRef]
Stump, C.S.; Short, K.R.; Bigelow, M.L.; Schimke, J.M.; Nair, K.S. Effect of Insulin on Human Skeletal Muscle Mitochondrial ATP Production, Protein Synthesis, and mRNA Transcripts. *Proc. Natl. Acad. Sci. USA* **2003**, *100*, 7996–8001. [CrossRef]
Tirosh, A.; Potashnik, R.; Bashan, N.; Rudich, A. Oxidative Stress Disrupts Insulin-Induced Cellular Redistribution of Insulin Receptor Substrate-1 and Phosphatidylinositol 3-Kinase in 3T3-L1 Adipocytes. A Putative Cellular Mechanism for Impaired Protein Kinase B Activation and GLUT4 Translocation. *J. Biol. Chem.* **1999**, *274*, 10595–10602. [CrossRef]
Huang, D.-W.; Shen, S.-C. Caffeic Acid and Cinnamic Acid Ameliorate Glucose Metabolism via Modulating Glycogenesis and Gluconeogenesis in Insulin-Resistant Mouse Hepatocytes. *J. Funct. Foods* **2012**, *4*, 358–366. [CrossRef]
Nair, A.; Preetha Rani, M.R.; Salin Raj, P.; Ranjit, S.; Rajankutty, K.; Raghu, K.G. Cinnamic Acid Is Beneficial to Diabetic Cardiomyopathy via Its Cardioprotective, Anti-Inflammatory, Anti-Dyslipidemia, and Antidiabetic Properties. *J. Biochem. Mol. Toxicol.* **2022**, *36*, e23215. [CrossRef] [PubMed]

12. Hafizur, R.M.; Hameed, A.; Shukrana, M.; Raza, S.A.; Chishti, S.; Kabir, N.; Siddiqui, R.A. Cinnamic Acid Exerts Anti-Diab Activity by Improving Glucose Tolerance in Vivo and by Stimulating Insulin Secretion in Vitro. *Phytomedicine* **2015**, *22*, 297– [CrossRef]
13. Huang, D.-W.; Shen, S.-C.; Wu, J.S.-B. Effects of Caffeic Acid and Cinnamic Acid on Glucose Uptake in Insulin-Resistant Mo Hepatocytes. *J. Agric. Food Chem.* **2009**, *57*, 7687–7692. [CrossRef]
14. Furukawa, S.; Fujita, T.; Shimabukuro, M.; Iwaki, M.; Yamada, Y.; Nakajima, Y.; Nakayama, O.; Makishima, M.; Matsuda, Shimomura, I. Increased Oxidative Stress in Obesity and Its Impact on Metabolic Syndrome. *J. Clin. Investig.* **2004**, *114*, 1752–1 [CrossRef]
15. Henriksen, E.J.; Diamond-Stanic, M.K.; Marchionne, E.M. Oxidative Stress and the Etiology of Insulin Resistance and Typ Diabetes. *Free Radic. Biol. Med.* **2011**, *51*, 993–999. [CrossRef] [PubMed]
16. Bhatti, J.S.; Bhatti, G.K.; Reddy, P.H. Mitochondrial Dysfunction and Oxidative Stress in Metabolic Disorders—A Step towa Mitochondria Based Therapeutic Strategies. *Biochim. Biophys. Acta Mol. Basis Dis.* **2017**, *1863*, 1066–1077. [CrossRef]
17. Małodobra-Mazur, M.; Lewoń, D.; Cierzniak, A.; Okulus, M.; Gliszczyńska, A. Phospholipid Derivatives of Cinnamic A Restore Insulin Sensitivity in Insulin Resistance in 3T3-L1 Adipocytes. *Nutrients* **2021**, *13*, 3619. [CrossRef]
18. Czarnecka, M.; Świtalska, M.; Wietrzyk, J.; Maciejewska, G.; Gliszczyńska, A. Synthesis and Biological Evaluation of Phosphati cholines with Cinnamic and 3-Methoxycinnamic Acids with Potent Antiproliferative Activity. *RSC Adv.* **2018**, *8*, 35744–35 [CrossRef] [PubMed]
19. Czarnecka, M.; Świtalska, M.; Wietrzyk, J.; Maciejewska, G.; Gliszczyńska, A. Synthesis, Characterization, and In Vitro Cancer C Growth Inhibition Evaluation of Novel Phosphatidylcholines with Anisic and Veratric Acids. *Molecules* **2018**, *23*, 2022. [CrossRef]
20. Ritz, P.; Berrut, G. Mitochondrial Function, Energy Expenditure, Aging and Insulin Resistance. *Diabetes Metab.* **2005**, *31*, 5S67–5 [CrossRef]
21. Lee, J.-Y.; Lee, D.-C.; Im, J.-A.; Lee, J.-W. Mitochondrial DNA Copy Number in Peripheral Blood Is Independently Associa with Visceral Fat Accumulation in Healthy Young Adults. *Int. J. Endocrinol.* **2014**, *2014*, 586017. [CrossRef] [PubMed]
22. Zheng, L.D.; Linarelli, L.E.; Liu, L.; Wall, S.S.; Greenawald, M.H.; Seidel, R.W.; Estabrooks, P.A.; Almeida, F.A.; Cheng, Z. Insu Resistance Is Associated with Epigenetic and Genetic Regulation of Mitochondrial DNA in Obese Humans. *Clin. Epigenet.* **2** *7*, 60. [CrossRef] [PubMed]
23. Gianotti, T.F.; Sookoian, S.; Dieuzeide, G.; García, S.I.; Gemma, C.; González, C.D.; Pirola, C.J. A Decreased Mitochondrial D Content Is Related to Insulin Resistance in Adolescents. *Obesity* **2008**, *16*, 1591–1595. [CrossRef]
24. Xu, F.X.; Zhou, X.; Shen, F.; Pang, R.; Liu, S.M. Decreased Peripheral Blood Mitochondrial DNA Content Is Related to HbA Fasting Plasma Glucose Level and Age of Onset in Type 2 Diabetes Mellitus. *Diabet. Med.* **2012**, *29*, e47–e54. [CrossRef]
25. Kelley, D.E.; He, J.; Menshikova, E.V.; Ritov, V.B. Dysfunction of Mitochondria in Human Skeletal Muscle in Type 2 Diabe *Diabetes* **2002**, *51*, 2944–2950. [CrossRef]
26. Ritov, V.B.; Menshikova, E.V.; He, J.; Ferrell, R.E.; Goodpaster, B.H.; Kelley, D.E. Deficiency of Subsarcolemmal Mitochondria Obesity and Type 2 Diabetes. *Diabetes* **2005**, *54*, 8–14. [CrossRef] [PubMed]
27. Hoek, J.B.; Cahill, A.; Pastorino, J.G. Alcohol and Mitochondria: A Dysfunctional Relationship. *Gastroenterology* **2002**, *1* 2049–2063. [CrossRef] [PubMed]
28. Jové, M.; Salla, J.; Planavila, A.; Cabrero, A.; Michalik, L.; Wahli, W.; Laguna, J.C.; Vázquez-Carrera, M. Impaired Expression NADH Dehydrogenase Subunit 1 and PPARgamma Coactivator-1 in Skeletal Muscle of ZDF Rats: Restoration by Troglitazon *Lipid Res.* **2004**, *45*, 113–123. [CrossRef]
29. Morino, K.; Petersen, K.F.; Dufour, S.; Befroy, D.; Frattini, J.; Shatzkes, N.; Neschen, S.; White, M.F.; Bilz, S.; Sono, S.; et al. Reduc Mitochondrial Density and Increased IRS-1 Serine Phosphorylation in Muscle of Insulin-Resistant Offspring of Type 2 Diabe Parents. *J. Clin. Investig.* **2005**, *115*, 3587–3593. [CrossRef]
30. Hansen, A.S.; Marckmann, P.; Dragsted, L.O.; Finné Nielsen, I.-L.; Nielsen, S.E.; Grønbæk, M. Effect of Red Wine and Red Gra Extract on Blood Lipids, Haemostatic Factors, and Other Risk Factors for Cardiovascular Disease. *Eur. J. Clin. Nutr.* **2005**, 449–455. [CrossRef]
31. Lee, M.-K.; Park, Y.B.; Moon, S.-S.; Bok, S.H.; Kim, D.-J.; Ha, T.-Y.; Jeong, T.-S.; Jeong, K.-S.; Choi, M.-S. Hypocholesterolemic a Antioxidant Properties of 3-(4-Hydroxyl)Propanoic Acid Derivatives in High-Cholesterol Fed Rats. *Chem.-Biol. Interact.* **2007**, *1* 9–19. [CrossRef] [PubMed]
32. Mnafgui, K.; Derbali, A.; Sayadi, S.; Gharsallah, N.; Elfeki, A.; Allouche, N. Anti-Obesity and Cardioprotective Effects of Cinnar Acid in High Fat Diet-Induced Obese Rats. *J. Food Sci. Technol.* **2015**, *52*, 4369–4377. [CrossRef] [PubMed]
33. Binienda, Z.K.; Sarkar, S.; Silva-Ramirez, S.; Gonzalez, C. Role of Free Fatty Acids in Physiological Conditions and Mitochondr Dysfunction. *Food Nutr. Sci.* **2013**, *2013*, 36092. [CrossRef]

nutrients

cle

)-5-hydroxy-7-methoxy-3-(2-hydroxybenzyl)-4-chromanone, a ajor Homoisoflavonoid, Attenuates Free Fatty Acid-Induced epatic Steatosis by Activating AMPK and PPARα Pathways in epG2 Cells

Eun Park [1] and Ji-Sook Han [2,*]

1 Department of Hotel Baking Technology, Busan Health University, Busan 49318, Republic of Korea; jepark@bhu.ac.kr
2 Department of Food Science and Nutrition, Pusan National University, Busan 46241, Republic of Korea
* Correspondence: hanjs@pusan.ac.kr; Tel.:+ 82-51-510-2836; Fax:+ 82-51-583-3648

Abstract: Background: (*E*)-5-hydroxy-7-methoxy-3-(2-hydroxybenzyl)-4-chromanone (HMC), a homoisoflavonoid isolated from *Portulaca oleracea*, has significant anti-adipogenesis potential; it regulates adipogenic transcription factors. However, whether HMC improves hepatic steatosis in hepatocytes remains vague. This study investigated whether HMC ameliorates hepatic steatosis in free fatty acid-treated human hepatocellular carcinoma (HepG2) cells, and if so, its mechanism of action was analyzed. Methods: Hepatic steatosis was induced by a free fatty acid mixture in HepG2 cells. Thereafter, different HMC concentrations (10, 30, and 50 μM) or fenofibrate (10 μM, a PPARα agonist, positive control) was treated in HepG2 cells.Results: HMC markedly decreased lipid accumulation and triglyceride content in free fatty acid-treated HepG2 cell; it (10 and 50 μM) markedly upregulated protein expressions of pAMP-activated protein kinase (AMPK) and acetyl-CoA carboxylase. HMC (10 and 50 μM) markedly inhibited the expression of sterol regulatory element-binding protein-1c, fatty acid synthase, and stearoyl-coA desaturase 1, which are the enzymes involved in lipid synthesis. Furthermore, HMC (10 and 50 μM) markedly upregulated the protein expression of peroxisome proliferator-activated receptor alpha (PPARα) and enhanced the protein expressions of carnitine palmitoyl transferase 1 and acyl-CoA oxidase 1. Conclusion: HMC inhibits lipid accumulation and promotes fatty acid oxidation by AMPK and PPARα pathways in free fatty acid-treated HepG2 cells, thereby attenuating hepatic steatosis.

Keywords: HMC; hepatic steatosis; AMPK; PPARα; HepG2

ion: Park, J.-E.; Han, J.-S. -hydroxy-7-methoxy-3-(2-oxybenzyl)-4-chromanone, a r Homoisoflavonoid, Attenuates Fatty Acid-Induced Hepatic osis by Activating AMPK and Rα Pathways in HepG2 Cells. ents 2024, 16, 3475. https://rg/10.3390/nu16203475

emic Editors: María Angeles in and Sonia Ramos Rivero

ived: 11 September 2024
sed: 7 October 2024
pted: 10 October 2024
shed: 14 October 2024

1. Introduction

Hepatic lipid levels are generally balanced by the influx of free fatty acids (FFAs) and lipid synthesis and consumption. When the influx of FFAs and lipogenesis increases in the liver, excessive lipid accumulation occurs, leading to hepatic steatosis. Hepatic steatosis has long been considered a symptom of alcoholic liver disease; however, steatosis has been found even in the absence of alcohol consumption. As the number of obese or overweight individuals increases, nonalcoholic fatty liver disease has become a highly prevalent chronic liver disease. Nonalcoholic fatty liver disease includes nonalcoholic fatty liver (simple hepatic steatosis), inflammatory nonalcoholic steatohepatitis, advanced fibrosis, and liver cirrhosis [1]. Nonalcoholic fatty liver is an early, reversible stage of nonalcoholic fatty liver disease characterized by the accumulation of lipids without an inflammatory response [2]. Although the etiology of nonalcoholic fatty liver disease is not fully understood, it is believed that lipid metabolic disorders in the liver may play a crucial role in its initiation and progression [3]. Thus, the inhibition of lipid accumulation by regulating hepatic lipid metabolism is important for the treatment and prevention of fatty liver disease.

The major regulators of hepatic lipid metabolism that inhibit lipid accumulati are AMP-activated protein kinase (AMPK) and peroxisome proliferator-activated recep alpha (PPARα) [4–6]. Acetyl-CoA carboxylase (ACC) and sterol regulatory element-bindi protein 1c (SREBP-1c) are modulated by AMPK [7]. The activation of AMPK inhibits AC and fatty acid synthase (FAS) by downregulating SREBP-1c and suppressing lipogenesis The PPARα, nuclear receptor protein activated by a ligand, modulates the expression lipid oxidation transcription factors such as carnitine palmitoyl transferase 1 (CPT1) a acetyl-CoA oxidase 1 (ACOX1). Activated PPARα increases the expression of ACOX1 a CPT1 [9]. A reduction in PPARα expression was observed in individuals with hepa steatosis, and a PPARα agonist was used to ameliorate this symptom [10,11]. Thus, AM and PPARα may be major targets for the prevention of hepatic steatosis.

The compound HMC ((*E*)-hydroxy-7-methoxy-3-(2′-hydroxybenzyl)-4-chromano was isolated from *P. oleracea*. It is homoisoflavonoid compound, a subclass of flavonoi with an additional carbon atom. Previous studies have revealed that HMC has antid betic, anti-obesity, and anti-inflammatory effects in cellular and diabetic animal mo systems [12–14]. Although several effects and mechanisms of HMC in diseases ha been elucidated, whether and how HMC improves nonalcoholic fatty liver in hepatocy remains unclear. In this study, we investigated whether HMC could attenuate hepa steatosis in FFA-stimulated human hepatocellular carcinoma (HepG2) cells. Furthermo we determined the mechanism of action of the compound.

2. Material and Methods

2.1. Materials

P. oleracea was collected from Hyosung-Food Inc. (Gangwon, Hongcheon, Repub of Korea). Extraction (using dichloromethane and methanol) and isolation of HMC fro *P. oleracea* were performed using a previously established method in our laboratory [1 HM-chromanone was isolated by reversed-phase HPLC (ODS-A, 75% aq. MeOH) usi 5% MeOH in the $CHCl_3$ fraction (0.35 g). The structure of the compound (Figure 1) w elucidated using a combination of spectroscopic methods, including HR/Mass and ^{1}H a ^{13}C NMR.

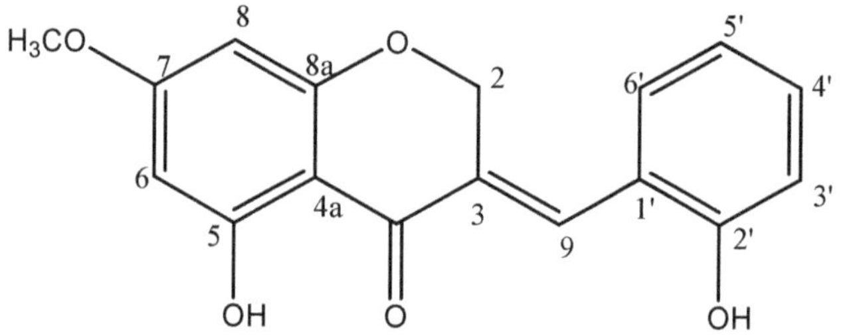

Figure 1. Chemical structure of HMC isolated from *P. oleracea*.

2.2. Cell Culture

The HepG2 cells (human liver cancer cell line, immortal cells) were purchased fro the Korean Cell Line Bank (KCLB No. 88065) and cultured in Dulbecco's modified eagl medium (DMEM; HyClone Laboratories, Logan, UT, USA) at 37 °C in 5% CO_2. The ce were then treated with a medium containing various concentrations of HMC and 1.0 m FFA mixture (oleate/palmitate = 2:1) for 24 h. Fenofibrate (10 μM) served as the positi control, and compound C (10 μM) served as the AMPK inhibitor.

2.3. Treatment of Free Fatty Acid Mixture

The FFA mixture was prepared by conjugating oleate/palmitate (2:1) with fatty ac free BSA [15]. Oleate and palmitate were dissolved in warmed BSA solution (45–50 °C) give a stock of 8 mM FFA mixture. The final molar ratio of FFA mixture to BSA was 6:1.

2.4. Cell Viability

HepG2 cells were seeded a density of 1×10^4 cells/well in 96-well plates and incubated at 37 °C in 5% CO_2. Thereafter, different HMC concentrations (10, 30, and 50 µM) or fenofibrate (10 µM, a PPARα agonist, positive control) with the FFA (1.0 mM)-treated cells were incubated for 24 h. To each well, the MTT solution (Sigma, St. Louis, MO, USA) was added and incubated for 4 h. The absorbance was measured using the microplate reader (540 nm).

2.5. Oil Red O Staining

Cells were seeded at a density of 3×10^5 cells/mL (1 mL) in 24-well plates for 24 h and then treated with the FFA mixtures. After 24 h of exposition, the HepG2 cells were washed with PBS and fixed in formalin for 2 h at room temperature. After washing with distilled H_2O and once with 60% isopropanol, the fixed adipocytes were completely dried, stained with Oil Red O working solution for 1 h, and washed immediately with distilled H_2O. Adipocytes were visualized using a microscope and image analysis. Intracellular triglyceride (TG) content was determined in cell lysates by an enzymatic colorimetric method and normalized by protein content, as previously described [16].

2.6. Western Blotting Analysis

The HepG2 cells were incubated with FFA stock solution (1.0 mM), followed by treatment with 10 and 50 µM HMC. The supernatants were quantified using Bio-Rad protein quantification reagent (Bio-Rad, Hercules, CA, USA). Samples were prepared by mixing buffer (Bio-Rad). The protein samples were separated by SDS-PAGE, and the acrylamide gel was transferred onto a nitrocellulose membrane. After blocking with TBS-T containing 5% skim milk for 1 h. The membrane was treated with the primary antibody (AMPK, PPARα, ACC, SREBP-1c, FAS and SCD-1, CPT1 and ACOX1) for 1 h. The secondary antibody (diluted 1:1000 with TBS-T) was incubated, and ECL solution was added to analyze protein expression in the dark using BioMax film. Relative protein expression was quantified by densitometric means using Multi Gauge v3.1.

2.7. Statistical Analysis

The results are expressed as mean ± standard deviation (SD, $n = 3$). Statistical analyses were performed using SPSS version 29.0 (IBM Corp., Armonk, NY, USA). The treatment groups were compared using one-way analysis of variance (ANOVA), followed by Tukey's HSD test.

3. Results

3.1. Cell Viability with HMC

To test the cytotoxicity of the FFA and HMC, various HMC concentrations were incubated with the FFA in the cells (Figure 2). In HepG2 cells treated with an FFA, the cell viability was markedly lower compared to that in the normal control cells. However, the HMC treatment markedly increased the viability rate lowered by the FFA treatment.

3.2. HMC Inhibits Lipid Accumulation

Lipid accumulation is an important factor in fatty liver disease in hepatocytes. To investigate whether HMC could inhibit lipid accumulation in hepatocytes, HepG2 cells were treated with 10, 30, and 50 µM of HMC or 10 µM fenofibrate (positive control), and then exposed to a 1.0 mM FFA mixture (oleate/palmitate = 2:1) for 24 h. As shown in Figure 3, HMC markedly reduced lipid droplet formation in HepG2 cells exposed to FFAs. Lipid accumulation markedly increased by 2.55-fold upon exposure to FFA in HepG2 cells compared with that in the control. However, 10, 30, and 50 µM of HMC markedly reduced lipid accumulation by 2.20-, 2.03- and 1.34-fold, respectively, in HepG2 cells. Fenofibrate reduced lipid accumulation by 1.28% when used at a concentration of 10 µM.

The 10 μM fenofibrate control and 50 μM HMC treatments showed similar values, with significant difference.

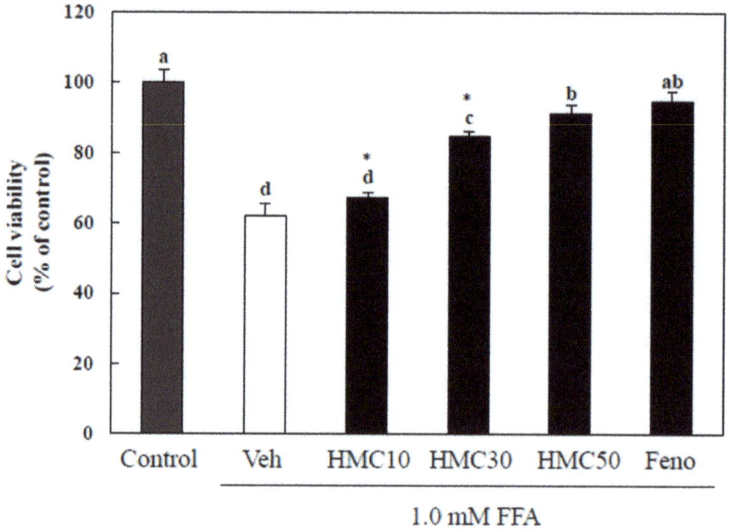

Figure 2. Cell Viability with HMC. Cell viability was stimulatied with 1.0 mM FFA stock soluti (oleate/palmitate = 2:1) and then treatment with HMC (10, 30, and 50 μM) for 24 h. Each valu expressed as the mean ± standard deviation (n = 3), and values with different superscript lett were markedly different with $p < 0.05$ as analyzed using ANOVA, followed by Tukey's HSD t *: Statistically significant differences.

3.3. HMC Reduces the Protein Expression of SREBP-1c, SCD-1, and FAS

In FFA-exposed HepG2 cells, treatment with 10 and 50 μM HMC markedly reduc SREBP-1c expression by 166.43% and 121.11%, respectively, which was increased to 221.1 by FFAs (Figure 4). The 10 μM fenofibrate reduced SREBP-1c expression to 117.84%, whi was not markedly different from the result of 50 μM HMC treatment. SCD-1 and FAS w also markedly increased to 259.98% and 297.45%, respectively, in HepG2 cells exposed FFAs compared to that in the normal control cells. However, treatment with 10 and 50 μ HMC markedly decreased SCD-1 to 200.31% and 139.52% and FAS to 234.61% and 127.88 respectively, compared to that in FFA-exposed HepG2 cells.

3.4. HMC Promotes AMPK and ACC in FFA-Exposed HepG2 Cells

To investigate whether HMC could induce AMPK in HepG2 cells exposed to FF HepG2 cells were treated with 10 and 50 μM of HMC and then exposed to FFAs. T AMPK phosphorylation markedly decreased by 48.51% in HepG2 cells exposed to FF compared to that in the control. However, treatment with 10 μM and 50 μM of HM markedly increased the phosphorylation levels of AMPK to 57.39% and 89.38%, respectiv (Figure 5). To confirm AMPK activation by HMC, ACC phosphorylation was investigat in HepG2 cells exposed to FFAs. The phosphorylation of ACC markedly decreased 32.21% in HepG2 cells exposed to FFAs compared to that in the control. However, 10 μ and 50 μM of HMC markedly increased pACC to 52.30% and 78.19%, respectively.

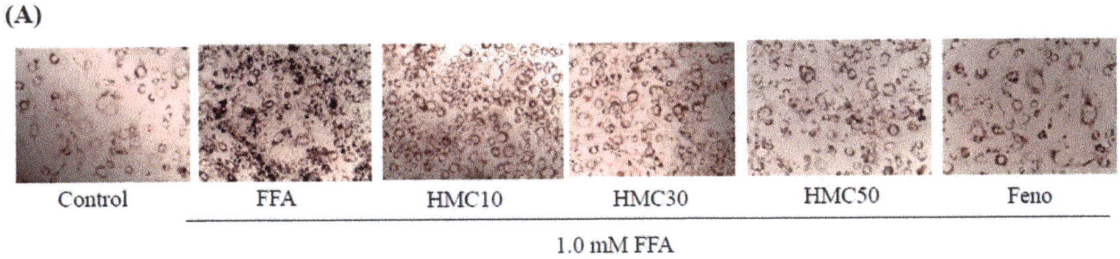

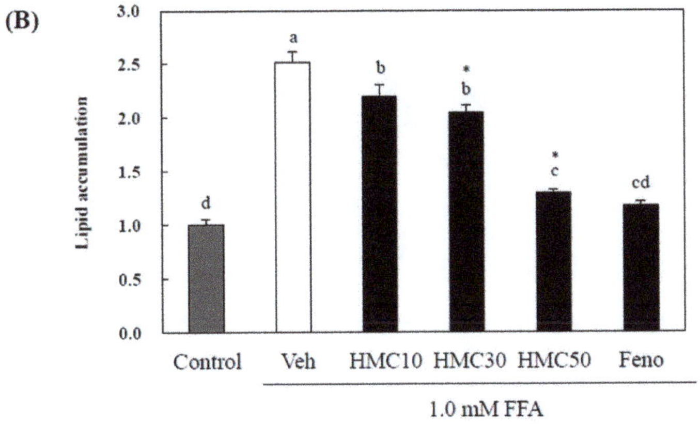

Figure 3. HMC inhibits lipid accumulation in FFA-induced hepatic steatosis. (**A**) Cells were fixed and stained with Oil Red O staining to visualize the lipid droplets by microscopy (magnification = 100X). Scale bar = 100 µM. (**B**) The effects of HMC on the inhibition of lipid accumulation. Each value is expressed as the mean ± standard deviation (n = 3), and values with different letters were markedly different with $p < 0.05$ as analyzed using ANOVA, followed by Tukey's HSD test. *: Statistically significant differences.

3.5. HMC Inhibits the Expression of SREBP-1c, SCD-1, and FAS via Activating AMPK

To determine whether AMPK phosphorylation mediates the effects of HMC on the inhibition of lipogenesis-related transcription factors and enzymes, FFA-exposed HepG2 cells were treated with an AMPK inhibitor, compound C, in addition to HMC treatment. Exposure to FFAs increased SREBP-1c to 221.10% in HepG2 cells compared to that in the control; however, this increase was markedly inhibited by treatment with HMC (Figure 6). At a concentration of 50 µM, HMC markedly inhibited SREBP-1c expression by 123.43% in FFA-exposed HepG2 cells, but this effect was blocked by compound C. We also examined SCD-1, FAS, and SREBP-1c expression. Treatment with 50 µM HMC markedly decreased the expression of SCD-1 and FAS to 138.26% and 124.58%, respectively, compared with that in FFA-exposed HepG2 cells. However, these effects were blocked by compound C. When cells were treated with HMC combined with compound C, the expression of SCD-1 and FAS markedly increased. These results show that HMC may inhibit the expression of SREBP-1c, SCD-1, and FAS by activating AMPK in HepG2 cells exposed to FFA.

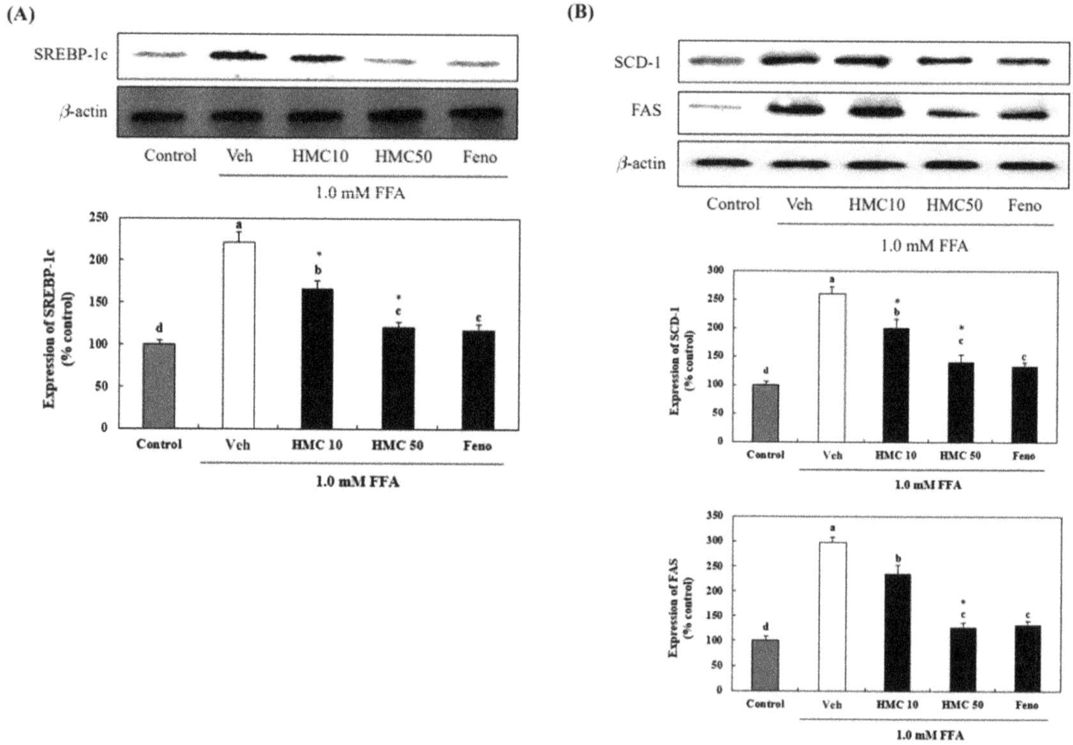

Figure 4. HMC inhibits the phosphorylation of SREBP-1c, SCD-1, and FAS in FFA-induced hepa steatosis. (**A**) The protein expression of SREBP-1c. (**B**) The protein expression of SCD-1 and FAS. Ea value is expressed as the mean ± standard deviation ($n = 3$), and values with different superscr letters were markedly different with $p < 0.05$ as analyzed using ANOVA, followed by Tukey's H test. *: Statistically significant differences.

3.6. HMC Increases the Phosphorylation of PPARα, CPT1, and ACOX1

The HMC markedly increased the expression of PPARα, CPT1, and ACOX1 in FF exposed HepG2 cells. The expression level of PPARα was markedly reduced to 42.86% FFAs in HepG2 cells compared to that in the control. However, the reduction was marke recovered by treatment with 10 μM and 50 Mm of HMC to 68.37% and 89.84%, respectiv (Figure 7). The PPARα expression level of the 10 μM fenofibrate control increased 92.33%, and there was no significant difference between the 50 μM HMC treatment a 10 μM fenofibrate control. The expression levels of CPT1 and ACOX1, the critical fatty ac oxidation enzymes, were also markedly decreased to 23.97% and 37.38%, respectively, FFA-exposed HepG2 cells compared to that in the control. However, treatment with 10 a 50 μM of HMC markedly increased the expression levels of CPT1 to 44.64% and 75.6 and ACOX1 to 63.75% and 92.33%, respectively. These results indicate that HMC co upregulate the expression of PPARα, CPT1, and ACOX1 in FFA-exposed HepG2 cells.

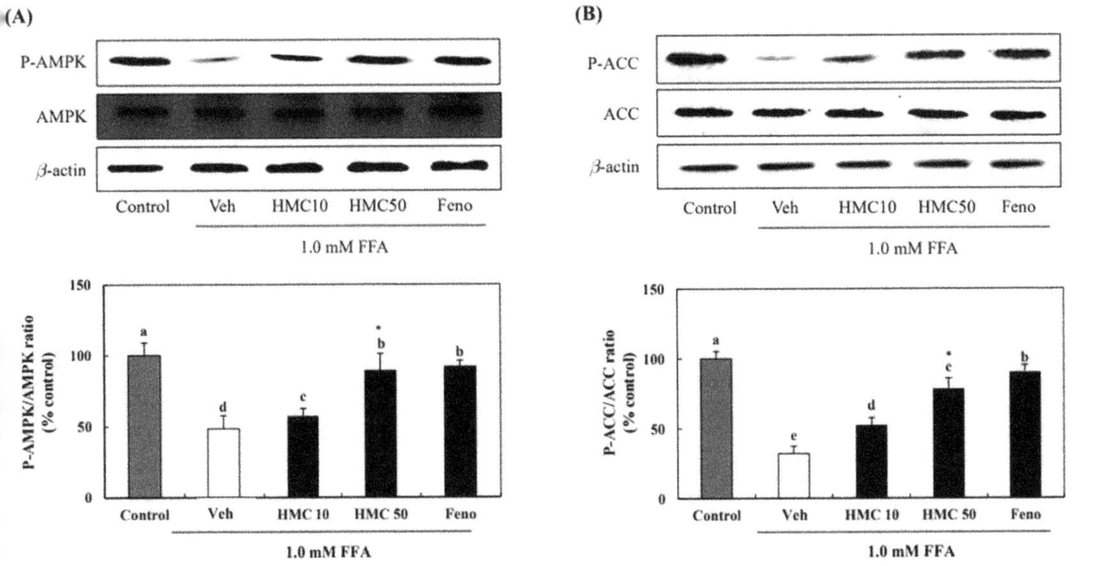

Figure 5. HMC promotes AMPK and ACC in FFA-induced hepatic steatosis. (**A**) The protein expression of AMPK. (**B**) The protein expression of ACC. Each value is expressed as the mean ± standard deviation ($n = 3$), and values with different superscript letters were markedly different with $p < 0.05$ as analyzed using one-way analysis of variance (ANOVA), followed by Tukey's HSD test. *: Statistically significant differences.

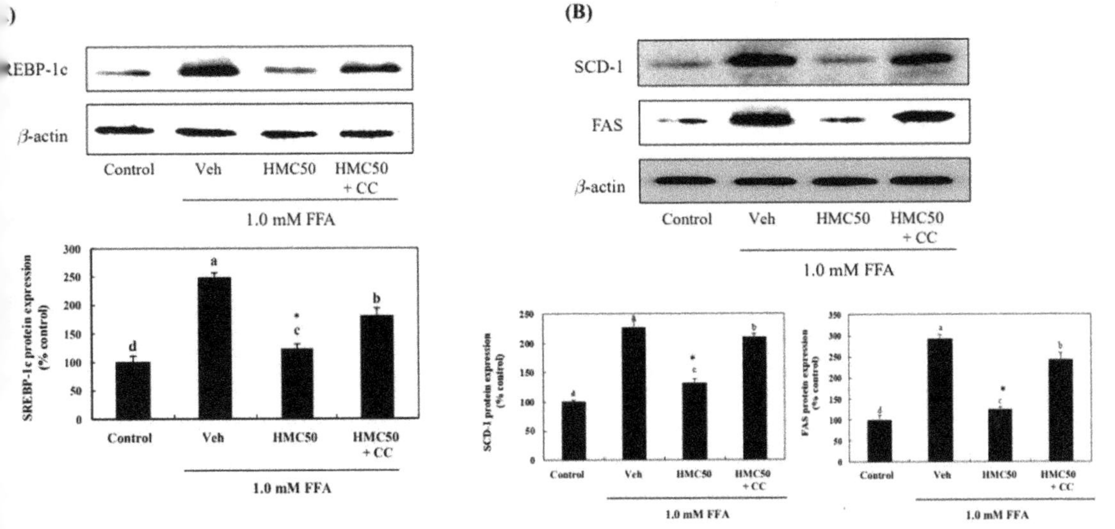

Figure 6. HMC inhibits the phosphorylation of SREBP-1c, SCD-1, and FAS via activating AMPK in FFA-induced hepatic steatosis. (**A**) The protein expression of SREBP-1c. (**B**) The protein expression of SCD-1 and FAS. Each value is expressed as the mean ± standard deviation ($n = 3$), and values with different superscript letters were markedly different with $p < 0.05$ as analyzed using one-way analysis of variance (ANOVA), followed by Tukey's HSD test. *: Statistically significant differences.

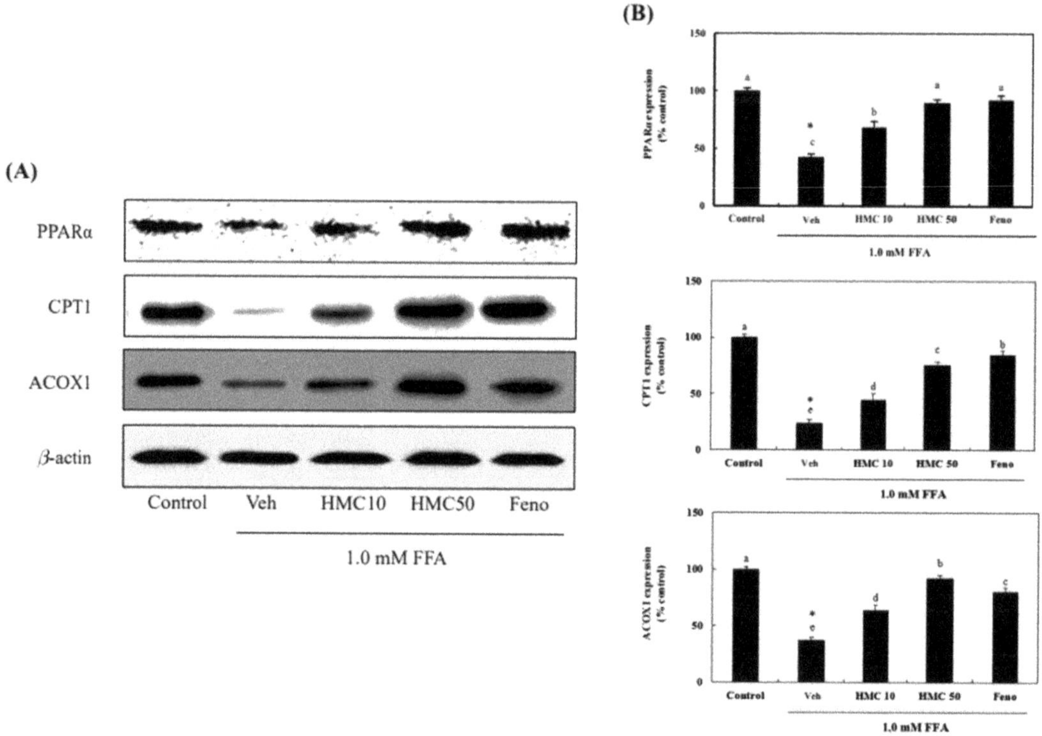

Figure 7. HMC increases the protein expression of PPARα, CPT1, and ACOX1. (**A**) The prot expression of PPARα, CPT1, and ACOX1. (**B**) Expression of PPARα, CPT1, and ACOX1. Each va is expressed as the mean ± standard deviation (n = 3), and values with different superscript lett were markedly different with $p < 0.05$ as analyzed using one-way analysis of variance (ANOV followed by Tukey's HSD test. *: Statistically significant differences.

4. Discussion

Hepatic steatosis begins with lipid accumulation in the liver and is considered a ma health problem. It could be caused by various factors, including the increased hepatic novo lipogenesis, FFA uptake into the liver, and damaged fatty acid β-oxidation [17,1 Insulin resistance also causes peripheral adipose lipolysis, which in turn promotes hepa steatosis by inducing hepatic TG production when FFA supply to the liver is increased [1 Patients with hepatic steatosis have approximately 60% of hepatic lipids induced from FF in adipose tissues. In contrast, healthy individuals obtain less than 5% of their hepatic lipi from de novo lipogenesis [20,21]. Thus, lipid accumulation due to excessive FFA infl and de novo lipogenesis is regarded as the main pathological factor in the development hepatic steatosis. Reducing lipid accumulation by modulating hepatic lipid metabolism an important strategy in the treatment of hepatic steatosis. We investigated whether HN could reduce hepatic lipid accumulation, and if so, by what mechanism, in FFA-stimulat HepG2 cells.

Increased FFA influx results in hepatic lipid accumulation, leading to hepatic steato Palmitic acid and oleic acid are common fatty acids. When used in vitro as a mixture can be an efficient model to investigate the effects of hepatic steatosis [22]. In this stu hepatic steatosis was induced using a FFA mixture (oleate/palmitate = 2:1) in Hep cells. The FFA mixture markedly increased lipid accumulation in HepG2 cell and HN markedly decreased the lipid accumulation. Reduction in hepatic lipid accumulati has been associated with several lipid metabolism proteins [23,24]. Activation of AMPK inhibits lipogenesis, which utilizes ATP. Activated AMPK decreases SREBP-1c and inhibits the activation of ACC, SCD-1, and FAS [25]. Changes in the expression of these proteins related to hepatic lipogenesis were investigated to elucidate the mechanism by which HMC reduces hepatic lipid accumulation.

Expression of phosphorylated AMPK and ACC was markedly increased by HMC in FFA-exposed HepG2 cells. Lipogenesis is inhibited when AMPK is activated by phosphorylation. AMPK phosphorylates and inactivates the lipogenic enzyme ACC [26]. Therefore, HMC appears to be involved in the inhibition of hepatic lipid accumulation by phosphorylating AMPK and ACC. Homoisoflavone compounds markedly increase AMPK, and this activation was attributed to the C-3 methine group in the compounds [27]. Consequently, AMPK activation by HMC is possibly due to the C-3 methine group in its 16-carbon skeleton structure.

A major lipogenic transcription factor that is abundant in the liver is SREBP-1c [24]. Increased FFA influx into hepatocytes induces the overexpression of SREBP-1c, leading to de novo lipogenesis [28]. Activated SREBP-1c is translocated to the nucleus and upregulates the expression of lipogenic enzymes [29], which are the main enzymes involved in fatty acid and triglyceride synthesis in the liver, respectively. The synthesis of mono-unsaturated fatty acid from saturated fatty acetyl Co-A is catalyzed by SCD-1, and its deletion prevents the development of fatty liver [30,31]. The last step of fatty acid biosynthesis is catalyzed by FAS [32]. In this study, HMC markedly reduced the expression of SREBP-1c, SCD-1, and FAS in FFA-exposed HepG2 cells. However, it failed to reduce the expression of SREBP-1c, SCD-1, and FAS increased by FFAs in HepG2 cells pretreated with the AMPK inhibitor. This indicated that HMC reduced the expression of SREBP-1c, SCD-1, and FAS by activating AMPK. This was consistent with a study by Li et al. [25] where activated AMPK reduced the expression of SREBP-1c and subsequently inhibited SCD-1 and FAS activation [6].

Flavonoids can decrease lipid accumulation by increasing AMPK activation while suppressing SREBP-1c expression in HepG2 cells with palmitic acid [33]. Nobiletin, a major flavonoid in citrus fruit, reduced SREBP-1c and FAS via AMPK and inhibited hepatic lipid accumulation [34]. Hesperidin also inhibited lipid accumulation by increasing AMPK and decreasing SREBP-1c, ACC, and FAS expression in oleic acid-treated HepG2 cell [35]. Reports have indicated that AMPK activated by flavonoids could reduce lipogenesis by inhibiting SREBP-1c expression, thereby decreasing ACC and FAS expression.

The activation of PPARα promotes fatty acid β-oxidation and decreases the lipid content [36]. Agonists of PPARα have lipid-lowering action, such as lowering TGs, and they have long been used to treat dyslipidemia [37]. It markedly decreased triglyceride content in the liver of an insulin-resistant animal model [38]. The alleviation of fatty liver by atorvastatin, a lipid-regulating drug, has been demonstrated to be due to an increase in hepatic PPARα expression in rats [39]. The enzyme CPT1 controls acetyl CoA inflow and fatty acid β-oxidation in the mitochondrial membrane. It is a major regulator of fatty acid β-oxidation in the liver. An increase in its expression decreases hepatic triglyceride content and ameliorates fatty liver [40]. The rate-regulating enzyme, ACOX1, is the first enzyme involved in fatty acid β-oxidation [41]. Administration of a PPARα agonist enhances fatty acid β-oxidation enzymes, such as CPT1 and ACOX1, in rodent hepatocytes [42]. Thus, HMC markedly increased PPARα expression and enhanced CPT1 and ACOX1 expression in FFA-exposed HepG2 cells.

Quercetin enhanced PPARα and CPT-1 expression and decreased hepatic steatosis by increasing fatty acid oxidation [43]. Genistein, a major isoflavone, increased *CPT-1* and *ACOX1* through the activation of PPARα [44]. These reports indicate that compounds such as flavonoids and isoflavones could activate PPARα and increase the expression of genes involved in fatty acid catabolism. This study suggests that HMC could also stimulate β-oxidation through the PPARα activation and the increase in CPT1 and ACOX1 expression in hepatocytes.

5. Conclusions

Our results demonstrate that HMC reduced lipid accumulation. This effect result from the activation of AMPK and PPARα signaling pathways, thereby suppressing SRE 1c, FAS, and SCD-1 expression and increasing CPT1 and ACOX1 expression. Thus, lipog esis was inhibited, and lipolysis was stimulated in the HepG2 cells (Figure 8).

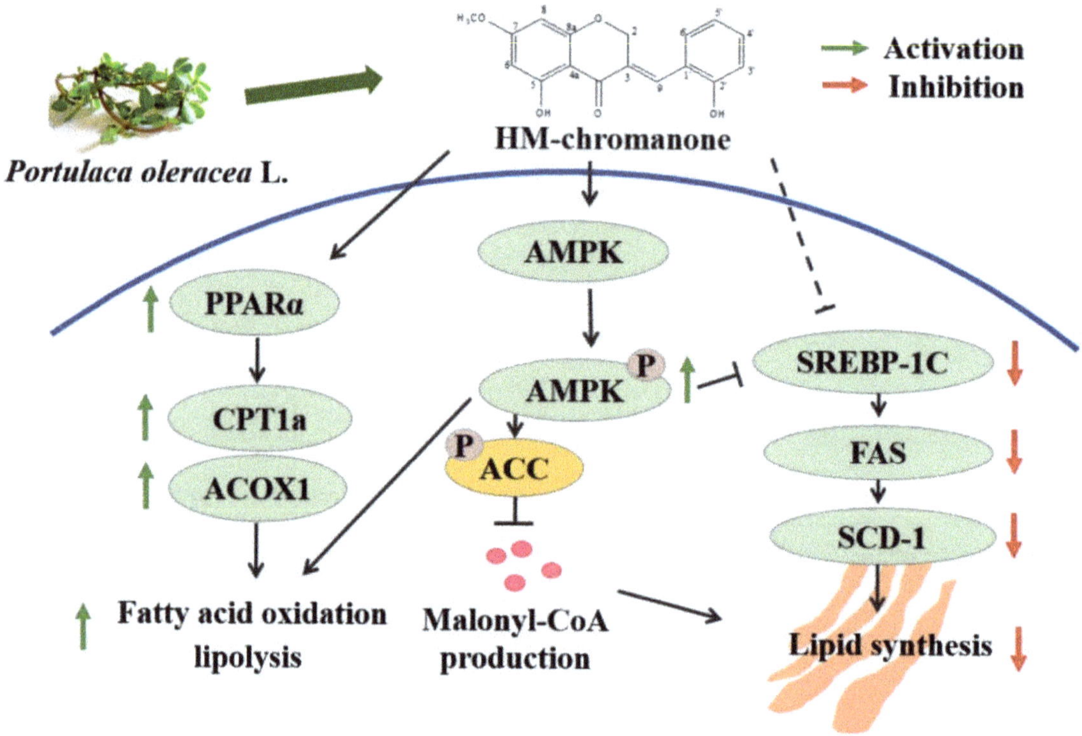

Figure 8. Proposed mechanism of attenuation of hepatic steatosis.

Author Contributions: Conceptualization, J.-E.P. and J.-S.H.; data curation, J.-E.P.; formal analy J.-E.P.; funding acquisition, J.-S.H.; investigation, J.-E.P.; methodology, J.-E.P.; project administrati J.-S.H.; resources, J.-S.H.; supervision, J.-S.H.; writing—original draft, J.-E.P.; writing—review a editing, J.-E.P. and J.-S.H. All data were generated in-house, and no paper mill was used. All auth agree to be accountable for all aspects of the work, ensuring its integrity and accuracy. All auth have read and agreed to the published version of the manuscript.

Funding: This research was funded by the National Research Foundation of Korea (NRF) (gra number: NRF-2020R1A2B5B01001547).

Institutional Review Board Statement: Not applicable.

Informed Consent Statement: Not applicable.

Data Availability Statement: Data are available upon reasonable request to the correspondi author.

Conflicts of Interest: The authors declare no conflicts of interest.

erences

Contos, M.J.; Choudhury, J.; Mills, A.S.; Sanyal, A.J. The histologic spectrum of nonalcoholic fatty liver disease. *Clin. Liver Dis.* **2004**, *8*, 481–500. [CrossRef] [PubMed]

Sozio, M.S.; Liangpunsakul, S.; Crabb, D. The role of lipid metabolism in the pathogenesis of alcoholic and nonalcoholic hepatic steatosis. *Semin. Liver Dis.* **2010**, *30*, 378–390. [CrossRef] [PubMed]

Torres, D.M.; Williams, C.D.; Harrison, S.A. Features, diagnosis, and treatment of nonalcoholic fatty liver disease. *Clin. Gastroenterol. Hepatol.* **2012**, *10*, 837–858. [CrossRef]

Brown, M.S.; Goldstein, J.L. The SREBP pathway: Regulation of cholesterol metabolism by proteolysis of a membrane-bound transcription factor. *Cell* **1997**, *89*, 331–340. [CrossRef] [PubMed]

Kang, O.H.; Kim, S.B.; Mun, S.H.; Seo, Y.S.; Hwang, H.C.; Lee, Y.M.; Lee, H.S.; Kang, D.G.; Kwon, D.Y. Puerarin ameliorates hepatic steatosis by activating the PPARα and AMPK signaling pathways in hepatocytes. *Int. J. Mol. Med.* **2015**, *35*, 803–809. [CrossRef]

Zhou, G.; Myers, R.; Li, Y.; Chen, Y.; Shen, X.; Fenyk-Melody, J.; Wu, M.; Ventre, J.; Doebber, T.; Fujii, N.; et al. Role of AMP-activated protein kinase in mechanism of metformin action. *J. Clin. Investig.* **2001**, *108*, 1167–1174. [CrossRef]

Long, Y.C.; Zierath, J.R. AMP-activated protein kinase signaling in metabolic regulation. *J. Clin. Investig.* **2006**, *116*, 1776–1783. [CrossRef]

Zeng, L.; Tang, W.J.; Yin, J.J.; Zhou, B.J. Signal transductions and nonalcoholic fatty liver: A mini-review. *Int. J. Clin. Exp. Med.* **2014**, *7*, 1624–1631. [PubMed]

Chitturi, S.; Abeygunasekera, S.; Farrell, G.C. NASH and insulin resistance: Insulin hypersecretion and specific association with the insulin resistance syndrome. *Hepatology* **2002**, *35*, 373–379. [CrossRef]

Seo, Y.S.; Kim, J.H.; Jo, N.Y. PPAR agonists treatment is effective in a nonalcoholic fatty liver disease animal model by modulating fatty-acid metabolic enzymes. *J. Gastroenterol. Hepatol.* **2008**, *23*, 102–109. [CrossRef]

Lei, X.; Li, J.; Liu, B.; Zhang, N.; Liu, H. Separation and Identification of Four New Compounds with Antibacterial Activity from *Portulaca oleracea* L. *Molecules* **2015**, *20*, 16375–16387. [CrossRef] [PubMed]

Je, J.Y.; Park, J.E.; Seo, Y.; Han, J.S. HMC inhibits adipogenesis by regulating adipogenic transcription factors and AMPK in 3T3-L1 adipocytes. *Eur. J. Pharmacol.* **2021**, *892*, 173689. [CrossRef] [PubMed]

Kang, E.; Park, J.E.; Seo, Y.; Han, J.S. (E)-5-hydroxy-7-methoxy-3-(2′-hydroxybenzyl)-4-chromanone isolated from *Portulaca oleracea* L. suppresses LPS-induced inflammation in RAW 264.7 macrophages by downregulating inflammatory factors. *Immunopharmacol. Immunotoxicol.* **2021**, *43*, 611–621. [CrossRef] [PubMed]

Park, J.E.; Park, J.Y.; Seo, Y.; Han, J.S. A new chromanone isolated from Portulaca oleracea L. increases glucose uptake by stimulating GLUT4 translocation to the plasma membrane in 3T3-L1 adipocytes. *Int. J. Biol. Macromol.* **2019**, *123*, 26–34. [CrossRef] [PubMed]

Sinha, S.; Perdomo, G.; Brown, N.F.; O'Doherty, R.M. Fatty acid-induced insulin resistance in L6 myotubes is prevented by inhibition of activation and nuclear localization of nuclear factor kappa B. *J. Biol. Chem.* **2004**, *279*, 41294–41301. [CrossRef]

Guo, H.X.; Liu, D.H.; Ma, Y. Long-term baicalin administration ameliorates metabolic disorders and hepatic steatosis in rats given a high-fat diet. *Acta Pharmacol. Sin.* **2009**, *30*, 1505–1512. [CrossRef]

Donnelly, K.L.; Smith, C.I.; Schwarzenberg, S.J.; Jessurun, J.; Boldt, M.D.; Parks, E.J. Sources of fatty acids stored in liver and secreted via lipoproteins in patients with nonalcoholic fatty liver disease. *J. Clin. Investig.* **2005**, *115*, 1343–1351. [CrossRef]

Postic, C.; Girard, J. Contribution of de novo fatty acid synthesis to hepatic steatosis and insulin resistance: Lessons from genetically engineered mice. *J. Clin. Investig.* **2008**, *118*, 829–838. [CrossRef] [PubMed]

Hooper, A.J.; Adams, L.A.; Burnett, J.R. Genetic determinants of hepatic steatosis in man. *J. Lipid Res.* **2011**, *52*, 593–617. [CrossRef] [PubMed]

Clark, J.M.; Brancati, F.L.; Diehl, A.M. Nonalcoholic fatty liver disease. *Gastroenterology* **2002**, *122*, 1649–1657. [CrossRef] [PubMed]

Hudgins, L.C.; Hellerstein, M.K.; Seidman, C.E.; Neese, R.A.; Tremaroli, J.D.; Hirsch, J. Relationship between carbohydrate-induced hypertriglyceridemia and fatty acid synthesis in lean and obese subjects. *J. Lipid Res.* **2000**, *41*, 595–604. [CrossRef] [PubMed]

Seo, M.S.; Hong, S.W.; Yeon, S.H. Magnolia officinalis attenuates free fatty acid-induced lipogenesis via AMPK phosphorylation in hepatocytes. *J. Ethnopharmacol.* **2014**, *157*, 140–148. [CrossRef] [PubMed]

Tessari, P.; Coracina, A.; Cosma, A.; Tiengo, A. Hepatic lipid metabolism and non-alcoholic fatty liver disease. *Nutr. Metab. Cardiovasc. Dis.* **2009**, *19*, 291–302. [CrossRef]

Musso, G.; Gambino, R.; Cassader, M. Recent insights into hepatic lipid metabolism in non-alcoholic fatty liver disease (NAFLD). *Prog. Lipid Res.* **2009**, *48*, 1–26. [CrossRef] [PubMed]

Li, Y.; Xu, S.; Mihaylova, M.M. AMPK phosphorylates and inhibits SREBP activity to attenuate hepatic steatosis and atherosclerosis in diet-induced insulin-resistant mice. *Cell Metab.* **2011**, *13*, 376–388. [CrossRef] [PubMed]

Bijland, S.; Mancini, S.J.; Salt, I.P. Role of AMP-activated protein kinase in adipose tissue metabolism and inflammation. *Clin. Sci.* **2013**, *124*, 491–507. [CrossRef]

Guo, H.; Zhao, H.; Kanno, Y. A dihydrochalcone and several homoisoflavonoids from Polygonatum odoratum are activators of adenosine monophosphate-activated protein kinase. *Bioorg. Med. Chem. Lett.* **2013**, *23*, 3137–3139. [CrossRef]

28. Méndez-Sánchez, N.; Arrese, M.; Zamora-Valdés, D.; Uribe, M. Current concepts in the pathogenesis of nonalcoholic fatty l disease. *Liver Int.* **2007**, *27*, 423–433. [CrossRef]
29. Horton, J.D.; Goldstein, J.L.; Brown, M.S. SREBPs: Activators of the complete program of cholesterol and fatty acid synthes the liver. *J. Clin. Investig.* **2002**, *109*, 1125–1131. [CrossRef]
30. Gómez-Lechón, M.J.; Donato, M.T.; Martínez-Romero, A.; Jiménez, N.; Castell, J.V.; O'Connor, J.E. A human hepatocell in vitro model to investigate steatosis. *Chem. Biol. Interact.* **2007**, *165*, 106–116. [CrossRef]
31. Shigiyama, F.; Kumashiro, N.; Tsuneoka, Y.; Igarashi, H.; Yoshikawa, F.; Kakehi, S.; Funato, H.; Hirose, T. Mechanisms of s deprivation-induced hepatic steatosis and insulin resistance in mice. *Am. J. Physiol. Endocrinol. Metab.* **2018**, *315*, E848–E [CrossRef]
32. Roder, K.; Zhang, L.; Schweizer, M. SREBP-1c mediates the retinoid-dependent increase in fatty acid synthase promoter acti in HepG2. *FEBS Lett.* **2007**, *581*, 2715–2720. [CrossRef] [PubMed]
33. Rajan, P.; Natraj, P.; Ranaweera, S.S.; Dayarathne, L.A.; Lee, Y.J.; Han, C.H. Anti-adipogenic effect of the flavonoids through activation of AMPK in palmitate (PA)-treated HepG2 cells. *J. Vet. Sci.* **2022**, *23*, e4. [CrossRef] [PubMed]
34. Yuk, T.; Kim, Y.; Yang, J.; Sung, J.; Jeong, H.S.; Lee, J. Nobiletin Inhibits Hepatic Lipogenesis via Activation of AMP-Activa Protein Kinase. *Evid. Based Complement. Altern. Med.* **2018**, *2018*, 7420265. [CrossRef] [PubMed]
35. Zhu, X.; Bian, H.; Wang, L. Berberine attenuates nonalcoholic hepatic steatosis through the AMPK-SREBP-1c-SCD1 pathway. *Radic. Biol. Med.* **2019**, *141*, 192–204. [CrossRef]
36. Escher, P.; Wahli, W. Peroxisome proliferator-activated receptors: Insight into multiple cellular functions. *Mutat. Res.* **2000**, 121–138. [CrossRef]
37. Francis, G.A.; Annicotte, J.S.; Auwerx, J. PPAR-alpha effects on the heart and other vascular tissues. *Am. J. Physiol. Heart C Physiol.* **2003**, *285*, H1–H9. [CrossRef] [PubMed]
38. Guerre-Millo, M.; Gervois, P.; Raspé, E. Peroxisome proliferator-activated receptor alpha activators improve insulin sensiti and reduce adiposity. *J. Biol. Chem.* **2000**, *275*, 16638–16642. [CrossRef]
39. Roglans, N.; Sanguino, E.; Peris, C. Atorvastatin treatment induced peroxisome proliferator-activated receptor alpha express and decreased plasma nonesterified fatty acids and liver triglyceride in fructose-fed rats. *J. Pharmacol. Exp. Ther.* **2002**, 232–239. [CrossRef] [PubMed]
40. Stefanovic-Racic, M.; Perdomo, G.; Mantell, B.S.; Sipula, I.J.; Brown, N.F.; O'Doherty, R.M. A moderate increase in carni palmitoyltransferase 1a activity is sufficient to substantially reduce hepatic triglyceride levels. *Am. J. Physiol. Endocrinol. Me* **2008**, *294*, E969–E977. [CrossRef]
41. Kondo, T.; Kishi, M.; Fushimi, T.; Kaga, T. Acetic acid upregulates the expression of genes for fatty acid oxidation enzyme liver to suppress body fat accumulation. *J. Agric. Food Chem.* **2009**, *57*, 5982–5986. [CrossRef] [PubMed]
42. Chou, C.J.; Haluzik, M.; Gregory, C. WY14,643, a peroxisome proliferator-activated receptor alpha (PPARalpha) agonist, impro hepatic and muscle steatosis and reverses insulin resistance in lipoatrophic A-ZIP/F-1 mice. *J. Biol. Chem.* **2002**, *277*, 24484–24 [CrossRef] [PubMed]
43. Panchal, S.K.; Poudyal, H.; Brown, L. Quercetin ameliorates cardiovascular, hepatic, and metabolic changes in diet-indu metabolic syndrome in rats. *J. Nutr.* **2012**, *142*, 1026–1032. [CrossRef] [PubMed]
44. Kim, S.; Shin, H.J.; Kim, S.Y. Genistein enhances expression of genes involved in fatty acid catabolism through activation PPARalpha. *Mol. Cell Endocrinol.* **2004**, *220*, 51–58. [CrossRef]

nutrients

MDPI

cle

otective Role of Lycopene in Subjects with Liver Disease: UTRIHEP Study

sella Donghia [1,*,†], Angelo Campanella [1,†], Caterina Bonfiglio [1], Francesco Cuccaro [2], Rossella Tatoli [1,†] and nluigi Giannelli [1]

1 National Institute of Gastroenterology—IRCCS "Saverio de Bellis", 70013 Castellana Grotte, Italy; angelo.campanella@irccsdebellis.it (A.C.); catia.bonfiglio@irccsdebellis.it (C.B.); rossella.tatoli@irccsdebellis.it (R.T.); gianluigi.giannelli@irccsdebellis.it (G.G.)
2 Local Health Unit—Barletta-Andria-Trani, 76121 Barletta, Italy; francescocuccaroepi@gmail.com
* Correspondence: rossella.donghia@irccsdebellis.it
† These authors contributed equally to this work.

Abstract: Background: Liver diseases are constantly increasing throughout the world and are often associated with other diseases, but above all they are caused by improper diet. Adherence to a diet with abundant vegetables has now been widely demonstrated to be important in combating this pathological condition. The aim of this study was to explore the protective role of lycopene (LYC) extracts from cooked and fresh tomato. Methods: The study cohort included 969 participants assessed in the NUTRIHEP cohort (2005–2006) and the associated follow-up (2014–2016), divided into two groups, based on liver condition: NAFLD, or AFLD and FLD. Results: The results indicated a statistical significance of LYC consumption, showing a protective role against liver disease, the best concentration being 9.50 mg/die, with an RR value of 0.59, $p = 0.01$, 0.39 to 0.90 at 95% C.I., and RRR = 0.40, $p = 0.002$, 0.22 to 0.71 at 95% C.I. Conclusions: The protective role of LYC extracts from tomato has not been amply demonstrated in humans. We conclude that this is one of the few papers in the literature to evaluate the protective effect of LYC against liver disease, as well as how this molecule could be used in future possible treatments. Utilizing lycopene as a supplement alone or in combination with other foods could be useful for developing treatments with reduced contraindications.

Keywords: lycopene; liver; antioxidant

ion: Donghia, R.; Campanella, onfiglio, C.; Cuccaro, F.; Tatoli, iannelli, G. Protective Role of pene in Subjects with Liver ise: NUTRIHEP Study. *Nutrients* *16*, 562. https://doi.org/ 90/nu16040562

emic Editors: Sonia Ramos o and María Angeles Martín

ved: 1 February 2024
sed: 16 February 2024
pted: 17 February 2024
shed: 18 February 2024

1. Introduction

Nonalcoholic fatty liver disease (NAFLD) is the most common cause of liver dysfunction worldwide. It affects an increasingly large segment of adults, reaching rates of 25% in Western countries [1], and up to 75% in patients with obesity or type 2 diabetes mellitus [2]. This condition is also increasingly diagnosed in children and teenagers [3]. NAFLD is a liver disorder defined by the accumulation of fat in more than 5% of hepatocytes, in the absence of significant alcohol consumption, drug use, or any other liver disease [4]. It covers a spectrum of hepatic diseases, ranging from simple steatosis to nonalcoholic steatohepatitis (NASH), which can evolve to cirrhosis, liver failure, and hepatocellular carcinoma [5].

Several studies have indicated chronic inflammation and oxidative stress as factors in the initiation and progression of NAFLD [6]. The "multiple hits" theory is the most accepted theory to explain the complex pathogenesis of NAFLD [7]. According to this theory, the close crosstalk between genetics and environmental factors leads to the occurrence of multiple hits, including inflammation, oxidative stress, lipid peroxidation, and mitochondrial dysfunction, promoting the onset of NAFLD and its progression to NASH.

Two different steps, or "hits", can be identified: the first is the accumulation triglycerides in hepatocytes, while the second is oxidative stress [8] that results in lip peroxidation, mitochondrial dysfunction, and inflammation [7–10].

Currently, no specific drug or therapy is available for the treatment of NAFLD, b international guidelines recommend a lifestyle-based approach, particularly a healthy d [https://www.iss.it/-/snlg-steatosi-epatica-non-alcolica, accessed on 21 November 202 It is now known how dietary factors can modulate hepatic steatosis and counteract liv damage, preventing the evolution of NAFLD into NASH [9]. Some studies have report the positive effects of the dietary intake of antioxidants from fruits and vegetables on t biomarkers of hepatic damage, in both human and animal models [11–13].

Recently, among dietary antioxidants, particular attention has been paid to LY a phytochemical belonging to the carotenoid family. LYC is abundantly found in r vegetables and fruits, being responsible for their characteristic color: tomatoes, papaya, g fruit, pink grapefruit, pink guava, carrots, and watermelon [14]. The level of its bioacti properties is influenced by many factors, such as bioavailability, metabolism, isomerizati or interactions with other carotenoids [15]. In nature, LYC occurs in its trans isofor Thermal treatments, such as cooking, increase the bioavailability of LYC due to its trans-cis isomerization [16]. It is assumed that thanks to the cis form, LYC is highly bioavailab in the human diet [17,18].

In addition, some studies suggest that an increase in the bioavailability of LYC as result of its trans-to-cis isomerization is due to the addition of sulfur compounds contain in garlic and onion [19]. Since it is a lipid-soluble substance, consumption together wi dietary fat sources amplifies its bioavailability [20]. The main sources of LYC in t diet are tomatoes and tomato-based products [21,22]. In the European population, LY consumption ranges between 5 and 7 mg/die [23], and more than 80% of its dieta daily intake comes from tomato products [24]. Several studies have demonstrated ma significant health benefits of LYC [25]. Due to its structure and lipophilic nature, LYC sho anti-inflammatory and antioxidant effects [26,27]. It is a much more potent antioxida than either alpha-tocopherol or beta-carotene [28].

The aim of this research is to evaluate the protective role of lycopene derived from t diet in a cohort from Southern Italy, demonstrating on large numbers the effect alrea discussed in the literature but based on in vitro studies.

2. Materials and Methods

2.1. Study Population

The NUTRIHEP study is a cohort extracted in 2005–2006 from the medical recor of general practitioners in the municipality of Putignano ($\geq$18 years) to minimize erro regarding the distribution of the population sample by age and gender. Furthermo Italian law requires the presence of a family doctor, allowing the overlap of medical a census data. Participants were interviewed at baseline in 2004–2005 by trained physicia and/or nutritionists to collect information on their sociodemographic characteristics, heal status, personal history, and lifestyle factors such as smoking, education level (Internatio Standard Classification of Education), marital status, and eating habits [29]. Weight w measured on an electronic scale (SECA©) and recorded to the nearest 0.1 kg. Height w measured with a wall-mounted stadiometer (SECA©) and recorded to the nearest 1 c From 2014 to 2018, all eligible subjects, starting from NUTRIHEP participants, were invit to take part in the follow-up. A total of 1426 subjects responded. The subjects underwe the same protocol as at the first enrollment.

NAFLD was diagnosed in subjects with hepatic steatosis on ultrasound, witho AFL (alcoholic fatty liver; the cutoffs for AFL were 30 g/day for men and 20 g/d for women [30]) or drug-induced fatty liver disease (e.g., corticosteroids, valproic ac amiodarone, hepatitis C or B viral infections, or other disorders) [31]. Meanwhile, fatty liv disease (FLD) was diagnosed as a common cause of chronic liver disease, encompassi several liver disorders, ranging from simple steatosis to steatohepatitis, liver fibrosis, cirrhosis, and hepatocellular carcinoma [32].

All participants signed informed consent after receiving complete information about their medical data to be studied. Those enrolled at baseline and follow-up were included in this analysis, investigating a total of 969 subjects. The study was approved by the Ethical Committee of the Minister of Health (DDG-CE-502/2005; DDG-CE-792/2014, on 20 May 2005 and 14 February 2014, respectively).

2.2. Dietary Assessments

The validated European Prospective Investigation into Cancer and Nutrition (EPIC) food frequency questionnaire was administered during the visit, and each food item (260 food items) was converted into an average daily intake in grams. LYC intake was calculated based on cooked and fresh tomato intake per day. The quantity of lycopene (mg/day) was obtained from the database (https://fdc.nal.usda.gov/ (accessed on 12 April 2019)).

2.3. Statistical Analysis

The patients' characteristics are reported as means and standard deviations (M ± SD) or medians and interquartile ranges for continuous variables, and as frequencies and percentages (%) for categorical variables.

We estimated a multiple multinomial logistic regression model using the three-category outcome variable (absent, NAFLD, and AFLD and FLD) as the dependent variable and LYC intake (both continuous and categorical) as predictors. The models were also adjusted for some covariates (gender, age, weight, glucose, triglycerides, AST, cholesterol, and HDL), and estimated coefficients were transformed into relative risk ratios (RRRs). To test the null hypothesis of non-association, the two-tailed probability level was set at 0.05.

The analyses were conducted with StataCorp 2023 Stata Statistical Software: Release 18 (College Station, TX, USA: StataCorp LLC.), while RStudio ("Prairie Trillium" Release) was used for the plots.

3. Results

Table 1 shows the epidemiological and clinical characteristics of the NUTRIHEP cohort at first recall and follow-up.

Table 1. Epidemiological and clinical characteristics of the cohorts, NUTRIHEP 1 and 2.

ameters *	Cohorts					
	NUTRIHEP 1 (2005–2006) (n = 969)			NUTRIHEP 2 (2014–2016) (n = 969)		
	Absent (n = 748)	NAFLD (n = 114)	AFLD and FLD (n = 107)	Absent (n = 470)	NAFLD (n = 368)	AFLD and FLD (n = 131)
(years)	41.59 ± 13.88	46.67 ± 11.81	54.08 ± 10.23	48.11 ± 13.14	58.41 ± 13.39	61.51 ± 12.41
Categories (%)						
50	424 (56.70)	45 (39.50)	13 (12.10)	283 (60.20)	96 (26.10)	25 (19.10)
50	324 (43.30)	69 (60.50)	94 (87.90)	187 (39.80)	272 (73.90)	106 (80.90)
der (M) (%)	324 (43.30)	69 (60.50)	94 (87.90)	203 (43.20)	179 (48.60)	105 (80.20)
oker (%)						
Never/Former	636 (85.10)	92 (81.40)	87 (81.30)	418 (88.90)	322 (87.70)	110 (84.60)
Current	111 (14.90)	21 (18.60)	20 (18.70)	52 (11.10)	45 (12.30)	20 (15.40)
rital Status (%)						
ingle	208 (28.00)	15 (13.20)	4 (3.70)	91 (19.40)	48 (13.00)	7 (5.40)
Married/Coupled	511 (68.70)	94 (82.50)	99 (92.50)	358 (76.30)	290 (78.80)	113 (86.90)
eparated/Divorced	15 (2.00)	0 (0.00)	1 (0.90)	13 (2.80)	7 (1.90)	4 (3.10)
Widower	10 (1.30)	5 (4.40)	3 (2.80)	7 (1.50)	23 (6.20)	6 (4.60)
pertension (Yes) (%)	72 (11.10)	24 (26.40)	29 (29.90)	74 (16.90)	155 (44.50)	55 (44.40)
betes (Yes) (%)	9 (1.20)	4 (3.50)	10 (9.30)	9 (2.10)	34 (9.80)	13 (10.50)
ight (kg)	66.07 ± 12.19	79.94 ± 17.14	83.93 ± 12.53	66.35 ± 12.06	78.63 ± 14.40	81.72 ± 15.50
I (kg/m^2)	24.43 ± 3.80	29.02 ± 5.64	29.66 ± 3.58	24.88 ± 3.62	30.16 ± 4.81	29.72 ± 4.99

Table 1. *Cont.*

Parameters *	Cohorts					
	NUTRIHEP 1 (2005–2006) (*n* = 969)			NUTRIHEP 2 (2014–2016) (*n* = 969)		
	Absent (*n* = 748)	NAFLD (*n* = 114)	AFLD and FLD (*n* = 107)	Absent (*n* = 470)	NAFLD (*n* = 368)	AFLD and FLD (*n* = 131)
Blood Parameters						
Glucose (mmol/L)	5.37 ± 0.87	5.84 ± 0.99	6.13 ± 1.17	4.98 ± 0.54	5.46 ± 1.11	5.79 ± 1.09
Total Cholesterol (mmol/L)	4.88 ± 0.97	5.19 ± 1.02	5.21 ± 0.99	4.85 ± 0.87	5.08 ± 0.95	4.84 ± 0.93
Triglycerides (mmol/L)	1.04 ± 0.69	1.64 ± 0.96	1.68 ± 1.13	0.89 ± 0.69	1.32 ± 0.85	1.41 ± 0.94
HDL (mmol/L)	1.40 ± 0.34	1.20 ± 0.30	1.17 ± 0.23	1.39 ± 0.33	1.27 ± 0.29	1.16 ± 0.28
WBC (K/mcL)	6.13 ± 1.57	6.81 ± 1.64	6.56 ± 1.63	5.63 ± 2.91	6.05 ± 1.62	6.36 ± 1.67
GGT (μkat/L)	0.19 ± 0.15	0.26 ± 0.12	0.27 ± 0.10	0.25 ± 0.14	0.32 ± 0.25	0.38 ± 0.26
AST (μkat/L)	0.18 ± 0.10	0.21 ± 0.08	0.21 ± 0.07	0.35 ± 0.12	0.38 ± 0.30	0.40 ± 0.10
ALT (μkat/L)	0.23 ± 0.21	0.35 ± 0.17	0.33 ± 0.16	0.33 ± 0.15	0.42 ± 0.44	0.43 ± 0.17
Blood Lipids (Yes) (%)	30 (4.60)	4 (4.40)	10 (10.30)	54 (12.30)	61 (17.60)	25 (20.20)
rMed ^ (Yes) (%)	8 (6–11)	8 (5–11)	8 (6–11)	9 (7–11)	9 (7–11)	8 (6–10)
Kcal (die)	2106.53 ± 770.46	2064.69 ± 838.32	2157.11 ± 761.54	2075.02 ± 707.08	1993.26 ± 775.25	2199.42 ± 775.94
Cooked Tomato (gr/die)	21.71 ± 24.84	22.10 ± 21.32	26.89 ± 24.09	11.98 ± 14.04	11.28 ± 14.03	16.14 ± 18.53
Fresh Tomato (gr/die)	45.36 ± 43.34	47.30 ± 66.78	51.23 ± 47.74	46.06 ± 46.66	49.27 ± 46.79	42.96 ± 34.48
LYC Intake (mg/die)	6.32 ± 5.43	6.78 ± 5.29	6.79 ± 5.31	4.82 ± 3.76	4.80 ± 3.74	5.48 ± 4.52

* As means and standard deviations (M ± SD) for continuous variables, and as frequencies and percentages for categorical variables; ^ as medians and interquartile ranges. Abbreviations: NAFLD, nonalcoholic fatty li disease; AFLD, alcoholic fatty liver disease; FLD, fatty liver disease; BMI, body mass index; HDL, high-den lipoprotein; WBC, white blood cells; GGT, gamma-glutamyl transferase; AST, aspartate aminotransferase; A alanine transaminase; rMED, relative Mediterranean diet score; LYC, lycopene.

As shown, only the subjects who returned in the second half were considered, the numbers are balanced. Furthermore, as described above, the cohort was divided ir patients without liver disease, those with NAFLD, and those with AFLD or FLD.

The multinomial logistic regression models are reported in Table 2.

Table 2. Multinomial logistic regression analysis of liver condition categories for NUTRIHEP 2 LYC intake at NUTRIHEP 1 as continuous and categorical variables inserted in the model ^.

Parameters	NAFLD ¥				AFLD and FLD ¥			
	RRR	se (RRR)	*p*	95% C.I.	RRR	se (RRR)	*p*	95% C.I.
LYC *(Continuous)*	0.96	0.02	0.008	0.93 to 0.99	0.94	0.02	0.006	0.90 to 0.9
LYC *(Categorical)*								
≥6.00 vs. <6.00	0.64	0.11	0.12	0.45 to 0.91	0.59	0.14	0.03	0.37 to 0.9
≥6.50 vs. <6.50	0.63	0.11	0.01	0.45 to 0.89	0.60	0.14	0.04	0.38 to 0.9
≥7.00 vs. <7.00	0.71	0.13	0.06	0.49 to 1.01	0.59	0.14	0.03	0.36 to 0.9
≥7.50 vs. <7.50	0.70	0.13	0.06	0.48 to 1.01	0.53	0.14	0.01	0.32 to 0.8
≥8.00 vs. <8.00	0.63	0.12	0.02	0.43 to 0.92	0.50	0.13	0.009	0.30 to 0.8
≥8.50 vs. <8.50	0.66	0.13	0.04	0.44 to 0.98	0.52	0.14	0.02	0.31 to 0.9
≥9.00 vs. <9.00	0.64	0.13	0.03	0.42 to 0.96	0.46	0.13	0.006	0.27 to 0.8
≥9.50 vs. <9.50	0.59	0.13	0.01	0.39 to 0.90	0.40	0.12	0.002	0.22 to 0.7
≥10.00 vs. <10.00	0.60	0.13	0.02	0.39 to 0.93	0.42	0.13	0.004	0.24 to 0.7
≥10.50 vs. <10.50	0.63	0.15	0.05	0.40 to 0.99	0.42	0.14	0.008	0.22 to 0.8

Abbreviations: NAFLD, nonalcoholic fatty liver disease; RRR, relative risk ratio; se(RRR), standard error of R 95% C.I., 95% confidence interval. ¥ Reference category: absent. ^ Models adjusted for gender, age, weight, gluc triglycerides, AST, cholesterol, and HDL in NUTRIHEP 2.

LYC intake (gr/die) was inserted as an explained variable in the adjusted models, both continuous and categorical form, to find the best concentration with a protective ro

As a continuous variable, LYC intake showed an RRR < 1 in the NAFLD and the AF and FLD groups (RRR = 0.96, $p = 0.008$, 0.93 to 0.99 at 95% C.I., and RRR = 0.94, p = 0.0 0.90 to 0.98 at 95% C.I., respectively). The protective role was better when using LYC a categorical variable (Figure 1).

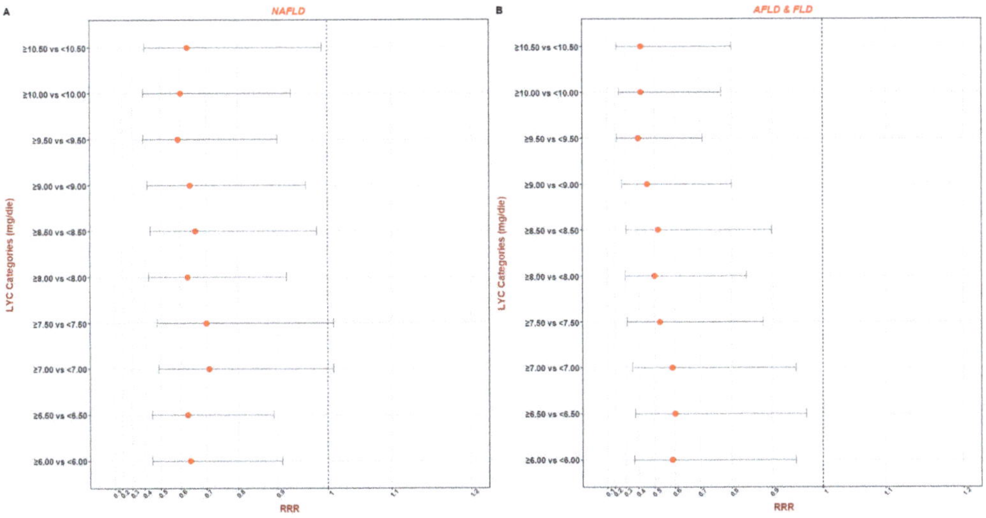

Figure 1. Forest plot of RRR value in the NAFLD (**A**) and AFLD and FLD (**B**) categories.

The RRR values reached increasingly lower concentrations, the lowest RRR value being at the concentration of 9.50 mg/die, with an RR value of 0.59, $p = 0.01$, 0.39 to 0.90 at 95% C.I., and RRR = 0.40, $p = 0.002$, 0.22 to 0.71 at 95% C.I.

4. Discussion

We studied the role of LYC, a phytochemical present in various fruits and vegetables, in modulating the risk of NAFLD, AFLD, and FLD. The innovative aspect of this research lies in its focus on LYC as a potential preventive agent against the progression of liver disease in humans. The study revealed a significant inverse association between LYC intake and the risk of NAFLD, suggesting a potential protective effect.

Furthermore, this study explored different concentrations of LYC, providing an understanding of its potential dose-dependent impact on NAFLD, AFLD, and FLD. At the same LYC concentrations, the protective effect was more evident against AFLD and FLD than against NAFLD. Previous research associated a high intake of LYC with low waist circumference, as well as with low subcutaneous and visceral fat mass, suggesting that LYC had an impact on adipocyte physiology [33].

Moreover, several studies have demonstrated that LYC exerts antioxidant and anti-inflammatory activities, primarily reducing steatosis and obesity-associated disorders in mice [34–36]. Other studies have investigated the potential impact of LYC on reducing the risk of cardiovascular disease and suggested mechanisms that involved the regulation of blood pressure and decreased inflammation [37,38]. The antioxidant properties of LYC were studied in relation to cancer prevention. Various studies have provided perspective on the ability of LYC to counteract free radicals, thus reducing the risk of developing various forms of cancer [39]. Other studies have suggested that LYC plays a protective role against ocular disorders and positively modulates the immune response, providing benefits in preventing infections and supporting the immune system [40].

Our findings are in line with the current understanding of the pathogenesis of NAFLD, AFLD, and FLD, along with the recognition of their global prevalence and lack of specific therapeutic options. Lifestyle interventions, particularly dietary interventions such as adherence to the Mediterranean diet, have beneficial effects for both the prevention and treatment of NAFLD. Several studies have confirmed the effectiveness of the Mediterranean diet in improving NAFLD, owing to a great variety of specific antioxidants helping to counteract the oxidative stress and inflammation associated with steatosis and, consequently, the extent of liver injury. Several anti-inflammatory and antioxidant mechanisms of L have been described that could explain the efficacy of LYC against multifactorial conditic such as NAFLD, AFLD, and FLD. Numerous studies have documented an increase superoxide dismutase (SOD) and catalase (CAT) activity following LYC administrati highlighting its strong antioxidant potential [41].

Additionally, the anti-inflammatory properties of LYC have been corroborated throu studies demonstrating a reduction in the expression of different types of cytokines (li TNF-α, IL1, IL6, and IL8), nitric oxide (NO), cyclooxygenase, and chemokines that cou modulate the immune system reaction. Furthermore, LYC exhibits a notable impact lipid metabolism, as evidenced by reductions in serum lipid levels and hepatic choleste levels. The precise mechanism underlying these effects is still not fully elucidated, b preclinical studies have striven to unravel the intricacies of LYC's actions.

However, there is a notable lack of clinical studies, emphasizing the need for furth exploration by the scientific community to comprehensively understand and harness t potential benefits of LYC in the prevention and treatment of NAFLD. LYC's lipophilic natu plays a crucial role in its anti-inflammatory effects. This property enables it to inter closely with cell membranes, regulating inflammatory mediator signaling pathways a activating antioxidant gene expression. LYC's ability to hinder the production of vario cytokines, chemokines, nitric oxide, and cyclooxygenase contributes to its modulati of the immune system. Notably, it inhibits the nuclear factor kappa B (NF-κB) signali pathway, a key player in inflammation, by binding to the inhibitor of NF-κB protein, th impeding its translocation to the nucleus.

LYC is found in numerous foods included in the Mediterranean diet, particular in red vegetables and fruits, such as tomatoes, cherries, pomegranates, pink grapefru peppers, carrots, and watermelon. Among these, the most closely studied is the toma serving as a vital source of vitamins, minerals, fiber, and carbohydrates, and playing significant antioxidant role, explained by LYC compounds [14]. Tomato byproducts such sauce and concentrates are part of the Mediterranean tradition and boast a greater quant of bioavailable LYC than fresh products. In particular, higher bioavailability is attributed the impact of food processing, cooking, and isomeric configurations [41].

Moreover, the bioavailability of LYC undergoes an increase when it is consum alongside oils or other fats, such as olive oil. The trans–cis isomerization process of t molecule plays a primary role. In nature, 90% of LYC is found in the trans form, which the most thermodynamically stable, while in the human body it is mainly present in the form, which is the most bioavailable. While the trans LYC form is the primary one in fru in cooked and processed products like tomato paste this percentage may decrease to 35

Despite the widespread prevalence of trans isomers, they exhibit lower bioavailabil compared to cis isomers. Several studies have indicated that the heightened antioxida activity of cis isomers may stem from their superior solubility and reduced tendency self-aggregate in polar media. It has been suggested that the peroxyl radical scavengi capacity of LYC isomers follows the order 5-cis > 9-cis > 13-cis > all-trans, likely owing the enhanced solubility of cis isomers in comparison to the all-trans form. In contrast, isomers of LYC prevail as the typical forms in tissues and plasma. This implies that the m jority of ingested trans LYC undergoes conversion to cis isomers through isomerization the enterocytes, liver, and stomach. Consequently, this process improves the bioavailabil and the absorption of LYC [42].

Our results showed that the amount of 9.50–10.00 mg/day was the most effective reducing the risk of steatosis. Although this intake is slightly higher than the actual avera intake in our cohort, it is an easily attainable dose based on consumption of LYC-ri foods. For example, to reach this equivalent dose, one would need to consume abo four medium-sized fresh tomatoes (calculated as 2.86 mg of LYC per 100 g of tomato; o medium-sized tomato weighs about 125 g) or 50 g of tomato sauce (about 20 mg of LY per 100 g of tomato sauce) per day [22]. While this study provides valuable insights, it h certain limitations.

The reliance on self-reported dietary data introduces the possibility of recall bias, potentially affecting the accuracy of LYC intake assessments. The observational nature of the study design limits the establishment of causal relationships. Additionally, the study cohort, extracted from a specific region in Italy, may not fully represent the diversity of global populations, impacting the generalizability of the findings. Furthermore, the potential influence of unmeasured confounding factors on the observed associations cannot be entirely ruled out.

5. Conclusions

In conclusion, our findings suggest that LYC, with its potent anti-inflammatory and antioxidant properties, may serve as a protective factor against the development and progression of steatosis and liver injury. Further clinical studies are needed to confirm these implications. Future studies should aim for more robust methodologies, such as randomized controlled trials, to confirm causation and validate any observed associations. This diet-centric approach could potentially pave the way for personalized nutrition strategies in the management of steatosis. Incorporating LYC-rich foods into dietary recommendations for individuals at risk may become a viable strategy.

Author Contributions: Conceptualization, R.D.; methodology, R.D. and C.B.; software, R.D. and C.B.; validation, G.G.; formal analysis, R.D. and C.B.; investigation, A.C., F.C. and R.T.; resources, G.G.; data curation, R.D. and C.B.; writing—original draft preparation, A.C. and R.T.; writing—review and editing, A.C. and R.T.; visualization, F.C. and G.G.; supervision, F.C.; project administration, R.D. and G.G.; funding acquisition, G.G. All authors have read and agreed to the published version of the manuscript.

Funding: This research was funded by the Italian Ministry of Health with Ricerca Corrente 2023.

Institutional Review Board Statement: The study was conducted in accordance with the Declaration of Helsinki and approved by the Ethics Committee of the National Institute of Gastroenterology and Research Hospital (DDG-CE-502/2005; DDG-CE-792/2014, on 20 May 2005 and 14 February 2014, respectively).

Informed Consent Statement: Informed consent was obtained from all subjects involved in the study.

Data Availability Statement: The original contributions presented in this study are included in the article. Further inquiries can be directed to the corresponding author.

Acknowledgments: The authors are grateful to the NUTRIHEP Group.

Conflicts of Interest: The authors declare no conflicts of interest.

erences

Keating, S.E.; Hackett, D.A.; George, J.; Johnson, N.A. Exercise and non-alcoholic fatty liver disease: A systematic review and meta-analysis. *J. Hepatol.* **2012**, *57*, 157–166. [CrossRef]
Tan, H.H.; Chang, J.P.E. Non-alcoholic Fatty Liver Disease. *Proc. Singap. Healthc.* **2010**, *19*, 1. [CrossRef]
Schwimmer, J.B.; McGreal, N.; Deutsch, R.; Finegold, M.J.; Lavine, J.E. Influence of gender, race, and ethnicity on suspected fatty liver in obese adolescents. *Pediatrics* **2005**, *115*, e561–e565. [CrossRef] [PubMed]
Maurice, J.; Manousou, P. Non-alcoholic fatty liver disease. *Clin. Med.* **2018**, *18*, 245–250. [CrossRef]
Vanni, E.; Bugianesi, E.; Kotronen, A.; De Minicis, S.; Yki-Järvinen, H.; Svegliati-Baroni, G. From the metabolic syndrome to NAFLD or vice versa? *Dig. Liver Dis.* **2010**, *42*, 320–330. [CrossRef]
Friedman, S.L.; Neuschwander-Tetri, B.A.; Rinella, M.; Sanyal, A.J. Mechanisms of NAFLD development and therapeutic strategies. *Nat. Med.* **2018**, *24*, 908–922. [CrossRef]
Tessari, P.; Coracina, A.; Cosma, A.; Tiengo, A. Hepatic lipid metabolism and non-alcoholic fatty liver disease. *Nutr. Metab. Cardiovasc. Dis.* **2009**, *19*, 291–302. [CrossRef] [PubMed]
Thorup, A.C.; Kristensen, H.L.; Kidmose, U.; Lambert, M.N.T.; Christensen, L.P.; Fretté, X.; Clausen, M.R.; Hansen, S.M.; Jeppesen, P.B. Strong and Bitter Vegetables from Traditional Cultivars and Cropping Methods Improve the Health Status of Type 2 Diabetics: A Randomized Control Trial. *Nutrients* **2021**, *13*, 1813. [CrossRef]
Liu, Y.; Wang, D.; Zhang, D.; Lv, Y.; Wei, Y.; Wu, W.; Zhou, F.; Tang, M.; Mao, T.; Li, M.; et al. Inhibitory effect of blueberry polyphenolic compounds on oleic acid-induced hepatic steatosis in vitro. *J. Agric. Food Chem.* **2011**, *59*, 12254–12263. [CrossRef]

10. Tilg, H.; Moschen, A.R. Evolution of inflammation in nonalcoholic fatty liver disease: The multiple parallel hits hypothe *Hepatology* **2010**, *52*, 1836–1846. [CrossRef]
11. Harrison, S.A.; Torgerson, S.; Hayashi, P.; Ward, J.; Schenker, S. Vitamin E and vitamin C treatment improves fibrosis in patie with nonalcoholic steatohepatitis. *Am. J. Gastroenterol.* **2003**, *98*, 2485–2490. [CrossRef]
12. Kugelmas, M.; Hill, D.B.; Vivian, B.; Marsano, L.; McClain, C.J. Cytokines and NASH: A pilot study of the effects of lifest modification and vitamin E. *Hepatology* **2003**, *38*, 413–419. [CrossRef] [PubMed]
13. Thong-Ngam, D.; Samuhasaneeto, S.; Kulaputana, O.; Klaikeaw, N. N-acetylcysteine attenuates oxidative stress and li pathology in rats with non-alcoholic steatohepatitis. *World J. Gastroenterol.* **2007**, *13*, 5127–5132. [CrossRef] [PubMed]
14. Khan, U.M.; Sevindik, M.; Zarrabi, A.; Nami, M.; Ozdemir, B.; Kaplan, D.N.; Selamoglu, Z.; Hasan, M.; Kumar, M.; Alshe M.M.; et al. Lycopene: Food Sources, Biological Activities, and Human Health Benefits. *Oxidative Med. Cell. Longev.* **2(** *2021*, 2713511. [CrossRef] [PubMed]
15. Przybylska, S.; Tokarczk, G. Lycopene in the Prevention of Cardiovascular Diseases. *Int. J. Mol. Sci.* **2002**, *23*, 1957. [CrossRe
16. Burton-Freeman, B.M.; Sesso, H.D. Whole food versus supplement: Comparing the clinical evidence of tomato intake a lycopene supplementation on cardiovascular risk factors. *Adv. Nutr.* **2014**, *5*, 457–485. [CrossRef] [PubMed]
17. Honda, M.; Maeda, H.; Fukaya, T.; Goto, M. Effects of Z-Isomerization on the Bioavailability and Functionality of Carotenoids Review. *Prog. Carotenoid Res.* **2018**, *228*, 139–159.
18. Saini, R.K.; Bekhit, A.E.D.A.; Roohinejad, S.; Rengasamy, K.R.R.; Keum, Y.S. Chemical Stability of Lycopene in Processed Produ A Review of the Effects of Processing Methods and Modern Preservation Strategies. *J. Agric. Food Chem.* **2020**, *68*, 712–7 [CrossRef]
19. Carvalho, G.C.; Camargo, B.A.F.; de Thayanne, J.; de Araújo, J.T.C.; Chrorilli, M. Lycopene: From tomato to its nutraceutical and its association with nanotechnology. *Trends Food Sci. Technol.* **2021**, *118*, 447–458. [CrossRef]
20. Shi, J.; Maguer, M. Lycopene in tomatoes: Chemical and physical properties affected by food processing. *Crit. Rev. Biotech* **2000**, *20*, 293–334. [CrossRef]
21. Colle, I.; Lemmens, L.; Van Buggenhout, S.; Loey, A.V.; Hendrick, M. Effect of thermal processing on the degradation, isomer tion, and bioaccessibility of lycopene in tomato pulp. *J. Food Sci.* **2010**, *75*, C753–C759. [CrossRef]
22. USDA National Nutrient Database for Standard Reference Legacy Release. FoodData Central. Available online: ht //agdatacommons.nal.usda.gov/articles/dataset/USDA_National_Nutrient_Database_for_Standard_Reference_Legacy_ Release/24661818 (accessed on 12 April 2019).
23. Petyaev, I.M. Lycopene Deficiency in Ageing and Cardiovascular Disease. *Oxidative Med. Cell. Longev.* **2016**, *2016*, 321 [CrossRef]
24. Maiani, G.; Castó, M.J.P.; Catasta, G.; Toti, E.; Cambrodón, I.G.; Bysted, A.; Granado-Lorencio, F.; Olmedilla-Alonso, B.; Knuths P.; Valoti, M.; et al. Carotenoids: Actual knowledge on food sources, intakes, stability and bioavailability and their protective r in humans. *Mol. Nutr. Food Res.* **2009**, *53*, S194–S218. [CrossRef]
25. Puah, B.P.; Jalil, J.; Attiq, A.; Kamisah, Y. New Insights into Molecular Mechanism behind Anti-Cancer Activities of Lycope *Molecules* **2021**, *26*, 3888. [CrossRef]
26. Arballo, J.; Amengual, J.; Erdman, J.J. Lycopene: A Critical Review of Digestion, Absorption, Metabolism, and Excreti *Antioxidans* **2021**, *10*, 342. [CrossRef] [PubMed]
27. Wang, H.; Liu, Z.; Shao, J.; Lin, L.; Jiang, M.; Wang, L.; Lu, X.; Zhang, H.; Chen, Y.; Zhang, R. Immune and Inflammation in Ac Coronary Syndrome: Molecular Mechanisms and Therapeutic Implications. *J. Immunol. Res.* **2020**, *2020*, 4904217. [CrossRef]
28. Pereira, B.L.; Reis, P.P.; Severino, F.E.; Felix, T.F.; Braz, M.G.; Nogueira, F.R.; Silva, R.A.; Cardoso, A.C.; Lourenço, M.A. Figueiredo, A.M.; et al. Tomato (*Lycopersicon esculentum*) or lycopene supplementation attenuates ventricular remodeling a myocardial infarction through different mechanistic pathways. *J. Nutr. Biochem.* **2017**, *46*, 117–124. [CrossRef]
29. Cozzolongo, R.; Osella, A.R.; Elba, S.; Petruzzi, J.; Buongiorno, G.; Giannuzzi, V.; Leone, G.; Bonfiglio, C.; Lanzilotta, E.; Mangh O.G.; et al. Epidemiology of HCV infection in the general population: A survey in a southern Italian town. *Am. J. Gastroente* **2009**, *104*, 2740–2746. [CrossRef]
30. European Association for the Study of the Liver (EASL); European Association for the Study of Diabetes (EASD); Europe Association for the Study of Obesity (EASO). EASL-EASD-EASO Clinical practice guidelines for the management of non-alcoho fatty liver disease. *J. Hepatol.* **2016**, *64*, 1388–1402. [CrossRef] [PubMed]
31. Chalasani, N.; Younossi, Z.; Lavine, J.E.; Charlton, M.; Cusi, K.; Rinella, M.; Harrison, S.A.; Brunt, E.M.; Sanyal, A.J. The diagnc and management of nonalcoholic fatty liver disease: Practice guidance from the American Association for the Study of Li Diseases. *Hepatology* **2018**, *67*, 328–357. [CrossRef] [PubMed]
32. Younossi, Z.M.; Koening, A.B.; Abdelatif, D.; Fazel, Y.; Henry, L.; Wymer, M. Global epidemiology of nonalcoholic fatty li disease-Meta-analytic assessment of prevalence, incidence, and outcomes. *Hepatology* **2016**, *64*, 73–84. [CrossRef]
33. Sluijs, I.; Beulens, J.W.J.; Grobbee, D.E.; van der Schouw, Y.T. Dietary carotenoid intake is associated with lower prevalence metabolic syndrome in middle-aged and elderly men. *J. Nutr.* **2009**, *139*, 987–992. [CrossRef]
34. Fenni, S.; Hammou, H.; Astier, J.; Bonnet, L.; Karkeni, E.; Couturier, C.; Tourniaire, F.; Landrier, J.F. Lycopene and tomato pow supplementation similarly inhibit high-fat diet induced obesity, inflammatory response, and associated metabolic disorders. *N Nutr. Food Res.* **2017**, *61*, 1601083. [CrossRef] [PubMed]

Li, C.C.; Liu, C.; Fu, M.; Hu, K.Q.; Aizawa, K.; Takahashi, S.; Hiroyuki, S.; Cheng, J.; von Linting, J.; Wang, X.D. Tomato Powder Inhibits Hepatic Steatosis and Inflammation Potentially Through Restoring SIRT1 Activity and Adiponectin Function Independent of Carotenoid Cleavage Enzymes in Mice. *Mol. Nutr. Food Res.* **2018**, *62*, e1700738. [CrossRef] [PubMed]
Singh, D.P.; Khare, P.; Zhu, J.; Kondepudi, K.K.; Singh, J.; Baboota, R.K.; Boparai, R.K.; Khardori, R.; Chopra, K.; Bishnoi, M. A novel cobiotic-based preventive approach against high-fat diet-induced adiposity, nonalcoholic fatty liver and gut derangement in mice. *Int. J. Obes.* **2016**, *40*, 487–496. [CrossRef]
Sesso, H.D.; Buring, J.E.; Norkus, E.P.; Gaziano, J.M. Plasma lycopene, other carotenoids, and retinol and the risk of cardiovascular disease in women. *Am. J. Clin. Nutr.* **2004**, *79*, 47–53. [CrossRef] [PubMed]
Hughes, D.A. Effects of lycopene and tomato products on cholesterol metabolism and atherosclerosis. *Pure Appl. Chem.* **2002**, *74*, 1431–1439.
Giovannucci, E. Tomatoes, tomato-based products, lycopene, and cancer: Review of the epidemiologic literature. *J. Natl. Cancer Inst.* **1999**, *91*, 317–331. [CrossRef] [PubMed]
Age-Related Eye Disease Study 2 Research Group. Lutein + zeaxanthin and omega-3 fatty acids for age-related macular degeneration: The Age-Related Eye Disease Study 2 (AREDS2) randomized clinical trial. *JAMA* **2013**, *15*, 2005–2015.
Abebavoli, L.; Procopio, A.C.; Paravati, M.R.; Costa, G.; Milić, N.; Alcaro, S.; Luzza, F. Mediterranean Diet: The Beneficial Effects of Lycopene in Non-Alcoholic Fatty Liver Disease. *J. Clin. Med.* **2022**, *11*, 3477. [CrossRef]
Stice, C.P.; Liu, C.; Aizawa, K.; Greenberg, A.S.; Ausman, L.M.; Wang, X.D. Dietary tomato powder inhibits alcohol-induced hepatic injury by suppressing cytochrome p450 2E1 induction in rodent models. *Arch. Biochem. Biophys.* **2015**, *572*, 81–88. [CrossRef] [PubMed]

nutrients

MDP

Article

Effect of a Phytochemical-Rich Olive-Derived Extract on Anthropometric, Hematological, and Metabolic Parameters

Anna Aiello [1,†], Luana Calabrone [2,†], Douglas M. Noonan [3,4], Paola Corradino [5], Sara Nofri [6], Simone Cristoni [2], Giulia Accardi [1], Giuseppina Candore [1], Calogero Caruso [1,‡], Angelo Zinellu [7] and Adriana Albini [5,*,‡]

1 Laboratory of Immunopathology and Immunosenescence, Department of Biomedicine, Neuroscience an Advanced Diagnostics, University of Palermo, 90134 Palermo, Italy; anna.aiello@unipa.it (A.A.); giulia.accardi@unipa.it (G.A.); giuseppina.candore@unipa.it (G.C.); calogero.caruso@unipa.it (C.C.)
2 ISB—Ion Source & Biotecnologie Srl, Rho, 20017 Milan, Italy; luana.calabrone@gmail.com (L.C.); simone.cristoni@gmail.com (S.C.)
3 Unit of Molecular Pathology, Biochemistry and Immunology, IRCCS MultiMedica, 20138 Milan, Italy; douglas.noonan@gmail.com
4 Department of Biotechnology and Life Sciences, University of Insubria, 21100 Varese, Italy
5 European Institute of Oncology (IEO), Istituto di Ricovero e Cura a Carattere Scientifico (IRCCS), 20141 Milan, Italy; paola.corradino@ieo.it
6 University of Florence, 50139 Florence, Italy; sara.nofri@googlemail.com
7 Department of Biomedical Sciences, University of Sassari, 07100 Sassari, Italy; azinellu@uniss.it
* Correspondence: adriana.albini@ieo.it
† These authors contributed equally to this work.
‡ These authors contributed equally to this work.

Abstract: Background: Extra virgin olive oil is a fundamental component of the Mediterranean d It contains several molecules that sustain human well-being by modulating cellular metabolism a exerting antioxidant, anti-inflammatory, and anti-ageing effects to protect normal tissues, and it c exert anti-angiogenic and pro-apoptotic effects on cancer cells. Metabolites found in different pa of the olive tree, including leaves, also possess properties that might help in cancer prevention a promote wellness in aging. Olive mill wastewater (OMWW), a liquid residue produced during ol oil extraction, represents an environmental issue. However, it is rich in phytochemicals with pot tial beneficial properties. Dietary supplements based on OMWW can be produced for nutritio supplementation with advantages to the ecology. Purpose: This work aims to measure hema chemical, anthropometric, and metabolomic parameters in volunteers taking an OMWW dieta supplement, Oliphenolia® (OMWW-OL). Methods: The supplementation of OMWW-OL 25 mL tw daily for 30 days was tested on a pilot cohort of volunteers with characteristics close to metabo syndrome. Hematochemical, anthropometric, serum biomarkers and serum metabolomic paramet were analyzed before the intervention, at 30 days, and 30 days after stopping consumption. Resu A total of 29 volunteers were enrolled, and 23 completed the study. The participants' parameters baseline were measured, and then twice daily at 30 days of treatment and 30 days after assumpti discontinuation. Although treatment was with an olive derivative, their weight did not increa Their body mass index, instead of augmenting, slightly decreased, particularly in the women. A hydration increased, especially in the women, while blood pressure, glycemia, and insulin decreas Cholesterol, high-density lipoproteins, and triglycerides were stable, and LDL levels decreased, wh vitamin D levels, alongside calcium, perceptibly increased. Albumin also increased. All the valu were in support of an equilibrium, with no damaging effects. By mass spectrometry analysis, also found favorable changes in the vitamin D/histamine and homocysteine/methionine ratios, increase in a new metabolite of unknown formula, and the vitamin D/unknown metabolite ra Conclusions: Supplementation of OMWW-OL has no detrimental effects and might imply the ben cial modulation of several biological parameters. Although this is a small pilot study, with limit potency, it preliminarily suggests that the OMWW extract use could be potentially valuable for peo at risk of metabolic syndrome. Some of these parameters could also be relevant in supporting healt ageing and in cancer prevention.

Citation: Aiello, A.; Calabrone, L.; Noonan, D.M.; Corradino, P.; Nofri, S.; Cristoni, S.; Accardi, G.; Candore, G.; Caruso, C.; Zinellu, A.; et al. Effect of a Phytochemical-Rich Olive-Derived Extract on Anthropometric, Hematological, and Metabolic Parameters. *Nutrients* **2024**, *16*, 3068. https://doi.org/10.3390/nu16183068

Academic Editors: Sonia Ramos Rivero and María Angeles Martín

Received: 31 July 2024
Revised: 4 September 2024
Accepted: 6 September 2024
Published: 11 September 2024

Keywords: extra virgin olive oil; olive mill wastewater; phytochemicals; polyphenols; antioxidants; antiaging; metabolic syndrome; cancer

1. Introduction

An integral part of the Mediterranean diet, extra virgin olive oil (EVOO) is high in monounsaturated fatty acids, which are regarded as advantageous for protection against diseases [1]. Since EVOO seems to improve omega-3 polyunsaturated fatty acids' oxidative stability, it might have positive effects on age-related conditions [2]. It is also a good source of antioxidants, including polyphenols and vitamin E, which may help protect against oxidative damage in the body, the cause of several chronic degenerative diseases [3,4]. EVOO reduces the risk of cardiovascular pathologies, type 2 diabetes, non-alcoholic fatty liver disease (NAFLD), neurodegenerative diseases, and some cancers by displaying anti-inflammatory and antioxidant effects [5–11]. Olive trees can produce useful metabolites in a variety of tissues, including the leaves. Olive mill wastewater (OMWW) is a large-scale liquid waste product released during olive oil extraction. It is an environmental concern on the one hand, but on the other, it is a rich source of phytochemicals that may have beneficial effects [6,12,13]. Hydroxytyrosol is the most common polyphenol in OMWW extracts aside from verbascoside and oleuropein, which are also abundant [14]. It has been shown that polyphenols control inflammation and oxidative stress [15], regulate gene expression in metabolic pathways, and have the potential to improve insulin sensitivity [16]. In this way, polyphenols could play an important role in the prevention of metabolic syndrome, which is characterized by hypertension, hyperglycemia, hyperinsulinemia, dyslipidemia, and obesity [17,18]. Thanks to pharmacometabolomics [19–21], the personalization of drug therapy is possible through the comprehensive analysis of metabolites in a patient's biological fluids. Metabolomics can provide important information on the effect of clinical treatments on the metabolism, and therefore, mass spectrometry (MS) has taken a central role in personalized and precision medicine [22–25].

In this work, we report the application of a dietary strategy to study the effects of an OMWW-derived supplement, Oliphenolia® (OMWW-OL), on different anthropometric and serum parameters at distinct time points. Our aim was to measure hematochemical, anti-inflammatory, antioxidant, and physiological parameters and to assess metabolic pathways before initiation (T0), after 30 days of nutritional intervention (T1), and 30 days after the end of the nutritional intervention (T2).

2. Materials and Methods

2.1. Study Design

This was a single-arm longitudinal interventional pilot study to evaluate the effect of OMWW-OL on a cohort of volunteers with characteristics close to metabolic syndrome as a pilot study. Participants were enrolled in Western Sicily within the project "Nutraceutical effects of olive products: Role in the achievement of longevity". The dietary intervention consisted in the consumption of 25 mL of OMWW-OL twice a day, 30 min before the main meals (lunch and dinner) for 30 days. Before starting the nutritional intervention, a 7-day washout period was carried out, during which the participants were required to abstain from EVOO and food supplements containing polyphenols. The flowchart of the study design is reported in Figure 1.

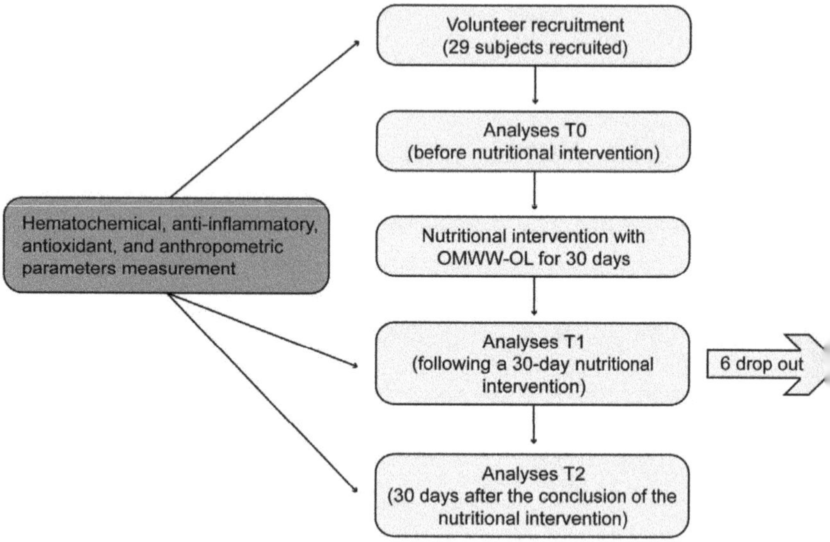

Figure 1. The flowchart of the study design. The study protocol was approved by the Eth Committee. During a 7-day washout period, the participants were required to abstain from ex virgin olive oil and food supplements containing polyphenols. The nutritional intervention consis of the consumption of the Oliphenolia® dietary supplement preparation (25 mL) twice a day, 30 m before the main meals (lunch and dinner), for 30 days, with no other restriction from the usual d

2.2. Study Participants

The subjects enrolled had at least one of the following inclusion criteria: slig dyslipidemia (TC 190–240 mg/dL, triglyceride level $\geq$ 150 mg/dL); a waist circumf ence $\geq$ 102 cm in men and $\geq$88 cm in women; and altered glucose tolerance (fasting blo glucose $\geq$ 100 mg/dL). The exclusion criteria were as follows: any treatment for speci diseases, including drugs to treat metabolic disorders; the diagnosis of severe system disease; a history of treatment with statins or similar drugs or other liposoluble or poglycemic drugs; restrictive dietary requirements (e.g., gluten-free diet). The onset gastrointestinal disorders during the intervention determined the prompt exclusion fro the study. No restriction related to gender was considered.

2.3. Ethical Considerations

Before enrollment, each recruited subject was fully informed about the purpos and procedures of the experiment, and their informed consent was obtained. To resp privacy, patients' names were anonymized by assigning each one an encoded alpha-nume identification code in accordance with the General Data Protection Regulation (GDPR, Regulation 2016/679) concerning the protection of individuals with regard to the processi of personal data, as well as the free circulation of such data. The study protocol w conducted in accordance with the 1964 Declaration of Helsinki and its later amendmer and the institutional Ethics Committee (Policlinico Paolo Giaccone, University Hospit approved the study protocol (Ruolo dei prodotti dell'olivo nella prevenzione delle malat età correlate/Role of olive products in the prevention of age-related diseases) on 18 Februa 2019, and the reference number is 02/2019.

2.4. Data Collection

The demographic, clinical, and anamnestic data of the probands were collected by group of well-trained biologists and physicians from the University of Palermo. Additio ally, a weekly food diary was provided to record data on their dietary intake. A database

Excel was created to collect and handle all participants' data and information. To obtain informative and reproducible analyses, anthropometric measures were obtained while subjects wore light clothes and no shoes. Body weight was measured using an electronic scale calibrated in kilograms. Height was measured with the subject in the supine position with a stadiometer. The bioelectrical impedance analysis (BIA) was also performed in the supine position, wearing light clothes and barefoot. The resistance and the reactance, both in ohms and the phase angle, were reported; the body compartment measurements (in percentage by weight and kg/m^2) were estimated using regression equations by Bodygram®Plus 1.1.4.4 software (AKERN®, Pisa, Italy). The body mass index (BMI, kg/m^2) was calculated using anthropometric measurements, such as height (m) and weight (kg).

2.5. Analysis of Hematochemical Parameters

Overnight fasting blood samples were collected in the morning. The blood was collected in specific tubes without additives in order to separate the serum samples, which were obtained after centrifugation at 2500 rpm for 15 min at 4 °C, rapidly frozen, and stored at −80 °C for further tests. All hematochemical analyses were measured by standard biochemical assays. The concentration of serum low-density lipoprotein (LDL) cholesterol was calculated using the Friedewald equation: LDL = total cholesterol-high-density lipoproteins (HDLs) − (triglycerides/5). Samples before initiation (T0), after 30 days of nutritional intervention (T1), and 30 days after the end of the nutritional intervention (T2) were also collected for spectrometry analysis.

2.6. Evaluation of the Antioxidative Effect

Serum samples from the patients were also used for the determination of human oxidized LDL (oxLDL) quantity, using a specific ELISA kit (Cusabio, Houston, TX, USA) following the manufacturer's instructions. The competitive inhibition enzyme immunoassay method was used in this assay. The microtiter plate was pre-coated with oxLDL. Standards or samples were added to the appropriate microtiter plate wells with a horseradish peroxidase-conjugated antibody preparation specific for oxLDL. The pre-coated oxLDL and oxLDL in the samples were used to start a competitive inhibition reaction. The absorption was determined with a microplate spectrophotometer (Tecan, Switzerland) at 450 nm. The results were interpreted by constructing a dose/response curve according to the standards provided in the kit.

2.7. Liquid Chromatography SANIST Mass Spectrometry

Serum metabolites were measured with an untargeted liquid chromatography (LC)-MS SANIST platform. The preliminary MS biomarker identification was obtained based on the accuracy of the derived molecular mass data (m/z, mass error: 10 ppm), and the ratios of the relative abundance of the different metabolites (i.e., "metabolite ratio") were calculated [26,27]. Thanks to the calculated ratios, metabolites were classified using an unsupervised method [26]. To 300 μL of serum, 600 μL of acetonitrile (CH_3CN, Sigma Aldrich, Milan, Italy) was added to precipitate proteins and centrifuged at 13,000× g for 10 min at room temperature. An aliquot of supernatant (600 μL) was collected and spiked with 100 μL of Ultramark polymer solution (10 ng/mL) to check the stability of the LC-MS signal during analysis. A SpeedVac (ThermoFisher, San Jose, CA, USA) was used to dry the remaining 300 μL; 100 μL of doubly distilled water/methanol (1:1, v/v; Sigma Aldrich, Milan, Italy) (1:1, v/v) solution was used to resuspend each sample, and 10 μL was added to the internal standard reserpine (10 μL, final concentration 10 ng/mL, Sigma Aldrich, Milan, Italy). The samples were directly injected into the LC-MS apparatus. A reverse-phase C-18 LC column (50 × 2.1 mm; particle size, 5 μm; pore size, 100 Å; Phenomenex, [Torrance, CA OR San Jose, CA], USA) was employed. The mobile phases consisted of (A) water containing 0.2% v/v formic acid (HCOOH, Sigma Aldrich, Milan, Italy), and (B) CH_3CN (100%), and the separation was achieved using an elution gradient. The flow rate was 0.25 mL/min, and the injection volume was 15 μL. Full-scan spectra were acquired in the mass range m/z 40–800. Tandem mass spectrometry (MS/MS) experiments on the seru samples were performed with collision-induced dissociation, using helium as the collisi gas. The precursor ions were isolated and fragmented (isolation windows, ±0.3 m collision energy, 30% of its maximum value, which was 5 V peak to peak), and then prod ions were obtained and analyzed by an Orbitrap mass analyzer with a highly accura m/z ratio (mass resolution m/z 15,000; mass error < 10 ppm). Compound identificati was made based on fragmentation patterns and a mass search database (NIST14) libra following the European Union database match mode (EU directive 2002/657/EC) [28]. T chromatographic data were elaborated using the Xcalibur Quan Browser (Version 4.1.31 The SANIST data elaboration tool was used to predict the metabolomic pathways alter in the serum samples of the participating subjects [29].

2.8. Statistical Analysis

Graphics and statistical analyses were performed using GraphPad Prism Softwa (GraphPad Software, Inc., San Diego, CA, USA, Ver. 10.0.2). The figures show the sing values at T0, T1, and T2. The Y-axis reports the values of the analyzed biomarker. Sta tics are performed in comparison between the average of T0/T1 ± SEM, T1/T2 ± SE T0/T2 ± SEM. Significance was calculated from the mean ± SD and analyzed using S dent's t-test and one-way Anova. A value of $p < 0.05$ was considered statistically significa * $p < 0.05$, ** $p < 0.01$, *** $p < 0.001$, and **** $p < 0.0001$.

3. Results

3.1. Anthropometric, Hematochemical, Inflammatory and Oxidative Parameters

A total of 29 participants were enrolled in this study; 17 were men, and the me age was 57 years (ranging from 30 to 72 years). Six participants dropped out of the stu after the 30-day time point for personal reasons, and 23 completed the study (average a 59 years). For better visualization, the parameters that showed significant differences, well as those with an interesting trend, upon the intervention are reported in graph (Figures 2–9; Figures S1 and S2).

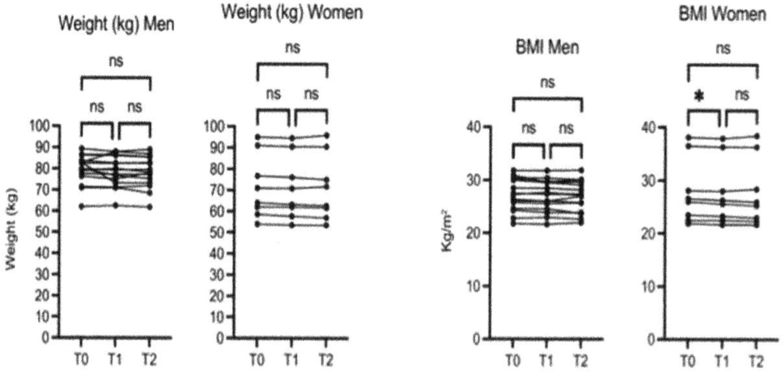

Figure 2. OMWW-OL consumption effects on weight and BMI in volunteers. The images show single values at T0, T1, and T2. There were no differences in the men's weight and BMI values; the women, the BMI showed a slight decrease at 30 days (T1), returning to the value of T0 af 60 days (T2). The Y-axis reports the values of the analyzed biomarker. Statistics are perform in comparison between the average of T0/T1 ± SEM, T1/T2 ± SEM, T0/T2 ± SEM. * $p < 0.$ ns: non-significant.

The weight (kg) and BMI did not increase and in female subjects actually decreas slightly but significantly at 30 days (T1) and went back to the original value at 60 days (T while no significant changes were observed in the men's weight and BMI (Figure 2).

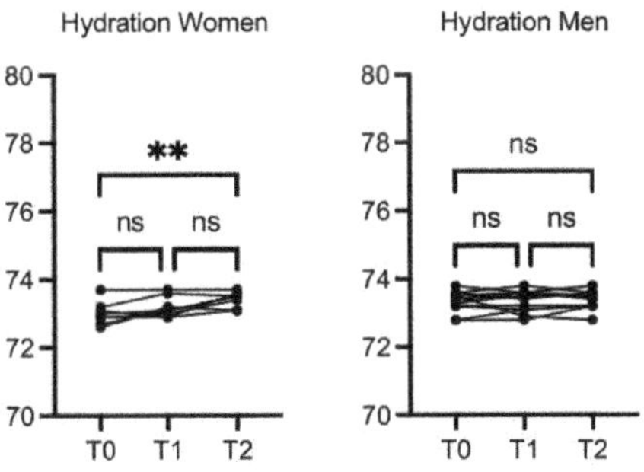

Figure 3. OMWW-OL consumption effects on hydration in volunteers. The images show the single values at T0, T1, and T2. In the women, hydration is increased at T2 vs. T0, while no significant differences were observed in the men. The *Y*-axis reports the values of the analyzed biomarker. Statistics are performed in comparison between the average of T0/T1 ± SEM, T1/T2 ± SEM T0/T2 ± SEM. ** $p < 0.01$, ns: non-significant.

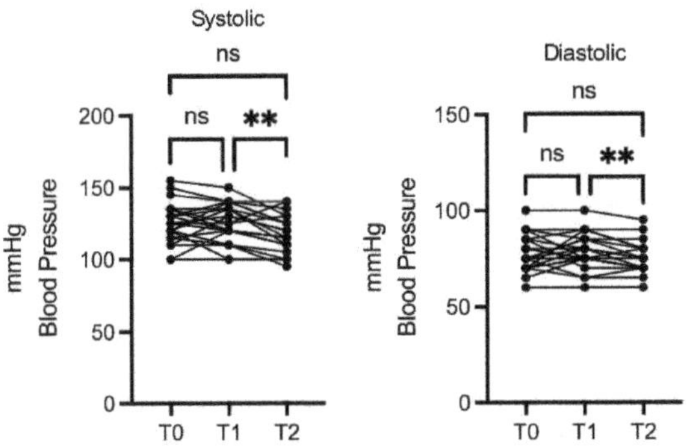

Figure 4. OMWW-OL consumption effects on blood pressure in volunteers. The images show the single values at T0, T1, and T2. Blood pressure values decreased significantly at T2 vs. T1, both in the maximum and minimum values. The *Y*-axis reports the values of the analyzed biomarker. Statistics are performed in comparison between the average of T0/T1 ± SEM, T1/T2 ± SEM, T0/T2 ± SEM. ** $p < 0.01$, ns: non-significant.

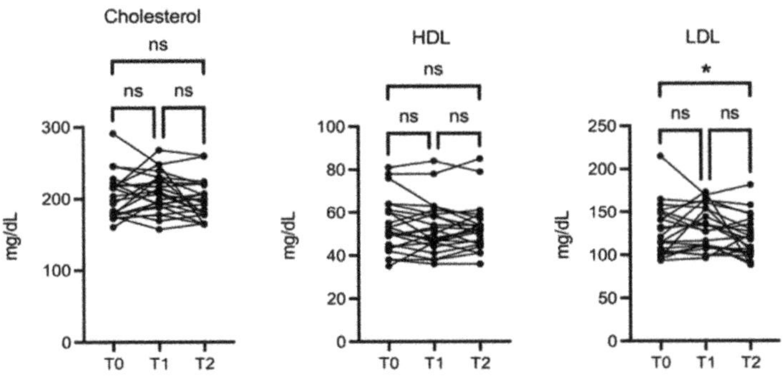

Figure 5. OMWW-OL consumption effects on cholesterol in volunteers. The images show the single values at T0, T1, and T2. There were no significant differences in total cholesterol and HDL, while the LDL decreased slightly but statistically significantly. The *Y*-axis reports the values of the analyzed biomarker. Statistics are performed in comparison between the average of T0/T1 ± SEM, T1/T2 ± SEM T0/T2 ± SEM. * $p < 0.05$, ns: non-significant.

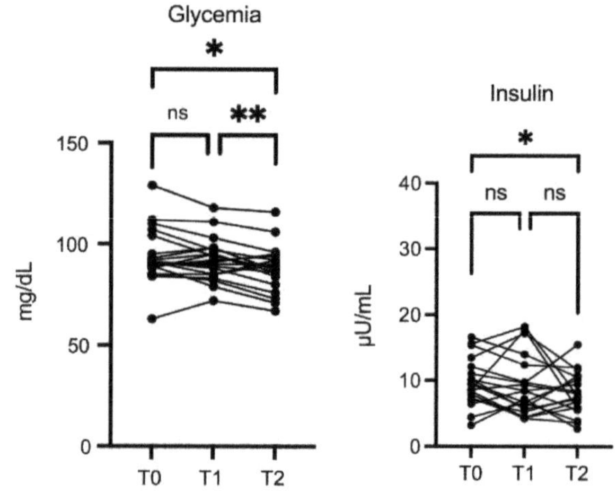

Figure 6. OMWW-OL assumption effects on glycemia and insulin in volunteers. The images sh the single values at T0, T1, and T2. Glycemia values decreased at T2 vs. T0, and at T2 vs. T1. Insu was also decreased at T2 vs. T0. The *Y*-axis reports the values of the analyzed biomarker. Statis are performed in comparison between the average of T0/T1 ± SEM, T1/T2 ± SEM T0/T2 ± SE * $p < 0.05$, ** $p < 0.01$, ns: non-significant.

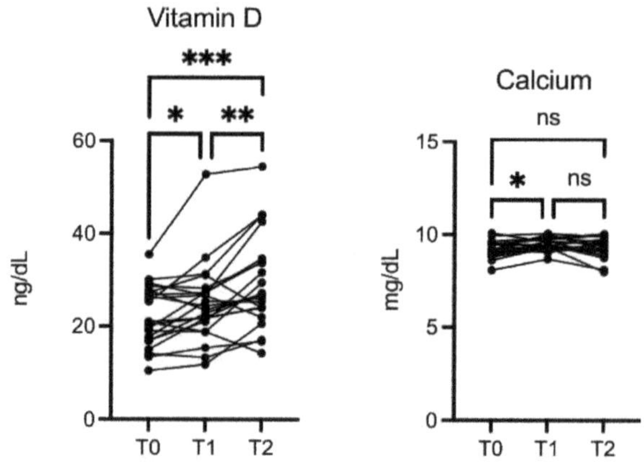

Figure 7. OMWW-OL consumption effects on vitamin D and calcium in volunteers. The images sh the single values at T0, T1, and T2. Vitamin D increases following OMWW intake in a statistica significant manner both at T1 and T2. Calcium was increased at T1 vs. T0. The *Y*-axis reports values of the analyzed biomarker. Statistics are performed in comparison between the average T0/T1 ± SEM, T1/T2 ± SEM T0/T2 ± SEM. * $p < 0.05$, ** $p < 0.01$, *** $p < 0.001$, ns: non-significa

The women's hydration was increased at T2 vs. T0 ($p \leq 0.01$), while hydration in men was increased slightly but not in a statistically significant manner (Figure 3).

Blood pressure decreased in a statistically significant manner, both for minimum a maximum values, at T2 vs. T1, while the T1 vs. T0 and T2 vs. T0 differences were r significant (Figure 4).

There was also a change, although not significant, in HDL, and a slight but significa decrease in LDL (Figure 5).

Glycemia values were lower at T2 vs. T0 ($p \leq 0.05$), and the decrease was mo significant at T2 vs. T1 ($p \leq 0.01$) (Figure 6). Insulin was also decreased at T2 vs. ($p = 0.05$) (Figure 6).

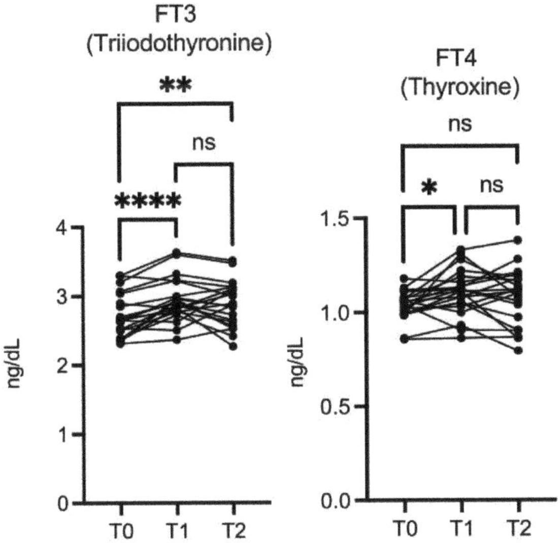

Figure 8. OMWW-OL consumption effects on free thyroid hormones (FT3 and FT4) in volunteers. The images show the single values at T0, T1, and T2. The FT3 level was increased significantly at T1 vs. T0 and at T2 vs. T0, while FT4 was significantly higher at T1 vs. T0. The *Y*-axis reports the values of the analyzed biomarker. Statistics are performed in comparison between the average of T0/T1 ± SEM, T1/T2 ± SEM T0/T2 ± SEM. * $p < 0.05$, ** $p < 0.01$, **** $p < 0.0001$, ns: non-significant.

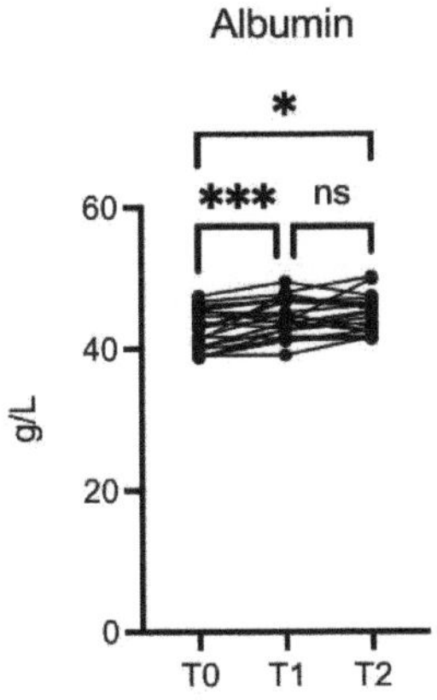

Figure 9. OMWW-OL consumption effects on albumin in volunteers. The images show the single values at T0, T1, and T2. The albumin values increase significantly at T1 vs. T0 and T2 vs. T0. The *Y*-axis reports the values of the analyzed biomarker. Statistics are performed in comparison between the average of T0/T1 ± SEM, T1/T2 ± SEM T0/T2 ± SEM. * $p < 0.05$, *** $p < 0.001$, ns: non-significant.

The vitamin D level was significantly increased at T1 vs. T0 ($p \leq 0.05$), T2 vs. T0 ($p \leq 0.001$), and T2 vs. T1 ($p \leq 0.01$) (Figure 7). Calcium was also increased at T1 vs. T0 ($p \leq 0.05$, Figure 7).

The free triiodothyronine (FT3) level was significantly higher at T1 vs. T0 and at T2 vs. T0, while free thyroxine (FT4) was significantly higher only at T1 vs. T0 (Figure 8).

The albumin level was significantly increased at T1 vs. T0 ($p \leq 0.001$) and T2 vs. T0 ($p \leq 0.05$) (Figure 9).

In our investigation, transferrin, ferritin, and sideremia (Figure S1), and mag sium (Figure S2) did not show variations. Potassium significantly decreased at T2 vs. (Figure S2).

3.2. Mass Spectrometry Serum Characterization

SANIST profiling revealed three chromatographic peak–area ratios, correlating th with enzyme functionality alteration. The detected potential biomarkers were identifi using the approach shown in Figure 10.

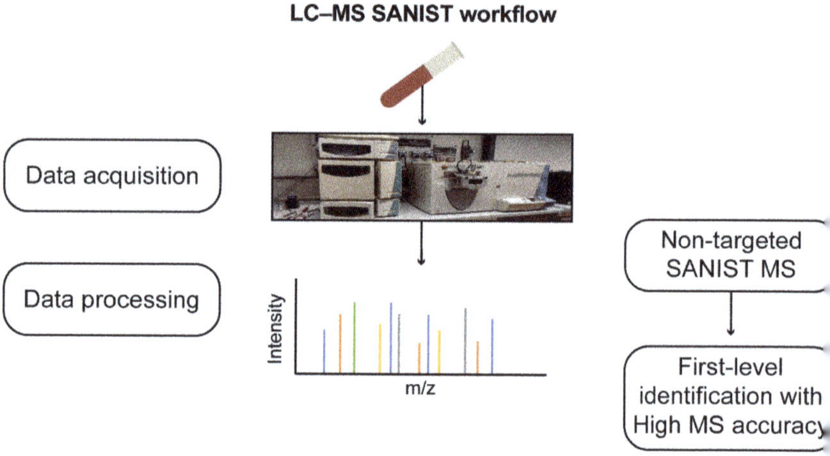

Figure 10. Workflow for non-targeted LC-MS SANIST metabolites profiling.

Mass spectra comparisons for the LC-MS data obtained with the NIST library shown in Figure 11. SANIST technology [26] was used to obtain and compare metabolom profiles. An Ultimate 3000 UPLC (ThermoFisher) LC system fitted with an Orbitrap m spectrometer (ThermoFisher) coupled to a surface-activated chemical ionization (SACI)/E source was used to analyze the metabolic profiling in both ESI-positive and ESI-negati ion modes.

The ion at m/z 150 → 133 (loss of a –OH group) (red) was identified by NIST methionine ($C_5H_{11}NO_2S$), present in the library spectrum (blue), with a 92% identificati score (direct match factor: 885; reverse match factor: 935) (Figure 11a). Methionine w confirmed by comparison to LC and MS/MS data acquired for the certified analyti standard, following the EU directive (EU directive 2002/657/EC). The ions m/z 4.0 → 3 and 112 → 95 were identified using the same approach and corresponding to calcitr and histamine, respectively (Figure 11b,c). We also identified homocysteine with ion m/z 136 → 90 (Figure 11d).

Furthermore, we observed a peak at m/z 371 but the corresponding metabolite w not identified using both an MS and an MS/MS database search. However, this unkno metabolite is very interesting because, according to the similarity approach in MS/M search, it appears to be highly similar to a peptide molecule. Due to the absence o commercial internal standard, verifying the data using EU Directive 2002/657/EC w not possible. In order to determine whether the metabolite at m/z 371 was present in t dietary supplement, OMWW-OL was analyzed using the LC-SANIST-MS/MS meth Nonetheless, it was evidently absent from the analysis of 10 OMWW-OL samples.

We observed a 10-fold increase in the homocysteine/methionine ratio (m/z 134/1 ($p = 4 \times 10^{-5}$), the vitamin 25 (OH) D3/histamine ratio (m/z 401/112) ($p = 5 \times 10^{-5}$), a the vitamin 25 (OH) D3/not identified metabolite ratio (m/z 401/371) ($p = 6 \times 10^{-5}$) wi respect to T0.

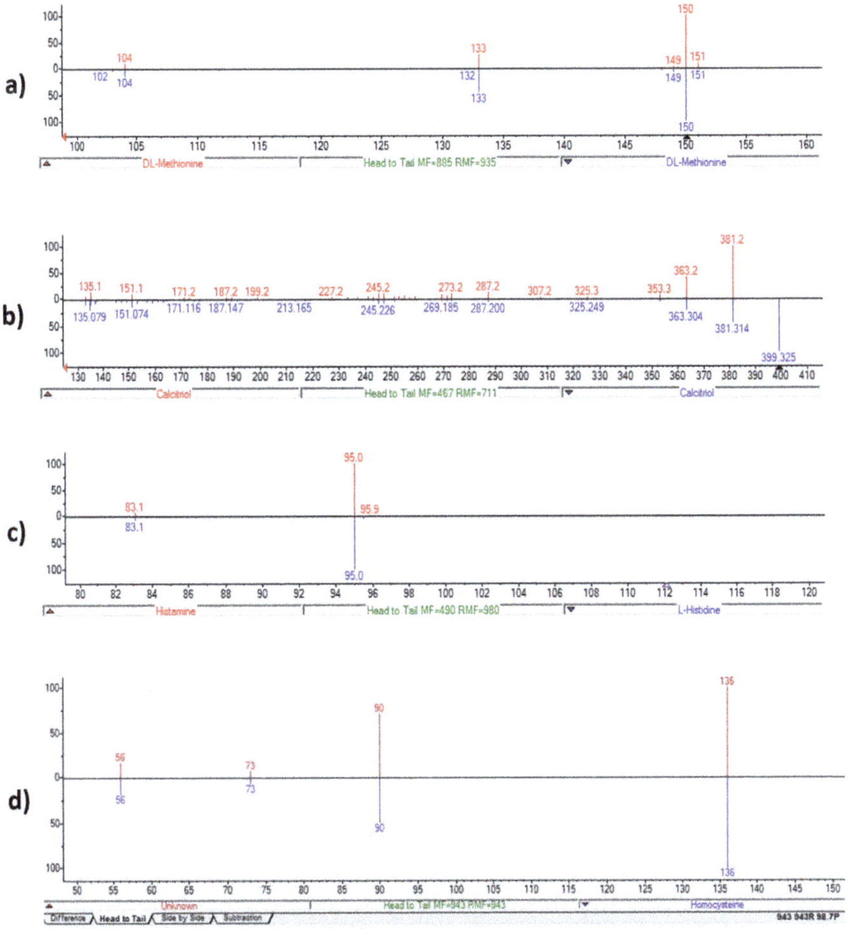

Figure 11. Biomarkers detected by NIST MS Search Program. Tandem mass spectra (MS/MS) library similarity match obtained by LC-MS data using NIST database: D-L-Methionine profile (**a**); Calcitriol (1α,25-dihydroxyvitamin D3) profile (**b**); Histamine profile (**c**), and Homocysteine profile (**d**).

The mass spectrometry data show very promising beneficial effects on metabolites, with the discovery of a new molecule induced during treatment.

4. Discussion

EVOO is an essential Mediterranean diet component rich in bioactive compounds [30]. Several studies suggest the role of its consumption in reducing the incidence of different diseases, such as cancer, cardiovascular pathologies, type 2 diabetes, and neurodegenerative disorders [31–35]. Every year, olive tree cultivation and oil extraction generate a large amount of byproducts, primarily OMWW, which can have a negative environmental impact. However, OMWW is rich in phenolic compounds displaying potent antioxidant, anti-inflammatory, vascular protective, and anti-aging properties [36–39], and it could potentially be useful in the pharmaceutical and food industries [30,38,40–45].

Here, we analyzed the effects of OMWW in a dietary supplement preparation (Oliphenolia®, OMWW-OL) on a cohort of volunteers with characteristics close to metabolic syndrome. Our data showed that OMWW-OL intake for 30 days had no negative effects on liver and kidney function and did not affect magnesium levels and iron absorption, transport, and storage. Several studies observed that diets, including daily EVOO, effective for weight loss [46–48]; however, data on women were still lacking. Interestin our work found that OMWW-OL was associated with a small decrease in the female p ticipants' weight and BMI, even in a short time. Furthermore, the finding that hydrati was significantly increased in females is a very promising result, as dehydration can s nificantly impair cognitive and physical abilities and has been associated with obes chronic diseases, and decreased longevity [49,50]. OMWW-OL might have the poten to ameliorate the state of hydration. Interestingly, hydration, weight, and BMI are m regulated in women than men, and in accordance with recent literature, including fr our group, gender differences in aging parameters were revealed [51]. Hence, althou supplementation with OMWW-OL was assessed in a pilot study and for a short time, disparity between men and women could have biological bases.

In agreement with previous observations, we found that OMWW-OL reduces blo pressure. This effect is probably related to its high polyphenol content [52,53], wh would also confer potential cardioprotective properties, as observed for EVOO [54]. also detected a glycemia decrease after OMWW-OL supplementation in our small coh This is consistent with the results obtained by Violi et al. [55], who showed that supplementation of EVOO at lunch is linked to lower post-prandial glycemic levels. Alb at a low significance level, cholesterol LDL showed a decreasing trend, while glycem and insulin decreased significantly. Besides its role in diabetes, insulin increases can risk by promoting the proliferation of normal, preneoplastic, and neoplastic cells [56]. decreasing glycemia and insulin levels, OMWW-OL could have some potential benefic effects in preventing metabolic syndrome and cancer. More significant differences might detectable upon supplementation for longer periods and in a larger cohort.

After nutritional intervention with OMWW-OL, we observed an increase in the F and FT4 serum levels, while TSH levels remained stable. Since low FT3 levels are associat with aging [57], it is possible to speculate about the ability of OMWW-OL to increase F and FT4 and to subsequently exert an anti-aging effect, as shown in a previous publicati of one of our institutions [58]. In this study, we also observed a significant increase albumin values. Albumin is involved in many bioactive functions, such as extracellu antioxidant defenses [59] and the transport of various ligands [60,61]. Decreases in albun levels are associated with an inflammatory state [62] and are observed in cancer, infectio malnutrition, liver illness, and kidney disease [63]. Conversely, high albumin concent tions may protect against early glycemic deterioration and type 2 diabetes progression [6 Therefore, the increased albumin levels observed with OMWW-OL could correlate w the anti-inflammatory activity of this olive byproduct and contribute to some extent metabolic syndrome prevention.

Since we observed a potassium reduction, this might be an interesting result for tients with chronic kidney disease, who are typically recommended to reduce potassiu consumption or eliminate it from their blood [65–67], also considering the absence of a relevant kidney toxicity in the short time of supplementation. Nevertheless, the effect kidney function should be evaluated upon supplementing OMWW-OL for longer tim frames. Several works have observed that vitamin D appears to be an effective agent prevent the production of reactive oxygen species and stop cytokine-mediated inflamr tion [68–71]. However, the literature does not provide much information on the impact natural compounds on endogenous vitamin D levels. Our finding that a significant increa in vitamin D, which was not due to seasonal variation, is achieved by supplementi with OMWW-OL supports a previous observation that in vitro digestion using olive resulted in higher absorption of the vitamin D precursor, calcifediol, than using other typ of edible oils [72]. Since this nutritional intervention also led to increased calcium lev based on our results it is possible to speculate that OMWW-OL may regulate the calciu balance either by inducing calcium absorption directly or by increasing vitamin D. Calciu is essential for preserving physiological functions and preventing aging, by promoti bone mineralization and sustaining heartbeat, blood clotting, and muscle and neur function [73–76]. Naturally, calcium levels decline as we age, and calcium deficiency has been increasingly linked to aging [74,75,77].

Further insights into the dietary intervention were achieved through MS, and differences were observed in the methionine cycle, which involves the regeneration of methionine from homocysteine [78]. The low homocysteine/methionine ratio found in our metabolomic analysis could be associated with active homocysteine metabolism, which is involved in the effective prevention of both oxidative stress and infectious diseases [79]. Elevated homocysteine levels increase the frequency of cardiovascular disease [80–84] with which metabolic syndrome is associated [85]. Therefore, also according to the decreased cardiovascular risk previously observed for olive oil [86], OMWW-OL could reduce cardiovascular risk by lowering homocysteine levels.

The absence of the metabolite at m/z 371.192 in OMWW-OL suggests that the detection of this metabolite in vivo is due to reduced vitamin D25 oxidation. Therefore, the potential antioxidant activity of OMWW-OL may preserve vitamin D25 and its functionality. Interestingly, the vitamin D/histamine ratio is also high, suggesting a decrease in histamine, an inflammatory mediator [87–89]. Since vitamin D regulates the immune system, particularly mast cells that carry histamine [90,91], the decrease in histamine observed with OMWW-OL intervention could be a secondary effect of the vitamin D increase.

In general, the data show that the vital parameters remain in the standard range, showing that, at least in the short time, there is no detrimental effect, with some positive encouraging trends to health benefits.

5. Conclusions

Here, we have shown that OMWW-OL has no detrimental effect on a variety of parameters, and also that in a preliminary way, the study indicates the potential to improve biological parameters in volunteers at risk for metabolic syndrome, according to a trend towards a favorable modulation of several biological parameters. The absence of significant variations in most hematochemical parameters indicates that regular Oliphenolia® ingestion has neither adverse effects on lipid levels nor on liver or kidney functions.

Some aging-associated parameters were modulated favorably. In addition to extending the period without needing treatment and slowing the aging process, it could be hypothesized that OMWW-OL could help prevent or postpone the onset of age-related disorders, improving quality of life and reducing healthcare costs. MS data revealed improved vitamin D/histamine, homocysteine/methionine, and vitamin D/unknown metabolite ratios.

Our data suggest that OMWW-OL may also have a promising effect in protecting cells from reactive oxygen species-induced injuries due to its high polyphenol content and vitamin D-boosting effect. The possibility of targeting aging instead of treating age-related pathologies is very attractive, as it can improve the quality of life and ensure longevity. This approach would align with positive biology, which researches the biological mechanisms involved in well-being [92]. Some of these parameters could also be relevant in cancer prevention or interception.

This study has some limitations, such as the small sample size, the single-arm protocol (thus lacking an external control group), and the short duration of the follow-up. Therefore, the information must be considered preliminary, although encouraging, and needs to be strengthened in further research with a longer supplementation time and follow-up period and an independent control group.

Future studies are needed to better elucidate the structure of the potential new metabolite and to investigate biological correlates.

Supplementary Materials: The following are available online at https://www.mdpi.com/article/10.3390/nu16183068/s1, Figure S1: Effects of OMWW-OL consumption on iron parameters in volunteers, Figure S2: Effects of OMWW-OL consumption on potassium and magnesium values in volunteers.

Author Contributions: A.A. (Anna Aiello): clinical studies, L.C.: mass spectrometry, L.C., D.M and P.C.: manuscript editing and illustration, S.N.: data analysis and editing, S.C.: mass spectro etry and metabolites identification, G.A., G.C., C.C. and A.Z.: study design and clinical stud A.A. (Adriana Albini): manuscript drafting and coordination. All authors have read and agreed the published version of the manuscript.

Funding: This work has been supported by the Italian Ministry of Health Ricerca Corrente–IRC IEO and IRCCS MultiMedica and through the project "Studi sulle proprietà degli estratti di acque vegetazione dell'olio di oliva. Approfondimenti di prevenzione e nutraceutica" to the IEO-MONZI Foundation and IRCCS IEO (A.A. (Adriana Albini)). The dietary supplement Oliphenolia® w provided at no cost by "Fattoria La Vialla di Gianni, Antonio e Bandino Lofranco-SAS" (Castigli Fibocchi, Arezzo, Italy).

Institutional Review Board Statement: The institutional Ethics Committee (Policlinico Paolo Gi cone, University Hospital) approved the study protocol (Ruolo dei prodotti dell'olivo nella prev zione delle malattie età correlate/Role of olive products in the prevention of age-related diseases) 18 February 2019, and the reference number is 02/2019.

Informed Consent Statement: Informed consent was obtained from all subjects involved in the stu

Data Availability Statement: The data that support the findings of this study are available on requ from the corresponding author.

Acknowledgments: The authors acknowledge Lara Vecchi, Valentina Attanasio, and Aashni Sh (Polistudium srl, Milan, Italy) for editorial assistance and Massimiliano Pianta (Polistudium, Milan, Italy) for graphics support. The authors also thank Danilo Morelli (INT, Istituto Nazionale Tumori Milano IRCCS) for helpful discussion.

Conflicts of Interest: The authors declare that the research was conducted in the absence of a commercial or financial relationships that could be construed as potential conflicts of interest.

References

1. Jimenez-Lopez, C.; Carpena, M.; Lourenço-Lopes, C.; Gallardo-Gomez, M.; Lorenzo, J.M.; Barba, F.J.; Prieto, M.A.; Simal-Gandar Bioactive Compounds and Quality of Extra Virgin Olive Oil. *Foods* **2020**, *9*, 1014. [CrossRef]
2. Therdyothin, A.; Prokopidis, K.; Galli, F.; Witard, O.C.; Isanejad, M. The Effects of Omega-3 Polyunsaturated Fatty Acids Muscle and Whole-Body Protein Synthesis: A Systematic Review and Meta-Analysis. *Nutr. Rev.* **2024**, nuae055. [CrossR [PubMed]
3. Lanza, B.; Ninfali, P. Antioxidants in Extra Virgin Olive Oil and Table Olives: Connections between Agriculture and Processi for Health Choices. *Antioxidants* **2020**, *9*, 41. [CrossRef] [PubMed]
4. Flynn, M.M.; Tierney, A.; Itsiopoulos, C. Is Extra Virgin Olive Oil the Critical Ingredient Driving the Health Benefits o Mediterranean Diet? A Narrative Review. *Nutrients* **2023**, *15*, 2916. [CrossRef] [PubMed]
5. Rossi, M.; Caruso, F.; Kwok, L.; Lee, G.; Caruso, A.; Gionfra, F.; Candelotti, E.; Belli, S.L.; Molasky, N.; Raley-Susman, K.M.; et Protection by Extra Virgin Olive Oil against Oxidative Stress In Vitro and In Vivo. Chemical and Biological Studies on the Hea Benefits Due to a Major Component of the Mediterranean Diet. *PLoS ONE* **2017**, *12*, e0189341. [CrossRef] [PubMed]
6. Cuffaro, D.; Bertolini, A.; Bertini, S.; Ricci, C.; Cascone, M.G.; Danti, S.; Saba, A.; Macchia, M.; Digiacomo, M. Olive M Wastewater as Source of Polyphenols with Nutraceutical Properties. *Nutrients* **2023**, *15*, 3746. [CrossRef]
7. Ke, Y.; Fan, H.; Zhao, Y.; Fu, X.; Wang, M.; Gao, Y.; Chen, Y.; Huo, W.; Zhang, W.; Wang, L.; et al. Olive Oil Intake a Cardiovascular Disease, Cancer, and All-Cause Mortality: A Systematic Review and Dose-Response Meta-Analysis of Prospect Cohort Studies. *Food Funct.* **2024**, *15*, 4668–4681. [CrossRef]
8. Zupo, R.; Castellana, F.; Crupi, P.; Desantis, A.; Rondanelli, M.; Corbo, F.; Clodoveo, M.L. Olive Oil Polyphenols Improve H Cholesterol and Promote Maintenance of Lipid Metabolism: A Systematic Review and Meta-Analysis of Randomized Controll Trials. *Metabolites* **2023**, *13*, 1187. [CrossRef]
9. Tsamos, G.; Kalopitas, G.; Evripidou, K.; Vasdeki, D.; Koufakis, T.; Kanavas, V.; Antza, C.; Germanidis, G.; Chourdakis, M. T Effects of Olive Oil Consumption on Biochemical Parameters and Body Mass Index of People with Nonalcoholic Fatty Liv Disease: A Systematic Review and Meta-Analysis of Randomized Controlled Trials. *Nutrients* **2024**, *16*, 857. [CrossRef]
10. Marrero, A.D.; Quesada, A.R.; Martínez-Poveda, B.; Medina, M.Á. Anti-Cancer, Anti-Angiogenic, and Anti-Atherogenic Potent of Key Phenolic Compounds from Virgin Olive Oil. *Nutrients* **2024**, *16*, 1283. [CrossRef]
11. Roberts, J.D.; Lillis, J.B.; Pinto, J.M.; Chichger, H.; López-Samanes, Á.; Coso, J.D.; Zacca, R.; Willmott, A.G.B. The Effect o Hydroxytyrosol-Rich Olive-Derived Phytocomplex on Aerobic Exercise and Acute Recovery. *Nutrients* **2023**, *15*, 421. [CrossR [PubMed]
12. Caponio, F.; Piga, A.; Poiana, M. Valorization of Food Processing By-Products. *Foods* **2022**, *11*, 3246. [CrossRef] [PubMed]

Dini, I.; Graziani, G.; Fedele, F.L.; Sicari, A.; Vinale, F.; Castaldo, L.; Ritieni, A. An Environmentally Friendly Practice Used in Olive Cultivation Capable of Increasing Commercial Interest in Waste Products from Oil Processing. *Antioxidants* **2020**, *9*, 466. [CrossRef]
Bellumori, M.; Cecchi, L.; Romani, A.; Mulinacci, N.; Innocenti, M. Recovery and Stability over Time of Phenolic Fractions by an Industrial Filtration System of Olive Mill Wastewaters: A Three-Year Study. *J. Sci. Food Agric.* **2018**, *98*, 2761–2769. [CrossRef]
Krawczyk, M.; Burzynska-Pedziwiatr, I.; Wozniak, L.A.; Bukowiecka-Matusiak, M. Impact of Polyphenols on Inflammatory and Oxidative Stress Factors in Diabetes Mellitus: Nutritional Antioxidants and Their Application in Improving Antidiabetic Therapy. *Biomolecules* **2023**, *13*, 1402. [CrossRef]
Williamson, G.; Sheedy, K. Effects of Polyphenols on Insulin Resistance. *Nutrients* **2020**, *12*, 3135. [CrossRef]
Cho, A.R.; Kwon, Y.J.; Kim, J.K. Pre-Metabolic Syndrome and Incidence of Type 2 Diabetes and Hypertension: From the Korean Genome and Epidemiology Study. *J. Pers. Med.* **2021**, *11*, 700. [CrossRef]
Vona, R.; Gambardella, L.; Cittadini, C.; Straface, E.; Pietraforte, D. Biomarkers of Oxidative Stress in Metabolic Syndrome and Associated Diseases. *Oxid. Med. Cell. Longev.* **2019**, *2019*, 8267234. [CrossRef]
Cristoni, S.; Rossi Bernardi, L.; Larini, M.; Natale, G.; Didomenico, N.; Varelli, M.; Conti, M.; Dorna, I.; Puccio, G. Predicting and Preventing Intestinal Dysbiosis on the Basis of Pharmacological Gut Microbiota Metabolism. *Rapid Commun. Mass Spectrom.* **2019**, *33*, 1221–1225. [CrossRef]
Balashova, E.E.; Maslov, D.L.; Lokhov, P.G. A Metabolomics Approach to Pharmacotherapy Personalization. *J. Pers. Med.* **2018**, *83*, 28. [CrossRef]
Maslov, D.; Balashova, E.; Lokhov, P.; Archakov, A.I. Pharmacometabonomics—The Novel Way to Personalized Drug Therapy. *Biomeditsinskaya Khimiya* **2017**, *63*, 115–123. [CrossRef] [PubMed]
Ang, M.Y.; Low, T.Y.; Lee, P.Y.; Wan Mohamad Nazarie, W.F.; Guryev, V.; Jamal, R. Proteogenomics: From Next-Generation Sequencing (NGS) and Mass Spectrometry-Based Proteomics to Precision Medicine. *Clin. Chim. Acta* **2019**, *498*, 38–46. [CrossRef]
Zinellu, E.; Zinellu, A.; Fois, A.G.; Carru, C.; Pirina, P. Circulating Biomarkers of Oxidative Stress in Chronic Obstructive Pulmonary Disease: A Systematic Review. *Respir. Res.* **2016**, *17*, 150. [CrossRef] [PubMed]
Qiu, S.; Cai, Y.; Yao, H.; Lin, C.; Xie, Y.; Tang, S.; Zhang, A. Small Molecule Metabolites: Discovery of Biomarkers and Therapeutic Targets. *Signal Transduct. Target. Ther.* **2023**, *81*, 132. [CrossRef]
Demicheva, E.; Dordiuk, V.; Polanco Espino, F.; Ushenin, K.; Aboushanab, S.; Shevyrin, V.; Buhler, A.; Mukhlynina, E.; Solovyova, O.; Danilova, I.; et al. Advances in Mass Spectrometry-Based Blood Metabolomics Profiling for Non-Cancer Diseases: A Comprehensive Review. *Metabolites* **2024**, *141*, 54. [CrossRef]
Albini, A.; Briga, D.; Conti, M.; Bruno, A.; Farioli, D.; Canali, S.; Sogno, I.; D'Ambrosio, G.; Consonni, P.; Noonan, D.M. SANIST: A Rapid Mass Spectrometric SACI/ESI Data Acquisition and Elaboration Platform for Verifying Potential Candidate Biomarkers. *Rapid Commun. Mass Spectrom.* **2015**, *29*, 1703–1710. [CrossRef]
Albini, A.; Bruno, A.; Bassani, B.; D'Ambrosio, G.; Pelosi, G.; Consonni, P.; Castellani, L.; Conti, M.; Cristoni, S.; Noonan, D.M. Serum Steroid Ratio Profiles in Prostate Cancer: A New Diagnostic Tool Toward a Personalized Medicine Approach. *Front. Endocrinol.* **2018**, *9*, 110. [CrossRef] [PubMed]
Cristoni, S.; Dusi, G.; Brambilla, P.; Albini, A.; Conti, M.; Brambilla, M.; Bruno, A.; Di Gaudio, F.; Ferlin, L.; Tazzari, V.; et al. SANIST: Optimization of a Technology for Compound Identification Based on the European Union Directive with Applications in Forensic Pharmaceutical and Food Analyses. *J. Mass Spectrom.* **2017**, *52*, 16–21. [CrossRef] [PubMed]
Cristoni, S.; Molin, L.; Lai, A.; Bernardi, L.R.; Pucciarelli, S.; Agostini, M.; Bedin, C.; Nitti, D.; Seraglia, R.; Repetto, O.; et al. MALDI-MS-NIST Library Approach for Colorectal Cancer Diagnosis. *Rapid Commun. Mass Spectrom.* **2009**, *23*, 2839–2845. [CrossRef]
Albini, A.; Albini, F.; Corradino, P.; Dugo, L.; Calabrone, L.; Noonan, D.M. From Antiquity to Contemporary Times: How Olive Oil By-Products and Waste Water Can Contribute to Health. *Front. Nutr.* **2023**, *10*, 1254947. [CrossRef]
Peri, S.; Ruzzolini, J.; Urciuoli, S.; Versienti, G.; Biagioni, A.; Andreucci, E.; Peppicelli, S.; Bianchini, F.; Bottari, A.; Calorini, L.; et al. An Oleocanthal-Enriched EVO Oil Extract Induces ROS Production in Gastric Cancer Cells and Potentiates the Effect of Chemotherapy. *Antioxidants* **2022**, *11*, 1762. [CrossRef] [PubMed]
Lăcătușu, C.M.; Grigorescu, E.D.; Floria, M.; Onofriescu, A.; Mihai, B.M. The Mediterranean Diet: From an Environment-Driven Food Culture to an Emerging Medical Prescription. *Int. J. Environ. Res. Public Health* **2019**, *16*, 942. [CrossRef] [PubMed]
Santangelo, C.; Vari, R.; Scazzocchio, B.; De Sanctis, P.; Giovannini, C.; D'Archivio, M.; Masella, R. Anti-Inflammatory Activity of Extra Virgin Olive Oil Polyphenols: Which Role in the Prevention and Treatment of Immune-Mediated Inflammatory Diseases? *Endocr. Metab. Immune Disord. Drug Targets* **2018**, *18*, 36–50. [CrossRef] [PubMed]
Nocella, C.; Cammisotto, V.; Fianchini, L.; D'Amico, A.; Novo, M.; Castellani, V.; Stefanini, L.; Violi, F.; Carnevale, R. Extra Virgin Olive Oil and Cardiovascular Diseases: Benefits for Human Health. *Endocr. Metab. Immune Disord. Drug Targets* **2018**, *18*, 4–13. [CrossRef] [PubMed]
Billingsley, H.E.; Carbone, S.; Lavie, C.J. Dietary Fats and Chronic Noncommunicable Diseases. *Nutrients* **2018**, *10*, 1385. [CrossRef]
Tsao, R. Chemistry and Biochemistry of Dietary Polyphenols. *Nutrients* **2010**, *2*, 1231–1246. [CrossRef]
Stagos, D. Antioxidant Activity of Polyphenolic Plant Extracts. *Antioxidants* **2019**, *9*, 19. [CrossRef]

38. Rudrapal, M.; Khairnar, S.J.; Khan, J.; Dukhyil, A.B.; Ansari, M.A.; Alomary, M.N.; Palai, S.; Deb, P.K.; Devi, R. Diet Polyphenols and Their Role in Oxidative Stress-Induced Human Diseases: Insights into Protective Effects, Antioxidant Potent and Mechanism(s) of Action. *Front. Pharmacol.* **2022**, *13*, 806470. [CrossRef]
39. Bhuyan, U.; Handique, J.G. Plant Polyphenols as Potent Antioxidants: Highlighting the Mechanism of Antioxidant Activity a Synthesis/Development of Some Polyphenol Conjugates. *Stud. Nat. Prod. Chem.* **2022**, *75*, 243–266. [CrossRef]
40. Selim, S.; Albqmi, M.; Al-Sanea, M.M.; Alnusaire, T.S.; Almuhayawi, M.S.; AbdElgawad, H.; Al Jaouni, S.K.; Elkelish, Hussein, S.; Warrad, M.; et al. Valorizing the Usage of Olive Leaves Bioactive Compounds, Biological Activities, and Fo Applications: A Comprehensive Review. *Front. Nutr.* **2022**, *9*, 1008349. [CrossRef]
41. Otero, C.; Miranda-Rojas, S.; Llancalahuén, F.M.; Fuentes, J.A.; Atala, C.; González-Silva, G.; Verdugo, D.; Sierra-Rosales Moreno, A.; Gordillo-Fuenzalida, F. Biochemical Characterization of Peumus Boldus Fruits: Insights of Its Antioxidant Proper through a Theoretical Approach. *Food Chem.* **2022**, *370*, 131012. [CrossRef] [PubMed]
42. Khwaldia, K.; Attour, N.; Matthes, J.; Beck, L.; Schmid, M. Olive Byproducts and Their Bioactive Compounds as a Valuable Sou for Food Packaging Applications. *Compr. Rev. Food Sci. Food Saf.* **2022**, *21*, 1218–1253. [CrossRef]
43. Vauzour, D.; Rodriguez-Mateos, A.; Corona, G.; Oruna-Concha, M.J.; Spencer, J.P. Polyphenols and Human Health: Preventior Disease and Mechanisms of Action. *Nutrients* **2010**, *2*, 1106–1131. [CrossRef] [PubMed]
44. Bahadoran, Z.; Mirmiran, P.; Azizi, F. Dietary Polyphenols as Potential Nutraceuticals in Management of Diabetes: A Revi *J. Diabetes Metab. Disord.* **2013**, *12*, 43. [CrossRef]
45. Cory, H.; Passarelli, S.; Szeto, J.; Tamez, M.; Mattei, J. The Role of Polyphenols in Human Health and Food Systems: A Mini-Rev *Front. Nutr.* **2018**, *5*, 87. [CrossRef]
46. Castañer, O.; Covas, M.I.; Khymenets, O.; Nyyssonen, K.; Konstantinidou, V.; Zunft, H.F.; de la Torre, R.; Muñoz-Aguayo, Vila, J.; Fitó, M. Protection of LDL from Oxidation by Olive Oil Polyphenols Is Associated with a Downregulation of CD40-Liga Expression and Its Downstream Products in Vivo in Humans. *Am. J. Clin. Nutr.* **2012**, *95*, 1238–1244. [CrossRef] [PubMed]
47. Mancini, J.G.; Filion, K.B.; Atallah, R.; Eisenberg, M.J. Systematic Review of the Mediterranean Diet for Long-Term Weight L *Am. J. Med.* **2016**, *129*, 407–415.e4. [CrossRef]
48. Flynn, M.M.; Reinert, S.E. Comparing an Olive Oil-Enriched Diet to a Standard Lower-Fat Diet for Weight Loss in Breast Can Survivors: A Pilot Study. *J. Womens Health* **2010**, *19*, 1155–1161. [CrossRef]
49. Nishi, S.K.; Babio, N.; Paz-Graniel, I.; Serra-Majem, L.; Vioque, J.; Fitó, M.; Corella, D.; Pintó, X.; Bueno-Cavanillas, Tur, J.A.; et al. Water Intake, Hydration Status, and 2-Year Changes in Cognitive Performance: A Prospective Cohort Stu *BMC Med.* **2023**, *21*, 82. [CrossRef]
50. Li, S.; Xiao, X.; Zhang, X. Hydration Status in Older Adults: Current Knowledge and Future Challenges. *Nutrients* **2023**, *15*, 2([CrossRef]
51. Caruso, C.; Accardi, G.; Virruso, C.; Candore, G. Sex, Gender, and Immunosenescence: A Key to Understand the Differ Lifespan between Men and Women? *Immun. Ageing* **2013**, *10*, 20. [CrossRef]
52. de Brito Alves, J.L.; de Sousa, V.P.; Cavalcanti Neto, M.P.; Magnani, M.; Braga, V.A.; da Costa-Silva, J.H.; Leandro, C.G.; Vi H.; Pirola, L. New Insights on the Use of Dietary Polyphenols or Probiotics for the Management of Arterial Hypertens *Front. Physiol.* **2016**, *7*, 448. [CrossRef] [PubMed]
53. Barona, J.; Jones, J.J.; Kopec, R.E.; Comperatore, M.; Andersen, C.; Schwartz, S.J.; Lerman, R.H.; Fernandez, M.L. A Mediterrane Style Low-Glycemic-Load Diet Increases Plasma Carotenoids and Decreases LDL Oxidation in Women with Metabolic Syndro *J. Nutr. Biochem.* **2012**, *23*, 609–615. [CrossRef] [PubMed]
54. Sarapis, K.; Thomas, C.J.; Hoskin, J.; George, E.S.; Marx, W.; Mayr, H.L.; Kennedy, G.; Pipingas, A.; Willcox, J Prendergast, L.A.; et al. The Effect of High Polyphenol Extra Virgin Olive Oil on Blood Pressure and Arterial Stiffness Healthy Australian Adults: A Randomized Controlled Cross-Over Study. *Nutrients* **2020**, *12*, 2272. [CrossRef]
55. Moreno-Luna, R.; Muñoz-Hernandez, R.; Miranda, M.L.; Costa, A.F.; Jimenez-Jimenez, L.; Vallejo-Vaz, A.J.; Muriana, Villar, J.; Stiefel, P. Olive Oil Polyphenols Decrease Blood Pressure and Improve Endothelial Function in Young Women with M Hypertension. *Am. J. Hypertens.* **2012**, *25*, 1299–1304. [CrossRef] [PubMed]
56. Violi, F.; Loffredo, L.; Pignatelli, P.; Angelico, F.; Bartimoccia, S.; Nocella, C.; Cangemi, R.; Petruccioli, A.; Monticolo, Pastori, D.; et al. Extra Virgin Olive Oil Use Is Associated with Improved Post-Prandial Blood Glucose and LDL Cholesterol Healthy Subjects. *Nutr. Diabetes* **2015**, *5*, e172. [CrossRef]
57. Vigneri, R.; Sciacca, L.; Vigneri, P. Rethinking the Relationship between Insulin and Cancer. *Trends Endocrinol. Metab.* **2020**, 551–560. [CrossRef]
58. Bertoli, A.; Valentini, A.; Cianfarani, M.A.; Gasbarra, E.; Tarantino, U.; Federici, M. Low FT3: A Possible Marker of Frailty in Elderly. *Clin. Interv. Aging* **2017**, *12*, 335–341. [CrossRef]
59. Alì, S.; Davinelli, S.; Accardi, G.; Aiello, A.; Caruso, C.; Duro, G.; Ligotti, M.E.; Pojero, F.; Scapagnini, G.; Candore, G. Heal Ageing and Mediterranean Diet: A Focus on Hormetic Phytochemicals. *Mech. Ageing Dev.* **2021**, *200*, 111592. [CrossRef]
60. Roche, M.; Rondeau, P.; Singh, N.R.; Tarnus, E.; Bourdon, E. The Antioxidant Properties of Serum Albumin. *FEBS Lett.* **2008**, 5 1783–1787. [CrossRef]
61. Belinskaia, D.A.; Voronina, P.A.; Shmurak, V.I.; Jenkins, R.O.; Goncharov, N.V. Serum Albumin in Health and Disease: Ester Antioxidant, Transporting, and Signaling Properties. *Int. J. Mol. Sci.* **2021**, *22*, 10318. [CrossRef] [PubMed]

Sitar, M.E.; Aydin, S.; Cakatay, U. Human Serum Albumin and Its Relation with Oxidative Stress. *Clin. Lab* **2013**, *59*, 945–952. [CrossRef] [PubMed]
Eckart, A.; Struja, T.; Kutz, A.; Baumgartner, A.; Baumgartner, T.; Zurfluh, S.; Neeser, O.; Huber, A.; Stanga, Z.; Mueller, B.; et al. Relationship of Nutritional Status, Inflammation, and Serum Albumin Levels during Acute Illness: A Prospective Study. *Am. J. Med.* **2020**, *133*, 713–722.e7. [CrossRef] [PubMed]
Rozga, J.; Piątek, T.; Małkowski, P. Human Albumin: Old, New, and Emerging Applications. *Ann. Transplant.* **2013**, *18*, 205–217. [CrossRef] [PubMed]
Jun, J.E.; Lee, S.E.; Lee, Y.B.; Jee, J.H.; Bae, J.C.; Jin, S.M.; Hur, K.Y.; Lee, M.K.; Kim, J.H. Increase in Serum Albumin Concentration Is Associated with Prediabetes Development and Progression to Overt Diabetes Independently of Metabolic Syndrome. *PLoS ONE* **2017**, *12*, e0176209. [CrossRef]
Morris, A.; Krishnan, N.; Kimani, P.K.; Lycett, D. Effect of Dietary Potassium Restriction on Serum Potassium, Disease Progression, and Mortality in Chronic Kidney Disease: A Systematic Review and Meta-Analysis. *J. Ren. Nutr.* **2020**, *30*, 276–285. [CrossRef]
Picard, K.; Barreto Silva, M.I.; Mager, D.; Richard, C. Dietary Potassium Intake and Risk of Chronic Kidney Disease Progression in Predialysis Patients with Chronic Kidney Disease: A Systematic Review. *Adv. Nutr.* **2020**, *11*, 1002–1015. [CrossRef]
Pourafshar, S.; Sharma, B.; Kranz, S.; Mallawaarachchi, I.; Kurland, E.; Ma, J.Z.; Scialla, J.J. Patterns of Fruit and Vegetable Intake in Adults with and without Chronic Kidney Disease in the United States. *J. Ren. Nutr.* **2023**, *33*, 88–96. [CrossRef]
Berridge, M.J. Vitamin D, Reactive Oxygen Species, and Calcium Signalling in Ageing and Disease. *Philos. Trans. R. Soc. Lond. B Biol. Sci.* **2016**, *371*, 20150434. [CrossRef]
Della Nera, G.; Sabatino, L.; Gaggini, M.; Gorini, F.; Vassalle, C. Vitamin D Determinants, Status, and Antioxidant/Anti-Inflammatory-Related Effects in Cardiovascular Risk and Disease: Not the Last Word in the Controversy. *Antioxidants* **2023**, *12*, 948. [CrossRef]
Masjedi, F.; Keshtgar, S.; Zal, F.; Talaei-Khozani, T.; Sameti, S.; Fallahi, S.; Kazeroni, M. Effects of Vitamin D on Steroidogenesis, Reactive Oxygen Species Production, and Enzymatic Antioxidant Defense in Human Granulosa Cells of Normal and Polycystic Ovaries. *J. Steroid Biochem. Mol. Biol.* **2020**, *197*, 105521. [CrossRef] [PubMed]
Quigley, M.; Rieger, S.; Capobianco, E.; Wang, Z.; Zhao, H.; Hewison, M.; Lisse, T.S. Vitamin D Modulation of Mitochondrial Oxidative Metabolism and mTOR Enforces Stress Adaptations and Anticancer Responses. *JBMR Plus* **2021**, *6*, e10572. [CrossRef]
McCourt, A.F.; O'Sullivan, A.M. Using Food Fortification to Improve Vitamin D Bioaccessibility and Intakes. *Proc. Nutr. Soc.* **2022**, *81*, 99–107. [CrossRef] [PubMed]
Pu, F.; Chen, N.; Xue, S. Calcium Intake, Calcium Homeostasis, and Health. *Food Sci. Hum. Wellness* **2016**, *5*, 8–16. [CrossRef]
Beto, J.A. The Role of Calcium in Human Aging. *Clin. Nutr. Res.* **2015**, *4*, 1–8. [CrossRef]
Veldurthy, V.; Wei, R.; Oz, L.; Dhawan, P.; Jeon, Y.H.; Christakos, S. Vitamin D, Calcium Homeostasis, and Aging. *Bone Res.* **2016**, *4*, 16041. [CrossRef]
Wicher, S.A.; Roos, B.B.; Teske, J.J.; Fang, Y.H.; Pabelick, C.; Prakash, Y.S. Aging Increases Senescence, Calcium Signaling, and Extracellular Matrix Deposition in Human Airway Smooth Muscle. *PLoS ONE* **2021**, *16*, e0254710. [CrossRef]
Shen, W.; Gao, C.; Cueto, R.; Liu, L.; Fu, H.; Shao, Y.; Yang, W.Y.; Fang, P.; Choi, E.T.; Wu, Q.; et al. Homocysteine-Methionine Cycle Is a Metabolic Sensor System Controlling Methylation-Regulated Pathological Signaling. *Redox Biol.* **2020**, *28*, 101322. [CrossRef] [PubMed]
Cristoni, S.; Bernardi, L.R.; Malvandi, A.M.; Larini, M.; Longhi, E.; Sortino, F.; Conti, M.; Pantano, N.; Puccio, G. A Case of Personalized and Precision Medicine: Pharmacometabolomic Applications to Rare Cancer, Microbiological Investigation, and Therapy. *Rapid Commun. Mass Spectrom.* **2021**, *35*, e8976. [CrossRef]
Brattström, L.; Wilcken, D.E. Homocysteine and Cardiovascular Disease: Cause or Effect? *Am. J. Clin. Nutr.* **2000**, *72*, 315–323. [CrossRef]
Dinavahi, R.; Falkner, B. Relationship of Homocysteine with Cardiovascular Disease and Blood Pressure. *J. Clin. Hypertens.* **2004**, *6*, 494–498. [CrossRef] [PubMed]
Yuan, S.; Mason, A.M.; Carter, P.; Burgess, S.; Larsson, S.C. Homocysteine, B Vitamins, and Cardiovascular Disease: A Mendelian Randomization Study. *BMC Med.* **2021**, *19*, 97. [CrossRef]
Guieu, R.; Ruf, J.; Mottola, G. Hyperhomocysteinemia and Cardiovascular Diseases. *Ann. Biol. Clin.* **2022**, *80*, 7–14. [CrossRef]
Kumar, A.; Palfrey, H.A.; Pathak, R.; Kadowitz, P.J.; Gettys, T.W.; Murthy, S.N. The Metabolism and Significance of Homocysteine in Nutrition and Health. *Nutr. Metab.* **2017**, *14*, 78. [CrossRef] [PubMed]
Nagayama, D.; Watanabe, Y.; Yamaguchi, T.; Suzuki, K.; Saiki, A.; Fujishiro, K.; Shirai, K. Issue of Waist Circumference for the Diagnosis of Metabolic Syndrome Regarding Arterial Stiffness: Possible Utility of a Body Shape Index in Middle-Aged Nonobese Japanese Urban Residents Receiving Health Screening. *Obes. Facts* **2022**, *15*, 160–169. [CrossRef]
Mehmood, A.; Usman, M.; Patil, P.; Zhao, L.; Wang, C. A Review on Management of Cardiovascular Diseases by Olive Polyphenols. *Food Sci. Nutr.* **2020**, *8*, 4639–4655. [CrossRef]
Branco, A.C.C.C.; Yoshikawa, F.S.Y.; Pietrobon, A.J.; Sato, M.N. Role of Histamine in Modulating the Immune Response and Inflammation. *Mediators Inflamm.* **2018**, *2018*, 9524075. [CrossRef]
Thangam, E.B.; Jemima, E.A.; Singh, H.; Baig, M.S.; Khan, M.; Mathias, C.B.; Church, M.K.; Saluja, R. The Role of Histamine and Histamine Receptors in Mast Cell-Mediated Allergy and Inflammation: The Hunt for New Therapeutic Targets. *Front. Immunol.* **2018**, *9*, 1873. [CrossRef]

89. Carthy, E.; Ellender, T. Histamine, Neuroinflammation, and Neurodevelopment: A Review. *Front. Neurosci.* **2021**, *15*, 6802 [CrossRef]
90. Murdaca, G.; Allegra, A.; Tonacci, A.; Musolino, C.; Ricciardi, L.; Gangemi, S. Mast Cells and Vitamin D Status: A Clinical a Biological Link in the Onset of Allergy and Bone Diseases. *Biomedicines* **2022**, *10*, 1877. [CrossRef]
91. Mehrani, Y.; Morovati, S.; Tieu, S.; Karimi, N.; Javadi, H.; Vanderkamp, S.; Sarmadi, S.; Tajik, T.; Kakish, J.E.; Bridle, B.W.; et Vitamin D Influences the Activity of Mast Cells in Allergic Manifestations and Potentiates Their Effector Functions agai Pathogens. *Cells* **2023**, *12*, 2271. [CrossRef] [PubMed]
92. Caruso, C.; Passarino, G.; Puca, A.; Scapagnini, G. "Positive biology": The centenarian lesson. *Immun. Ageing* **2012**, *9*, 5. [CrossR [PubMed]

nutrients

MDPI

cle

fects of Evodiamine on Behavior and Hippocampal Neurons rough Inhibition of Angiotensin-Converting Enzyme and odulation of the Renin Angiotensin Pathway in a Mouse odel of Post-Traumatic Stress Disorder

xing Wang [1,2,†], Chengcai Lai [1,2,†], Baoying Shen [2,†], Bowei Li [2], Junru Chen [2], Xin Shen [2], Zhengping Huang [3], nqi Yang [2] and Yue Gao [1,2,*]

1 Medical College, Qinghai University, Xining 810016, China; wzx98748@163.com (Z.W.); asa2057516@163.com (C.L.)

2 Department of Pharmaceutical Sciences, Beijing Institute of Radiation Medicine, Beijing 100850, China; shenby25511@163.com (B.S.); weiboli1029@163.com (B.L.); 15088094928@163.com (J.C.); shenxin9204@126.com (X.S.); ycq1qaz@outlook.com (C.Y.)

3 Department of Neurology, Fujian Medical University, Quanzhou 362000, China; 15750848187@163.com

* Correspondence: gaoyue@bmi.ac.cn

† These authors contributed equally to this work.

on: Wang, Z.; Lai, C.; Shen, B.; Chen, J.; Shen, X.; Huang, Z.; C.; Gao, Y. Effects of amine on Behavior and ocampal Neurons through tion of Angiotensin-Converting ne and Modulation of the Renin tensin Pathway in a Mouse l of Post-Traumatic Stress der. *Nutrients* **2024**, *16*, 1957. //doi.org/10.3390/nu16121957

emic Editors: Sonia de al-Teresa, María les Martín and Sonia Ramos o

ved: 25 April 2024
ed: 12 June 2024
oted: 14 June 2024
shed: 19 June 2024

Abstract: Post-traumatic stress disorder (PTSD) is a persistent psychiatric condition that arises following exposure to traumatic events such as warfare, natural disasters, or other catastrophic incidents, typically characterized by heightened anxiety, depressive symptoms, and cognitive dysfunction. In this study, animals subjected to single prolonged stress (SPS) were administered evodiamine (EVO) and compared to a positive control group receiving sertraline. The animals were then assessed for alterations in anxiety, depression, and cognitive function. Histological analysis was conducted to examine neuronal changes in the hippocampus. In order to predict the core targets and related mechanisms of evodiamine intervention in PTSD, network pharmacology was used. The metabolic markers pre- and post-drug administration were identified using nontargeted serum metabolomics techniques, and the intersecting Kyoto Encyclopedia of Genes and Genomes (KEGG) pathways were screened. Finally, the core targets were validated through molecular docking, enzyme-linked immunosorbent assays, and immunofluorescence staining to confirm the anti-PTSD effects and mechanisms of these targets. As well as improving cognitive impairment, evodiamine reversed anxiety- and depression-like behaviors. It also inhibited the reduction in the number of hippocampal neuronal cells and Nissl bodies in SPS mice inhibited angiotensin converting enzyme (ACE) levels in the hippocampus of SPS mice, and modulated the renin angiotensin pathway and its associated serum metabolites in brain tissue. Evodiamine shows promise as a potential candidate for alleviating the symptoms of post-traumatic stress disorder.

Keywords: post-traumatic stress disorder; evodiamine; renin angiotensin system; angiotensin converting enzyme; single prolonged stress

1. Introduction

Post-traumatic stress disorder (PTSD) is a mental illness triggered by trauma, characterized by intrusive reliving of past events, hyperarousal, avoidance behavior, and intense fear [1]. This chronic condition affects many individuals who show limited improvement with current therapies and is associated with an increased risk of suicide, psychological disorders, cardiovascular disease, and cognitive impairment [2]. Studies during the Corona Virus Disease 2019 (COVID-19) pandemic revealed a high estimated PTSD incidence rate of up to 17.52% across 24 countries [3]. Currently, only selective serotonin reuptake inhibitors (SSRIs) such as sertraline and paroxetine are approved for treating PTSD [4]. However,

SSRIs have notable side effects, including initial neuroticism, emotional blunting, gastrc testinal issues, insomnia and sexual dysfunction, as well as drawbacks such as slow on of action and uncertain effectiveness [5]. The development of new drugs and formulatic for the treatment of PTSD is urgently needed.

Evodiamine (EVO) is a major alkaloid compound found in the dry unripened fr Evodia fructus [6] that possesses a variety of pharmacological activities. With respect to pharmacological effects of EVO, increasing attention has been given to its beneficial effe including its anti-inflammatory [7], immune-modulatory [8], and antitumor effects [9], well as its ability to retard the development of atherosclerosis [10]. Additionally, evc amine has been found to be effective in treating cerebral ischemic injury [6] and chro unpredictable mild stress-induced depression [11].

Although its mechanisms are not well understood, the renin angiotensin system (R has been identified as a potential therapeutic target for PTSD [12]. Due to its involvem in sympathetic arousal, the renin angiotensin system has been directly linked to PT and cardiovascular disease (CVD) [13]. Renin is released by this hormone system, wh regulates blood pressure. Stress activates the sympathetic nervous system, which relea renin into the bloodstream and activates angiotensinogen. The angiotensin converti enzyme (ACE) converts angiotensin I to angiotensin II, a peptide that increases blood p sure and vascular resistance. Additionally, angiotensin II inhibits norepinephrine reupta and increases norepinephrine release, further enhancing sympathetic activity [14]. The fore, given the well-established sympathetic overactivation in PTSD, the renin angioten system is a crucial target to better understand the increased CVD risk in PTSD patients [1

Neuroprotective and cognitive protective effects of evodiamine have been dem strated in an animal model of intracerebroventricular streptozotocin (ICV-STZ)-induc sporadic Alzheimer's [15]. Despite its notable antidepressant-like activity and ability improve cognition, the pharmacological potential of evodiamine has not been explor in a mouse model of PTSD. Therefore, molecular docking pharmacological network ar ysis was used to predict that evodiamine interferes with the renin angiotensin pathw through relevant targets for treating single prolonged stress (SPS)-induced PTSD. Throu experimental validation, it was confirmed that EVO treatment attenuated the sympto of SPS-induced PTSD in vivo, and the relevant bioactive constituents were identifi through nontargeted serum metabolomics. Mechanistically, EVO interferes with the re angiotensin pathway to alleviate PTSD mainly by inhibiting the angiotensin converti enzyme (ACE) in brain tissue, which may be a novel therapeutic target for PTSD.

This study investigated the histological and pharmacological alterations in the h pocampus of a mouse model of post-traumatic stress disorder (PTSD), along with behavioral changes associated with PTSD. The study revealed that EVO exhibited an olytic and antidepressant properties, as well as improved memory and cognitive functi in a post-stress behavioral task. Furthermore, the results demonstrated the neuroprotect effects of EVO on hippocampal neurons in stressed mice, including the prevention hippocampal neuronal apoptosis. The renin angiotensin system (RAS) and other canoni systems have been documented as significant contributors to stress, particularly in t context of post-traumatic stress disorder. This paper provides additional insights i the involvement of ACE and RAS in stress-induced PTSD, as well as the potential therapeutic intervention in stress and stress-related conditions.

2. Results

2.1. Effects of EVO on Behavioral Deficits in SPS-Induced PTSD Mice

The results of the Open Field Trial (OFT) showed that EVO and sertraline (SE treatments significantly increased the percentage of distance in the center, total distar and average speed, and shortened incubation period in the center compared to the S group (Figure 1B–E). Moreover, in the elevated plus-maze test (EPMT), compared with S EVO and SER led to significant increases in distance in the open arms and time spent the open arms (Figure 1I,J). In mice, SPS-induced PTSD inhibits locomotor activity a spontaneous exploration, while EVO and SER reverse these effects. The results of the tail suspension test (TST) revealed that SPS notably prolonged the immobility time of the mice compared to that of the control group, whereas the EVO and SER treatments partially reduced the immobility time (Figure 1G). Additionally, compared with SPS, EVO and SER significantly decreased the immobilization time in the forced swim test (FST) (Figure 1H). These results suggest that EVO treatment has the potential to alleviate SPS-induced despair behavior in mice.

In order to test spatial learning, mice were trained in the Morris water maze (MWM), which entailed three trials per day for five consecutive days. The study revealed a significant decrease in escape latency across all groups during the training period. Particularly in the spatial navigation task, the latency to escape consistently decreased as the training days progressed for all groups of mice. This decrease was statistically significant in terms of the time and handling factors, with no significant interaction between these factors. Further analysis through a two-by-two comparison at different time points revealed that on day 4, the control group had a significantly shorter escape latency than the SPS group (Figure 1L). These findings suggest that mice in the control group effectively learned the task, while no notable difference in learning ability was observed among the experimental groups.

In the spatial exploration experiment, compared with those in the SPS group, the distance traveled to explore the target quadrant and the number of shuttles to the platform area in the EVO- and SER-treated mice were markedly greater (Figure 1M,N). These results indicate that SPS may mildly impair the memory capacity of mice, albeit without a significant effect, and that EVO has a substantial positive impact on this symptom.

The results demonstrated that administration of EVO significantly improved the depression-like behavior of SPS-exposed mice. Additionally, EVO treatment was expected to alleviate SPS-induced memory and cognitive deficits.

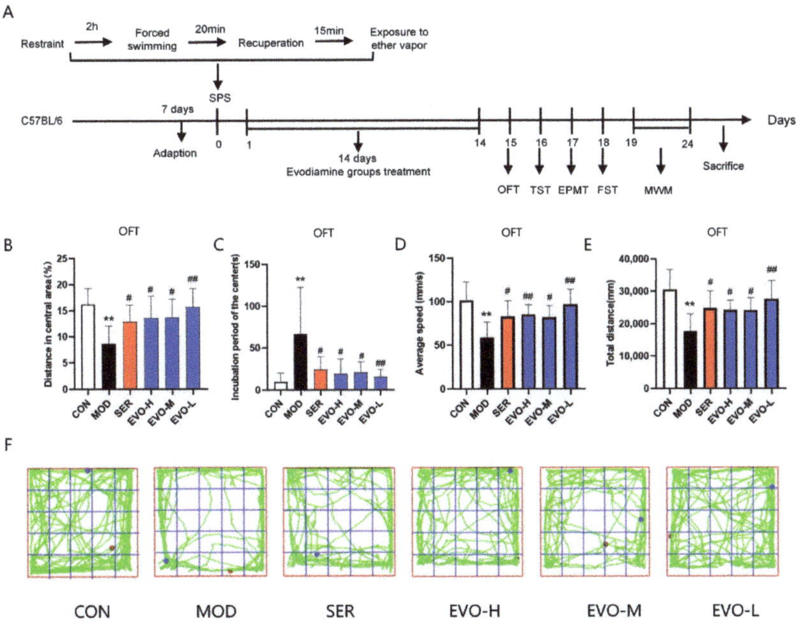

Figure 1. *Cont.*

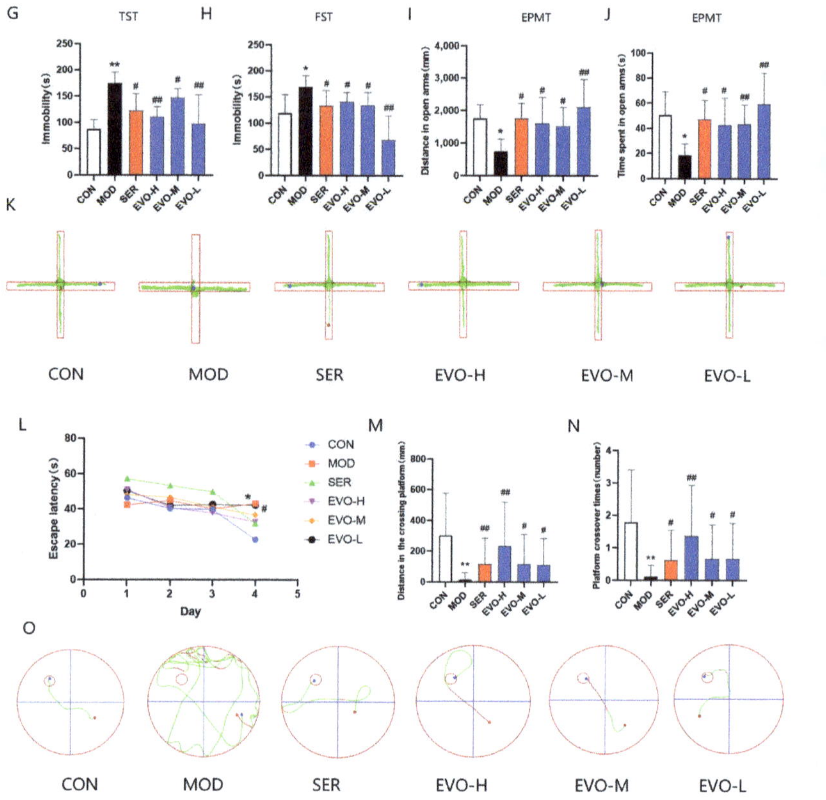

Figure 1. Effects of EVO on behavioral deficits in SPS-induced PTSD mice. (**A**) Experimental fl chart of the mouse model and treatment. (**B**) Percentage of distance in the center. (**C**) Incubati period in the center of the OFT. (**D**) Average speed in the OFT. (**E**) Total distance in the OFT. (**F**) R resentative trajectory diagram of mice in the OFT. In the figure, the mice started at the red point a ended when they reached the blue point, while the green line indicates the trajectory the mice trave (**G**) Immobilization time in the TST. (**H**) Immobilization time in the FST. (**I**) Distance in the open ar in the EPMT. (**J**) Time spent in the open arms in the EPMT. (**K**) Representative trajectory diagram mice in the EPMT. (**L**) Escape latency in the MWM test. (**M**) Distance in the crossing platform in MWM test. (**N**) Number of platform crossover times in the MWM test. (**O**) Representative traject plots of mice in the MWM test. Forty-eight mice were randomly allocated into six groups: the cont group (CON), model group (MOD), sertraline group (SER), evodiamine high dose group (EVO- evodiamine medium dose group (EVO-M), and evodiamine low dose group (EVO-L). The data expressed as the means ± SDs ($n = 8$). * $p < 0.05$, ** $p < 0.01$ compared to the CON group; # $p < 0.$ ## $p < 0.01$ compared to the SPS group.

2.2. *Effects of EVO on Hippocampal Morphology in SPS-Induced PTSD Mice*

Morphological changes in the hippocampal region were assessed by HE staini (Figure 2A). Compared with the control group, the number of cells in the hippocampal C CA2, CA3, and DG regions was significantly reduced in the model group (Figure 2B– The brain tissue structure was mildly abnormal, and a small number of neuronal ce in the hippocampal area in the field of view were degenerated with deepened stainir and no clear cytosol was seen (Figure 2A). On the contrary, compared with the moc group, the hippocampal tissues in the EVO and sertraline-treated group tended to normalized, including lighter staining, normal structure, more aligned, and more regu morphology with clear nuclei in the hippocampal CA1 region and DG region (Figure 2A), and a moderate dose of EVO significantly suppressed the reduction of neuronal cells in the hippocampus of the DG region in SPS mice (Figure 2E).

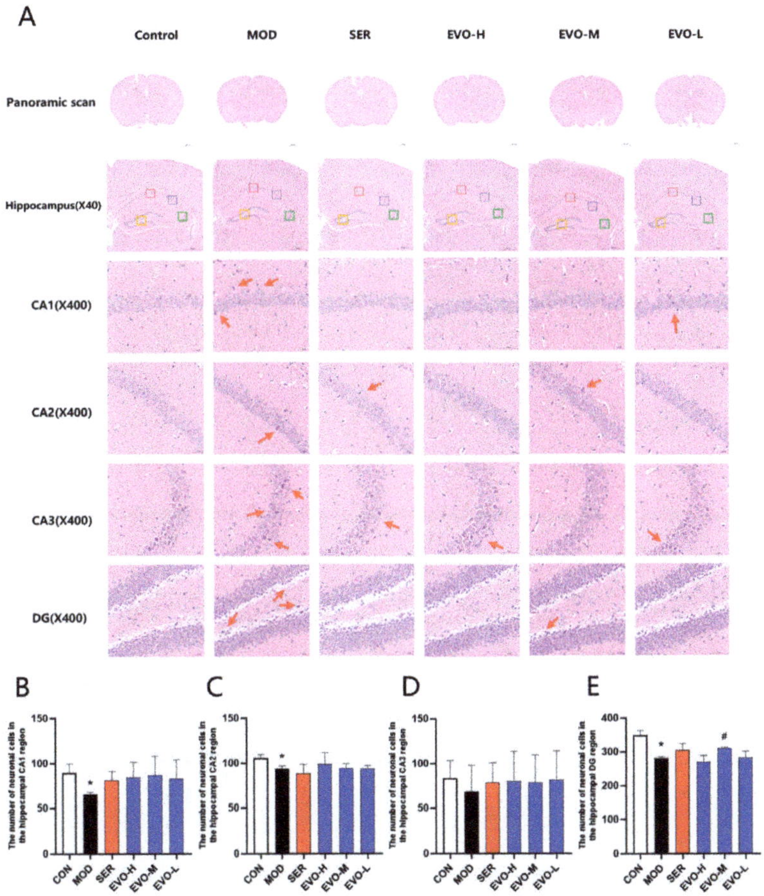

Figure 2. Effects of EVO on hippocampal morphology in SPS-induced PTSD mice. (**A**) The hematoxylin–eosin (H&E) staining was conducted on hippocampal tissue. Subsequent to staining, panoramic scanning was executed. The provided image displays HE staining of hippocampal regions CA1, CA2, CA3, and DG at a magnification of 400× (scale bar: 50 μm). The CA1 region is highlighted with a red box, CA2 with a blue box, CA3 with a green box, and the DG region with an orange box. (**B**) Quantification of neurons in the CA1 region of the hippocampus across each group of mice. (**C**) Evaluation of neurons in the CA2 region of the hippocampus within each group of mice. (**D**) Enumeration of neurons in the CA3 region of the hippocampus among all groups of mice. (**E**) Enumeration of neurons in the DG region of the hippocampus across each group of mice. Data are expressed as the means ± SDs (n = 3). Compared with the CON group, * $p < 0.05$; compared with the SPS group, # $p < 0.05$.

2.3. *Effect of EVO on Nissl Bodies in the Hippocampus of PTSD Mice Induced by SPS*

Nissl staining revealed that exposure to SPS resulted in a decrease in the number of Nissl bodies. However, the administration of EVO significantly ameliorated this effect (Figure 3). Specifically, in the hippocampal CA2 region, the Nissl bodies exhibited less pigmentation and a significant reduction in quantity in the model group than in the control group. Conversely, in the group treated with EVO, the coloration of neuronal Nissl bodie the CA2 region of the hippocampus deepened, and the number of such vesicles increas The results demonstrated that EVO could increase the number of Nissl bodies in t hippocampus of mice, providing a protective effect on the hippocampal neurons of m with SPS-induced PTSD.

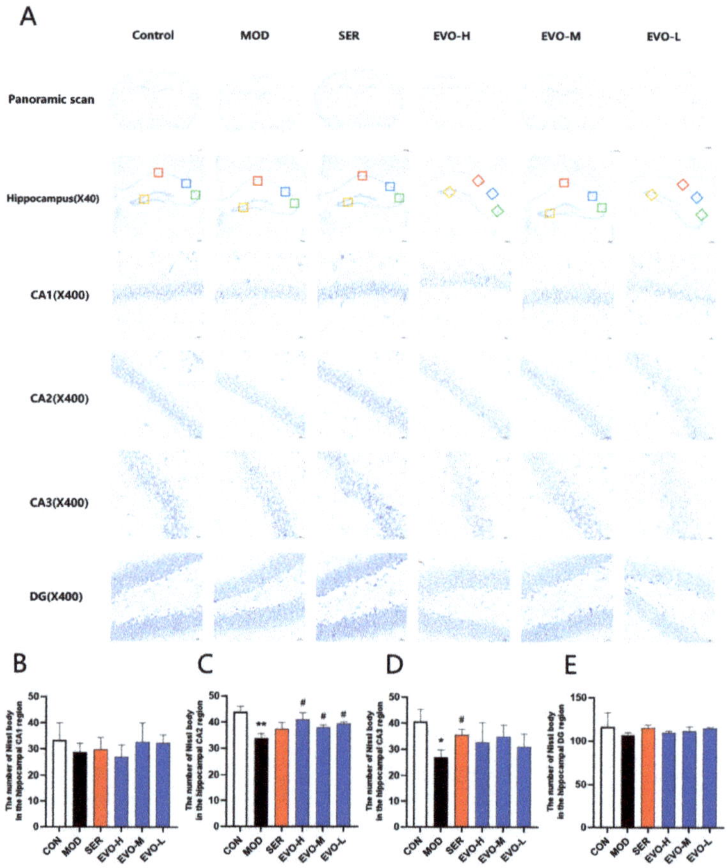

Figure 3. Effect of EVO on Nissl bodies in the hippocampus of PTSD mice induced by SPS. (**A**) H- pocampal Nissl staining representative images. Nissl staining images of the hippocampal CA1, C CA3, and DG regions are shown in the upper panel, which is a panoramic scan of the brain tiss after Nissl staining, 400× magnification (scale bar is 50 μm). A red box represents the CA1 regior blue box represents the CA2, a green box represents the CA3 region, and an orange box represents DG region. (**B**) Quantification of Nissl bodies in the CA1 region of the mouse hippocampus acr each group. (**C**) Assessment of Nissl bodies in the CA2 region of the mouse hippocampus wit each group. (**D**) Enumeration of Nissl bodies in the CA3 region of the mouse hippocampus amc all groups. (**E**) Enumeration of Nissl bodies in the DG region of the mouse hippocampus across e group. Data are expressed as the means ± SDs (n = 3). Compared with the CON group, * p < 0. ** $p < 0.01$; compared with the SPS group, # $p < 0.05$.

2.4. Network Pharmacology Reveals Potential Targets and Pathways Associated with Evodiamine Intervention in PTSD

To identify potential targets for treating PTSD, various disease target databases, such as GeneCards, OMIM, PharmGkb, TTD, and DrugBank, were consulted. A total of 2460 potential targets were identified from these databases. Further analysis using a Venn diagram identified 68 key targets that were common to both evodiamine and PTSD (Figure 4A). These targets, including ACE, REN, ADRB2, PTGS2, and INS, were identified as potential core targets of evodiamine in its therapeutic action on PTSD. The STRING database was used to construct a protein-protein interaction (PPI) network representing potential targets for evodiamine intervention in the treatment of PTSD (Figure 4B). The core of the PPI network was determined using the Cyto NCA plug-in, with filtering conditions set at values greater than the median. After two rounds of filtering, a total of five key targets were identified (ACE, REN, ADRB2, PTGS2, and INS), suggesting their significance in the therapeutic effects of EVO on PTSD (Figure 4C).

Gene Ontology (GO) functional enrichment analysis of the shared targets of evodiamine and PTSD revealed 1285 biological processes, 85 cellular components, and 103 molecular functions. Based on enrichment analysis results, evodiamine treatment for PTSD primarily affected tube diameter regulation, blood vessel diameter maintenance, and tube size control. Additionally, the molecular functions were primarily associated with synaptic membranes, postsynaptic membranes, and axon termini. In addition to neurotransmitter receptors, postsynaptic neurotransmitter receptors, and G protein-coupled amine receptors were all major components of the cell. These findings suggest that post-traumatic stress disorder (PTSD) involves various biological processes in living organisms and that evodiamine potentially exerts a therapeutic effect on PTSD by regulating these biological processes. The selection of the top 10 entries for the enrichment results bar and bubble plots was based on a significance level of $p < 0.05$ (Figure 4D).

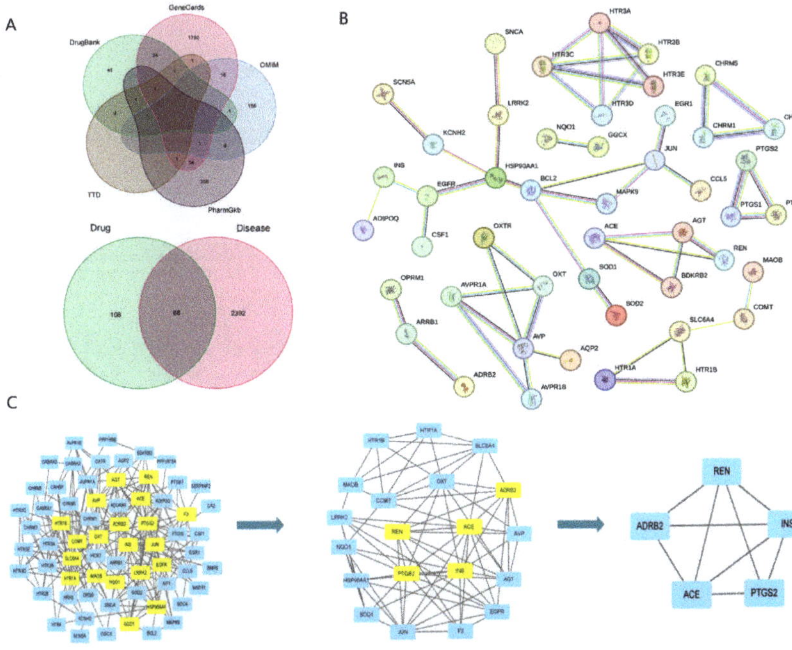

Figure 4. *Cont.*

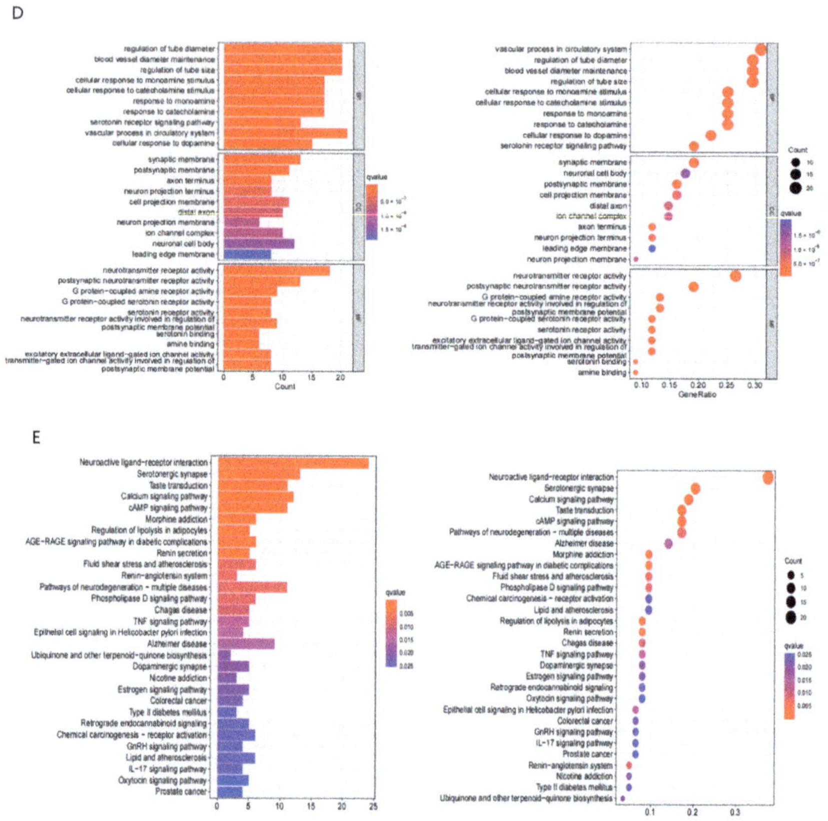

Figure 4. Network pharmacological analysis suggested potential anti-PTSD components in EV (**A**) Venn map of PTSD-associated genes combined with the Venn map and EVO-PTSD intersecti target genes. (**B**) Protein interaction network of EVO for PTSD. (**C**) Finding the core of the I network. (**D**) Results of GO enrichment plotted as bars and bubbles. (**E**) KEGG enrichment results bar and bubble plots. The length of the bar and the size of the bubble indicate the number of targ enriched, and the color of the bar and node from red to blue indicates the *p* value from large to sm Therefore, the darker the red color is the greater the significance of the signaling pathway, indicati the greater importance of the signaling pathway.

Evodiamine and PTSD share 41 signaling pathways with a *p* value of 0.05, according KEGG pathway enrichment analysis. The top 30 KEGG pathways, as determined throu screening, were chosen for the creation of bar and bubble plots (Figure 4E). Among the pathways, renin secretion and the renin–angiotensin system were notably relevant to t core targets.

2.5. Effect of EVO on Serum Metabolites in SPS-Induced PTSD

Serum samples were collected from mice in the normal group, SPS model group, a EVO low-dose administration group for nontargeted metabolomic analysis based on t behavioral results. PLS-DA results show differences between the control group and t model group, as well as between the model group and the low-dose EVO treatment gro (Figure 5A). This mode helps exclude irrelevant noise related to classification, thereby i proving the analytical robustness and credibility of the model. To further identify potent biomarkers that play a significant role in metabolic differentiation, an OPLS-DA model w established for both the ESI+ and ESI− modes (Figure 5B). The parameters of the OPLS-I replacement assay plot for the comparison between the OPLS-DA control and model groups in the ESI− mode were R2Y = 0.999 and Q2 = 0.256, while for the model group versus the EVO low-dose treatment group, the parameters were R2Y = 0.998 and Q2 = 0.497. These satisfactory parameters (R2Y and Q2) indicate reliable predictive ability for both aspects of the ESI model. In conclusion, the study findings validate the model constructed in negative ion mode. Significant differences were observed between the control and model groups and between the model and low-dose EVO treatment groups in negative ion mode. Moreover, EVO intervention significantly impacted the composition of serum metabolites in SPS mice. By applying thresholds of VIP > 1.0, multiplicity of difference (FC) > 1.5 or FC < 0.66, and p value < 0.05, the results of differential expression of metabolites in negative ion mode indicated the presence of one upregulated metabolite (Figure 5C). The model group exhibited one upregulated metabolite and two downregulated metabolites compared to those in the control group. Additionally, compared with the model group, the low-dose EVO treatment group displayed eight upregulated metabolites and twelve downregulated metabolites. Specifically, compared to the model group, the low-dose EVO treatment group exhibited eight upregulated and twelve downregulated metabolites. Notably, neuraminic acid was upregulated in the model group. Among the examined metabolites, neuraminic acid was upregulated in the model group and downregulated following EVO treatment. Cluster analysis of negative ion modeling for mid-differentially abundant metabolites in both the normal and model groups, as well as the administered and model groups, revealed that samples from the same group clustered together, suggesting a reasonably accurate screening for potential differentially abundant metabolites. (Figure 5D), enabling visualization of the contribution of metabolite differences across different subgroups. HMDB annotation and LIPID annotation analyses of differentially abundant metabolites in the negative ion model (Figure 5E) revealed the involvement of multiple small molecule metabolite pathways, including lipid and lipid-like molecules, organic acids and derivatives, organoheterocyclic compounds, benzenoids, phenylpropanoids, and polyketides. Furthermore, a variety of lipid structures and annotations were identified, such as steroids, bile acids and derivatives, steroid conjugates, sterols, secosteroids, and isoprenoids.

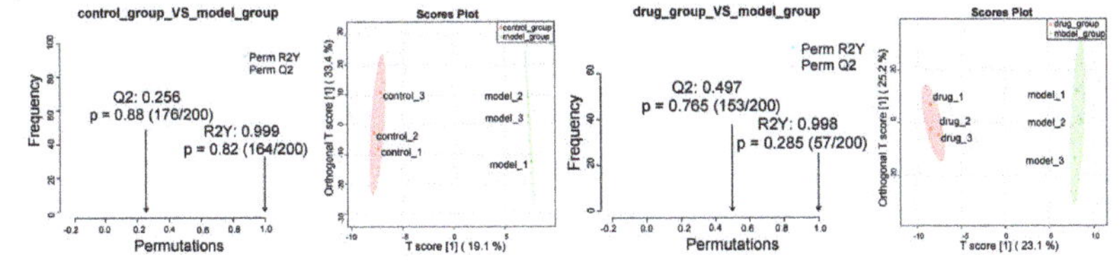

Figure 5. *Cont.*

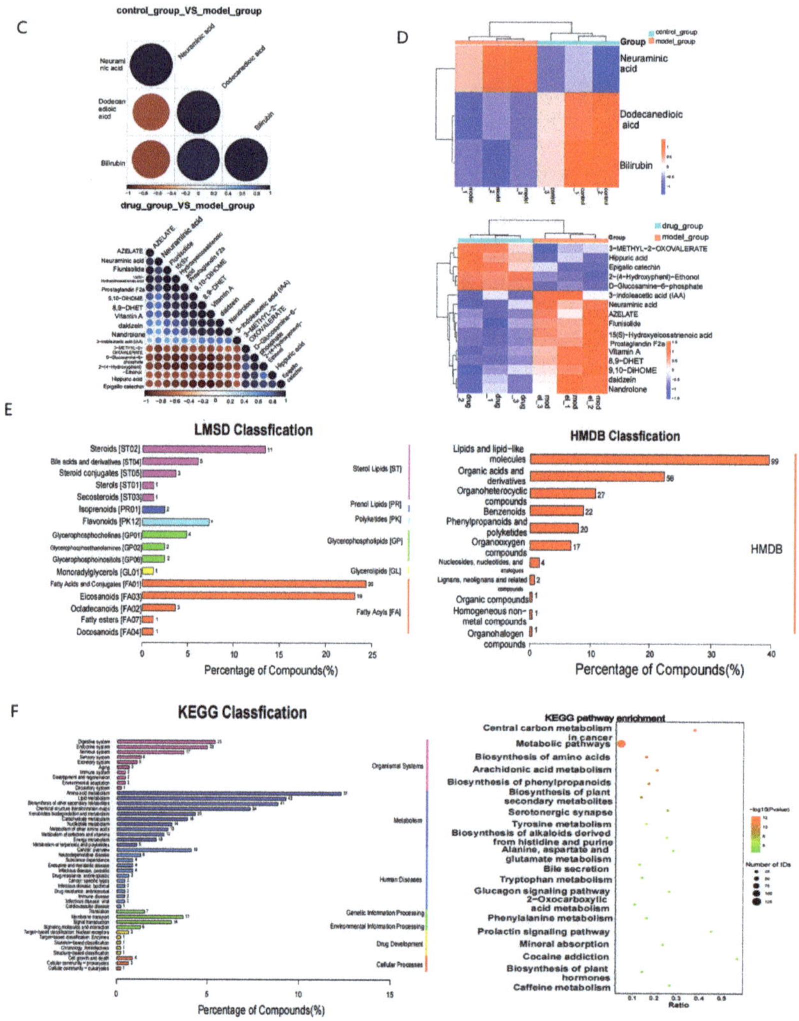

Figure 5. Nontargeted serum metabolomics analysis of the effects of EVO intervention on S stimulated mice. (**A**) Score plots of PLS-DA for the normal and model groups, as well as administered and model groups under negative ion mode. The horizontal axis represents the fi principal component, the vertical axis represents the second principal component and the numb following each principal component denotes the explained variance. (**B**) OPLS-DA substituti detection and score plots for the normal and model groups, and the administration and mo groups under negative ion mode. R2Y and Q2 indicate the explained variance and predictabi of the model, respectively. The horizontal axis of the plots represents the Spearman correlati coefficients between postdisruption and predisruption sample groupings. A correlation coeffici closer to 1 signifies a better fit. Typically, a Q2 value ≥ 0.2 is considered indicative of model reliabi (**C**) Correlation plots displaying differentially abundant metabolites between the normal and mo groups, as well as the administered and model groups in negative ion mode. For positive correlati the linear relationship tends toward 1 between two metabolites, whereas for negative correlati it tends toward -1. Significance testing for metabolite correlation analysis was conducted, wit chosen threshold of p value < 0.05 indicating a significant correlation. (**D**) Cluster analysis plots w generated for comparing metabolites with significant differences between the normal group and

model group, and between the administration group and the model group, in negative ion mode. The horizontal axis represents sample numbers and their clustering relationship, while the vertical axis indicates the clustering relationship of metabolites. (**E**) Statistical plot of HMDB and LIPID statistics in negative ion mode. The horizontal axis displays the percentage of each metabolite type relative to the total number of entries, and the left vertical axis represents subclass entries. (**F**) KEGG statistical plot and pathway enrichment bubble plot in negative ion mode. The KEGG statistics plot illustrates the percentage of each metabolite type relative to the total number of entries on the horizontal axis, with subclasses (subclass) on the left vertical axis and major classes (class) on the right vertical axis. The KEGG pathway enrichment bubble plot presents the ratio on the horizontal axis, individual pathway names on the vertical axis, and colors representing the degree of enrichment (−log10(p value)), with circle size indicating the number of metabolic pathways.

KEGG classification analysis revealed that the disease was associated with 45 metabolic pathway categories (Figure 5F). These pathways were primarily involved in the digestive system, endocrine system, nervous system, amino acid metabolism, and lipid metabolism. Further evaluation of the enrichment of KEGG-specific pathways revealed a total of 250 pathways. Among these pathways, central carbon metabolism in cancer, metabolic pathways, biosynthesis of amino acids, arachidonic acid metabolism, and biosynthesis of phenylpropanoids exhibited significant changes and were highly enriched. The results indicated that the metabolic marker neuraminic acid was upregulated in the model group and downregulated after EVO treatment. The KEGG classifications were primarily distributed across the digestive system, endocrine system, and nervous system.

2.6. *Intersection of Network Pharmacology and LC-MS/MS KEGG Pathways*

Through the intersection of 41 KEGG pathways acquired from network pharmacology with 250 KEGG pathways obtained from the negative ion mode of nontargeted metabolomics, a total of 24 pathways were identified (Figure 6A). Among these pathways, the renin secretion pathway was included. Furthermore, by considering the relationships between the potential targets identified from network pharmacology and the KEGG pathways, a more detailed analysis of the renin secretion pathway and the renin–angiotensin system was conducted, and the results were visually represented (Figure 6B). Additionally, the core targets within these pathway maps were identified and labeled accordingly (Figure 6C). The results demonstrated that EVO treatment for PTSD was associated with the renin secretion and renin–angiotensin system.

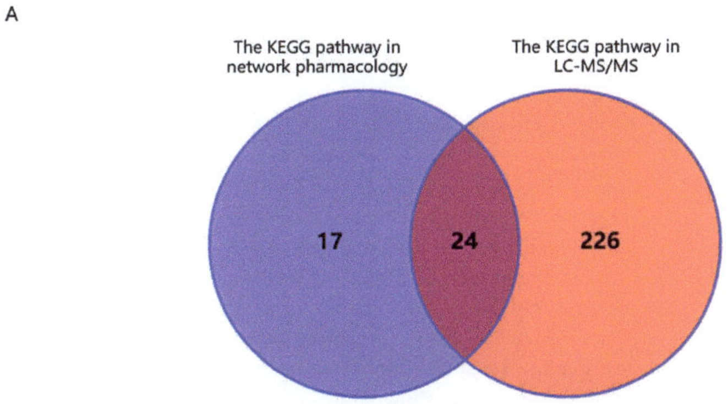

Figure 6. *Cont.*

B

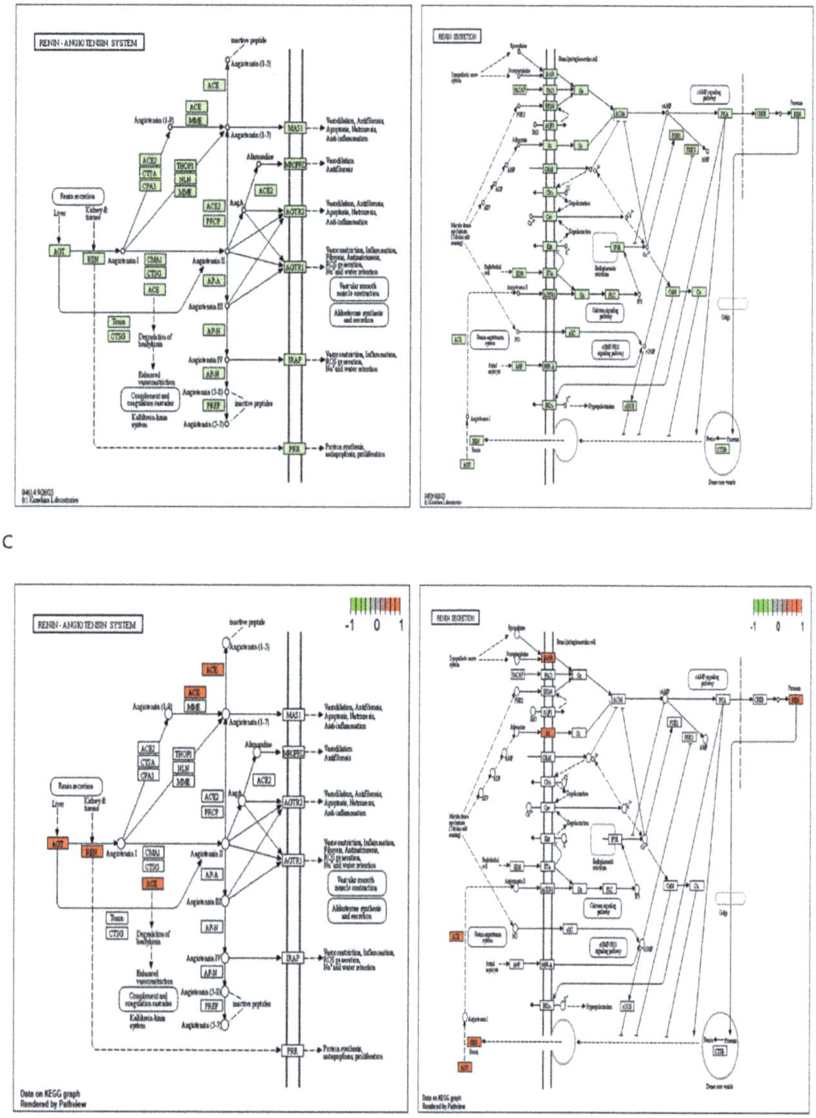

Figure 6. Network pharmacology and untargeted metabolomics of EVO intervention in PTSD ass ated with KEGG pathways. (**A**) Venn diagram of network pharmacology and off-target metabolom intersecting KEGG pathways. KEGG pathways 1–24: Dopaminergic synapse, cAMP signaling pa way, morphine addiction, parathyroid hormone synthesis, secretion and action, taste transducti regulation of lipolysis in adipocytes, longevity regulating pathway—multiple species, renin secreti serotonergic synapse, neuroactive ligand–receptor interaction, endocrine resistance, Type II diabe mellitus, GnRH signaling pathway, relaxin signaling pathway, estrogen signaling pathway, calci signaling pathway, phospholipase D signaling pathway, oxytocin signaling pathway, choliner synapse, sphingolipid signaling pathway, ubiquinone and other terpenoid–quinone biosynthe retrograde endocannabinoid signaling, ovarian steroidogenesis. (**B**) The renin secretion and ren angiotensin system pathway diagram. (**C**) Diagram of core target locations in the renin secretion a renin–angiotensin system pathways.

2.7. Molecular Docking Results for EVO and Potential Targets

Molecular docking was performed to determine whether EVO plays a significant role in regulating PTSD through relevant potential targets. The docking results show that EVO exhibits robust affinities for ACE, REN, ADRB2, and PTGS2, with binding energies of −8.5, −8.4, −8.5, and −9.7 kcal/mol, respectively. In ACE, the amino group of EVO forms hydrogen bonds with GLN-444 (Figure 7A). Similarly, in REN, the carbonyl group of EVO forms hydrogen bonds with TYR-83 (Figure 7B). In addition, the carbonyl group of EVO forms hydrogen bonds with THR-46 (Figure 7C). Moreover, the carbonyl group of EVO engages in hydrogen bonds with ARG-376, while its amino group forms hydrogen bonds with ASN-375 and GLY-225 (Figure 7D). After molecular docking, EVO exhibited a strong affinity for ACE, REN, ADRB2, and PTGS2.

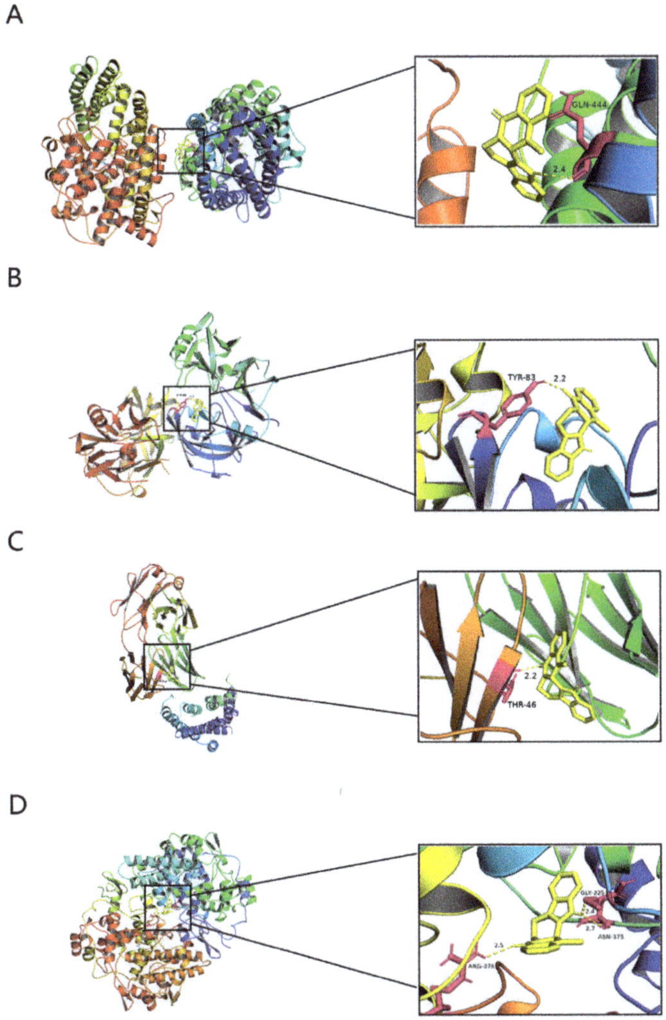

Figure 7. Molecular docking results. (**A**) A schematic diagram showing EVO's binding mode and binding site details with ACE. (**B**) A schematic diagram showing EVO's binding mode and binding site details with REN. (**C**) A schematic diagram showing EVO's binding mode and binding site details with ADRB2. (**D**) A schematic diagram showing EVO's binding mode and binding site details with PTGS2.

2.8. Effects of EVO on Core Target Levels and the Renin–Angiotensin Pathway in the Hippocampus of SPS-Induced PTSD Mice

Network pharmacology is core target screening and prediction through databas Molecular docking is to see whether EVO can be molecularly docked with the predict targets. We initially identified the core targets corresponding to EVO using netwo pharmacology and molecular docking techniques. To further confirm these potential co targets, we performed ELISA experiments. The results showed that ACE levels we significantly elevated in the model group, which was reversed by EVO treatment, b the expression of REN, ADRB2, and PTGS2 remained unchanged (Figure 8A). To furth confirm whether the changes in ACE expression were accurate, we further perform immunofluorescence experiments. The analysis demonstrated a significant elevation ACE expression in the CA1, CA2, CA3, and DG regions of the hippocampus in SPS-treat mice, while EVO and SER treatments effectively counteracted these effects (Figure 8B, Finally, we found that ACE is indeed a core target, and EVO treats PTSD by affecting AC expression and simultaneously affecting the RAS pathway.

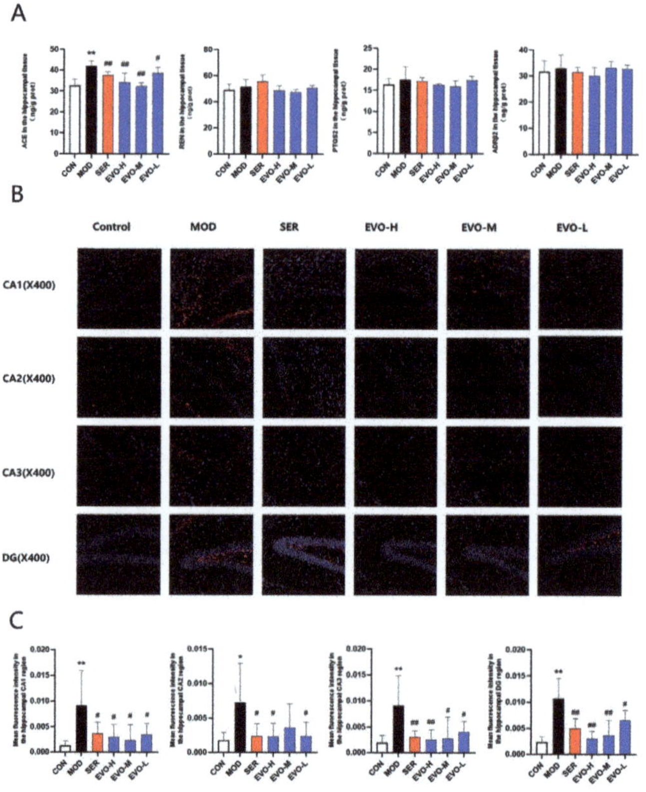

Figure 8. Effects of EVO on core target levels and the renin angiotensin pathway in the hippocamp of SPS-induced PTSD mice. (**A**) Evaluation of ACE, REN, ADRB2, and PTGS2 expression in hippocampus of mice across each group. (**B**) Immunofluorescence staining of ACE in the CA1, C CA3, and DG regions of the hippocampus in each group of mice; magnification ×400 (scale bar: 50 μ Immunofluorescence staining of mouse brain slices in the hippocampal region of mice stained bl and red for ACE expression. (**C**) Mean optical densities of the CA1, CA2, CA3, and DG regions the hippocampus of mice in each group. Data are presented as means ± standard deviations (n = Compared with the CON group, * $p < 0.05$, ** $p < 0.01$; compared with the SPS group, # p < 0 ## $p < 0.01$.

3. Discussion

This study is the first to evaluate the therapeutic effect of EVO in a PTSD model in SPS mice. The single-prolonged stress (SPS) model, as introduced by Liberzon et al., represents a significant advancement in the study of post-traumatic stress disorder (PTSD) by successfully replicating alterations in the hypothalamic–pituitary–adrenal (HPA) axis observed in PTSD patients. This animal model incorporates three distinct stressors—2 h of restraint, forced swimming, and ether anesthesia—to induce psychological, physiological, and endocrine stress responses, respectively. The combination of these stressors results in a significant increase in serum corticosteroid levels, mimicking the symptom severity of PTSD in experimental settings [16]. Network pharmacology and nontargeted serum metabolomics verified that EVO has the potential to act as an anti-PTSD agent by reducing ACE levels in the hippocampus of SPS mice. This reduction subsequently modulates the renin–angiotensin pathway. PTSD leads to cognitive impairment, reduced recognition [17], and spatial memory impairment [18]. Various pharmacological activities of EVO have been reported, including antinociceptive and cognitive enhancement effects [19]. Further, evodiamine induces antidepressant-like activity by increasing the expression of the serotonin transporter (5-HTT) [20]. Recent studies examining the renin–angiotensin system in trauma-exposed individuals have offered additional evidence supporting its potential involvement in both PTSD and CVD. For instance, elevated circulating levels of renin have been observed in the trauma-exposed group, with the highest levels present in the PTSD group compared to healthy controls [21]. Inhibition of the renin–angiotensin system through the use of ACE inhibitors (ACE Is) and angiotensin receptor blockers (ARBs) has been linked to decreased anxiety and stress [22]. Experimental trials on isolated rat aortae revealed that EVO demonstrated endothelium-dependent vasodilatory effects [23]. Moreover, EVO reduces blood pressure by inhibiting aldosterone secretion via Ang II-related pathways [24]. However, the mechanism by which EVO exerts its anti-PTSD effects through ACE interference exhibits both similarities and differences compared to those of ACE inhibitors.

In this study, mice subjected to SPS exhibited prolonged time to OFT central zone motor latency, reduced central zone distance traveled (Figure 1B), and elevated anxiety indices in the EPM test (Figure 1G). The EPM test, a widely used paradigm, offers an unbiased assessment of anxiety-like behaviors in rodents, measured by the percentage of time spent entering or exploring the open arms [25]. Rodents displaying anxiety-like behaviors typically avoid entering the open arms in the EPM test, while anxiolytic medications typically increase their exploration of these open arms [26]. According to the findings of this study, SPS-exposed animals displayed a significant reduction in both the percentage of time spent in the EPM and the duration of time spent exploring the open arms, indicating an anxious state. However, EVO administration increased both the distance traveled and the time spent entering the open arms (Figure 1F), suggesting potential anxiolytic and antidepressant effects. The observed increase in immobility time in both the TST and FST reflects anxious behaviors and learned helplessness in the animals following trauma exposure (Figure 1D,E). EVO administration significantly decreased immobility time in both tests, indicating substantial improvement in depressive-like behavior among SPS-exposed mice, similar to the positive control sertraline.

Cognitive impairment and depression often cooccur in humans, and the relationship between these conditions is substantial [27]. Thus, in this SPS model, we evaluated the impact of EVO on cognitive performance using the MWM test. Both the hippocampus and amygdala are implicated in anxiety-like and depressive-like behaviors [28], with the hippocampus being particularly vital for learning and memory processes [29]. Animals rely on various maze cues to remember the location of hidden platforms in a pool, thereby completing the hippocampus-dependent spatial learning and memory challenge known as the MWM test [30]. Memory was quantified as the duration spent in the platform area during the test session after platform removal, while learning was measured by the decrease in time taken to locate the hidden platform across sessions [31]. The findings of this study indicate that prolonged exposure of mice to SPS stimuli in the MWM task l to heightened transfer latencies and reduced quadrant dwell times, demonstrating th EVO substantially mitigated hippocampus-dependent spatial learning and memory defic (Figure 1H). Nonetheless, in the drug-treated group, escape latency consistently decreas with the progression of training days, exhibiting significant improvement by the fou day (Figure 1H). Furthermore, mice treated with EVO showed a significant increase in t number of journeys and entries within the quadrant, a key measure of spatial memo retention (Figure 1H). Therefore, the findings of this study suggest that EVO treatme holds promise for mitigating SPS-induced memory and cognitive deficits.

The single prolonged stress (SPS) model is widely regarded as an appropriate anim model for PTSD and has been linked to hippocampal atrophy in confirmed patients [3 Although PTSD is commonly understood as a prolonged, maladaptive stress respor that does not directly affect the central nervous system [33], neuroimaging studies ha revealed abnormal increases in activity within the cerebral cortex, hippocampus, a corpus callosum, along with alterations in gray matter volume and density in some PT patients [34]. Untargeted serum metabolomics results showed that the levels of neurami acid in serum were elevated in the model group compared to the normal group, and E treatment reduced neuraminic acid levels in the SPS model group. Studies have shown th neuraminic acid belongs to the family of salivary acids, which are based on neuraminic ac (Neu), which is usually found in the form of *N*-acetyl (NeuAc) or *N*-ethanolyl (NeuG or 2-keto-3-deoxynonanoic acid (Kdn) [35]. Additional studies have also shown th the term sialic acid (neuraminic acid) is used to refer to any member of this family [3 Serum sialic acid (SSA) levels are abnormally high in pathological states exhibiting tiss destruction, tissue proliferation, or inflammation [37]. Whereas post-traumatic stre disorder (PTSD) is associated with increased inflammation. c-reactive protein (CRP) a marker of systemic inflammation, and single nucleotide polymorphisms (SNPs) in t CRP gene have recently been associated with elevated blood CRP protein levels a increased disease severity in patients with PTSD [38]. There is also a significant correlati between serum salivary acid levels and cardiovascular disease (CVD) risk factors. patients with CVD, elevated levels of SSA may be due to the release of free and bou sialic acid from damaged cardiomyocytes or vascular endothelium [39]. Since the ren angiotensin system is involved in sympathetic arousal, it is directly associated with PT and CVD [13]. The EVO treatment group downregulated serum sialic acid and modulat the renin–angiotensin system, thus serum sialic acid has the potential to be a marker the effectiveness of EVO in the treatment of PTSD, but further validation and resear are needed. The hippocampus and prefrontal cortex, recipients of inputs from multip sensory systems and higher cortical regions, are integral to the processes of learning a memory, while also playing crucial roles in regulating emotions and behaviors. Stud utilizing magnetic resonance imaging have demonstrated that individuals diagnosed w PTSD exhibit a reduction in hippocampal volume compared to unaffected controls, wi reductions reaching up to 8%. Furthermore, the extent of this reduction is found to directly correlated with the severity of PTSD symptoms [40]. In this paper, we found significant reduction in the number of neuronal cells in the CA1, CA2, CA3, and DG regio of the hippocampus in the SPS model through pathological observation, mild abnormalit in brain organization, and a small number of neuronal cells in the hippocampus in the fie of view were degenerated with deepened staining, and clear cytosolic bodies were n seen. While asymmetrical alterations in activity and augmented prefrontal cortex volu have been noted in certain studies involving PTSD patients, contrasting findings ha documented cortical thinning and white matter atrophy in others [41]. Therefore, stress thought to exert partial influence on the hippocampus and prefrontal cortex, implicati them in the pathogenesis of PTSD. Thus, the current study investigates behavioral chang and hippocampal neuron alterations in mice subjected to SPS [42]. Studies utilizing bra imaging techniques in PTSD patients have unveiled a reduction in hippocampal volume phenomenon that correlates with the severity of exposure to traumatic environments [4

Moreover, mice exposed to SPS displayed a notable reduction in the Nissl bodies, while EVO treatment resulted in an elevation of Nissl bodies in the CA2 region of the hippocampus post-SPS exposure (Figure 3). Ultimately, hippocampal neuronal apoptosis is intricately linked to PTSD symptoms.

The core targets ACE, REN, ADRB2, PTGS2, and INS were screened by network pharmacology for the treatment of PTSD with EVO. ACE, REN, and ADRB2 were enriched in renin secretion and the renin–angiotensin pathway (Figure 6B). However, in addition to INS, we further investigated ACE, REN, ADRB2, and PTGS2. To validate the KEGG pathway identified through network pharmacology and assess EVO absorption into the bloodstream post-administration for its anti-PTSD effects in mice, we measured serum metabolites in EVO-treated mice from the normal, SPS-stimulated, and low-dose groups. We identified 43 differentially expressed metabolites that are frequently associated with PTSD-related pathways, including amino acid biosynthesis and the arachidonic acid metabolic pathway (Figure 5). Subsequently, by integrating the results of metabolomics and network pharmacology and comprehensively exploring the underlying mechanisms [44], we observed that both KEGG pathway analyses included the renin secretion pathway (Figure 6A). Additionally, serum metabolomics revealed greater endocrine correlations in lipid metabolism pathways in both the SPS model group and the EVO-treated group. In conclusion, EVO treatment for PTSD is associated with renin secretion and the renin–angiotensin system.

AT1R and ACE signaling hyperactivation exacerbates cognitive impairment, cell death, and inflammation in neurons [45]. AT1R and ACE signaling exacerbate oxidative stress in areas critical for cognitive function, such as the cortex, hippocampus, and basal ganglia. Following ischemic injury, activation of AT1Rs increases cholinergic and noncholinergic cell death in the rodent cortex and hippocampus [46]. Acetylcholine release from cholinergic neurons is decreased when ACE expression is upregulated [47]. Cholinergic cell death and dysfunction are frequently observed features of cognitive impairment [48]. Transgenic mice with three copies of the ACE gene were previously utilized to characterize the RAS system [49]. These transgenic mice exhibited high ACE activity, no change in blood pressure, and impaired short- and long-term memory, as measured by the new object recognition (NOR) test [50]. The molecular docking results demonstrated that EVO had a strong affinity for ACE, REN, ADRB2, and PTGS2 (Figure 6A). This finding was further verified by ELISA, which revealed that the hippocampal ACE level was significantly increased in SPS mice. Both the EVO group and the sertraline group showed a decrease in the ACE level. Similarly, immunofluorescence analysis revealed that the integrated optical density (IOD) of ACE in the CA1, CA3, and DG regions was greater in the SPS group than in the control group. However, EVO treatment reversed this phenomenon. These results suggest that EVO can modulate the renin–angiotensin pathway by downregulating hippocampal ACE levels in SPS mice and exerting a therapeutic effect on PTSD.

4. Materials and Methods

4.1. Animals

Healthy male C57BL/6 mice, weighing between 18 and 20 g, were obtained from Beijing Vitality River Laboratory Animal Technology Co., Ltd, Beijing, China (license number SCXK [Beijing] 2021-0006). These mice were individually housed in cages maintained at a temperature of 24–25 °C and a humidity level of 50 ± 5%, following a 12 h day-night cycle. Prior to the experiment, the mice were acclimatized for one week.

4.2. Stimulation of the SPS Model

With slight modifications, the SPS model was generated as previously described [51]. The SPS procedure comprised three stages: restraint (2 h), forced swimming (20 min), and ether anesthesia. Initially, mice were confined in a restraint tube for 2 h, ensuring unimpeded breathing through the smaller end of the cone. Subsequently, mice were subjected to forced swimming in a transparent acrylic cylinder (diameter: 240 mm, height: 500 mm) filled to two-thirds capacity with water at 24 °C, for a duration of 20 min. Following a 15 min respite, mice were anesthetized with diethyl ether until reaching a state unconsciousness, after which they were returned to their respective cages.

4.3. Experimental Design

The mice were randomly allocated into six groups (n = 8): the control group, S stimulation model group, SER treatment group (15 mg/kg/day, i.p.), EVO low-dose gro (5 mg/kg/day, i.p.), EVO medium-dose group (10 mg/kg/day, i.p.), and EVO high-d group (20 mg/kg/day, i.p.). Sertraline (15 mg/kg, i.g.) was administered as a positi control, based on previous PTSD studies [52]. The three doses of EVO (high, medium, a low) were selected through pre-experimentation screening. At the conclusion of the dosi period, the mice were euthanized, and samples were collected for subsequent stud Behavioral experiments were conducted on days 15, 16, 17, 18, and 19 (Figure 1A).

4.4. Behavioral Tests

4.4.1. Open Field Test (OFT)

A well-established test for assessing mice's locomotor activity, exploratory hab and depression is the open field test [53]. A black square-bottomed open box, measuri 50 cm by 50 cm by 35 cm, supplied by Shanghai Xinruan Information Technology Co., L Shanghai, China, was utilized for the test. The mice were gently placed in the center the square for a duration of 6 min. This period consisted of a 1 min acclimatization pha followed by a 5 min observation phase. During the 5 min observation phase, the time sp by the mice in the center and their movement patterns were recorded. Subsequently 20% alcohol solution was employed to eliminate any lingering odors from the box af each test.

4.4.2. Tail Suspension Test (TST)

The TST is a commonly used behavioral test to assess depressive-like behaviors mice [54]. Using tape placed approximately 0.5 cm from the tail tip, each mouse's t was suspended from the edge of a shelf (Shanghai Xinruan Information Technology C Ltd., Shanghai, China). Video cameras were placed in front of the animals to track th movements while they were immobilized for five minutes. The TST experiment reflec anxious behavior and learned helplessness by immobility time.

4.4.3. Elevated Plus Maze Test (EPMT)

The EPMT evaluates exploratory and anxiety behaviors in mice [55]. The maze consist of two open arms, two closed arms, and a central area (dimensions: 35 cm × 5 cm × 10 supplied by Shanghai Xinruan Information Technology Co., Ltd., Shanghai, China). M were positioned in the central area, facing an open arm, and permitted to explore fo min. Measurements included the distance traveled and frequency of entries into the op arm. Cleaning between sessions was conducted using a 20% ethanol solution. The EI experiment reflects depressive behavior by detecting distance in the open arms and ti spent in the open arms.

4.4.4. Forced Swimming Test (FST)

The FST assesses the despair behavior of mice [56]. The mice were gently plac in a glass cylinder filled with water at 25 °C to a depth of 20 cm (Shanghai Xinru Information Technology Co., Ltd., Shanghai, China). In the last 5 min of the 6 min test, t duration of immobilization was recorded. Software was used to analyze the moveme of the mice using a video camera. The FST experiment reflected depressive behavior immobility time.

4.4.5. Morris Water Maze (MWM) Test

The MWM evaluates spatial learning and memory abilities [57]. It consisted o circular pool with a diameter of 120 cm and a depth of 50 cm (Shanghai Xinruan Informati

Technology Co., Ltd., Shanghai, China). In the pool, a submerged platform (9 cm in diameter) was located. The water temperature was maintained at room temperature (25 ± 1 °C). Mouse swimming activities were monitored and recorded using a video analysis system. Each mouse underwent three training sessions per day with a 30 s rest between sessions. On days 1–5, the mice were released into the water from the two farthest quadrants from the platform. Navigating was allowed for 60 s for each mouse.

4.5. Histopathologic Analysis

Prior to HE staining, mouse brain tissues were fixed in 4% paraformaldehyde. The tissues were then dehydrated in varying concentrations of alcohol, embedded, and sectioned. Following dewaxing, the paraffin sections were stained with hematoxylin for nuclei and eosin for cytoplasm, dehydrated again, and sealed. The morphological changes in mouse brain tissue were assessed with an orthogonal optical microscope (Nikon Eclipse ci, Tokyo, Japan).

4.6. Nissl Staining and Neuron Counting

In order to evaluate neuronal damage in the hippocampal area, Nissl staining was performed. In brief, dewaxed hippocampal slices were stained using Nissl stain solution (toluidine blue method) according to the manufacturer's instructions. The slices were then imaged using an orthogonal optical microscope (Nikon Eclipse ci, Tokyo, Japan) after dehydration, hyalinization, and mounting. Under a microscope, cellular morphological changes were observed in hippocampal CA1, CA2, CA3, and DG regions.

4.7. Network Pharmacological Analysis

4.7.1. Collection and Screening of EVO Targets

Traditional Chinese Medicine System Pharmacology (TCMSP) database (https://tcmspw.com/tcmsp.php, accessed on 16 July 2023) was used to identify potential targets of EVO. During screening, the criteria used were an oral bioavailability (OB) $\geq$ 30% and a drug-likeness (DL) $\geq$ 0.18. Furthermore, the Bioinformatics Analysis Tool for Traditional Chinese Medicine Molecular Mechanisms (BATMAN-TCM) database (http://bionet.ncpsb.org/batman-tcm/, accessed on 16 July 2023) was used for screening, with a score of $\geq$20 and $p \leq 0.05$ as the screening conditions. For further processing and analysis, detailed information on the relevant potential targets was compiled.

4.7.2. Collection of Disease Targets

The disease targets were obtained through the GeneCards (https://www.genecards.org/, accessed on 16 July 2023), OMIM (https://www.omim.org/, accessed on 16 July 2023), PharmGkb (https://www.pharmgkb.org/, accessed on 16 July 2023), TTD (http://db.idrblab.net/ttd/, accessed on 16 July 2023), and DrugBank (https://www.drugbank.ca/, accessed on 16 July 2023) databases. The disease genes from the databases were merged using R language (R version 4.4.0), redundant gene entries were eliminated, and ultimately, PTSD-related genes were obtained.

4.7.3. Constructing the PPI Network Diagram and Obtaining Core Targets

The gene symbols of the EVO and PTSD targets were uploaded to the STRING database (https://string-db.org/, accessed on 16 July 2023) to construct the protein interaction network. The species was specified as "Homo sapiens", and the minimum interaction threshold was set to "highest confidence". The network's core was identified using the Cyto NCA plug-in by applying a filtering condition greater than the median value, leading to the identification of core targets after two rounds of filtering.

4.7.4. Analysis of GO and KEGG Pathway Enrichment

The target proteins of EVO and PTSD were analyzed using R for GO function and KEGG pathway enrichment. GO function enrichment analysis encompassed BP, CC, and

MF, with the top 10 entries selected based on $p < 0.05$ for visual analysis. KEGG pathw enrichment analysis identified the top 30 signaling pathways with $p < 0.05$ for visu analysis, shedding light on the mechanisms of pain and diarrhea treatment in PTSD. Th analysis was crucial for understanding the biological functions and associated signali pathways of EVO in treating PTSD.

4.8. LC-MS/MS-Based Serum Metabolomics Studies

After the behavioral analysis, eyeballs were extracted from each group of mice, a blood samples were obtained. Blood from the Con, Mod, and EVO_L (0.5 mg/kg/ groups underwent centrifugation at 12,000 rpm for 15 min at 4 °C. Serum samples (50 μ from these groups were then transferred to 1.5 mL EP tubes placed on ice, followed by t addition of 200 μL of cold methanol containing an internal standard. Following 2 min vortexing, the samples were cooled for 10 min and subsequently centrifuged at 14,000× for 15 min at 4 °C. The resulting supernatant (200 μL) was transferred to a new EP tu then either lyophilized or concentrated via low-temperature centrifugation, and stor at −20 °C. Before online analysis, the concentrated metabolite extract was dissolved 100 μL of 20% methanol/water solution until completely dissolved, and the supernata was collected after centrifugation for analysis in both positive and negative ion modes.

4.9. Molecular Docking

The process of molecular docking is initiated with the retrieval of 2D structures small molecule ligands from the PubChem database (https://pubchem.ncbi.nlm.nih.go accessed on 17 July 2023). Converting these 2D structures into 3D formats follow using ChemOffice 14.0. Subsequent to this, the protein structures of the core targets we sourced from the PDB database (http://www.rcsb.org/, accessed on 17 July 2023) a visualized utilizing PyMol 2.2.0. Downloading the core target protein structures fro the PDB database, the core protein was then isolated from the proto-ligand employi PyMol 2.4.0. Following isolation, the protein underwent hydrogenation, and Autodo 1.5.6 software was used to configure its atom types. Retrieving small molecule structu from the TCMSP database, they were saved as "mol2" files and subsequently employed AutoDock and Vina for conducting the docking simulations.

4.10. Enzyme-Linked Immunosorbent Assay (ELISA)

ELISA kits for ACE, REN, ADRB2, and PTGS2 were procured from Jiangsu Enzy Immunity located in Nanjing, China. The levels of these proteins in the hippocampal sues of each mouse group were assessed using ELISA kits according to the manufacture instructions. Optical density was measured at 450 nm with an enzyme-linked immunos bent assay (ELISA) reader. Subsequently, the levels of ACE, REN, ADRB2, and PTGS2 hippocampal tissues were determined based on the standard curve.

4.11. Immunofluorescence

For antigen retrieval, mouse brain tissue sections were subjected to immersion in a b containing EDTA antigen repair buffer (pH 8.0) within a microwave oven. Following th step, the sections were gently dried by shaking and a circular boundary was marked arou the tissue using a histochemical pen to prevent antibody runoff. Next, an autofluorescen quencher was applied to the delineated circle for 5 min, followed by rinsing under runni water for 10 min and subsequent incubation in serum for 30 min. Subsequently, rab anti-ACE (1:100, Proteintech, Wuhan, China) was added to the sections, which were th left to incubate overnight at 4 °C. Following this incubation, the sections underwent furth incubation with a secondary antibody for 50 min at room temperature in the absence light. Finally, the sections were stained with diaminophenylindole (DAPI) and observ under an orthogonal light microscope (Nikon Eclipse ci, Tokyo, Japan).

4.12. Statistical Analysis

The study's data are expressed as mean values accompanied by their respective standard deviations. To compare data between two groups, a *t*-test was employed, while for comparisons among multiple groups, one-way analysis of variance (ANOVA) was utilized. Graphs were generated using GraphPad Prism 8.0.2 software. Statistical significance was defined as a *p* value below 0.05.

5. Conclusions

The present study pioneered the demonstration of the therapeutic effects of EVO in a PTSD model in SPS mice. Behavioral studies confirmed that EVO has anxiolytic and antidepressant effects and enhances memory and cognition in a behavioral task after SPS. The protective effect of EVO on hippocampal neurons and the attenuation of hippocampal neuron apoptosis in SPS mice were confirmed by HE staining and Nysted staining. In terms of mechanistic studies, mutual validation via network pharmacology and nontargeted serum metabolomics demonstrated that the improvements in these behavioral indices after SPS may be related to the modulatory effect of EVO on the renin angiotensin pathway. Additionally, using molecular docking, ELISA, and immunofluorescence staining, it was verified that EVO could regulate the renin angiotensin pathway by decreasing the ACE level in the hippocampus of SPS mice, thus exerting an anti-PTSD effect. This study is limited by the lack of clarification regarding the relationship between the upstream and downstream pathways of EVO in the treatment of PTSD. The specific mechanism by which EVO exerts its therapeutic effects through regulating the ACE-regulated RAS pathway has not been experimentally validated in both in vivo and in vitro settings, necessitating further research to comprehensively elucidate the mechanism of EVO's anti-PTSD effects via the ACE and renin angiotensin pathways. Furthermore, the specific site or sites within hippocampal tissue or blood where EVO primarily exerts its effects remain to be elucidated, a critical consideration for the advancement of drug development (Figure 9).

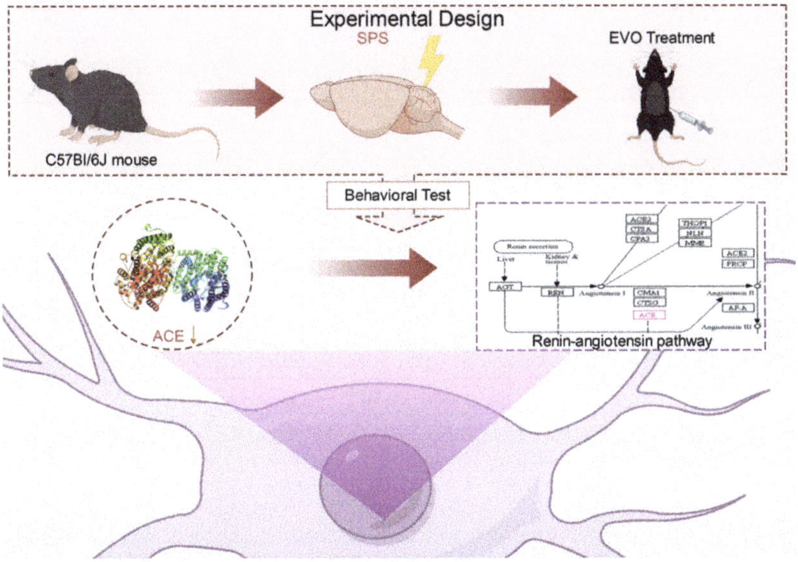

Figure 9. Mechanisms of evodiamine intervention in behavioral and hippocampal neurons in a mouse model of post-traumatic stress disorder.

Author Contributions: Z.W. and C.L. were responsible for the study design, statistical analysis, a manuscript preparation. Z.W., B.S., B.L., J.C., X.S. and Z.H. were responsible for performing experiments. Z.W., Y.G., C.Y. and B.L. were involved in evolving the ideas and editing the manuscr Y.G. and C.L. were responsible for providing the funding for the study. All authors have read a agreed to the published version of the manuscript.

Funding: This work was supported by the National Natural Science Foundation of China (82204751) and the Innovation Team and Talents Cultivation Program of the National Administrati of Traditional Chinese Medicine (No. ZYYCXTD-D-202207).

Institutional Review Board Statement: All animal experimental protocols were reviewed and proved by the Institutional Animal Care and Use Committee of the Beijing Institute of Radiati Medicine (permit number IACUC-DWZX-2022-688). Full efforts were made to reduce animal and improve animal welfare.

Informed Consent Statement: Informed consent was obtained from all subjects involved in study.

Data Availability Statement: The data presented in this study are available on request from corresponding author.

Conflicts of Interest: There are no conflicts of interest to declare.

References

1. Lee, B.; Choi, G.M.; Sur, B. Silibinin prevents depression-like behaviors in a single prolonged stress rat model: The possible role serotonin. *BMC Complement Med. Ther.* **2020**, *20*, 70. [CrossRef] [PubMed]
2. Feduccia, A.A.; Jerome, L.; Yazar-Klosinski, B.; Emerson, A.; Mithoefer, M.C.; Doblin, R. Breakthrough for Trauma Treatme Safety and Efficacy of MDMA-Assisted Psychotherapy Compared to Paroxetine and Sertraline. *Front. Psychiatry* **2019**, *10*, 6 [CrossRef] [PubMed]
3. Yunitri, N.; Chu, H.; Kang, X.L.; Jen, H.J.; Pien, L.C.; Tsai, H.T.; Kamil, A.R.; Chou, K.R. Global prevalence and associated r factors of posttraumatic stress disorder during COVID-19 pandemic: A meta-analysis. *Int. J. Nurs. Stud.* **2022**, *126*, 1041 [CrossRef] [PubMed]
4. Sartori, S.B.; Singewald, N. Novel pharmacological targets in drug development for the treatment of anxiety and anxiety-rela disorders. *Pharmacol. Ther.* **2019**, *204*, 107402. [CrossRef] [PubMed]
5. Stübner, S.; Grohmann, R.; Greil, W.; Zhang, X.; Müller-Oerlinghausen, B.; Bleich, S.; Rüther, E.; Möller, H.J.; Engel, R.; Falkai et al. Suicidal Ideation and Suicidal Behavior as Rare Adverse Events of Antidepressant Medication: Current Report from AMSP Multicenter Drug Safety Surveillance Project. *Int. J. Neuropsychopharmacol.* **2018**, *21*, 814–821. [CrossRef] [PubMed]
6. Zhao, T.; Zhang, X.; Zhao, Y.; Zhang, L.; Bai, X.; Zhang, J.; Zhao, X.; Chen, L.; Wang, L.; Cui, L. Pretreatment by evodiamin neuroprotective in cerebral ischemia: Up-regulated pAkt, pGSK3β, down-regulated NF-κB expression, and ameliorated B permeability. *Neurochem. Res.* **2014**, *39*, 1612–1620. [CrossRef] [PubMed]
7. Yuan, S.M.; Gao, K.; Wang, D.M.; Quan, X.Z.; Liu, J.N.; Ma, C.M.; Qin, C.; Zhang, L.F. Evodiamine improves congnitive abili in SAMP8 and APP(swe)/PS1(ΔE9) transgenic mouse models of Alzheimer's disease. *Acta Pharmacol. Sin.* **2011**, *32*, 295–3 [CrossRef] [PubMed]
8. Chang, C.P.; Chang, J.Y.; Wang, F.Y.; Tseng, J.; Chang, J.G. The effect of Evodia rutaecarpa extract on cytokine secretion by hum mononuclear cells in vitro. *Am. J. Chin. Med.* **1995**, *23*, 173–180. [CrossRef] [PubMed]
9. Shi, L.; Yang, F.; Luo, F.; Liu, Y.; Zhang, F.; Zou, M.; Liu, Q. Evodiamine exerts anti-tumor effects against hepatocellular carcino through inhibiting beta-catenin-mediated angiogenesis. *Tumour. Biol.* **2016**, *37*, 12791–12803. [CrossRef]
10. Wei, J.; Ching, L.C.; Zhao, J.F.; Shyue, S.K.; Lee, H.F.; Kou, Y.R.; Lee, T.S. Essential role of transient receptor potential vanill type 1 in evodiamine-mediated protection against atherosclerosis. *Acta Physiol.* **2013**, *207*, 299–307. [CrossRef]
11. Wu, Y.; Pan, X.; Xu, Y.; Lu, X.; He, S.; He, R.; Gong, M. Optimization of combinations of ginsenoside-Rg1, ginsenoside-R evodiamine and rutaecarpine for effective therapy of mouse migraine. *J. Nat. Med.* **2016**, *70*, 207–216. [CrossRef]
12. Smith, H.C.; Yu, Z.; Iyer, L.; Marvar, P.J. Sex-dependent effects of angiotensin type 2 receptor expressing medial prefrontal cor (mPFC) interneurons in fear extinction learning. *Biol. Psychiatry Glob. Open Sci.* 2023; *in press*. [CrossRef]
13. Seligowski, A.V.; Duffy, L.A.; Merker, J.B.; Michopoulos, V.; Gillespie, C.F.; Marvar, P.J.; Stein, M.B.; Ressler, K.J. The ren angiotensin system in PTSD: A replication and extension. *Neuropsychopharmacology* **2021**, *46*, 750–755. [CrossRef]
14. Osborn, J.W.; Fink, G.D. Region-specific changes in sympathetic nerve activity in angiotensin II-salt hypertension in the rat. *E Physiol.* **2010**, *95*, 61–68. [CrossRef]
15. Wang, D.; Wang, C.; Liu, L.; Li, S. Protective effects of evodiamine in experimental paradigm of Alzheimer's disease. *Co Neurodyn.* **2018**, *12*, 303–313. [CrossRef]
16. Yamamoto, S.; Morinobu, S.; Takei, S.; Fuchikami, M.; Matsuki, A.; Yamawaki, S.; Liberzon, I. Single prolonged stress: Toward animal model of posttraumatic stress disorder. *Depress Anxiety* **2009**, *26*, 1110–1117. [CrossRef]

Keller, S.M.; Schreiber, W.B.; Stanfield, B.R.; Knox, D. Inhibiting corticosterone synthesis during fear memory formation exacerbates cued fear extinction memory deficits within the single prolonged stress model. *Behav. Brain Res.* **2015**, *287*, 182–186. [CrossRef]
Schoenfeld, T.J.; Rhee, D.; Martin, L.; Smith, J.A.; Sonti, A.N.; Padmanaban, V.; Cameron, H.A. New neurons restore structural and behavioral abnormalities in a rat model of PTSD. *Hippocampus* **2019**, *29*, 848–861. [CrossRef]
Heo, S.-K.; Yun, H.-J.; Yi, H.-S.; Noh, E.-K.; Park, S.-D. Evodiamine and rutaecarpine inhibit migration by LIGHT via suppression of NADPH oxidase activation. *J. Cell. Biochem.* **2009**, *107*, 123–133. [CrossRef] [PubMed]
Hu, Y.; Ehli, E.A.; Hudziak, J.J.; Davies, G.E. Berberine and evodiamine influence serotonin transporter (5-HTT) expression via the 5-HTT-linked polymorphic region. *Pharmacogenomics J.* **2012**, *12*, 372–378. [CrossRef] [PubMed]
Terock, J.; Hannemann, A.; Janowitz, D.; Freyberger, H.J.; Felix, S.B.; Dorr, M.; Nauck, M.; Volzke, H.; Grabe, H.J. Associations of trauma exposure and post-traumatic stress disorder with the activity of the renin-angiotensin-aldosterone-system in the general population. *Psychol. Med.* **2019**, *49*, 843–851. [CrossRef]
Zanchetti, A.; Elmfeldt, D. Findings and implications of the Study on COgnition and Prognosis in the Elderly (SCOPE)—A review. *Blood Press* **2006**, *15*, 71–79. [CrossRef]
Chiou, W.F.; Chou, C.J.; Shum, A.Y.; Chen, C.F. The vasorelaxant effect of evodiamine in rat isolated mesenteric arteries: Mode of action. *Eur. J. Pharmacol.* **1992**, *215*, 277–283. [CrossRef]
Hung, P.H.; Lin, L.C.; Wang, G.J.; Chen, C.F.; Wang, P.S. Inhibitory effect of evodiamine on aldosterone release by Zona glomerulosa cells in male rats. *Chin. J. Physiol.* **2001**, *44*, 53–57.
Carobrez, A.P.; Bertoglio, L.J. Ethological and temporal analyses of anxiety-like behavior: The elevated plus-maze model 20 years on. *Neurosci. Biobehav. Rev.* **2005**, *29*, 1193–1205. [CrossRef]
Malik, H.; Usman, M.; Arif, M.; Ahmed, Z.; Ali, G.; Rauf, K.; Sewell, R.D.E. Diosgenin normalization of disrupted behavioral and central neurochemical activity after single prolonged stress. *Front. Pharmacol.* **2023**, *14*, 1232088. [CrossRef]
Zhang, J.P.; Zhang, K.Y.; Guo, L.; Chen, Q.L.; Gao, P.; Wang, T.; Li, J.; Guo, G.Z.; Ding, G.R. Effects of 1.8 GHz Radiofrequency Fields on the Emotional Behavior and Spatial Memory of Adolescent Mice. *Int. J. Environ. Res. Public Health* **2017**, *14*, 1344. [CrossRef]
Belzung, C.; Turiault, M.; Griebel, G. Optogenetics to study the circuits of fear- and depression-like behaviors: A critical analysis. *Pharmacol. Biochem. Behav.* **2014**, *122*, 144–157. [CrossRef]
Bird, C.M.; Burgess, N. The hippocampus and memory: Insights from spatial processing. *Nat. Rev. Neurosci.* **2008**, *9*, 182–194. [CrossRef]
Rossato, J.I.; Zinn, C.G.; Furini, C.; Bevilaqua, L.R.; Medina, J.H.; Cammarota, M.; Izquierdo, I. A link between the hippocampal and the striatal memory systems of the brain. *Acad Bras. Cienc.* **2006**, *78*, 515–523. [CrossRef] [PubMed]
Kenney, J.W.; Gould, T.J. Modulation of hippocampus-dependent learning and synaptic plasticity by nicotine. *Mol. Neurobiol.* **2008**, *38*, 101–121. [CrossRef]
Bossini, L.; Tavanti, M.; Lombardelli, A.; Calossi, S.; Polizzotto, N.R.; Galli, R.; Vatti, G.; Pieraccini, F.; Castrogiovanni, P. Changes in hippocampal volume in patients with post-traumatic stress disorder after sertraline treatment. *J. Clin. Psychopharmacol.* **2007**, *27*, 233–235. [CrossRef]
Daniels, J.K.; McFarlane, A.C.; Bluhm, R.L.; Moores, K.A.; Clark, C.R.; Shaw, M.E.; Williamson, P.C.; Densmore, M.; Lanius, R.A. Switching between executive and default mode networks in posttraumatic stress disorder: Alterations in functional connectivity. *J. Psychiatry Neurosci.* **2010**, *35*, 258–266. [CrossRef]
Carrion, V.G.; Weems, C.F.; Watson, C.; Eliez, S.; Menon, V.; Reiss, A.L. Converging evidence for abnormalities of the prefrontal cortex and evaluation of midsagittal structures in pediatric posttraumatic stress disorder: An MRI study. *Psychiatry Res.* **2009**, *172*, 226–234. [CrossRef]
Schauer, R. Sialic acids: Fascinating sugars in higher animals and man. *Zoology* **2004**, *107*, 49–64. [CrossRef]
Schnaar, R.L.; Gerardy-Schahn, R.; Hildebrandt, H. Sialic acids in the brain: Gangliosides and polysialic acid in nervous system development, stability, disease, and regeneration. *Physiol. Rev.* **2014**, *94*, 461–518. [CrossRef]
Yerlikaya, F.H.; Toker, A.; Cicekler, H.; Aribas, A. The association of total sialic acid and malondialdehyde levels with metabolic and anthropometric variables in obesity. *Biotech. Histochem.* **2015**, *90*, 31–37. [CrossRef]
Otsuka, T.; Hori, H.; Yoshida, F.; Itoh, M.; Lin, M.; Niwa, M.; Ino, K.; Imai, R.; Ogawa, S.; Matsui, M.; et al. Association of CRP genetic variation with symptomatology, cognitive function, and circulating proinflammatory markers in civilian women with PTSD. *J. Affect Disord.* **2021**, *279*, 640–649. [CrossRef]
Bansal, P.; Bansal, P.; Verma, R. Association of serum sialic acid concentration with diabetic complications and cardiovascular risk factors in an Indian population. *Arch. Med. Sci. Atheroscler. Dis.* **2021**, *6*, e14–e17. [CrossRef]
Lee, T.; Jarome, T.; Li, S.J.; Kim, J.J.; Helmstetter, F.J. Chronic stress selectively reduces hippocampal volume in rats: A longitudinal magnetic resonance imaging study. *Neuroreport* **2009**, *20*, 1554–1558. [CrossRef]
Geuze, E.; Westenberg, H.G.; Heinecke, A.; de Kloet, C.S.; Goebel, R.; Vermetten, E. Thinner prefrontal cortex in veterans with posttraumatic stress disorder. *Neuroimage* **2008**, *41*, 675–681. [CrossRef]
Gao, J.; Wang, H.; Liu, Y.; Li, Y.Y.; Chen, C.; Liu, L.M.; Wu, Y.M.; Li, S.; Yang, C. Glutamate and GABA imbalance promotes neuronal apoptosis in hippocampus after stress. *Med. Sci. Monit.* **2014**, *20*, 499–512. [CrossRef]

43. Lee, A.L.; Ogle, W.O.; Sapolsky, R.M. Stress and depression: Possible links to neuron death in the hippocampus. *Bipolar. Disc* **2002**, *4*, 117–128. [CrossRef]
44. Wang, F.; Song, J.; Yan, Y.; Zhou, Q.; Li, X.; Wang, P.; Yang, Z.; Zhang, Q.; Zhang, H. Integrated Network Pharmacology Analy and Serum Metabolomics to Reveal the Anti-malaria Mechanism of Artesunate. *ACS Omega* **2022**, *7*, 31482–31494. [CrossRef
45. Labandeira-Garcia, J.L.; Rodriguez-Perez, A.I.; Garrido-Gil, P.; Rodriguez-Pallares, J.; Lanciego, J.L.; Guerra, M.J. Brain Ren Angiotensin System and Microglial Polarization: Implications for Aging and Neurodegeneration. *Front. Aging Neurosci.* **20** *9*, 129. [CrossRef]
46. Zhang, T.L.; Fu, J.L.; Geng, Z.; Yang, J.J.; Sun, X.J. The neuroprotective effect of losartan through inhibiting AT1/ASK1/MKK4/JN pathway following cerebral I/R in rat hippocampal CA1 region. *CNS Neurosci. Ther.* **2012**, *18*, 981–987. [CrossRef]
47. Tota, S.; Kamat, P.K.; Saxena, G.; Hanif, K.; Najmi, A.K.; Nath, C. Central angiotensin converting enzyme facilitates mem impairment in intracerebroventricular streptozotocin treated rats. *Behav. Brain Res.* **2012**, *226*, 317–330. [CrossRef]
48. Kumaran, D.; Udayabanu, M.; Kumar, M.; Aneja, R.; Katyal, A. Involvement of angiotensin converting enzyme in cereb hypoperfusion induced anterograde memory impairment and cholinergic dysfunction in rats. *Neuroscience* **2008**, *155*, 626–6 [CrossRef]
49. Gadelha, A.; Vendramini, A.M.; Yonamine, C.M.; Nering, M.; Berberian, A.; Suiama, M.A.; Oliveira, V.; Lima-Landman, M Breen, G.; Bressan, R.A.; et al. Convergent evidences from human and animal studies implicate angiotensin I-converting enzy activity in cognitive performance in schizophrenia. *Transl. Psychiatry* **2015**, *5*, e691. [CrossRef]
50. Schwengel, K.; Namsolleck, P.; Lucht, K.; Clausen, B.H.; Lambertsen, K.L.; Valero-Esquitino, V.; Thone-Reineke, C.; Muller, Widdop, R.E.; Denton, K.M.; et al. Angiotensin AT2-receptor stimulation improves survival and neurological outcome af experimental stroke in mice. *J. Mol. Med.* **2016**, *94*, 957–966. [CrossRef]
51. Zhang, X.; Zhao, Y.; Du, Y.; Sun, H.; Zhang, W.; Wang, A.; Li, Q.; Li, C.; Wang, Y.; Du, Z.; et al. Effect of ketamine on mo dysfunction and spatial cognition deficits in PTSD mouse models via HCN1-BDNF signaling. *J. Affect Disord.* **2021**, *286*, 248–2 [CrossRef]
52. Zhang, Z.S.; Qiu, Z.K.; He, J.L.; Liu, X.; Chen, J.S.; Wang, Y.L. Resveratrol ameliorated the behavioral deficits in a mouse mode post-traumatic stress disorder. *Pharmacol. Biochem. Behav.* **2017**, *161*, 68–76. [CrossRef]
53. Kurosawa, N.; Shimizu, K.; Seki, K. The development of depression-like behavior is consolidated by IL-6-induced activation locus coeruleus neurons and IL-1β-induced elevated leptin levels in mice. *Psychopharmacology* **2016**, *233*, 1725–1737. [CrossR
54. Li, W.; Liu, X.; Qiao, H. Downregulation of hippocampal SIRT6 activates AKT/CRMP2 signaling and ameliorates chro stress-induced depression-like behavior in mice. *Acta Pharmacol. Sin.* **2020**, *41*, 1557–1567. [CrossRef]
55. Lo Iacono, L.; Ielpo, D.; Accoto, A.; Di Segni, M.; Babicola, L.; D'Addario, S.L.; Ferlazzo, F.; Pascucci, T.; Ventura, R.; Andoli D. MicroRNA-34a Regulates the Depression-like Behavior in Mice by Modulating the Expression of Target Genes in the Dor Raphe. *Mol. Neurobiol.* **2020**, *57*, 823–836. [CrossRef]
56. Taniguchi, E.; Tashiro, A.; Hattori, A.; Furuse, M.; Yasuo, S. Photoperiodic changes in hippocampal neurogenesis and plas metabolomic profiles in relation to depression-like behavior in mice. *Behav. Brain Res.* **2021**, *403*, 113136. [CrossRef]
57. Morris, R. Developments of a water-maze procedure for studying spatial learning in the rat. *J. Neurosci. Methods* **1984**, *11*, 47– [CrossRef]

MDPI AG
Grosspeteranlage 5
4052 Basel
Switzerland
Tel.: +41 61 683 77 34

Nutrients Editorial Office
E-mail: nutrients@mdpi.com
www.mdpi.com/journal/nutrients

www.ingramcontent.com/pod-product-compliance
Lightning Source LLC
LaVergne TN
LVHW070459061126
851858LV00011B/262
* 9 7 8 3 7 2 5 8 3 3 6 1 0 *